Essential Anatomy

Essential Anatomy

and some clinical applications

J. S. P. Lumley MS FRCS FMAA(Hon) FGA

Professor of Vascular Surgery, Assistant Director of Surgical Professorial Unit, and Honorary Consultant Surgeon, St Batholomew's Hospital, London; Examiner in Anatomy for the Royal College of Surgeons of England

J. L. Craven BSc MD FRCS

Consultant Surgeon, District General Hospital, York; Examiner in Pathology for the Royal College of Surgeons of England

Both Formerly Assistant Lecturers at University College, London

J. T. Aitken MD

Emeritus Professor of Anatomy, University College, London; Past Examiner in Anatomy for the Royal College of Surgeons of England

FOURTH EDITION

CHURCHILL LIVINGSTONE

EDINBURGH LONDON MELBOURNE AND NEW YORK 1987

CHURCHILL LIVINGSTONE
Medical Division of Longman Group UK Limited

Distributed in the United States of America by
Churchill Livingstone Inc., 1560 Broadway, New York,
N.Y. 10036, and by associated companies, branches
and representatives throughout the world.

First edition 1965
Second edition 1975
Third edition 1980
Fourth edition 1987

ISBN 0-443-03573-3

British Library Cataloguing in Publication Data
Lumley, J. S. P.
Essential anatomy : and some clinical
applications.—4th ed.
1. Anatomy, Human
I. Title II. Craven, J. L. III. Aitken,
J. T.
611 QM23.2

Library of Congress Cataloging in Publication Data
CIP Data available

Produced by Longman Singapore Publishers (Pte.) Ltd.
Printed in Singapore

Preface to the Fourth Edition

The anatomy content of the medical preclinical course has changed greatly over the last decade. Less time is now allocated to dissection, and emphasis is given to clinical application rather than topographical detail. This clinical emphasis is reflected in the present edition, there being a general streamlining of the contents and the inclusion of a chapter on the anatomy of clinical examination. The aim throughout has been to provide a comprehensive revision work for the undergraduate student and to include enough detail for the postgraduate student preparing for a primary fellowship examination.

Radiographs have been revised and pictures of obsolete procedures omitted. Computer Tomography and Magnetic Resonance Imaging have gained prominence, this being in keeping with the clinical development of these imaging techniques. With this has developed the need for knowledge of the anatomy of the body cavities as seen in transverse sections.

We are indebted to Drs. M. Charlesworth, J. Dacie, R. Resnick and J. Webb for providing typical radiographs and to Professor Don Mayor of Southampton for examining the text and advising on its future direction.

London and York
1987

J. S. P. L.
J. L. C.
J. T. A.

Preface to the First Edition

Preclinical medical and dental students labour to-day under a weight of ever expanding subjects and books, so that some justification needs to be given for producing yet another textbook.

As teachers we know that the profusion of data in anatomy often obscures many important and essential facts, and we have seen students, lacking a critical and methodical approach, perform badly in examinations. This concise book omits much detail and avoids unnecessary repetition. Short notes on relevant histology and embryology have been included and as far as possible the New York (1963) nomenclature has been used.

We have tried to emphasise a methodical approach to the subject by the description of structures in a consistent manner, and to this end the book-mark may prove a useful *aide mémoire*. The diagrams have been chosen with care and their simplicity should make them easy to understand.

This combination of brevity, method and regional description will, we hope, help the student to be selective both in his dissection and in his revision for the second professional examination.

We are greatly indebted to Professor J. Z. Young for his encouragement, guidance and the facilities of his department. Many colleagues have also helped in the process of sifting and emphasising. We are grateful to Mrs J. Astafiev, departmental artist, who has made our drawings and diagrams comprehensible. Mr Charles Macmillan and the staff of E. & S. Livingstone Ltd. have been, as always, very helpful in all stages of the preparation of this work.

1965

J. S. P. L.
J. L. C.
J. T. A.

Method of Description

Joints

1. Classification:
   (a) bony
   (b) fibrous
   (c) cartilaginous
   (d) synovial
2. Bones:
   (a) names
   (b) shapes
   (c) articular covering
3. Ligaments:
   (a) capsule
   (b) capsular thickenings
   (c) accessory
4. Intracapsular structures:
   (a) synovial membrane
   (b) discs
   (c) ligaments, tendons, fat pads
5. Functional aspects:
   (a) movements (muscles involved)
   (b) stability (bones, ligaments and muscles)
   (c) relation to line of weight
6. Blood and nerve supply
7. Relations
8. Bursae
9. Radiology

Organs

1. Definition
2. General outline:
   (a) shape
   (b) size
   (c) surfaces
   (d) situation
   (e) colour
   (f) consistency
3. Relations
4. Blood supply
5. Nerve supply
6. Lymph drainage
7. Histology
8. Embryology
9. Functional aspects
10. Radiology

Vessels and Nerves

1. Origin
2. Course
3. Termination
4. Relations
5. Branches

Muscles

1. Attachments
2. Action
3. Nerve supply

Contents

PART ONE

General introduction

— 1 —

The structure of the body

The body is formed of innumerable cells and varying amounts of intercellular substances. They may be grouped into (a) the tissues, (b) the cells of the body fluids, and (c) the mobile cells of the macrophage (reticulo-endothelial) system. Aggregations of one or more tissues form the functional units (the organs) of the body. All the cells, tissues and organs subserving a particular function are classified as a system.

The cell

Cells are composed of a protoplasmic gel, the protoplasm of the nucleus being called karyoplasm and that of the cell body, cytoplasm. The resting nucleus is surrounded by an elastic nuclear membrane and contains a large amount of basophil staining chromatin which, in the dividing cell, is grouped to form the chromosomes. In dividing cells, the chromosomes are in pairs, differing in size and shape from other pairs; they contain the genes which control the chemical processes of the body.

The cytoplasm is limited by a thin cell membrane and within it are found a number of intracellular inclusions, the organelles. These comprise the mitochondria, Golgi apparatus, endoplasmic reticulum, centrosome, lysosomes, fibrils and tubules. The mitochondria are often small rod-shaped particles with inward projecting folds, the cristae. They are rich in enzymes and are essential for the metabolism of the cell. The Golgi apparatus is situated near the nucleus and consists of a number of vesicles (membrane-bound cavities) and a smooth-walled membranous component. It is probably concerned with cell secretion. The endoplasmic reticulum is a membranous lattice network scattered throughout the cytoplasm; it is concerned with protein synthesis. The centrosome is a spherical area of clear cytoplasm near the nucleus and is concerned with cell division. The lysosomes are small membranous vesicles containing autolytic enzymes.

Occasionally the growth of a group of cells in the body may become uncontrolled with the resulting development of a local swelling known as a tumour, a growth or a neoplasm. This abnormal process is known as neoplasia. The suffix '-oma' is added to the name of the tissue in describing the neoplasm, e.g. bone tumours are called 'osteomas'. A benign (simple) neoplasm is usually a slow growing, localised collection of cells. If infiltration or spread of the neoplastic tissue around the body occurs, the condition is said to be malignant, a cancer. A growth of this malignant tissue at a second site away from the primary site is called a secondary growth (a metastasis). Cancers of epithelial tissues are known as carcinomas, and of mesodermal tissues, sarcomas. A change in the form or function of a tissue or organ is referred to as a lesion. A serious disorder of this nature is known as a disease. Death of a tissue is called necrosis.

THE SYSTEMS AND ORGANS

The systems and organs of the body are composed of epithelial, connective, muscular and nervous tissues.

EPITHELIAL TISSUE

This forms a protective covering over the internal and external surfaces of the body and is characterised by having a minimal amount of intercellular substance and a tendency to form sheets. It may be simple, transitional or stratified.

1. Simple epithelium

This consists of a single layer of cells on a basement membrane. It is described as squamous (pavement), cubical or columnar, depending on the shape of its cells. Squamous cells are found lining the alveoli of the lungs, the blood vessels (endothelium) and the serous cavities (mesothelium). Cubical cells line the ducts of many glands. Columnar cells are often ciliated and may be modified as mucus-forming goblet cells; they line much of the alimentary, respiratory and reproductive tracts. Mucus, a glycoprotein, accumulates in the cell and is discharged from its free surface.

2. Transitional epithelium

This contains two or three layers of cells, most of which are attached to the basement membrane and are nucleated. It lines most of the urinary tract, is stretchable and does not desquamate. It contains few glands.

3. Stratified squamous epithelium

This also has two or more layers of cells. Those in contact with the basement membrane are columnar cells. The more superficial cells are flattened and the surface cells have no nuclei (enucleate) and are continually being rubbed away (desquamated). This form of epithelium covers the exterior of the body, lines both ends of the alimentary tract and is particularly suited to areas exposed to wear and tear. In the upper respiratory tract the differing lengths of the columnar cells gives the appearance of a double layer and this is known as pseudostratified columnar epithelium.

SKIN

This consists of two layers, an outer epidermis and an inner dermis (corium). The epidermis is composed of keratinised stratified squamous epithelium. Hair follicles, sweat and sebaceous glands and nails are modifications of the epidermis. The scales on the surface of the skin consist mainly of keratin, a sulphur-containing fibrous protein largely responsible for the protective and barrier properties of the skin.

The dermis is a layer of vascular connective tissue moulded tightly to the epidermis and merging in its deeper part with the subcutaneous tissues. Lying in the dermis are the coiled tubular sweat ducts opening on to the skin surface, the compound alveolar sebaceous glands opening into the hair follicles and the arrectores pili muscles which are attached to the hair follicles. The roots of the hairs and the sweat glands extend into the subcutaneous tissue.

MUCOUS AND SEROUS MEMBRANES

These line the wet internal surfaces of the body and consist of two layers, an epithelium and a corium. The epithelium of mucous membranes is usually of a simple variety with many mucous or serous cells but the urinary tract is lined by transitional epithelium, and the respiratory tract by pseudostratified columnar ciliated epithelium with mucous cells. The serous membranes line most of the closed body cavities. The corium is more deeply placed and composed of connective tissue. In the alimentary tract it contains a thin sheet of smooth muscle—the muscularis mucosa.

GLANDS

These are epithelial ingrowths modified to produce secretions. These secretions may pass on to the epithelial surface (exocrine glands) or into the blood stream (endocrine glands). Exocrine glands may be unicellular (goblet) or multicellular. The latter may be simple (containing one duct) or compound (branched) where numerous small ducts open into a single main duct. The secretory part of the gland may be long and thin (tubular), globular (acinar), oval (alveolar) or intermediate, e.g. tubulo-alveolar. The secretions of the exocrine glands may be formed by disintegration of the whole cell (holocrine, e.g. sebaceous glands), disintegration of the free end of the cell (apocrine, e.g. mammary glands) or without cellular damage (merocrine or epicrine, e.g. most other glands). Most endocrine glands are of the last type. If the duct of an exocrine gland becomes blocked and the gland continues to secrete, the fluid accumulates and a cyst is formed. A generalised enlargement of glands is termed adenopathy.

CONNECTIVE TISSUE

This is characterised by having a large amount of intercellular substance. It forms areolar tissue, the packing material of the body, and supporting tissues, cartilage, bone and blood. Embryonic connective tissue is called mesenchyme.

AREOLAR TISSUE

The intercellular substance is semisolid and composed of proteins and mucopolysaccharides. Three types of fibres are found: coarse collagen fibres which are white (in bulk), flexible, inelastic and arranged in bundles; elastic fibres which are yellowish (in bulk), less frequent and branching; reticular fibres which form a very fine silver-staining network throughout the tissues.

The cells are of five main varieties: large slender poorly-staining fibroblasts, closely concerned with the production of the three types of tissue fibres; tissue macrophages which are phagocytic i.e. engulf particulate matter; oval plasma cells with their cartwheel-like staining nucleus, concerned with antibody production; granular basophilic mast cells, concerned with histamine and heparin production; the cyst-like fat-containing cells.

The relative amounts of cells and intercellular substances vary throughout the body. Subcutaneous tissue contains a variable amount of fat and fibrous tissue, and forms layers called fascia. Superficially, fat is usually predominant but deeper, the fascia forms a well-defined fibrous sheet over the muscles. In other places condensations of nonelastic fibrous tissue form ligaments, tendons and aponeuroses, and retinacula. Ligaments are usually attached to the bones on each side of a joint, maintaining its stability; tendons join the muscles to the bones by blending with the periosteum; aponeuroses are thin flattened tendinous sheets through which muscles gain wider attachments. Retinacula are usually thickenings of the deep fascia related to joints.

SUPPORTING TISSUES

1. Cartilage

This is an avascular firm tissue composed of cells (chondrocytes) in an abundant intercellular substance (matrix). It is formed from an overlying fibrous layer, the perichondrium, and classified, according to its predominant fibres, into hyaline cartilage, fibrocartilage and yellow elastic cartilage.

(a) Hyaline cartilage contains many cells, a few fine collagen-like fibres and is found in the rib cartilages and over most articular surfaces. It also forms the precursor in cartilaginous ossification.

(b) Fibrocartilage contains many dense fibrous bundles, fewer cells and is present in the intervertebral disc, over the articular surface of bones which ossify in membranes, e.g. the mandible, and in intra-articular cartilages, e.g. the menisci of the knee.

(c) Yellow elastic cartilage contains elastic fibres and is found in the auricular, epiglottic and arytenoid cartilages of the head.

2. Bone

This is a hard supporting tissue composed mainly of inorganic calcium salts impregnating a network of collagen fibres. The basic unit is composed of concentric layers around a central vessel, the arrangement being known as a Haversian system. The bone cells (osteocytes) lie within spaces (lacunae) between the layers and their processes pass into canaliculi in the bone. Compact bone is dense and strong and forms the outer part of most bones. Inside, the cancellous (spongy) bone consists of a network of thin partitions (trabeculae) around inter-communicating spaces; the osteocytes lie within lacunae in the trabeculae. The outer surface of a bone is covered by a thick fibrous layer, the periosteum, many of the cells of which are the granular, bone-forming osteoblasts. These cells become osteocytes when enclosed in the hard intercellular substance. The main blood supply of bone comes from the muscles attached to the periosteum and bone.

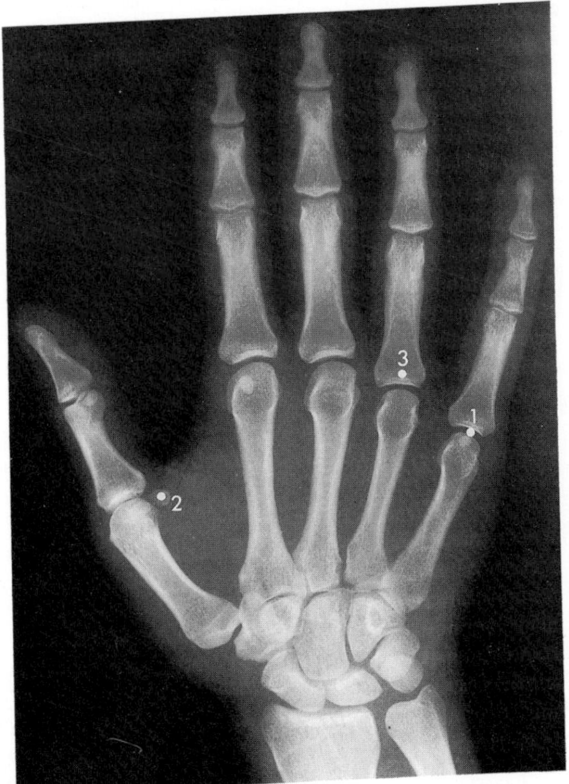

Fig. 1.1 Radiograph of adult hand. Note the joint lines (1), the sesamoid bones (2), and the bony trabeculae (3).

The shape of the bones of the body, the proportion of compact to cancellous tissue and the architecture of the trabeculae, are arranged to give maximum strength along with economy of material. Both genetic and local factors influence the shape and size of a bone. Adjacent muscles or organs (e.g. the brain) mould the bone to some extent. Many of these properties of bone can be investigated in the living person by means of X-rays (Fig. 1.1).

Bones are classified as long, short, flat or irregular.

Long bones are present in the limbs. The body (shaft) is a cylinder of compact bone surrounding a medullary cavity which is filled with some spongy bone and a large amount of yellow fatty marrow. The two ends are formed of spongy bone with a thin outer shell of compact bone. The trabeculae in the cancellous bone are laid down along the lines of force. In the developing long bone, blood-forming tissue is found in the ends, but not normally in the adult.

The short bones are found in the carpus and tarsus. They consist of spongy bone covered with a thin layer of compact bone.

The flat bones, e.g. the scapula, give attachment to muscles and form a protective covering (the bones of the skull vault). They consist of two layers of compact bone with a thin intervening spongy layer (the diploë in the skull).

The remaining irregular bones may contain red, blood-forming, marrow (the vertebrae), or air spaces (sinuses) being then called pneumatic bones (some bones of the face).

Excessive force applied to a bone may cause it to break (fracture) and in such injuries the adjacent soft tissues may be damaged both by the force and by the broken ends of the bone. Knowledge of the anatomical relations of the bone enables the doctor to predict the likely association of nerve, artery and muscle injury with a bone fracture. A break in the overlying skin is a serious complication of fractures, as it allows entry to infecting organisms. In these circumstances the fracture is said to be compound.

Ossification

Bone may develop either (i) in a condensed fibrous tissue model and the process is called membranous (mesenchymal) ossification, or (ii) in a cartilage model which has replaced the mesenchyme and the process is called cartilaginous (endochondral) ossification.

Mesenchymal ossification usually starts in the fifth to sixth week of intra-uterine life, and is found in the bones of the skull vault, the bones of the face and the clavicle (partly).

Cartilaginous ossification occurs in all long bones except the clavicle. A primary centre appears for the body about the eighth week of intra-uterine life and secondary centres for each end between birth and puberty. Fusion of the body and these centres occurs about the 18th year in males. Secondary centres appear and fuse up to a year earlier in the female. Further ossification centres may develop at puberty in areas of major muscle attachment. Examples are the processes of the vertebrae and the crests of the scapula and the hip. They fuse with the rest of the bone about the 25th year.

Development of a long bone (Fig. 1.2)
A long bone develops from a cartilaginous model possessing an outer perichondrium and an irregular cartilaginous matrix; the deep layers of the perichondrium have pronounced bone forming properties.

The first changes occur in the cartilage cells of the middle of the body (the **diaphysis**). They become greatly enlarged and the matrix is correspondingly reduced and calcified. The cells die and undergo shrinkage, leaving spaces known as primary alveoli. The deeper layer of perichondrium around the middle of the body starts to produce bone and is then known as the periosteum. Blood vessels and bone cells (the bone-forming osteoblasts and the bone-removing osteoclasts) pass inwards from the periosteum to the calcified zone. The multinucleated osteoclasts initiate absorption of the calcified material producing larger spaces, the secondary alveoli. Osteoblasts come to line the secondary alveoli and layers of bone are deposited. Some osteoblasts are incorporated into the bone, becoming

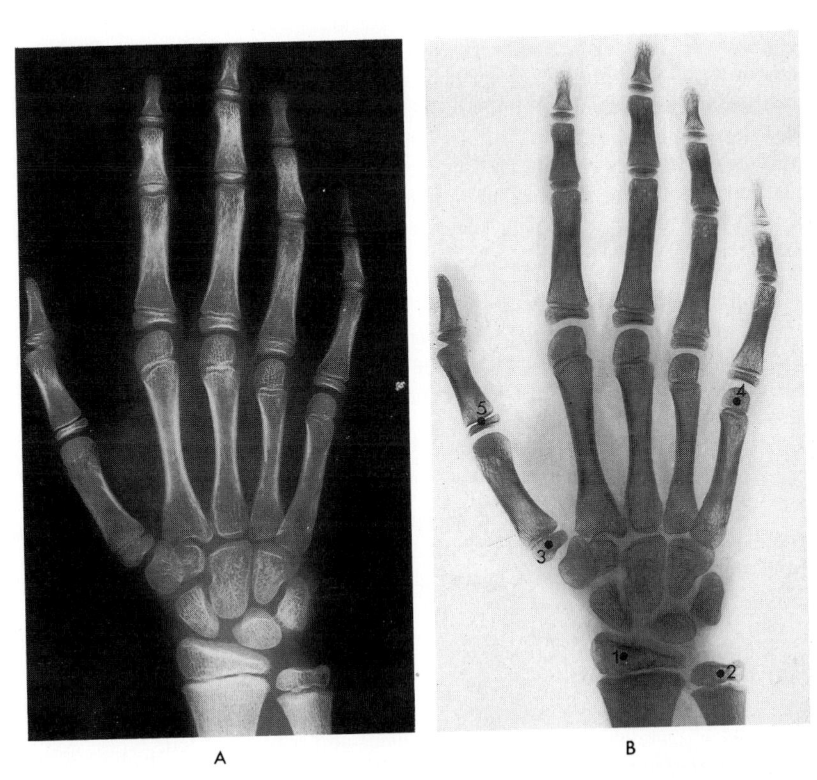

Fig. 1.2 Radiograph of young person showing unfused epiphyses (1, 2, 3 and 4). Note the clear zone (epiphyseal cartilage, 5) which separates the body of the bone from the epiphysis. Compare A and B and note that the method of printing has been reversed. In A the dense tissues appear white but in B they are dark. Both methods of reproduction are used in the text in an effort to obtain a clear picture.

the bone cells (osteocytes). Ossification extends up and down the body from this primary centre. The cells of the adjacent cartilage come to lie in parallel longitudinal rows and are subsequently replaced in the manner already described. This form of ossification is known as endochondral (cartilaginous).

Secondary centres of ossification (the **epiphyses**) appear later in life. Osteogenic cells invade the calcified cartilage after the cells have undergone hypertrophy, death and shrinkage. The layer of cartilage left between the epiphysis and the diaphysis is known as the **epiphyseal plate**. The part of the diaphysis bordering the plate is known as the **metaphysis**; the cartilage adjacent to the metaphysis is continually being ossified. New cartilage cells are formed in the epiphyseal plate. Growth in length of the bone continues until the cartilage cells stop multiplying and fusion of the diaphysis and epiphysis then occurs. The internal architecture of the bone is remodelled by osteoclastic and osteoblastic activity. Simultaneous laying down of layers of bone around the body by the periosteum increases the girth of the bone and is known as subperiosteal ossification; it is a form of mesenchymal ossification.

Epiphyses may be classified into three types: pressure epiphyses seen at the ends of weight-bearing bones, traction epiphyses occurring at the site of muscle attachments, and 'atavistic' epiphyses being functionless skeletal remnants which may show on an X-ray and be mistaken for disease or injury.

Injury in a young person can dislodge the epiphysis from the metaphysis, e.g. a fall on the outstretched arm may produce a slipped epiphysis of the lower end of the radius. This injury may interfere with further growth at that end of the bone (see Fig. 19.4, p. 263).

Summary: most primary centres of long bones appear by the end of the second month of intra-uterine life and most epiphyseal centres before puberty. The epiphyses at the knee joint appear just before birth and are an indication of the age of the fetus. Most long bones cease growing in length between the 18th and 20th years in men and a year or so earlier in women.

The skeleton

The skeleton is divisible into an axial part (the bones of the head and trunk) and an appendicular part (the bones of the limbs). The upper limb is joined to the trunk by the mobile pectoral girdle, and the lower limb by the stable pelvic girdle.

JOINTS (Figs. 1.1, 1.2)

These are unions between two or more bones and may be of four types: bony, fibrous, cartilaginous or synovial.

1. Bony

The three elements of the hip bone are joined by bony union, as are the occipital and the sphenoid in the skull after completion of the second dentition.

2. Fibrous

The bony surfaces are united by fibrous tissue. They comprise the skull sutures, the articulations of the roots of the teeth, and the inferior tibiofibular joint.

3. Cartilaginous

These may be primary or secondary. In the primary cartilaginous joints the bony surfaces are united by hyaline cartilage as seen in the union of the body and the ends in a developing long bone.

In the secondary cartilaginous joints (the symphyses) the bony surfaces are covered with hyaline cartilage and united by a fibrocartilaginous disc. These joints all lie in the midline and comprise the intervertebral discs, the symphysis pubis and the manubriosternal and xiphisternal junctions.

4. Synovial

The bony articular surfaces (facets) are covered with hyaline cartilage (with the exception of the temporomandibular and sternoclavicular joints where they are covered with fibrocartilage). A fibrous capsule is attached near to the articular margins of the bones. The surfaces of the interior of the joint, except those covered by cartilage, are lined by a delicate vascular synovial membrane which secretes a watery synovial fluid into the joint cavity.

The capsule may possess ligamentous thickenings, and accessory extracapsular and intracapsular ligaments may pass across the joint. Fibrous intra-articular discs are present in some joints, occasionally completely dividing their cavity (e.g. temporomandibular joint). Tendons occasionally enter the joint cavity by piercing the capsule (biceps brachii) and fat pads may be present between the capsule and the synovial lining (knee joint).

Muscles or tendons crossing superficial to the joint may be protected by a synovial sheath or sac whose fluid prevents excessive friction. The sacs are known as **bursae** and the cavity may communicate with that of the joint.

Functional aspects of joints

MOVEMENT

Bony, fibrous and primary cartilaginous joints are immobile, secondary cartilaginous joints are slightly mobile, and synovial joints are freely mobile.

Synovial joints are subdivided into a number of varieties according to the movements possible at the joint. These varieties are listed below and the movements are best understood by examining the examples given.

(a) Hinge—elbow, ankle and interphalangeal joints.
(b) Pivot—proximal radio-ulnar joint and the dens articulation of the atlanto-axial joint.
(c) Condyloid—knee and temporomandibular joints. These are modified hinge joints.
(d) Ellipsoid—radiocarpal (wrist) and the metacarpophalangeal joints.

(e) Saddle—carpometacarpal joint of the thumb.
(f) Ball and socket—hip and shoulder joints.
(g) Plane—intercarpal joints and joints between the articular processes of adjacent vertebrae.

STABILITY

This may be due to bony, ligamentous or muscular factors and is usually inversely related to the mobility of the joint.

Many of these functional aspects of joints may be assessed in an X-ray examination of the part. Displacement of the articulating surfaces of a joint is known as dislocation. Dislocation of a joint may follow severe injury and is always associated with damage to the capsular and accessory ligaments. There may also be fractures of the bony structures of the joint and occasionally damage to closely related nerves and vessels. Chronic inflammatory processes are prone to affect the bone ends (osteoarthritis) and synovial membrane (rheumatoid arthritis), and joints thus affected may be deformed and painful with marked limitation of movement.

MUSCULAR TISSUE

This is a contractile tissue. There are striated, smooth, and cardiac varieties.

1. Striated (voluntary) muscle

This acts mainly on the bony skeleton or as a diaphragm, but it is also found around the pharynx and larynx and forms some sphincters. It is composed of unbranched fibres of sarcoplasm limited by a membrane, the sarcolemma, and contains many nuclei. Each fibre has a motor end plate and contains many contractile units, the myofibrils, which have alternating dark (A) and light (I) bands. A dark line (the Z disc) crosses the middle of the I band. The bands of adjacent myofibrils coincide, giving the muscle fibre its striated appearance. Each fibre is enveloped and attached to its neighbour by a fibrous endomysium and bundles of fibres are enclosed by a fibrous perimysium. A muscle composed of many bundles is surrounded by an epimysium. The electron microscope has greatly increased our knowledge of the details of the structure and function of muscle.

The motor nerve supply of the muscle comes from the anterior horn cells of the spinal cord and the motor nuclei of the brain stem. The sensory supply arises in the more specialised spindles and tendon organs as well as the simpler touch and pain endings. Impulses from the sensory endings pass into the posterior horns of the spinal cord. Muscles have a very rich blood supply.

MUSCLES OF THE SKELETON

These are formed of voluntary fibres. The muscles are attached at each end, usually to bone, either directly or through tendons and aponeuroses and cross one or more joints. Occasionally two muscles meet at a common stretchable union known as a raphé, e.g. mylohyoid muscles.

Muscle fibres are arranged either parallel to the direction of the action (sartorius) or obliquely to it (rectus femoris). In muscles of equal volume, a parallel arrangement of fibres gives greater movement but less power than an oblique arrangement. The less moving attachment of a muscle is often called its origin, and the more moving attachment its insertion. In some situations these roles are reversed, so reference will be made to muscle attachments in the text, rather than to origin or insertion.

When a movement occurs at a joint, the muscles concerned in producing it are known as the prime movers and those opposing it, as antagonists. Muscles contracting to steady the joint across which movement is occurring, and other joints involved are known as synergists. A further type of action is known as the action of paradox, in which a muscle gradually relaxes against the pull of gravity, e.g. bending forwards by relaxing the back muscles.

2. Smooth (unstriated, involuntary) muscle

This is present in the walls of most vessels and hollow organs of the body. It is composed of unbranched spindle-shaped fibres with a single central nucleus and containing many unstriated myofibrils. The fibres are arranged in interlacing bundles and are supplied by the autonomic nervous system.

3. Cardiac muscle

This is found in the heart. It consists of short branched cylindrical fibres joined end to end. The adherent ends of adjacent fibres form the dark intercalated discs. Each fibre contains a single central nucleus and striated myofibrils resembling those of voluntary muscle and is supplied by the autonomic nervous system.

NERVOUS TISSUE

This is capable of excitability and conductivity. It consists of excitable cells (neurons) and supporting cells which in the central nervous system are the neuroglial cells, and in the peripheral nervous system, the Schwann cells.

The **neuron** consists of a cell body, the perikaryon, and processes, usually an axon and one or more dendrites. The cell bodies are situated in the central nervous system or in peripheral ganglia. They possess a large nucleus and well-marked cellular inclusions. The axon (fibre) begins at a small axon hillock on the cell body and carries impulses away from the cell body. This often long slender process ends by dividing into many branches which have small terminal knobs, the boutons, related to the cell bodies or branches of other neurons. The relationship is known as a synapse and it may be either facilitatory or inhibitory, depending on the neuron of origin and possibly on the receptor area of the second neuron. (A rather specialised synapse is formed when a nerve ends on a muscle fibre at a motor end plate.)

Axons may give off one or more short collateral branches. Many axons are surrounded by a myelin sheath. This sheath is interrupted about every millimetre or so and the constriction is called a node of Ranvier. In the peripheral nervous system each internodal segment of sheath is produced by a Schwann cell, the nucleus of which is seen on its surface. These cells play an important role in peripheral nerve regeneration. Fibres of the peripheral nervous system are also covered by a thin fibrous membrane, the neurilemma. In the central nervous system oligodendroglia takes the place of the Schwann cells. Dendrites are usually short unmyelinated processes carrying action potentials to the cell body. The volume over which the dendrites of a single cell extend is known as the dendritic field.

Afferent neurons carry information towards the central nervous system and efferent neurons carry instructions away from it. Within the central nervous system afferent and efferent neurons are often connected by many intercalated (internuncial or intermediate) neurons.

The neurons are organised to form the central and peripheral nervous systems. The former comprises the brain and spinal cord (p. 437) and the latter the cranial nerves (p. 401), spinal nerves and the autonomic nervous system (see below). A group of neurons in the central nervous system is called a *nucleus*, and when the group is outside, a *ganglion*.

Neuroglia consists of astrocytes, oligodendroglia and microglia, and makes up almost half of the brain substance.

Astrocytes are stellate cells with large nuclei and numerous processes which may be of the thick protoplasmic variety, as found mainly in the grey matter, or the thin fibrous variety found mainly in the white matter. Some of the processes end on blood vessels and the astrocytes are thought to be concerned with the fluid balance in the central nervous system and with the nutrition of the neurons. Oligodendroglia are oval dark-staining cells possessing few processes. They produce the myelin of the central nervous system. Microglia are small mobile phagocytic cells and form part of the macrophage system.

Ependymal cells are columnar in shape and line the cavities of the brain and spinal cord. In certain regions the ependyma is modified to form the choroid plexuses of the brain which produce the cerebrospinal fluid (pp. 441 and 486).

CRANIAL NERVES (p. 437)

Twelve pairs of nerves are attached to the brain. Some are mainly sensory, some mainly motor and others are mixed sensory and motor.

SPINAL NERVES (Fig. 4.3)

The 31 paired spinal nerves (8 cervical, 12 thoracic, 5 lumbar, 5 sacral and 1 coccygeal) are formed within the vertebral canal, each by the union of a ventral and a dorsal root. The roots are formed from a number of rootlets which emerge from the anterolateral and posterolateral sulci of the spinal cord. The ventral root carries efferent (motor) fibres from the cord and the dorsal root, afferent (sensory) fibres to the cord. The cell bodies of the sensory fibres are situated in a ganglion on the dorsal root. The spinal nerves are therefore a mixture of motor and sensory fibres. Each nerve leaves the vertebral canal through an intervertebral foramen and soon divides into a large ventral and smaller dorsal ramus (branch).

The adjacent ventral rami of most regions communicate to form plexuses (cervical, brachial and lumbosacral) while those of the thoracic region become the intercostal and subcostal nerves (p. 59). The dorsal rami pass backwards into the postvertebral muscles and divide into medial and lateral branches. These rami supply the muscles and skin over the posterior aspect of the body but give no branches to the limbs.

Tumours within the vertebral canal or a protrusion from a degenerate intervertebral disc may compress a spinal nerve and produce segmental sensory and motor dysfunction. Knowledge of the anatomical distribution of the individual nerves will enable the site of the disease to be identified.

AUTONOMIC NERVOUS SYSTEM

This system supplies glands and smooth and cardiac muscle. Its fibres form a fine network on the blood vessels and in the nerves. All its fibres arise from neurons of the visceral columns of the brain and spinal cord (p. 440) and synapse with peripheral ganglion cells before reaching the organs they supply. This fine network is divisible into two complementary parts, sympathetic and parasympathetic, which leave the central nervous system at different sites. They usually have opposing effects on the structure they supply through endings which are mainly adrenergic or cholinergic. Evidence is accumulating for other types of endings producing different transmitter substances. Most viscera are innervated by both parts, the sympathetic preparing the body for fight or flight, and the parasympathetic controlling its vegetative functions.

Sympathetic system

Each ventral ramus from the 1st thoracic to the 2nd lumbar nerves gives a bundle of myelinated (preganglionic) fibres to the sympathetic trunk. The bundles arise near the formation of a ramus and are called rami communicantes; they form the sympathetic outflow of the central nervous system. Each ventral ramus later receives a bundle of unmyelinated (postganglionic) fibres from the sympathetic trunk—a ramus communicans. The peripheral ganglia of the system lie within the two parallel sympathetic trunks alongside the vertebral column. The trunks extend from the base of the skull to the coccyx and have 3 cervical, 12 thoracic, 4 lumbar and 5 sacral ganglia. The preganglionic fibres from the thoracolumbar outflow may synapse: (i) in the adjacent ganglion, (ii) in other ganglia higher or lower in the chain, (iii) in the collateral ganglia situated in the plexuses around the aorta (e.g. coeliac). Each preganglionic fibre may synapse with 15 or more ganglionic cells, thus giving rise to widespread activity. A number of preganglionic fibres end in the medulla of the suprarenal gland.

Postganglionic (unmyelinated) fibres may: (i) return to a spinal nerve in a ramus communicans to be distributed to peripheral smooth muscle, e.g. arterial walls, (ii) pass along the major arteries and their branches to be distributed to the organs these supply, (iii) form named nerves, e.g. cardiac, running to the viscus concerned.

The cell bodies of the sympathetic fibres supplying the upper and lower limbs are situated in ganglia in the cervicothoracic and lumbosacral regions respectively. Chemical blockage or surgical removal of these ganglia may be undertaken to improve the blood supply to the tissues of the limb, and less commonly, to reduce excessive sweating. The visceral branches supply the circular smooth muscle, including the sphincters of the viscera.

Parasympathetic system

This system receives preganglionic fibres from four cranial nerves (oculomotor, facial, glossopharyngeal and vagus) and the 2nd, 3rd and 4th sacral nerves (craniosacral outflow). The peripheral ganglia of this system are near the organs they supply, usually in its walls. There are, however, four well-defined, isolated, parasympathetic ganglia associated with the cranial nerves (p. 431). The postganglionic fibres are usually short and unmyelinated. The visceral branches usually supply the smooth muscle responsible for emptying the organ and also produce dilatation of the blood vessels. There is little evidence of parasympathetic supply to the limbs.

Afferent (e.g. pain) fibres from the viscera are present in both sympathetic and parasympathetic systems and pass to the central nervous system without synapsing. Their subsequent paths are similar to those of somatic pain (p. 479). Afferent (reflex) fibres from the lungs, heart, bladder, etc, and visceral sensations of nausea, hunger, rectal distension, etc. also reach the central nervous system, probably along parasympathetic pathways.

Most transmitter chemicals can be classified as adrenergic (for the sympathetic system) or cholinergic (for the parasympathetic system). Other chemicals are being identified and they are usually grouped as purinergic transmitters.

THE CELL-CONTAINING BODY FLUIDS

These comprise mainly the blood and lymph. They are usually classified with the connective tissues.

Blood consists of cells and plasma, slightly less than half its volume being cellular.

Blood cells—these are the red and white blood corpuscles and the platelets.

The red blood corpuscles are enucleate biconcave discs. They contain the pigment haemoglobin and are concerned with the transport of blood gases. The white blood cells are less numerous, nucleated and larger. Their cytoplasm may be granular (neutrophils, eosinophils and basophils) or nongranular (lymphocytes and monocytes). Neutrophils are the most numerous of the granular cells. They have a lobulated nucleus and fine granular cytoplasm staining with neutral dyes. Eosinophils usually have a lobulated nucleus and eosinophilic cytoplasmic granules. Basophils have a kidney-shaped nucleus and basic-staining coarse granules.

Lymphocytes have a large darkly staining nucleus and pale basophilic cytoplasm; monocytes are similar but slightly larger. Platelets are small subcellular elements concerned in blood clotting.

The red blood corpuscles, granular white corpuscles and platelets are all formed in the red bone marrow, the lymphocytes in the lymph nodes and the monocytes in the macrophage system.

Bone marrow is a pulp-like tissue contained within all bones; it may be red or yellow.

Red marrow is present in parts of most bones at birth. It is composed of a reticular network of fine collagen fibres and contains red and granular white cells and their precursors, large multinucleate megakaryocytes which give rise to the platelets, and a few plasma cells and monocytes. Yellow marrow is less vascular and contains many fat cells. It gradually replaces the red marrow and, in the adult, the latter is present only in the vertebrae, ribs, sternum and flat bones.

Anaemia is a deficiency of circulating haemoglobin and may be caused by chronic blood loss, excess red cell destruction, neoplasia affecting the bone marrow or by insufficient blood precursors such as iron and vitamin B_{12}. In most chronic and severe anaemias there is a compensatory increase in the amount of red marrow found in the bones.

THE CARDIOVASCULAR SYSTEM

This comprises the heart (p. 65) and blood vessels, the latter being the arteries, veins, capillaries and sinusoids. The arteries and veins passing to organs and muscles are usually accompanied by the nerves and together form a compact neurovascular bundle. The region of entry of the bundle into an organ is called the **hilus**.

The walls of the arteries possess three coats: an intima, composed of an endothelial lining and a small amount of connective tissue; a media, composed mainly of elastic tissue in the larger arteries, and almost entirely of smooth muscle in the smaller (arterioles) and medium sized arteries; an outer fibrous adventitia. The coats of the veins correspond to those of the arteries but the media contains less smooth muscle and elastic fibres. In the larger veins the adventitia is thicker than the media. In most veins, valves are present. They are formed of paired folds of endothelium, and help to determine the direction of flow. The capillaries, which unite the arteries and veins, have their walls formed of a single endothelial layer of large angular flattened cells.

The direct union between two vessels is called an anastomosis. Sinusoids are thin-walled, dilated channels uniting arteries and veins and are found in the bone marrow, liver, spleen and suprarenal glands. Their endothelial lining contains many mobile cells of the macrophage system.

In some situations blood passes through two capillary beds before returning to the heart; this constitutes a portal circulation. The passage of blood from the stomach, intestine, pancreas and spleen through the liver exemplifies such a system. Short vessels passing through foramina in the skull and joining venous channels (sinuses) inside and veins outside are called emissary veins. Blood may pass in either direction in different circumstances.

Reduction of the blood supply to a region is known as ischaemia and this is of particular importance in the heart and brain. One important degenerative arterial disease which can affect the vessels is arteriosclerosis and this is very prevalent in developed countries. The arterial narrowing produced by the disease may cause local intravascular clotting (thrombosis) to occur. A local thrombus may also be flushed into the blood stream (forming an embolus) and block distal smaller vessels. Local death of an area of tissue or organ due to reduction of its blood supply is known as an infarction. In situations where bacteria can get into the dead tissue (infarcted area) it undergoes putrification, a condition known as gangrene. In some instances it is possible surgically to bypass arterial blockages, thus re-establishing the distal blood supply and preventing infarction and gangrene. The body responds to an injury, e.g. invading bacteria, by the process known as inflammation. The capillaries dilate and white blood cells pass out of the circulation to phagocytose the offending organisms. The area becomes red and hot because of the increased blood supply, and swollen with increased tissue fluid; it is also painful. A collection of dead tissue and dead blood cells is called an abscess, or if under the skin, a boil.

THE LYMPH SYSTEM

Lymph consists of cells and plasma. The cellular and protein proportions are less than in blood.

Lymph cells—mainly lymphocytes.

This system collects tissue fluid and conveys it to the blood stream. It comprises the lymph capillaries and vessels, the lymph nodes and aggregations of lymph tissue in the spleen, thymus and around the alimentary tract. The system forms an extensive network over the body and its fine vessels are not easily identified.

The lymph capillaries are larger than those of the blood; they are composed of a single layer of endothelial cells. The lymph vessels resemble veins and possess many paired valves. The larger collecting vessels open into the venous system near the formation of the brachiocephalic veins.

A lymph node is an aggregation of lymph tissue along the course of a lymph vessel. It is bean-shaped with a number of afferent vessels (conveying lymph to the node) entering its convex surface, and an efferent vessel (carrying lymph away from the node) leaving its hilus (opening). It is surrounded by a fibrous capsule from which fibrous trabeculae pass inwards. It is filled with a reticular network of fine collagen fibres, and the cells are either primitive (lymphocyte precursors) or fixed macrophages. Numerous lymphocytes and a few monocytes lie freely within the meshwork but they are absent peripherally, leaving a subcapsular lymph space. The cells of the outer part of the node (cortex) are densely packed and form oval masses called follicles. The centre of the follicle and the hilar (medullary) regions of the node contain loosely packed lymphocytes.

Lymph aggregations elsewhere in the body consist of a mixture of follicles and loosely packed lymphocytes.

Bacterial infections produce inflammatory responses in the regional lymph nodes. In many malignant diseases neoplastic cells spread via the lymph vessels to the regional lymph nodes and there develop to such an extent as to completely replace the normal tissue of the lymph node and occlude lymph flow. The stagnation of lymph within the tissues due to obstruction of flow produces a swelling of the tissues known as lymphoedema. The term lymphadenopathy is used to describe a generalised enlargement of the lymph nodes though they are not glands in the strict definition of the term (see p. 5).

THE MOBILE CELLS OF THE BODY

These form the macrophage (reticulo-endothelial) system. They are capable of ingesting particulate material (phagocytosis) and are an important defence against micro-organisms. They are found in the lymph tissue (macrophages), blood (monocytes), connective tissue (tissue macrophages), nervous tissue (microglia), and the sinusoids of the bone marrow, liver, spleen and suprarenals (reticulo-endothelial cells).

ORIENTATION

The anatomical position, about which anatomical relations are orientated, is one in which the person stands upright with feet together, eyes looking forward and arms straight down the side of the body, with the palms facing forward.

Structures in front are termed anterior (ventral) and those behind, posterior (dorsal). Structures above are superior (cranial, rostral) and those below, inferior (caudal). Structures may be nearer to (medial) or further from (lateral) the midline and those in the midline are called median. A sagittal plane passes vertically anteroposteriorly through the body and a coronal plane passes vertically at right angles to a sagittal plane. Transverse (horizontal) planes pass horizontally through the body.

Proximal and distal are terms used to indicate the relation of a structure to the centre of the body. The ankle is distal to the knee joint; the shoulder is proximal to the elbow joint. Blood flows distally (peripherally) in the arteries and proximally (centrally) in the veins.

Movements. Forward movement in a sagittal plane is usually flexion and backward movement, extension. Owing to rotation of the lower limb during development, backward movement of the leg extends the hip and flexes the knee; downward movement of the toes is flexion. Upward movement at the ankle joint is dorsiflexion (extension) and downward movement is plantar flexion (flexion). Movement away from the midline in the coronal plane is abduction and movement towards the midline is adduction. Side to side movement of the neck and trunk is termed lateral flexion. Circumduction is the movement when the distal end of a bone describes the base of a cone whose apex is at the proximal end. Rotation occurs in the long axis of a bone; in the limbs it may be medial, towards the midline or lateral, away from it.

RADIOGRAPHIC AND RADIO-ISOTOPIC EXAMINATION OF THE BODY

The study of normal and disordered anatomy in the living human subject by means of X-rays has developed enormously in the last 60 years. Plain radiographs will demonstrate most bony features, for bone is markedly radio-opaque in comparison with the soft tissues of the body. The use of ingested or injected substances, with an absorptive capacity for X-rays different from the body tissues, has allowed many organs to be examined. For example, the ingestion of barium sulphate by mouth, or its introduction per anum by an enema, allows the upper or lower parts of the gastro-intestinal tract to be examined. The gall bladder can be examined radiographically by some iodine-containing substances (taken by mouth or by intravenous injection) which are excreted by the liver into the bile. Certain contrast materials are excreted by the kidneys and thus the structure of the kidney and the ureters can be examined. Direct injection of radio-opaque substances into arteries, veins, lymph vessels, the subarachnoid space, bladder and joints also allows radiographic examination of these structures (Figs. 1.1, 10.1, 11.3, 12.1). Tomographic X-ray techniques enable a thin segment of the body to be examined. More recently, information from such segments has been computer analysed to provide detailed information on all the tissues included in a segment. The apparatus is known as the CAT (computerised axial tomography) scanner. (see Figs. 11.3, 12.3, 32.7).

Some radioisotopes when ingested or injected are preferentially localised in certain organs. The gamma-camera allows a fairly precise localisation of the radioisotope by measuring its emissions and in this way the thyroid, liver, kidney, pancreas and brain can be indirectly studied (Fig. 11.2).

— 2 —

Embryology

(To be read in conjunction with the descriptions of different organs or tissues.)

Embryology is the study of development from the time of fertilisation of the mature female ovum by the male spermatozoon to the time of birth. During this period of about nine calendar months the embryo undergoes a process of growth and differentiation. Interruption, or delay, of the process will almost certainly give rise to congenital abnormalities. Thus a clear knowledge of normal human development will enable the mechanisms of the production of embryological defects to be understood (Fig. 2.1).

The mature ovum is about 120 μm in diameter and is surrounded by a thick coat known as the **zona pellucida**. As it is expelled from the ovary it is caught by fimbriae of the uterine tube and drawn into the ampulla, probably by suction. Here it may be fertilised by a spermatozoon but its life cycle without fertilisation is only 12 to 24 hours. Although only one spermatozoon is necessary for fertilisation many others are probably necessary to aid penetration of the zona pellucida by the fertilising sperm. The mature female and male germ cells contain only half (haploid) the adult number of chromosomes. Fusion of the two germ cells to form the **zygote** restores the adult number (diploid) of chromosomes. During the 3–6 days after fertilisation the zygote is passed to the uterine cavity by contraction of the muscular walls of the uterine tube and by the cilia of the epithelial lining. Failure of this transport can give rise to an ectopic pregnancy in which the embryo develops in the uterine tube or even in the abdominal cavity or on the ovary.

During the passage to the uterus the zygote undergoes cellular division, each cell being called a blastomere. Once the 12–16 cell stage has been reached the structure is known as the **morula** (mulberry) and two distinct cell types are present. The smaller outer cells line the zona pellucida and are nutritive in function, and the larger inner cells (**inner cell mass**) are destined to become the future embryo.

Once the morula has reached the uterus, fluid from the uterine cavity enters its intercellular spaces. Confluence of these spaces produces a central cavity and the structure is now known as a **blastocyst**. The zona pellucida disappears at this stage and allows enlargement of the blastocyst and its implantation into the

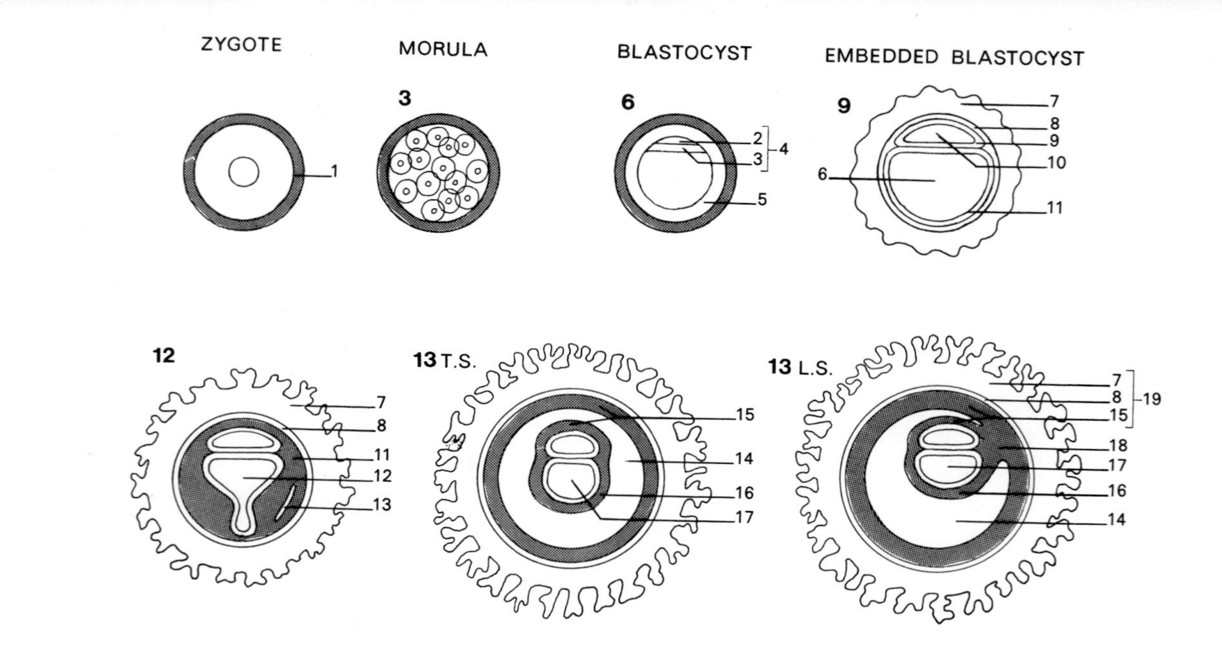

Fig. 2.1 Early development of fertilised egg. The bold numbers indicate the age in days after fertilisation. 1. Zona pellucida 2. Ectodermal germ layer 3. Endodermal germ layer 4. Embryonic disc 5. Trophoblast 6. Exocoelomic cavity 7. Syncytiotrophoblast 8. Cytotrophoblast 9. Amnioblast 10. Amniotic cavity 11. Extra-embryonic mesoderm (e.e.m) 12. Developing yolk sac 13. Exocoelomic cyst 14. Extra-embryonic coelome 15. Somatopleuric e.e.m 16. Splanchnopleuric e.e.m. 17. Yolk sac 18. Body stalk 19. Chorion

uterine wall. This occurs between 7 and 11 days. The cells which surround the cyst are known as the **trophoblast**, while the cells of the slower dividing inner cell mass come to lie at one end of the cyst and are known as the embryoblast. By the 8th day the embryoblast consists of two layers of cells, a flattened layer, adjacent to the cavity, known as the **endodermal germ layer** and a higher columnar layer, adjacent to the trophoblast, known as the **ectodermal germ layer**. At this stage another cavity develops between the ectodermal layer and the trophoblast; this is the **amniotic cavity** (and the cells lining the trophoblast side of the cavity are known as amnioblasts). The ectodermal and endodermal layers together form the **embryonic disc**. By the 9th day a further layer of cells is developing within the blastocyst known as the **extra-embryonic mesoderm** (Heuser's membrane or primary mesenchyme). It lines the trophoblast and is probably formed from it, being continuous with the periphery of the embryonic endoderm. At this stage the cavity of the blastocyst is known as the **exocoelomic cavity (primitive yolk sac)**. Between the 9th and 14th days the exocoelomic cavity is gradually obliterated by an increase of extra-embryonic mesoderm and by the development of the **secondary (definitive) yolk sac**. The yolk sac is lined by endoderm but it is uncertain whether its formation is due to a cavity developing within the endoderm or whether a layer of endoderm grows out from the periphery of the germ disc and nips off part of the exocoelomic cavity to form the sac. By the 13th day the exocoelome is represented only by one or more exocoelomic cysts. The extra-embryonic mesoderm splits to form a further cavity, the **extra-embryonic coelome**. In this way the extra-embryonic mesoderm is divided into two parts; an inner surrounding the yolk sac, embryonic disc and the enlarging amniotic cavity, (the extra-embryonic splanchnopleuric mesoderm), and an outer layer lining the trophoblast, the extra-embryonic somatopleuric mesoderm. The two layers are continuous as a narrow **body stalk** at one end of the embryonic disc and this stalk will develop into the **umbilical cord**. The trophoblast and somatopleure together make up a structure known as the **chorion** and this forms a vital link between the developing embryo and the uterus.

While the embryo is passing along the uterine tube, the uterine lining, influenced by hormones produced by the corpus luteum, undergoes the so-called **decidual reaction**. Menstruation is inhibited, the endometrial glands proliferate and the endometrial vascularity increases. After losing the zona pellucida the blastocyst becomes at first adherent and then embedded in the uterine wall, usually posteriorly. The endometrial covering of the blastocyst is now called the decidua capsularis, that between the embryo and the uterine muscle the decidua basalis, and that over the remainder of the uterus the decidua parietalis. After implantation the trophoblast undergoes rapid proliferation. Cell definition is lost in its outer layer, the nuclei lying in a common cytoplasm called the **syncytiotrophoblast**. An inner layer of cells remains definable and is known as the **cytotrophoblast**. Syncytiotrophoblast with the underlying cytotrophoblast extends in finger-like processes known as primary villi into the decidua. As organisation progresses the extra-embryonic mesoderm of the chorion grows into

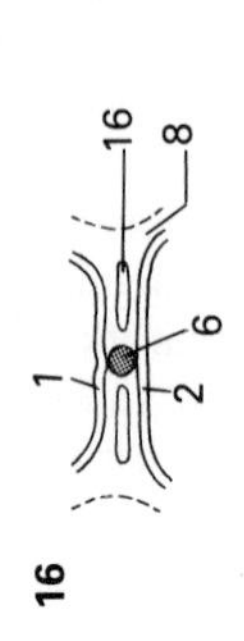
16
1
2
6
16
8

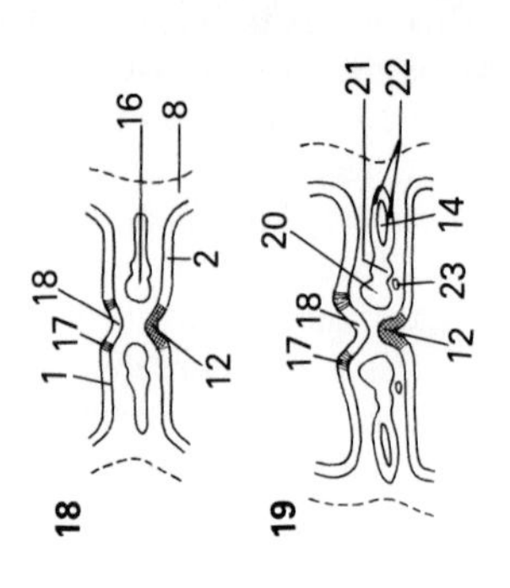
18
1
17
18
12
2
16
8
19
17
18
12
20
23
14
21
22

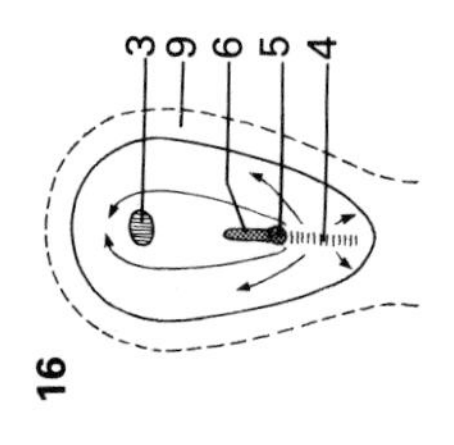
16
3
9
6
5
4

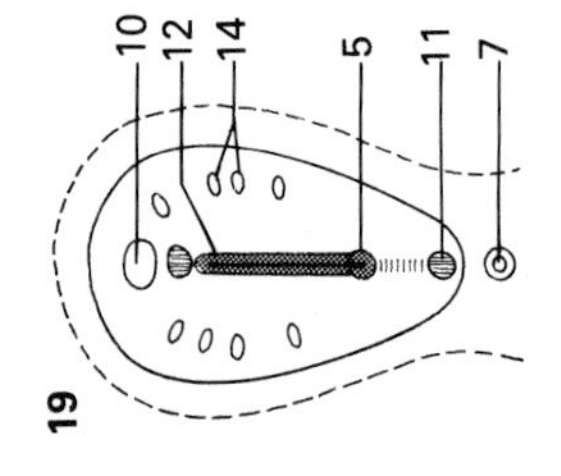
19
10
12
14
5
11
7

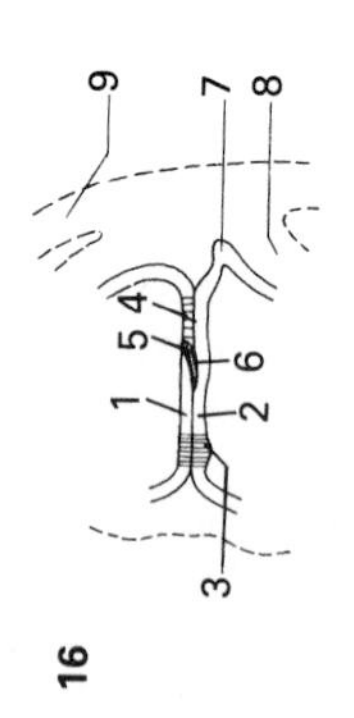
16
3
1
2
5
4
6
9
7
8

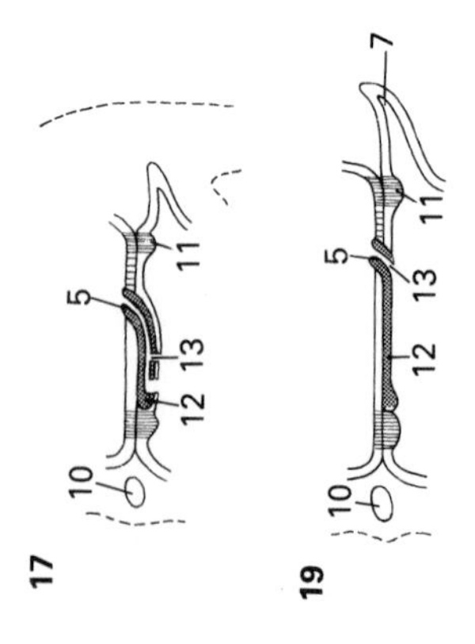
17
10
12
13
5
11
19
10
12
5
13
11
7

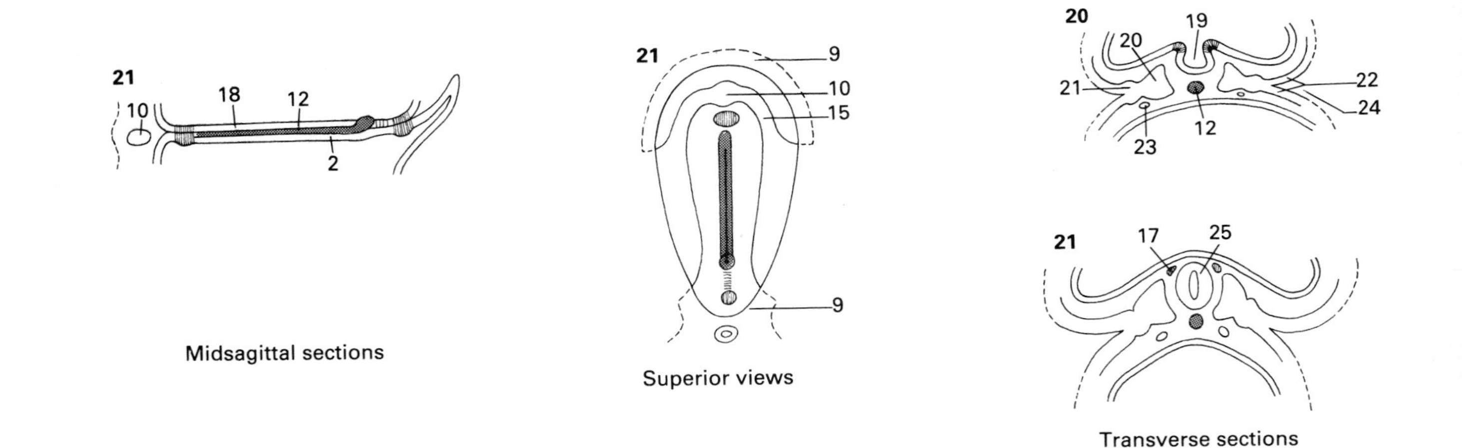

Fig. 2.2 Development of the embryonic disc. The bold numbers indicate the age in days after fertilisation.
1. Ectoderm 2. Endoderm 3. Prochordal plate (oropharyngeal membrane) 4. Primitive streak 5. Primitive knot with central pit 6. Head process 7. Allantois 8. Splanchnopleuric extra-embryonic mesoderm (e.e.m) 9. Somatopleuric e.e.m. 10. Pericardial cavity 11. Cloacal membrane 12. Notochord 13. Neurenteric canal 14. Clefts in lateral plate mesoderm (beginning of intra-embryonic coelome) 15. Pericardioperitoneal canal 16. Intra-embryonic mesoderm 17. Neural crest 18. Neuroectoderm 19. Neural groove 20. Somatic (paraxial) mesoderm 21. Intermediate cell mass mesoderm 22. Lateral plate mesoderm 23. Dorsal aorta 24. Communication between intra- and extra-embryonic coelomes 25. Neural tube

the primary villi and they are then known as secondary villi. When branches of the future umbilical artery and vein have developed within the mesoderm the villi are called tertiary villi. The maternal tissue between the villi becomes a collection of blood lakes, the **intervillous spaces**.

GENERAL BODY FORM (Fig. 2.2)

By the beginning of the 3rd week (15th day) the oval-shaped embryonic disc, made up of ectodermal and endodermal layers, lies between the amniotic and yolk sacs (Fig. 2.2). Both the disc and the sacs are surrounded by splanchnopleure. The body stalk of extra-embryonic mesoderm joins these structures to the chorion. About the 14th day a midline endodermal area of thickening develops at the end of the disc away from the body stalk, this is the **prochordal plate** (future mouth area). In this region the endoderm and ectoderm are closely adherent and remain so throughout development. The area later becomes the **oropharyngeal membrane**. The prochordal plate is at the cranial (cephalic) end of the disc, and the body stalk is at the caudal end. The embryo can thus be oriented. During the 15th and 16th days a midline differentiation occurs in the caudal half of the disc to form the **primitive streak**. From this region cells stream cranially between the ectoderm and endoderm on either side of the midline to form a third germinal layer known as the **mesodermal germ layer** (secondary mesenchyme). The mesoderm extends cranially on each side of the prochordal plate reaching the midline beyond it to form the **cardiogenic mesoderm** which will later contain the **pericardial cavity**.

At the cephalic end of the primitive streak there is a thickening known as the **primitive knot (Hensen's node)** in which there is a central pit, the **primitive pit**. About the 17th day a midline collection of cells, the **head process**, grows cranially from Hensen's node between the ectodermal and endodermal layers to reach the prochordal plate. For a short time the head process replaces the endoderm being in direct contact with the yolk sac. About the 18th day the head process becomes canalised to form the **notochordal (neurenteric) canal** which temporarily unites the amniotic and yolk sac cavities. The head process soon folds along its longitudinal axis to form a solid column of cells known as the **notochord**, this being the precursor of the vertebral column. The neurenteric canal becomes obliterated and the endoderm reunites beneath the notochord to separate it from the yolk sac. This process is completed by the end of the 3rd week.

The midline area of ectoderm caudal to the primitive streak is closely related to the endodermal layer. Together they form the **cloacal membrane** which will later become the **urogenital** and **anal membranes**. A diverticulum arises from the yolk sac adjacent to the cloacal membrane, and extends into the body stalk. It is known as the **allantois**. Although the allantois of lower animals plays an important role in the developing urogenital system it remains rudimentary in man. By the end of the 3rd week of development three embryonic germ layers have appeared and some differentiation has already taken place, the embryonic disc becoming elongated and pear-shaped with a recognisably wider cephalic end. The ectoderm and endoderm remain in contact at the prochordal plate and at the cloacal membrane, and the precursors of the vertebral column (notochord) and the pericardial cavity are already discernible. Development now enters the **embryonic period** (4th–8th weeks) during which time longitudinal and side-to-side folding of the disc turns it into a recognisable body form, and the main organ systems are laid down.

The ectodermal surface of the embryonic disc is carried by the longitudinal (**head** and **tail folds**) and side-to-side folding of the disc over the outer surface of the embryo. The head fold not only swings the oropharyngeal membrane (prochordal plate) to the ventral surface of the embryo, it also brings the pericardial region to the ventral surface, caudal to the oropharyngeal membrane. The tail fold brings the cloacal membrane to the ventral aspect of the embryo, and both the head and tail folds tend to narrow the attachment of the embryo to the body stalk. The simultaneous ventral folding of the sides of the embryonic disc enclose an endoderm-lined tube of yolk sac which extends from the oropharyngeal membrane to the cloacal membrane. This tube is the future alimentary tract and its connection to the yolk sac is soon only a narrow stalk, the **vitelline duct**. The further development of the embryo can now be considered as being from three concentric cylinders (representing the three germ layers). They are curved longitudinally to form a 'C', the opening of which is attached to the chorion by the body stalk, which in turn contains the remains of the yolk sac. The outer cylinder (representing the ectoderm) is not completed ventrally until:

(a) the pericardial region has come to lie caudal to the oropharyngeal membrane.
(b) some communications between the extra-embryonic coelome and a new cavity within the mesoderm (the intra-embryonic coelome) have been closed.
(c) the development of the alimentary and urogenital systems allows completion of the anterior abdominal wall (Fig. 2.3)

As the ectoderm comes to surround the embryo, the amniotic cavity, which started as a small cavity adjacent to the ectoderm, also expands and surrounds the embryo. The embryo is now enclosed in the amniotic cavity, the body stalk being narrowed down to the thin elongated umbilical cord. The three primitive germ layers become differentiated during the folding process into the tissues and organs of the body. As an understanding of mesodermal differentiation is the key to this process it will be considered first.

MESODERMAL GERM LAYER (Figs. 2.2, 2.4)

The primitive mesoderm is formed from the primitive streak and extends forwards on each side of the notochord and oropharyngeal membrane; it also forms the midline pericardial region, the latter coming to lie caudal to the oropharyngeal membrane after development of the head fold. The mesoderm on each side of the midline differentiates into a medial block of **paraxial mesoderm** and the laterally placed **lateral plate mesoderm**. From the end of the third week the paraxial mesoderm on each side of the midline becomes transversely divided into the segmented blocks known as **somites**. This process starts adjacent to the cranial end of the notochord and progresses in caùdal sequence. It is customary to express the age of the embryo between the 4th and 6th weeks by the number of somites that have appeared, these being discernible on external examination. By the end of the 4th week approximately 30 pairs of somites have appeared and by the end of the 5th week 42–44 pairs are present. Some of the most cranial somites are incorporated in the skull, and the most caudal somites disappear, so that the number of somites corresponds to the number of spinal nerves. By the 6th week the embryo has taken up a recognisable form and the crown to rump measurement (**crown-rump length**) is thereafter used to assess growth.

Each somite differentiates into a ventromedially placed **sclerotome** and a dorsolateral **dermomyotome**. The sclerotome in turn differentiates into young connective tissue (mesenchyme) which forms the bone, cartilage and packing tissue of the trunk. During the 4th week each sclerotome migrates medially, and surrounds the notochord, contributing to the vertebral bodies. Each adult vertebral body is in fact intersegmental in position, i.e. it is formed from parts

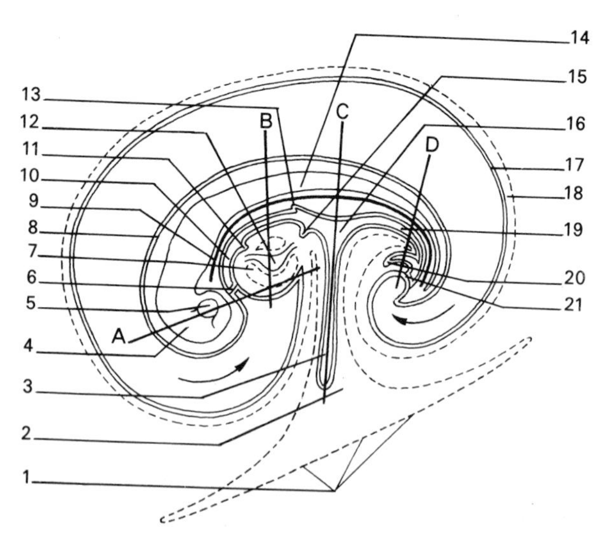

Fig. 2.3 Longitudinal sagittal section of a 30 day embryo. Transverse sections of the embryo at sites A–D are shown in Figure 2.4. The two arrows are in the enlarging amniotic cavity.
1. Placental site 2. Body stalk (umbilical cord) 3. Yolk sac (vitello-intestinal duct) 4. Forebrain 5. Eye rudiment 6. Oropharyngeal membrane 7. Pericardial sac 8. Head fold 9. Notochord 10. Foregut 11. Lung bud 12. Heart tube 13. Dorsal pancreatic bud 14. Spinal cord 15. Ventral pancreatic bud 16. Midgut 17. Amniotic ectoderm 18. Extra-embryonic mesoderm 19. Hindgut 20. Allantois 21. Cloacal membrane

of two adjacent somites. The nucleus pulposus of the adult intervertebral disc is produced by mucoid degeneration of the notochord. The skull is also partly developed from the sclerotomes. Initially two distinct parts, the **neurocranium** surrounding the brain, and the **viscerocranium** forming the face, are present.

The dermomyotome has two distinct elements, an outer **dermatome** and an inner **myotome**. The dermatome migrates deep to the ectoderm to form the dermis and subcutaneous tissues of the body. The myotome differentiates into young muscle cells, myoblasts which form the striated somatic musculature of the body. A smaller dorsal block of the myotome (epimere) is innervated by the dorsal ramus of a spinal nerve while the larger ventral part of the myotome (hypomere) is innervated by its ventral ramus. Not all striated muscle, however, is somatic in origin. The muscles derived from the pharyngeal arches, e.g. those of mastication, of facial expression and of the pharynx and larynx develop from the mesenchymal core of the arch. It is debated whether the limb musculature develops from the mesenchymal core of the limb bud or whether the appropriate

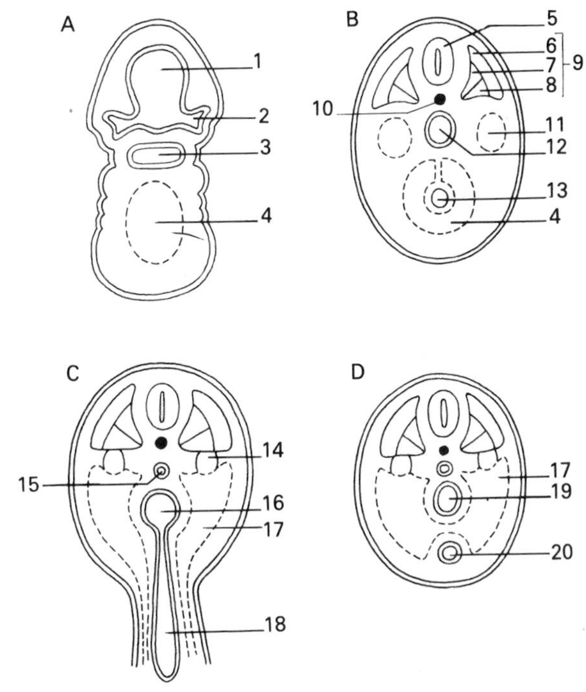

Fig. 2.4 Transverse sections of 30 day embryo cut at sites shown in Figure 2.3. 1. Forebrain vesicle 2. Eye rudiment 3. Stomodeum 4. Pericardial cavity 5. Neural tube 6. Dermatome 7. Myotome 8. Sclerotome 9. Somitic (paraxial) mesoderm 10. Notochord 11. Pericardioperitoneal canal 12. Foregut 13. Primitive heart tube 14. Intermediate cell mass mesoderm 15. Dorsal aorta 16. Midgut 17. Peritoneal cavity bounded by lateral-plate mesoderm 18. Yolk sac in body stalk 19. Hindgut 20. Allantois

somatic myotomes are carried into the limb bud with the ventral rami of the respective spinal nerves. Further uncertainty exists about the development of the extrinsic eye muscles and the tongue muscles. It is thought that the eye muscles may be derived from the so-called pre-otic myotomes situated on either side of the prochordal plate, and the tongue muscles by the ventral migration of occipital myotomes. Each myotome acquires its appropriate cranial nerve innervation.

At the end of the 3rd week clefts appear within the lateral plate mesoderm. These clefts coalesce to form a longitudinal cavity on each side of the midline, the **intra-embryonic coelome**. The lateral plate mesoderm is thus divided into a layer lining the ectoderm, known as somatopleuric mesoderm, and a layer covering the endoderm, known as splanchnopleuric mesoderm. Coincident with the development of the intra-embryonic coelome within the lateral plate mesoderm a further space is forming within the pericardial mesoderm—the pericardial cavity. The latter unites with the coelome on each side to form a single ∩-shaped cavity. The longitudinal cavities adjacent to the pericardial sac become the **pericardioperitoneal canals** from which develop the future pleural sacs. Whilst these cavities are uniting, the clefts in the lateral plate mesoderm extend into the extra-embryonic mesoderm and communication soon exists between intra-embryonic and extra-embryonic coelomic cavities. The intra-embryonic somatopleuric mesoderm becomes continuous with the extra-embryonic mesoderm over the amniotic cavity, and the intra-embryonic splanchnopleuric mesoderm becomes continuous with the extra-embryonic mesoderm over the yolk sac (Figs. 2.2, 2.3, 2.4). The communication between intra-embryonic and extra-embryonic coelomic cavities is limited and does not extend cranially into the pericardial cavity or into the short length of intra-embryonic coelome immediately adjacent to it on each side, the pericardioperitoneal canals (the future pleural cavities).

The communications are thought to have a nutritional function for the embryo but as the umbilical and placental circulation develops, the intra-embryonic coelome becomes cut off and the extra-embryonic coelome is obliterated by the enlarging amniotic sac (Fig. 2.3). The enclosed coelomic cavity surrounds the gut in such a way that the splanchnopleuric mesoderm from each side unites dorsally and ventrally and forms the **dorsal** and **ventral mesenteries** of the gut. This mesoderm also forms the smooth muscle, connective tissue and blood vessels of the gut and its derivatives.

The pericardial, pleural and peritoneal cavities are formed from the intra-embryonic coelome by septa developing cranial and caudal to the pericardioperitoneal canals. The separation of the midline pericardium from the pleural cavities is by two tranverse folds known as the **pleuropericardial membranes**. The folds overlie the common cardinal veins which remain in position as the heart and pericardium descend and the lungs expand, laterally and upwards. The pericardioperitoneal canals are stretched over the enlarging lungs. The folds separating the pleural cavities from the peritoneal cavity later form part of the diaphragm. The diaphragm is formed largely from a single midline ventral mass of mesenchyme, the **septum transversum**, lying between the pericardium and the cranial lip of the yolk sac. Additionally, paired transverse dorsolateral folds, the **pleuroperitoneal membranes**, grow out to fuse with the septum transversum in

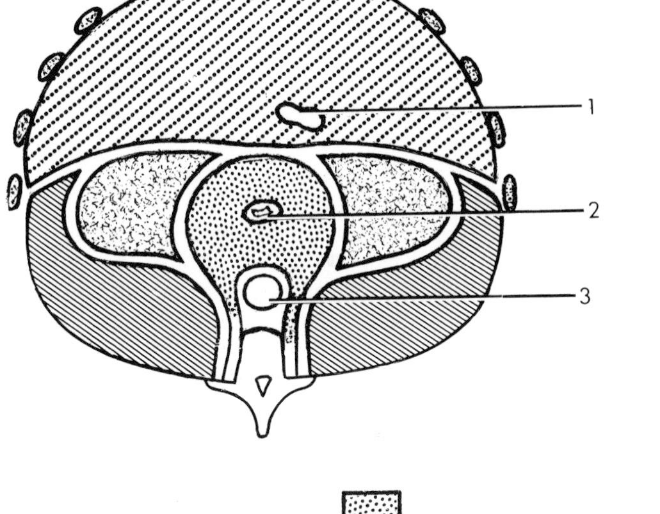

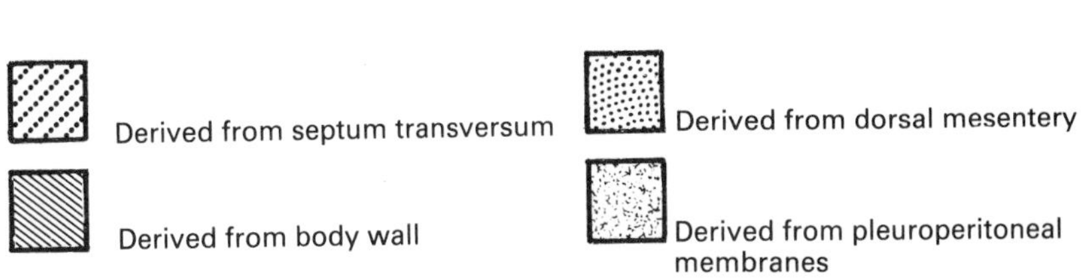

Fig. 2.5 Development of the diaphragm. Contributions from the septum transversum and the body wall are attached to the xiphoid process, ribs and costal cartilages. The mesenteric contribution (the crura) is attached to the vertebrae. The pleuroperitoneal membranes have no bony attachment.
1. Inferior vena cava 2. Oesophagus 3. Aorta

front, and the mesentery of the lower oesophagus behind. The diaphragm is completed by peripheral contributions from the body wall (Fig. 2.5). In the abdomen the dorsal mesentery is retained from the lower oesophagus to the upper rectum, being most easily recognisable as the small gut mesentery and being greatly expanded and folded to form the greater omentum. The ventral mesentery is absent below the level of the first part of the duodenum which means that the two lateral compartments of the coelomic cavity can unite to form a single peritoneal cavity (p. 115).

UROGENITAL DEVELOPMENT (Figs. 2.2 to 2.9)

A longitudinal nonsegmented condensation of mesoderm, the intermediate cell mass (nephrogenic cord), develops on each side of the midline between the paraxial and the lateral plate mesoderm. This region develops craniocaudally into

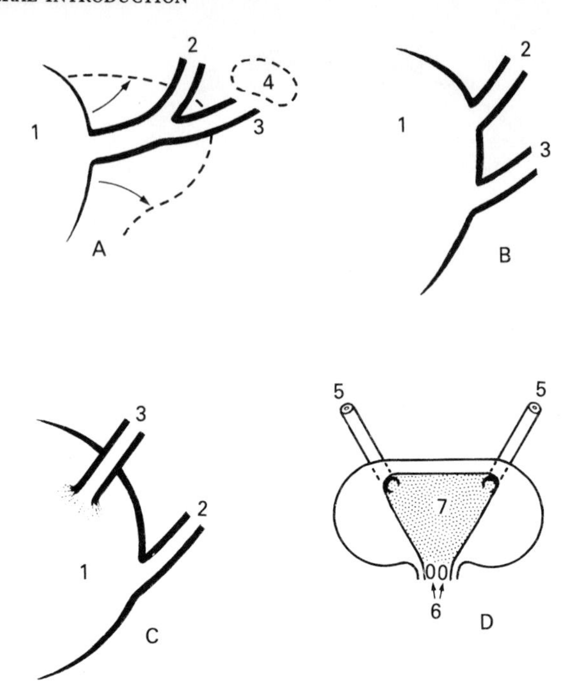

Fig. 2.6 Development of the urinary bladder. **A.** Earliest stage with mesonephric duct of left side entering the urogenital sinus. **B.** The sinus enlarges incorporating the mesonephric duct and the ureteric bud which contribute to the bladder wall. **C.** Differential growth and increasing size bring about a reversal in the positions of ureter and duct. In the male, the duct becomes the ductus deferens and, joining with the duct of the seminal vesicle, forms the ejaculatory duct. In the female, the mesonephric duct regresses. **D.** The situation at birth. The triangle between the ureters and the urethra forms the trigone (stippled) and is derived from the mesenchyme of the two mesonephric ducts. The rest of the bladder is derived from the endoderm of the urogenital sinus.
1. Urogenital sinus 2. Mesonephric duct 3. Ureteric bud 4. Metanephric cap 5. Ureter 6. Ejaculatory ducts 7. Trigone

the **pronephros**, then the **mesonephros** (primitive kidney) and lastly into the **metanephros** (definitive kidney). Of the first two components, only the **mesonephric (Wolffian) duct** normally persists. The tubules of the metanephros form the secreting part of the adult kidney and a ureteric bud from the caudal end of the mesonephric duct forms the collecting duct system and joins up with the tubules of the metanephros (Fig. 2.6A). Failure of these parts to unite may produce a condition called a polycystic kidney.

A second longitudinal duct, the **paramesonephric (Mullerian) duct** appears in the 6th week on each side, lying parallel and lateral to the mesonephric duct. All four ducts pass caudally and open close together into the terminal part of the alimentary tract called the **cloaca**. During the 4th week, mesenchymal tissue

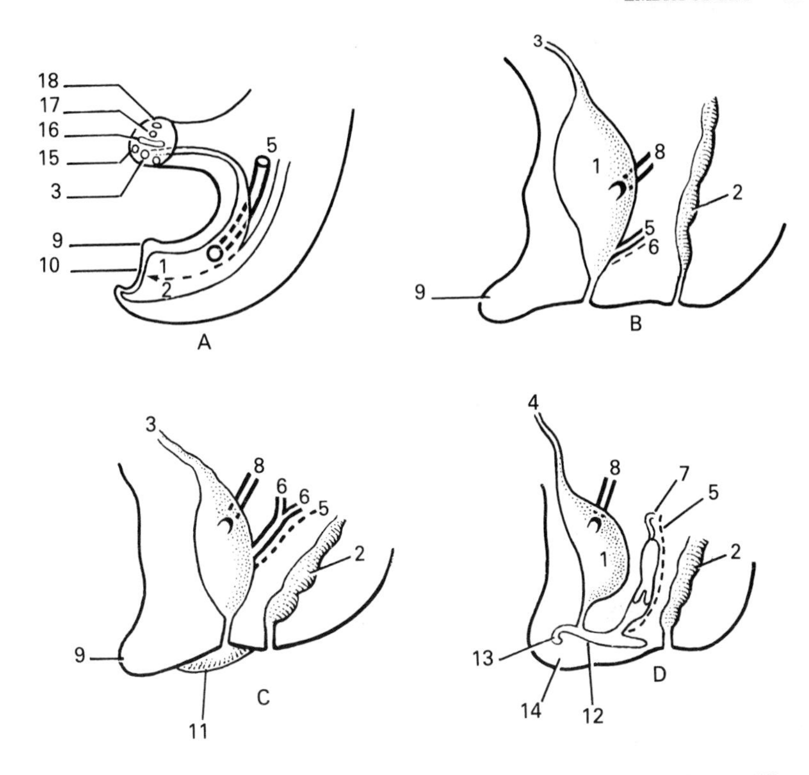

Fig. 2.7 Development of the urogenital system. **A.** Early undifferentiated stage. The dotted arrow represents the urorectal septum which divides the cloaca into the urogenital sinus (1) in front and the recto-anal canal (2) behind. The cloacal membrane (10) is likewise divided. **B.** Early male stage (see also Fig. 2.9). **C.** Early female stage. **D.** Later female stage (see also Fig. 2.9).
1. Urogenital sinus 2. Recto-anal canal 3. Allantois 4. Urachus 5. Mesonephric duct 6. Paramesonephric duct 7. Uterine tube 8. Ureter 9. Cloacal eminence (genital tubercle) 10. Cloacal membrane 11. Genital fold 12. Labia minora 13. Clitoris 14. Labia majora 15. Right and left umbilical arteries 16. Coelomic cavity 17. Vitello-intestinal duct 18. Left umbilical vein

grows into each side of the cloaca and forms a **urorectal septum**. This divides the cloaca into a larger ventrally situated **urogenital sinus**, receiving the four ducts, and a smaller dorsally placed **recto-anal canal**. The cloacal membrane is similarly divided into urogenital and anal parts (Fig. 2.7A).

The portion of the urogenital sinus which is cranial to the points of entry of the mesonephric ducts is the definitive bladder and proximal urethra; the portion distal to the ducts becomes the more distal urethra. As development proceeds, the distal ends of the mesonephric ducts are incorporated into the wall of the sinus so that the ureters come to open independently into the upper part of the sinus (Figs. 2.6B, C; 2.7B). That part of the sinus between the openings of the ureters and the urethra becomes the **trigone** of the bladder; it is derived from

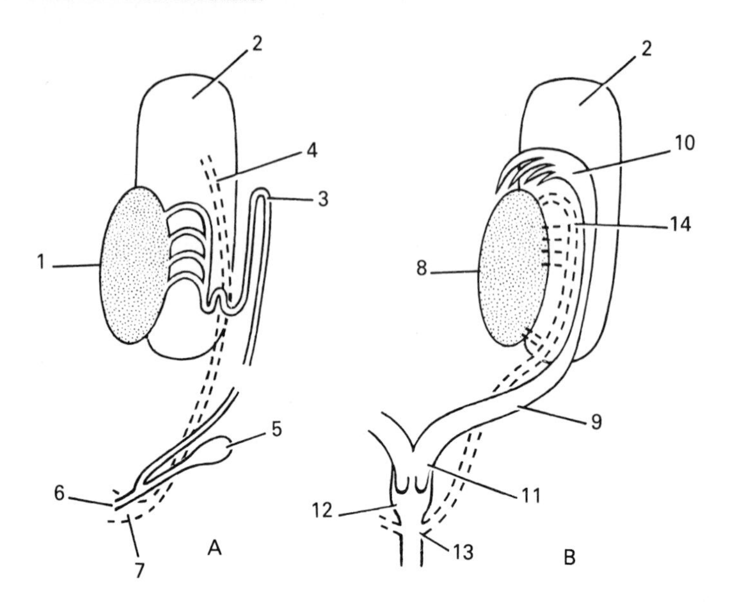

Fig. 2.8 Differentiation and development of the internal genitalia. **A.** Male **B.** Female.
1. Testis 2. Mesonephros 3. Mesonephric duct 4. Vestigial paramesonephric duct 5. Seminal vesicle 6. Ejaculatory duct 7. Prostatic utricle 8. Ovary 9. Paramesonephric duct forming the uterine tube and its fimbriated extremity (10), the uterus (11) and the upper vagina (12) 13. Opening of vestigial mesonephric duct into upper vagina 14. Vestigial mesonephric duct in future broad ligament

mesenchyme (Fig. 2.6D). The apex of the urogenital sinus (the future bladder) is connected to the allantois by a narrow duct called the **urachus**. This usually is obliterated before birth, forming the median umbilical ligament. However, it may remain patent and urine may then leak from the umbilicus after birth.

In the male, the mesonephric duct persists as the ductus (vas) deferens. Its cranial end becomes associated with the developing testis medial to the mesonephros (see Figs. 2.8A and p. 115). The paramesonephric duct regresses and largely disappears though occasionally cysts of the testis arise from it. The prostate gland is formed from urethral outgrowths.

In the female, the paramesonephric ducts persist and the cranial end, opening directly into the peritoneal cavity, becomes the uterine tube and overhangs the developing ovary (see Figs. 2.7C and D; 2.8B and p. 185). The more caudal portions of the two ducts cross ventral to the mesonephric ducts, fuse together and form the definitive uterus and upper vagina. The part of the urogenital sinus cranial to the point of entry of the paramesonephric ducts contributes to the urethra. The caudal part of the sinus forms the vestibule and a small portion of the vagina. Both urethra and vagina open into the vestibule. The mesonephric duct regresses and largely disappears though occasional cysts in the broad liga-

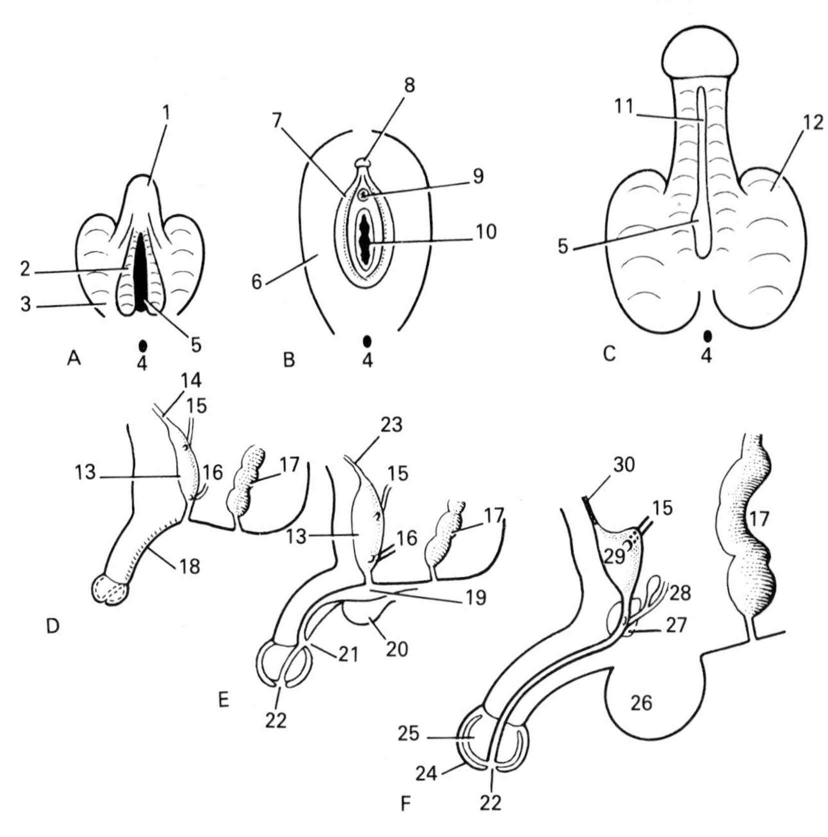

Fig. 2.9 Development of the external genitalia. **A.** The indifferent stage—about three weeks. **B.** Towards female. **C.** Towards male. **D.**, **E.** and **F.** Later stages in the development of the male genitalia.
1. Cloacal eminence (genital tubercle) 2. Genital folds 3. Cloacal swellings 4. Anus 5. Urogenital membrane 6. Labia majora 7. Labia minora 8. Clitoris 9. Urethral orifice 10. Vaginal orifice 11. Urethral groove 12. Lateral scrotal swelling 13. Urogenital sinus 14. Allantois 15. Ureter 16. Mesonephric duct 17. Rectal canal 18. Urethral groove 19. Urethral fold 20. Scrotal fold 21. Primary urethral orifice 22. Secondary urethral orifice 23. Urachus 24. Prepuce 25. Glans penis 26. Scrotum 27. Prostate 28. Ductus deferens 29. Bladder 30. Median umbilical ligament

ment arise from it. Incomplete fusion of the paramesonephric ducts leads to varying degrees of divided uterus.

The external genitalia develop from swellings around the urogenital and anal membranes. (Fig. 2.9) At the end of the 3rd week mesenchyme from the primitive streak migrates and surrounds the cloacal membrane forming a midline **cloacal eminence** adjacent to the allantois, and, on each side, a medial **genital fold** and a lateral **cloacal (labioscrotal) swelling**. After the separation of the urogenital and anal membranes the part of the cloacal swelling adjacent to the anal membrane comes to surround the latter as the **anal fold**. In the female the genital folds border the urogenital membrane and then become the labia minora, the cloacal swellings becoming the labia majora. The cloacal eminence becomes known as the **genital tubercle** and later forms the clitoris. In the male the genital folds become the **urethral folds** and form the body of the penis. A groove along the ventral aspect becomes enclosed and forms the bulbous and spongy urethra. The genital tubercle forms the glans penis. The cloacal swellings become the **scrotal swellings** which fuse in the midline and form the adult scrotum.

The gonad develops in the 4th week as the **gonadal ridge** of mesenchyme situated on each side of the midline, between the mesonephros and the dorsal mesentery. The primitive germ cells are first formed from the endoderm of the yolk sac adjacent to the allantois but they migrate dorsally and along the dorsal mesentery to reach the gonadal ridges in the 6th week (see pp. 115 and 191).

CARDIOVASCULAR DEVELOPMENT (See also Figs. 2.10, 2.11 and p. 73)

The cardiovascular system is an early mesenchymal derivative distributing nutriment through the embryo. The blood vessels are derived from clusters of primitive mesenchymal cells known as angioblasts. Confluence of intercellular clefts within the clusters form blood channels. The angioblasts lining the channels become the flattened endothelium of the vessels, while the clumps of cells enclosed within the channels become the red blood cells. White cells develop later. This process is first seen in the walls of the yolk sac but is rapidly followed in the chorion and within the embryo itself. The heart starts to develop at the end of the 3rd week as two tubes; initially lying ventral to the pericardium, they come to lie dorsal to it as the head fold develops. About this time the tubes fuse and invaginate the pericardium to produce the dorsal mesentery. Early in the 4th week the heart becomes disproportionally elongated and buckles in ∽-shaped fashion within the pericardium. The main arteries and veins thus become more closely related over the craniodorsal aspect of the pericardium (see also p. 73). The dorsal mesentery of the pericardium breaks down and the defect, situated between the arteries cranially and the veins caudally, becomes the transverse sinus. The first heart beat occurs about the 4th week. As the circulation is not established at this time, the main effect is to produce some ebb and flow of blood and also possibly to produce movement of fluid between the intra- and extra-embryonic coelomic cavities.

The arterial end of the heart tube (**conus arteriosus**) divides spirally and unequally into a **ventral aorta** and a **pulmonary trunk**. The right and left **dorsal aortae** develop on each side of the notochord in the 4th week. They are joined at their cranial ends to the ventral aorta by a series of arches, the pharyngeal arch arteries (Fig. 2.10). The first arch arteries have largely disappeared before the last (6th) appear. Caudal to the 6th arch vessels, the dorsal aortae later fuse and form a single vessel. The paired vitelline branches from the dorsal aortae to the yolk sac and gut also fuse and form the coeliac, and superior and inferior

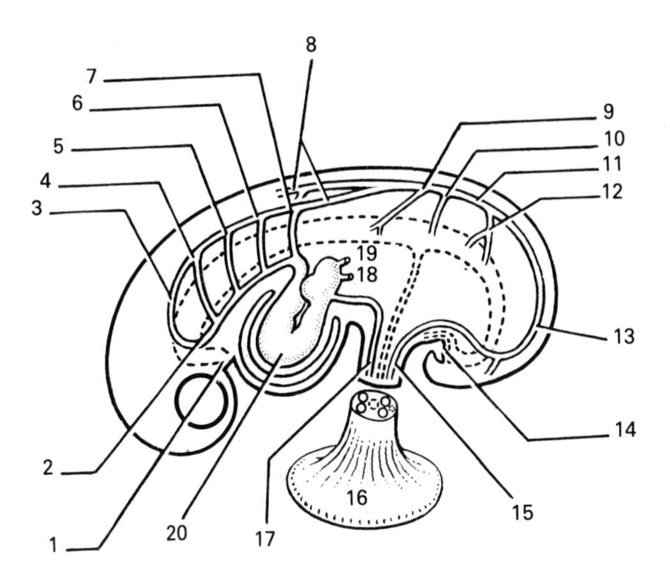

Fig. 2.10 Diagram indicating the early circulation. The gut is shown in dotted lines. The vitello-intestinal duct and the allantois are both shown entering the umbilical cord. Blood returning to the heart tube in the umbilical veins (17), the vitelline veins (18) and the common cardinal veins (19), enters the sinus venosus, passes into the auricle and then the ventricle (20). Blood is expelled through the conus arteriosus into the ventral aorta (2).
1. Oropharyngeal membrane 2. Ventral aorta 3, 4, 5, 6, 7. Pharyngeal arch arteries 8. Dorsal aortae 9. Coeliac trunk 10. Superior mesenteric artery 11. Abdominal aorta 12. Inferior mesenteric artery 13. Left common iliac artery 14. Anal membrane 15. Left umbilical artery 16. Placenta 17. Left umbilical vein 18. Vitelline vein 19. Common cardinal vein 20. Ventricle of heart tube

mesenteric arteries to the abdominal foregut, midgut and hindgut respectively. The umbilical arteries are terminal branches of the dorsal aorta. Each is represented in the adult by the common and internal iliac and superior vesical arteries. The terminal (distal) remnant of the umbilical artery, the lateral umbilical ligament, passes from the superior vesical artery towards the umbilicus. Other visceral branches of the dorsal aorta supply the urogenital ridge and its derivatives (kidney and gonad). Somite branches pass in an intersegmental manner to the spinal cord, the body wall and to the limbs.

The venous drainage parallels the arterial supply in many ways (Fig. 2.11A, B, C). The cranial end of the embryo is drained by paired anterior cardinal veins which join with posterior cardinal veins and form the left and right common cardinals. These drain into the **sinus venosus** (the venous end of the primitive heart tube). The anterior cardinal veins become the jugular, brachiocephalic and superior vena caval complex. The posterior cardinal veins have associated with them subcardinal and supracardinal veins. These together eventually form the inferior vena cava and azygos systems of veins with some of their tributaries. The left umbilical vein persists but the right disappears. The vitelline and umbilical

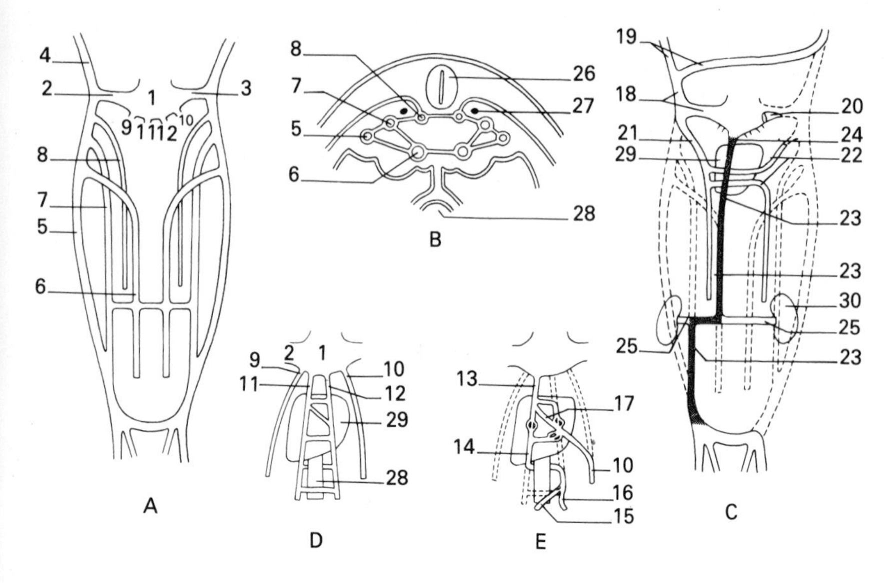

Fig. 2.11 Development of the systemic venous system (**A.**, **B.** and **C.**) and of the portal venous system (**D.** and **E.**). The dotted lines represent vessels which disappear.
1. Sinus venosus 2. Right common cardinal vein 3. Left common cardinal vein 4. Anterior cardinal vein 5. Posterior cardinal vein 6. Subcardinal vein 7. Supracardinal vein 8. Azygos vein 9. Right umbilical vein 10. Left umbilical vein 11. Right vitelline vein 12. Left vitelline vein 13. Hepatic vein joining IVC 14. Portal vein 15. Superior mesenteric vein 16. Splenic vein 17. Ductus venosus 18. Superior vena cava 19. Brachiocephalic veins 20. Coronary sinus 21. Adult azygos vein 22. Hemiazygos and accessory hemiazygos veins 23. Definitive inferior vena cava (shaded) 24. IVC link with hepatic veins 25. Renal veins 26. Neural tube 27. Sympathetic chain 28. Primitive midgut 29. Liver 30. Kidney

veins become involved in the development of the liver and the hepatoportal vein (Fig. 2.11D and E). The ductus venosus by-passes the liver by joining the left umbilical vein to the inferior vena cava just before the latter enters the heart. The hepatoportal vein is formed by parts of the two vitelline veins and their communicating vessels around the upper part of the midgut.

ENDODERMAL GERM LAYER (Figs. 2.2, 2.3, 2.4, 2.10, 9.5)

With the longitudinal and side-to-side folding of the embryo the primitive endodermal lining of the yolk sac is fashioned into a longitudinal tube extending between the oropharyngeal and the cloacal membranes. This forms the primitive gut. The gut remains attached to the yolk sac by a stalk of diminishing diameter, the vitelline duct. A series of paired vitelline arteries form around the duct and communicate with the dorsal aortae, but when the latter fuse, the vitelline arteries also fuse and become reduced to three midline arteries, the coeliac and the superior and inferior mesenteric arteries.

The **primitive foregut** extends from the oropharyngeal membrane to approximately the level of origin of the liver bud. The oropharyngeal membrane breaks down in the 3rd week. During the 4th and 5th weeks the cranial end of the foregut acquires a series of paired lateral pharyngeal grooves which divide the adjacent mesenchyme into six bars of tissue known as the pharyngeal (branchial) arches. The derivatives of the arches are considered on pages 374, 388 and 418. In the region immediately caudal to the last pair of pharyngeal pouches, and preceding their full development, a ventral midline **respiratory (tracheobronchial) diverticulum** appears which grows caudally. In the 6th week the diverticulum bifurcates to form the paired **lung buds**. The lung buds grow into the pericardioperitoneal canals and acquire a covering of splanchnopleuric mesoderm which develops into the connective tissue, smooth muscle and blood vessels of the adult organ. The lining of the canal becomes the pulmonary and parietal pleura.

Beyond the respiratory diverticulum, the **foregut** tube continues as the oesophagus, passes through a mass of mesenchyme, the **septum transversum** (future diaphragm) and then elongates and enlarges to form the stomach and the proximal part of the duodenum. The latter lies on the posterior abdominal wall and gives origin to the liver and pancreas (see pp. 151 and 157). The abdominal gut has a **dorsal mesentery** in most of its length but the ventral mesentery is restricted to the foregut. The **ventral mesentery** is attached to the anterior abdominal wall above the body stalk and the lower free edge contains the umbilical veins in front, and the liver diverticulum near the duodenum. The liver enlarges and distorts the ventral mesentery. The part between the anterior abdominal wall and the liver is the future falciform ligament and the part between the liver and the stomach is the future gastrohepatic (lesser) omentum. In the adult, the free edge of the falciform ligament passes from the umbilicus to the liver and contains the ligamentum teres (obliterated left umbilical vein). The free edge of the lesser omentum (its right border) passes between the liver and the duodenum and contains the bile duct, hepatic artery and the portal vein.

The spleen develops in the left side of the dorsal mesentery which also elongates to form the greater omentum, a long fold hanging from the greater curvature of the stomach. The pancreas develops from a ventral bud adjacent to the liver bud and a more proximal dorsal bud. The buds come together dorsally and the duct system is formed (see p. 157). The abdominal foregut and its derivatives are supplied by the branches of the coeliac artery.

The **midgut** extends from the origin of the liver diverticulum on the duodenum to approximately the junction of the middle and distal thirds of the adult transverse colon. The midgut has only a dorsal mesentery, and running in this are the branches of the superior mesenteric artery. This part of the gut later differentiates into most of the small intestine and about two thirds of the large intestine (see p. 118 for further developments).

The **hindgut** extends from the distal third of the transverse colon to the anal membrane. Some parts have a dorsal mesentery, others lie on the posterior abdominal wall. This region of the gut tube, supplied by the inferior mesenteric artery and its branches, differentiates into the distal part of the transverse colon, the descending and sigmoid colon, the rectum and much of the anal canal. The terminal hindgut (cloaca) partakes in the development of both the urogenital system (bladder) and the alimentary system (rectum), and has been studied above. The anal membrane usually disappears after the 9th week but may be present at birth (imperforate anus) and has to be ruptured operatively. The details of the further development of the gastro-intestinal tract are described on pages 118 and 182.

ECTODERMAL GERM LAYER (Figs. 2.2, 2.3, 2.4, 31.2)

The primitive ectoderm initially forms the dorsal aspect of the embryonic disc, being continuous at its periphery with the ectoderm lining of the amniotic cavity. At the beginning of the 3rd week the midline cranial portion of ectoderm becomes differentiated, into the **neural plate**, the future nervous system. The midline grooving, folding and tube formation of this region to form the central nervous system is considered on page 438. The neural crest, a dorsal longitudinal plate of cells on each side of the neural tube, becomes differentiated into the ganglia of the dorsal roots and the cranial nerves, their peripheral and central branches, the ganglia of the autonomic nervous system and also the medulla of the suprarenal glands.

The remainder of the ectoderm extends, surrounds the developing embryo and forms the epidermis and skin derivatives, under which develops the mesenchymal dermis. The ectoderm also contributes to the formation of the cutaneous and mammary glands. In early development the ectoderm over the prochordal plate is depressed to form a pit, the **stomodeum**. Mesenchymal condensations around the stomodeum during the 5th week raise ectodermal processes (pharyngeal arches) which contribute to the developing face (p. 356). The early breakdown of the oropharyngeal membrane (3rd to 4th week) has created some uncertainty as to which structures in the mouth are ectodermal and which are endodermal in origin. Probable ectodermal derivatives are the gums, enamel organs of the teeth, nasal sinuses and the parotid salivary glands. A midline ectoderm diverticulum from the stomodeum known as **Rathke's pouch** extends dorsally to the base of the brain and contributes to the development of the pituitary gland. An area of ectoderm on each side of the head adjacent to the developing optic cup (itself an outgrowth from the brain) becomes the lens of the eye. The ectoderm covering the pharyngeal arches has transverse clefts (pharyngeal clefts) which mostly disappear. The first, however, contributes to the formation of the external acoustic meatus (p. 388).

The ectoderm over the cloacal membrane becomes depressed and forms a pit, the **proctodeum**. The migration of mesenchyme raises ectodermal folds around the membrane which have already been referred to. The folds form the external genitalia and also the lower part of the anal canal.

By the end of the 2nd month of development the embryo has taken on the familiar appearance with a large head, short limbs and wide body stalk. The period extending from the 3rd month to birth is known as the **fetal period** during which time the age of the fetus is expressed by its crown-rump or crown-heel length. During this period the head becomes relatively smaller, the limbs longer and the umbilical cord thin and elongated. The external genitalia also develop during this period.

LIMB DEVELOPMENT

The limb buds appear at the end of the 5th week, the upper limbs preceding the lower. The debate as to whether the muscles are derived from somatic or core mesenchyme has already been referred to but the nerves, arising from the ventral rami, have a well marked segmental organisation. The digits are developed by the formation of four longitudinal grooves at the distal extremity of each limb. Initially the limbs project laterally but with the formation of the joints they come to lie more ventrally. A 90° axial rotation of the limbs (the direction being reversed in the upper and lower limbs) brings the palms to point ventrally, and the sole of the foot to face dorsally, and later downwards.

PLACENTA

Up to the 2nd month of development the uterine decidual reaction around the embedded trophoblast (the decidua capsularis and the decidua basalis) is uniform. Subsequently, however, with the increasing size of the fetus the decidua capsularis becomes thinned and the decidua basalis thickens to take up the disc-like form of the placenta. By the 3rd month the uterine cavity is obliterated with the progressive enlargement of the amniotic cavity. The extra-embryonic coelome is also obliterated, the amniotic membrane fusing with the chorion. Both membranes are continuous with the placenta at its margins. In the early stages of development, the chorion is evenly covered by villi. By the 3rd month, however, most villi have atrophied except those near the site of attachment of the body stalk. These villi increase in size and the whole area becomes the future placenta. The maternal surface of the full-term placenta is divided into about 20 smaller lobes (cotyledons). Each consists of 3 or 4 villous stems which anchor the placenta to the uterine lining. The uterine vessels open on to the spaces between the major villous stems. The lining of these lakes is continuous with the maternal endothelium. Many other villi, carrying fetal blood vessels also occupy the blood-filled spaces between the major stem villi. The fetal blood is contained in the capillaries of the villi and separated from the maternal blood by fetal endothelium, mesenchyme and the trophoblast (syncytiotrophoblast outside and the incomplete cytotrophoblast inside). The fetal surface of the placenta is covered by the smooth transparent amnion, and the umbilical vessels can be seen ramifying and passing into the placental substance. At birth, the placenta is 15–20 cm in diameter, 3–4 cm thick and weighs approximately 500 g. The full-term fetus measures about 50 cm from crown to heel and weighs about 3 kg.

UMBILICAL CORD (Figs. 2.7A, 2.10)

At birth the umbilical cord is about 60 cm long and 1.5 cm in diameter. It joins the placenta to the anterior abdominal wall. Due to the presence of the muscular vessels the cord shows a continuous spiral. The outer covering is amnion and the structures inside are embedded in a clear jelly. The two umbilical arteries are accompanied by the left umbilical vein (the right having become obliterated). The remnant of the vitello-intestinal duct can be identified and occasionally the urachus and a small extension of the abdominal cavity. The umbilicus is the scar which remains on the abdominal wall when the remnant of the cord atrophies (wastes away).

— 3 —

The vertebral column

This curved flexible pillar forms the central axis of the skeleton. It supports the weight of the head, trunk and upper limbs and transmits it through the pelvic girdle to the lower limbs. In the male it is about 70 cm long. It is composed of 33 bony segments, the vertebrae, united by cartilaginous discs and ligaments, and within its canal it contains and protects the spinal cord.

The vertebrae are classified into five groups from above downwards, the cervical (7), thoracic (12), lumbar (5), sacral (5) and coccygeal (4). In the adult the sacral and coccygeal vertebrae fuse to form two bones, the sacrum and the coccyx.

A TYPICAL VERTEBRA

A typical vertebra has a body in front and a vertebral arch behind, together enclosing the vertebral foramen. The arch consists of paired anterior rounded pedicles and posterior flattened laminae. Two transverse processes are attached in the region of union of the pedicles and laminae, and the spinous process to the union of the laminae. There are two superior and two inferior articular processes to contact with similar processes on adjacent vertebrae.

The **body** is short and cylindrical, slightly constricted in its middle and possessing flat upper and lower articular surfaces. Its posterior aspect is perforated by foramina for large basivertebral veins. The **pedicles** are short, thick, rounded processes extending backwards from the posterolateral aspects of the body and are grooved above and below. With adjacent pedicles they bound intervertebral foramina which are traversed by the spinal nerves and vessels. The **laminae** are flat and fuse with each other posteriorly in the midline forming a backwardly projecting **spinous process**. The laminae of adjacent vertebrae overlap each other. The expanded tip of the spinous process may be palpable posteriorly in the midline. Superior and inferior pairs of **articular processes** arise near the junction of the pedicles and laminae. Each bears an articular facet for articulation with the adjacent vertebra. The stout **transverse processes** project laterally from the junction of the pedicles and laminae.

REGIONAL VARIATIONS

Cervical, thoracic and lumbar groups of vertebrae are recognisable by characteristics which are most marked in their central members, whilst the vertebrae at each end of the group tend to resemble those of the adjacent region.

Cervical region

A typical cervical vertebra (Fig. 3.1) has a small oval body, the superior articular surface of which bears lateral lips which articulate by small synovial joints with the reciprocally bevelled inferior articular surface of the vertebra above. The vertebral foramen is relatively large. Each short wide transverse process encloses a **foramen transversarium** and ends laterally as anterior and posterior tubercles. The articular processes are large and rounded; the superior articular surface faces backwards and upwards, the inferior in the opposite direction. The spine is bifid and placed deeply between the extensor muscles of the neck. The 1st, 2nd and 7th cervical vertebrae show some variations from this pattern.

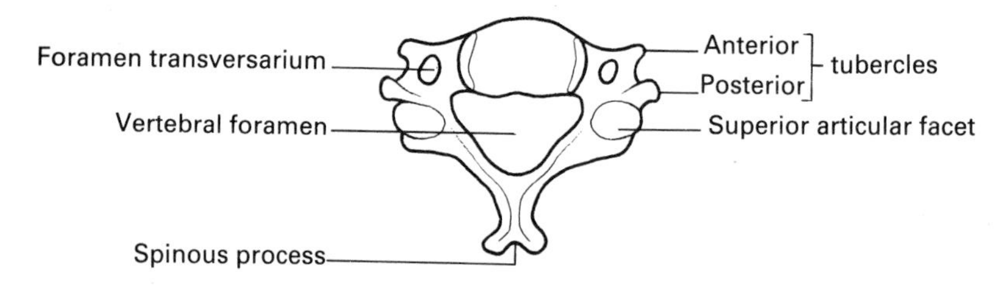

Fig. 3.1 A typical cervical vertebra, superior aspect

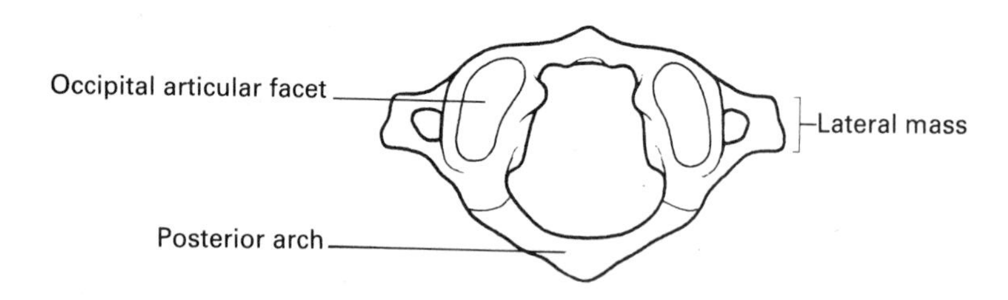

Fig. 3.2 The atlas, superior aspect

The 1st cervical vertebra (the **atlas**, Fig. 3.2) is modified to allow free movement of the head. It consists of a ring of bone with neither body nor spine. The two bulky articular **lateral masses** are united by a shorter anterior and a longer posterior arch. The upper articular facets on the lateral masses are kidney-shaped, concave in their long axis and articulate with the occipital condyles. The lower articular facets are flat, circular and articulate with the 2nd cervical vertebra. Medially the masses bear a tubercle for the attachment of the transverse ligament of the atlas. The **anterior arch** has a small midline tubercle anteriorly for attachment of the anterior longitudinal ligament and bears posteriorly a facet for articulation with the dens of the axis. Its superior border gives attachment to the anterior atlanto-occipital membrane. The **posterior arch** bears a small midline tubercle for the attachment of the ligamentum nuchae. Its superior surface is grooved on each side laterally by the vertebral artery and between the grooves gives attachment to the posterior atlanto-occipital membrane. The transverse processes are long, wide, possess no tubercles and the tip is palpable behind the angle of the mandible where it is crossed by the accessory nerve.

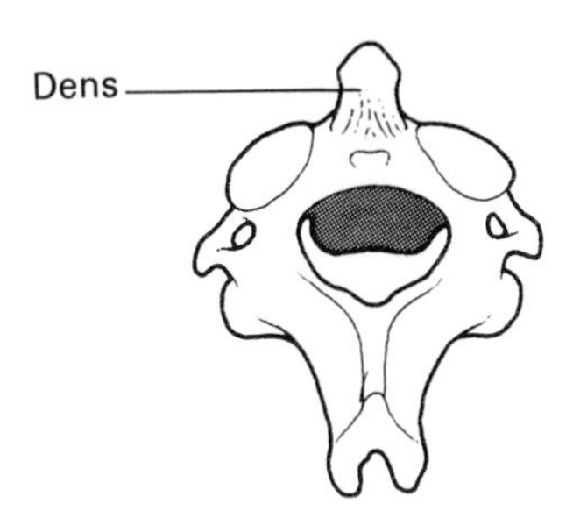

Fig. 3.3 The axis, superior aspect.

The 2nd cervical vertebra (the **axis**, Fig. 3.3) bears a conical projection, the **dens**, on the upper surface of its body. (This represents the body of the atlas.) It articulates with the back of the anterior arch of the atlas and is held in position by the transverse ligament of the atlas. The superior articular facets face upwards and are large, flat and circular.

The 7th cervical vertebra (the **vertebra prominens**) has a long nonbifid spine which is distinctly palpable at the lower end of the nuchal furrow at the back of the neck.

Thoracic region

A typical thoracic vertebra (Fig. 3.4) has a wedge-shaped body which is deeper posteriorly. The sides of each body articulate with paired ribs at a **superior** and a smaller **inferior costal facet** near the back of the upper and lower borders. The transverse processes are directed laterally and backwards, and their expanded ends bear articular facets for the tubercles of the corresponding ribs. The superior articular facets face backwards and laterally. The broad laminae and downward projecting spines overlap those of the vertebra below.

The bodies of the upper vertebrae are smaller and narrower and their superior articular facets tend to face more superiorly. The bodies of the lower vertebrae are larger and broader, and their spines are more horizontal. The costal facets on the transverse processes are absent in the last two vertebrae. Each transverse process of the 12th thoracic vertebra is small and irregular.

The body of the 1st thoracic vertebra has a larger upper facet for the 1st rib

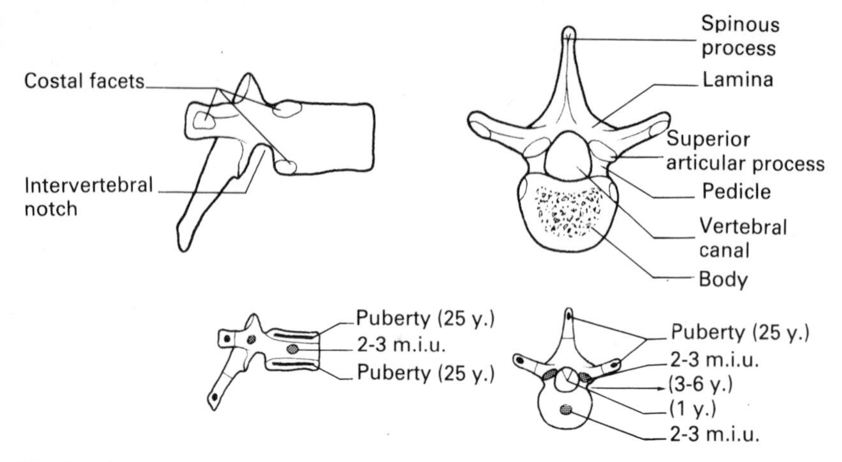

Fig. 3.4 A typical thoracic vertebra, lateral and superior aspects. Inset: ossification (m.i.u. = months in utero; y = years).

and a smaller inferior facet for the 2nd rib. The costal facets on the bodies of the 10th, 11th and 12th vertebrae are normally single and complete.

Lumbar region

A typical lumbar vertebra (Fig. 3.5) possesses a large kidney-shaped body, wider from side to side than from before backwards. The bulky transverse processes project directly laterally, and posteriorly at their bases bear a small tubercle, the **accessory process**. The superior articular facets face backwards and medially, and on the posterior surface of each process is a rounded **mamillary process**. The broad flat spinous process projects horizontally backwards.

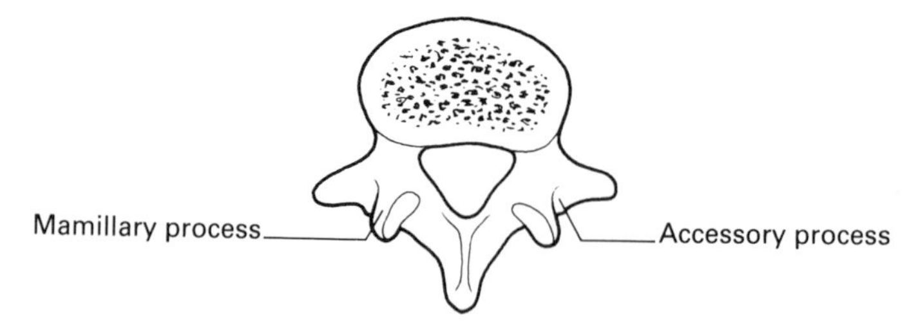

Fig. 3.5 A typical lumbar vertebra, superior aspect.

The 5th lumbar vertebra is atypical—its body is wedge-shaped, being thicker anteriorly than posteriorly. Its transverse process is small and conical and arises from the side of the body and the pedicle. The forward facing inferior articular facet prevents it sliding anteriorly on the sloping upper surface of the sacrum.

The sacrum and coccyx are described on pages 174 and 175 respectively.

JOINTS AND LIGAMENTS

The articular surfaces of the bodies of adjacent vertebrae are covered by a thin layer of hyaline cartilage and united by a thick fibrocartilaginous **intervertebral disc**. The centre of the disc (**nucleus pulposus**) is gelatinous and is surrounded by a peripheral fibrous part, the **anulus fibrosus**. The disc acts as a shock absorber. It is normally under pressure and occasionally the semisolid nucleus protrudes through a defect in the anulus and may press on the spinal cord or its nerves producing symptoms and signs of compression. The discs contribute about a quarter to the length of the column and are thicker in the cervical and lumbar regions. The vertebral bodies are also united by **anterior** and **posterior longitudinal ligaments**. The anterior ligament is a flat band extending from the occipital bone to the front of the sacrum. It is firmly attached to each vertebral body and disc. The posterior ligament is a broad band which extends from the back of the body of the axis to the sacrum. It is also attached to each intervertebral disc.

Adjacent vertebral arches articulate by two plane synovial joints between paired articular processes. The joints possess a weak capsular ligament and additional support is provided by accessory ligaments:

(i) the **ligamenta flava** unite the adjacent laminae. They contain a large amount of yellow elastic tissue. They maintain the curvatures of the vertebral column, support the column when in the flexed position and help to restore it to the extended erect position.

(ii) the **supraspinous, interspinous** and **intertransverse ligaments**.

VARIATIONS

In the cervical region the vertebral bodies are also united by small synovial joints between the reciprocally lipped and bevelled lateral margins of their bodies.

The atlanto-occipital and atlanto-axial joints are modified to allow free movement of the head. X-ray pictures taken through the wide open mouth show these joints.

The **atlanto-occipital joints** are condyloid synovial joints between the convex occipital condyles and the concave upper surface of the lateral masses of the atlas. They possess weak capsular ligaments and two accessory ligaments, the **anterior** and **posterior atlanto-occipital membranes** which pass between the corresponding arch of the atlas and the margins of the foramen magnum.

The **atlanto-axial joints** comprise two lateral synovial plane joints between the lateral masses of the two vertebrae, and a median synovial pivot between the dens and a ring formed by the anterior arch and the transverse ligament of the atlas. This latter has a weak capsular ligament and accessory ligaments, viz:

(i) the **membrana tectoria** is the upward continuation of the posterior longitudinal ligament. It ascends from the back of the axis to the occipital bone just within the foramen magnum.

(ii) the **cruciate ligament** lies anterior to the membrana tectoria. It consists of the transverse ligament of the atlas passing between the lateral masses of the atlas, and a longitudinal ligament passing from the body of the axis to the anterior margin of the foramen magnum.

(iii) the two alar ligaments diverge from the upper part of the dens to the medial side of the occipital condyles.

(iv) the apical ligament is a thin ligament and lies anterior to the cruciate ligament. It ascends from the apex of the dens to the anterior margin of the foramen magnum.

FUNCTIONAL ASPECTS

Curvatures and mobility

In the fetus the column is flexed throughout its length producing a primary curvature. After birth two secondary curvatures develop; extension of the cervical region is produced by the muscles raising the head, and extension of the lumbar region accompanies the adoption of the erect posture. The thoracic and sacral regions retain the primary curvature. The curves and intervertebral discs confer a certain resilience upon the column.

MUSCLES

See page 415.

MOVEMENT

Whilst only limited movements between adjacent vertebrae are possible, these augment each other and the column as a whole will move extensively. Flexion is most marked in the cervical region, rotation in the thoracic, and extension and lateral flexion in the lumbar region. Flexion: rectus abdominis aided by the prevertebral muscles. When flexion occurs from the anatomical position, gravity is an important factor. Extension: by the long muscles of the back. Lateral flexion: sternocleidomastoid and trapezius, the oblique abdominal wall muscles and quadratus lumborum. Rotation: sternocleidomastoid and internal and external oblique abdominal muscles. Movements of the head: the head can be rotated, flexed, extended and laterally flexed. Rotation occurs at the atlanto-axial joints and the other movements at the atlanto-occipital joints. Movement of the head is usually accompanied by synergic movements of the neck. Rotation: sternocleidomastoid, trapezius, splenius capitis and suboccipital muscles. Flexion: longus capitis and the muscles depressing the fixed mandible, aided by gravity. Extension: postvertebral muscles. Lateral flexion: sternocleidomastoid and trapezius.

STABILITY

This depends almost entirely on the muscles surrounding the column aided by the ligamenta flava, for neither bony nor ligamentous factors alone could withstand the large forces frequently acting on the column. The majority of fractures of the vertebral bodies do not produce instability of the vertebral column, as the strong ligaments connecting the vertebrae are not easily torn. If however some ligamenta flava and interspinous ligaments are ruptured the vertebral column is unstable and subsequent displacement may produce compression of the spinal cord and irreversible damage.

Vertebral canal

This extends from the foramen magnum to the sacral hiatus. It is a bony-ligamentous canal and opens below on to the back of the lower sacral segments and the coccyx. The spinal cord ends above the level of the 2nd lumbar vertebra, the dural sac at the level of the 2nd sacral vertebra and the filum terminale is attached to the back of the coccyx. The supracristal plane, through the highest points of the iliac crests, cuts across the spine of the 4th lumbar vertebra. The umbilicus is about the level of the lower border of the body of the 3rd lumbar vertebra. The dural sac with its enclosed spinal cord, spinal nerves, and cerebrospinal fluid is separated from the body-ligamentous wall by a fat-filled extradural space in which lie the emerging spinal nerves and the internal vertebral venous plexus.

By means of an injection through the sacral hiatus, it is possible to anaesthetise the lower spinal nerves, especially those supplying the perineal region (see also p. 175).

Blood supply

The vertebral column is supplied by segmental spinal branches of the ascending cervical, vertebral, intercostal, lumbar and lateral sacral arteries. The veins drain to the external vertebral venous plexuses which communicate throughout their length with the internal vertebral venous plexus within the vertebral canal. The external vertebral venous plexus drains to the internal iliac, lumbar, azygos and vertebral veins. The internal vertebral venous plexuses communicate with pelvic plexuses below and intracranial venous sinuses above. The basivertebral veins carry blood from the red marrow of the vertebral bodies into the internal vertebral plexus. Blood borne metastases may develop in a vertebral body and lead to its collapse.

PART TWO

Thorax

— 4 —

The thoracic wall and diaphragm

The thorax has a bony-cartilaginous skeleton and contains the principal organs of respiration and circulation. Its walls are formed of 12 thoracic vertebrae posteriorly, a midline sternum anteriorly, and 12 pairs of ribs, with costal cartilages and intercostal muscles laterally. It is conical in shape, possessing a narrow inlet superiorly and a wide outlet inferiorly.

The thoracic inlet is about 10 cm wide and 5 cm anteroposteriorly. It slopes downwards and forwards and is bounded by the 1st thoracic vertebra posteriorly, the superior border of the manubrium anteriorly and the 1st rib and costal cartilage laterally. It transmits the oesophagus, the trachea, the large vessels and nerves of the head and neck, and the vessels of the upper limbs. On each side is the apex of a lung, covered by the dome of the pleura.

The thoracic outlet also is widest from side to side. It is bounded by the 12th thoracic vertebra posteriorly, the 12th and 11th ribs laterally and the costal cartilages of the 10th, 9th, 8th and 7th ribs anteriorly. The costal margin ascends to meet the sternum, forming the infrasternal angle. The outlet is closed by the diaphragm which separates thoracic and abdominal cavities.

Sternum (Fig. 4.1)

The sternum is a flat bone forming the median part of the anterior thoracic wall. It has three parts, the manubrium, the body and the xiphoid process from above downwards. It is palpable throughout most of its length.

The **manubrium** is a thick plate of bone about 5 cm long, wider above than below. Its upper border bears a central jugular notch between two lateral facets for articulation with the ends of the clavicles. The lateral borders are marked above by a pit for the 1st costal cartilage and below by a half facet for the upper part of the 2nd costal cartilage. The narrow inferior border articulates with the body at the palpable **sternal angle**.

The **body** of the sternum is longer (10 cm) and thinner than the manubrium. It is formed of four pieces (sternebrae) whose lines of fusion are marked by three faint transverse ridges on the adult bone. The lateral border possesses a half facet for the 2nd costal cartilage and five shallow pits for 3rd–7th cartilages. The body tapers inferiorly and articulates with the xiphoid process.

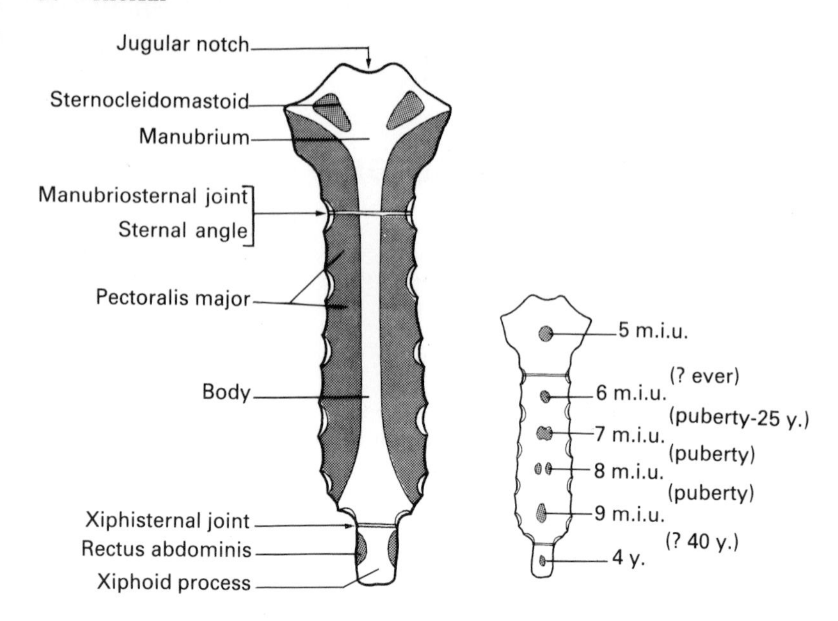

Fig. 4.1 The sternum, anterior aspect. Inset: ossification (m.i.u., months in utero; y., years).

The **xiphoid process** is thin, flat, and irregularly shaped and is cartilaginous in early life. It lies in the rectus sheath and gives attachment to the diaphragm posteriorly and the rectus abdominis anteriorly.

Ribs

The ribs articulate posteriorly with the thoracic vertebrae and anteriorly with their costal cartilages. They form a large part of the thoracic wall. There are usually 12 on each side but this number may be increased by the development of a cervical or lumbar rib. The upper seven articulate through their cartilages with the sternum and are known as true ribs; the 8th–10th articulate through their cartilages with the cartilage immediately above and are called false ribs; the 11th and 12th have free anterior ends and are known as floating ribs.

A typical rib consists of a head, neck, tubercle and body from behind forwards. The **head** bears two facets for articulation with adjacent vertebrae, separated by a transverse crest which is attached to the intervertebral disc. The constricted **neck** is flattened but roughened and its surfaces give attachment to the costotransverse ligaments. The **tubercle** lies posteriorly at the junction of the neck and body. It bears a medial facet for articulation with the transverse process. The **body** is flattened in a vertical plane and curved to conform to the chest wall. Its upper border is rounded whilst its lower is sharp and limits the costal groove on its internal surface. The body bends forward lateral to the tubercle at the **angle** which marks the lateral limit of the attachment of the

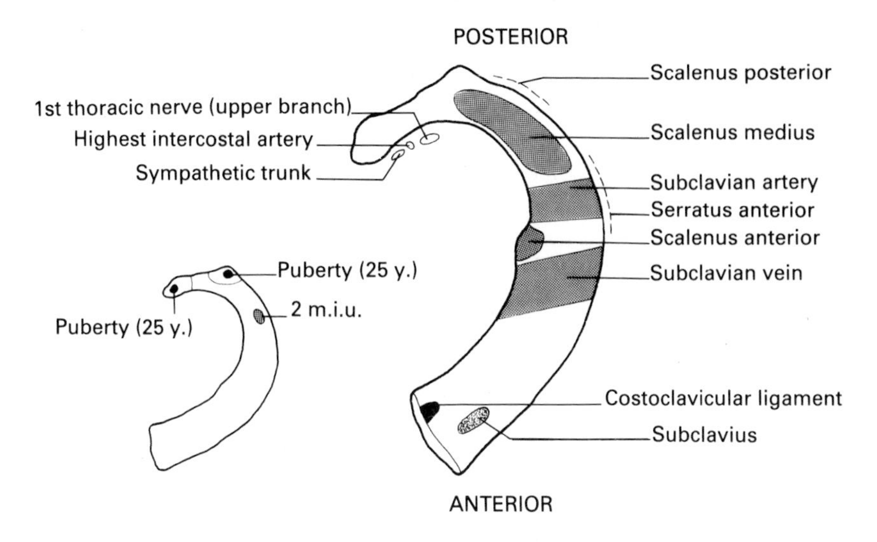

Fig. 4.2 The upper surface of the 1st rib, showing some of its muscular attachments and immediate relations. Inset: ossification (m.i.u., months in utero; y., years).

erector spinae muscles. The anterior end of the rib possesses a small depression for the costal cartilage.

The 1st, 10th, 11th and 12th ribs are atypical.

The **1st rib** (Fig. 4.2) is short, wide, flattened, and lies in an oblique plane. It has superior and inferior surfaces, and medial and lateral borders. Its head bears a single facet for articulation with the 1st thoracic vertebra and posteriorly its tubercle and angle coincide. The lower surface is smooth and lies on the pleura. A small **scalene tubercle** on its medial border marks the attachment of scalenus anterior. On the upper surface the tubercle separates an anterior groove for the subclavian vein and a posterior groove for the subclavian artery and the lower trunk of the brachial plexus. Scalenus medius is attached to a roughened area behind the artery. Anteriorly the upper surface gives attachment to the subclavius muscle and the costoclavicular ligament. Between the pleura and the front of the neck of the rib (from medial to lateral) lie the sympathetic trunk, the superior intercostal artery and the large branch of the 1st thoracic nerve to the brachial plexus.

The heads of the 10th, 11th and 12th ribs bear a single articular facet. The 11th and 12th ribs are short and possess neither necks nor tubercles. The 12th also lacks an angle and a subcostal groove, and gives attachment to quadratus lumborum muscle and the lateral arcuate ligament of the diaphragm.

The **costal cartilages** are flexible bars of hyaline cartilage which contribute to the elasticity of the thoracic walls. They extend forwards from the anterior ends of the ribs. The 1st–7th articulate directly with the sternum, the 8th–10th with the lower borders of the preceding cartilages. The 11th and 12th possess free pointed ends which lie in the abdominal wall musculature.

Thoracic vertebrae
See page 45.

JOINTS OF THE THORACIC CAGE

These comprise the intervertebral, rib and sternal joints.

INTERVERTEBRAL JOINTS

See page 47.

RIB JOINTS

A typical rib articulates behind with the vertebral column in two places and anteriorly with its own costal cartilage.

Costovertebral joints
The two facets on the head of most ribs articulate with adjacent vertebrae at two synovial joints; the lower facet with its own vertebra and the upper with the vertebra above. The crest on the head is attached to the intervening disc by an intra-articular ligament. The capsule of these joints is strengthened by ligaments which run from the head of the rib to the vertebral bodies. The heads of the 1st, 10th, 11th and 12th ribs usually have only one facet and articulate with only their own vertebra.

Costotransverse joints
These are synovial joints on the upper 10 ribs. The upper joint surfaces are reciprocally curved and the lower are more flattened. The capsular ligament is strengthened by accessory costotransverse ligaments. The lower two ribs are attached to the transverse processes only by ligaments.

The joints between the ribs and their costal cartilages (**costochondral**), and between the 1st costal cartilage and the sternum are primary cartilaginous joints. Joints between the 2nd–7th costal cartilages and the sternum, and those between adjacent costal cartilages have a synovial cavity.

STERNAL JOINTS

The manubriosternal joint is a secondary cartilaginous joint. The fibrocartilaginous disc allows a slight hinge movement of the body on the manubrium during inspiration. The xiphisternal joint is also of the secondary cartilaginous variety but usually becomes a bony union in later life.

Movement: of the ribs, see page 62.

THE THORACIC CAGE

Note that:

(*a*) the vertebral column intrudes into the cage posteriorly, making the thoracic cavity kidney-shaped in transverse section;
(*b*) the ribs increase in length from the 1st to the 7th and thereafter become shorter;
(*c*) they lie at an angle of about 45° to the vertebral column, the obliquity increasing down to the 9th rib where it reaches its maximum;
(*d*) the anterior ends of the ribs fall increasingly short of the sternum and their costal cartilages are correspondingly lengthened;
(*e*) from the back it is seen that the rib angles lie nearer the midline superiorly.

The cage gives attachment to:

(*a*) some of the muscles supporting the weight of the upper limb;
(*b*) muscles of the abdominal wall, thereby helping to support the abdominal viscera;
(*c*) extensor muscles of the back as far laterally as the angles of the ribs;
(*d*) scalene muscles which help to raise the 1st ribs;
(*e*) diaphragm and intercostal muscles (see below).

THE INTERCOSTAL SPACES

These are bounded by adjacent ribs and costal cartilages. They contain the intercostal muscles and the intercostal vessels and nerves. Deep to them lies the pleura.

The **external intercostal muscle** descends obliquely forwards from the lower border of the rib above to the upper border of the rib below. Anteriorly between the costal cartilages its fibres are replaced by a fibrous sheet, the external intercostal membrane.

The **internal intercostal muscle** lies deep to the external intercostal muscle. It descends obliquely backwards from the subcostal groove of the rib above to the upper border of the rib below. Posteriorly, behind the angle of the rib, its fibres are replaced by a fibrous sheet, the internal intercostal membrane.

The **innermost intercostal muscle** forms an incomplete sheet, lying between the ribs and the pleura. Muscle fibres are attached to the inner surfaces of the sternum, costal cartilages and ribs.

Action
The intercostal muscles move the ribs. When the 1st rib is anchored by the scalene muscles, the approximation of the ribs raises the sternum and air is drawn into the lungs. The external intercostal may then be most active. When the lower ribs are anchored by the abdominal muscles, approximation of the ribs lowers the sternum and air is forced out of the lungs. The internal intercostal muscles may then be most active. Contraction of the intercostal muscles also helps to maintain the rigidity of the chest wall, thus preventing the tissues between the ribs from being pushed out or pulled in by changes in the intra-thoracic pressure produced mainly by action of the diaphragm (see p. 62).

Nerve supply
Intercostal nerves.

THE INTERCOSTAL VESSELS AND NERVES

These form a neurovascular bundle which passes forwards around the chest wall in the subcostal groove deep to the internal intercostal muscle and lying on the innermost intercostal muscle and the pleura.

The intercostal arteries

Each space is supplied by a single posterior artery and paired anterior intercostal arteries. The lower two spaces have only posterior arteries.

The 1st and 2nd posterior intercostal arteries arise from the superior intercostal artery (a branch of the costocervical trunk from the subclavian artery). The remainder come from the descending thoracic aorta. They pass forwards in the space above the corresponding nerve, and give spinal branches, branches to the lateral chest wall and a small collateral branch which passes forwards in the space with a collateral branch of the nerve.

The anterior intercostal arteries of the upper six spaces are branches of the internal thoracic artery, and the arteries of the next three spaces, of its terminal branch, the musculophrenic artery. They pass laterally to anastomose with the posterior artery and its collateral branch.

The internal thoracic artery

This arises from the subclavian artery (p. 421), descends just lateral to the sternal border and ends in the 6th intercostal space by dividing into musculo-phrenic and superior epigastric arteries. Near its origin it is crossed over by the phrenic nerve and in the chest it lies between the costal cartilages superficially and the innermost intercostal muscle and pleura deeply.

Branches

(i) *anterior intercostal arteries*—to the upper six spaces.

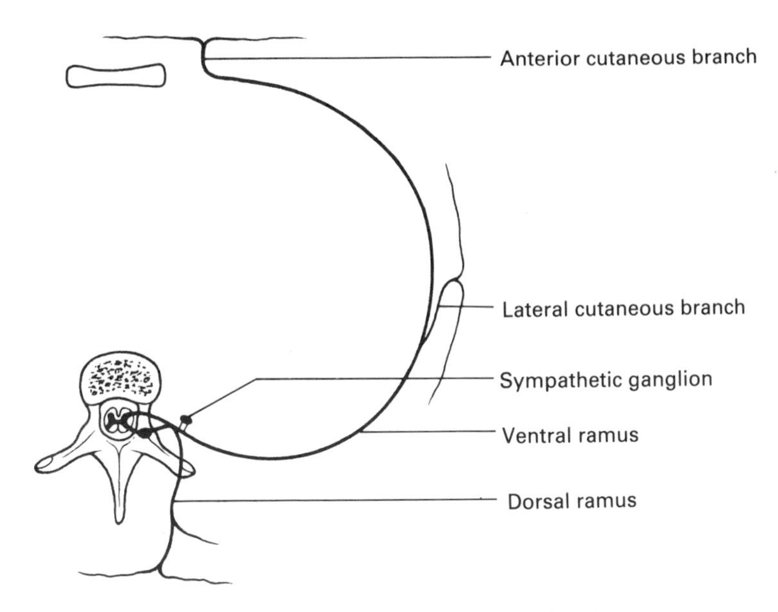

Fig. 4.3 The distribution of a typical intercostal nerve. Muscle branches are not shown. Usually the dorsal root with its ganglion and the ventral root join and form the spinal nerve in the intervertebral foramen.

(ii) *perforating cutaneous branches*—in the female these are larger and supply the breast.

(iii) **superior epigastric artery**—descends between the costal and xiphoid origins of the diaphragm into the rectus sheath and anastomoses with the inferior epigastric artery (branch of the external iliac artery). It supplies the contents of the rectus sheath.

(iv) **musculophrenic artery**—descends laterally behind the costal margin, gives anterior intercostal branches to the 7th, 8th and 9th spaces and supplies the diaphragm and pericardium.

The intercostal veins

The intercostal spaces drain by two anterior veins and a single posterior intercostal vein. The anterior pass to the musculophrenic and internal thoracic veins; the posterior pass in an irregular manner to the brachiocephalic vein (1st space) and the azygos system of veins (the other spaces).

The intercostal nerves (Fig. 4.3)

These are the ventral rami of the upper 11 thoracic spinal nerves. Each spinal nerve is attached to the spinal cord by ventral and dorsal roots. The latter possess a ganglion which is situated in the intervertebral foramen. Before emerging from the foramen the two roots unite and almost immediately divide into dorsal and ventral rami. Each dorsal ramus supplies the extensor muscles of the back before ending in medial and lateral cutaneous branches which supply the skin of the back. Each ventral ramus forms the intercostal nerve.

A typical intercostal nerve (3rd or 4th) passes laterally into the subcostal groove. It continues around the chest wall between internal intercostal muscle externally and the innermost intercostal muscles and pleura internally. At the anterior end of the space it turns forwards and pierces the internal intercostal muscle, the anterior intercostal membrane and pectoralis major. It ends as an anterior cutaneous branch.

Branches

(i) collateral branch—arises near the angle of the rib and runs forwards with a similar branch of the artery below the parent trunk.

(ii) muscular branches.

(iii) lateral cutaneous nerve—also arises near the angle of the rib, pierces the intercostal muscles in the midaxillary line and ends up by dividing into anterior and posterior branches.

(iv) anterior cutaneous nerve—divides into a small medial and a large lateral branch.

The 1st intercostal nerve has no lateral cutaneous branch but sends a large branch to the brachial plexus. The lateral branch of the 2nd and sometimes the 3rd intercostal nerve forms the intercostobrachial nerve which supplies the skin of the axilla. The 12th thoracic (subcostal) nerve runs forwards below the 12th rib to the lower abdominal wall and supplies its muscles. (Its lateral cutaneous branch supplies the skin over the anterior superior iliac spine.)

Surgical access to a pleural cavity (thoracotomy) is gained by an incision which follows the line of a rib and enters the chest via the bed of the rib or via the intercostal space. An alternative approach to the mediastinum and heart is through a midline sternotomy dividing the sternum vertically.

Rib fractures usually tear the pleura and may be accompanied by haemorrhage from intercostal vessels; the bleeding may be into the pleural cavity producing a haemothorax. Occasionally the underlying lung may be torn by a fractured rib and the air leak into the pleural cavity is known as a pneumothorax. A pneumothorax may also result from penetrating wounds or may occur spontaneously. In most cases there is a partial or complete collapse of the lung, until the pleural wound heals. Multiple rib fractures may allow a segment of the chest wall to become unstable thus interfering with the normal respiratory movements. The tidal respiratory volume may then be dangerously diminished.

THE DIAPHRAGM

This musculotendinous septum separates the abdominal and thoracic cavities. It has a central tendinous part and a peripheral muscular part. During quiet respiration, the central tendon moves little so that the position of the heart varies only slightly during the respiratory cycle.

Attachments

PERIPHERAL

(i) *sternal part*—from the back of the xiphoid process by two muscular slips.

(ii) *costal part*—from the inner surfaces of the lower six ribs and costal cartilages, interdigitating with transversus abdominis.

(iii) *vertebral part*—from the sides of the bodies of the upper lumbar vertebrae by two crura, and from the medial and lateral arcuate ligaments on each side. The **right crus** is attached to the bodies of the first three vertebrae and the **left crus** to the first two. The larger right crus passes forwards and to the left and surrounds the oesophageal opening. The **medial arcuate ligament** is the thickened upper edge of psoas fascia and passes in front of psoas from the body of the 1st lumbar vertebra to its transverse process. The **lateral arcuate ligament** is anterior to quadratus lumborum and passes from the transverse process of the 1st lumbar vertebra to the 12th rib.

There is often a gap between the vertebral and costal attachments and the pleura is then only separated by the perirenal tissues from the peritoneum.

CENTRAL

The fibres are attached to the periphery of a trilobed central tendon.

Action

It is the principal muscle of inspiration. During contraction, the dome of the diaphragm descends and flattens, and the vertical diameter of the chest is increased. The abdominal wall is relaxed to accommodate the downward movement of the liver and other viscera. In expiration, the diaphragm relaxes. In expulsive efforts, e.g. micturition and defaecation, the intra-abdominal pressure is increased by simultaneous contraction of the diaphragm and abdominal wall muscles.

Nerve supply

The right and left halves are supplied by the corresponding phrenic nerves. The periphery receives additional sensory branches from intercostal nerves.

Openings

There are three large openings in the diaphragm, from before backwards (see Fig. 2.5):

(i) *for the inferior vena cava*—in the central tendon to the right of the midline at the level of the 8th thoracic vertebra. It also transmits the right phrenic nerve.

(ii) *for the oesophagus*—in the muscle derived from the right crus, to the left of the midline at the level of the 10th thoracic vertebra. It also transmits the anterior and posterior gastric branches of the vagus nerves, and oesophageal branches of the left gastric vessels.

(iii) *for the aorta*—between the crura of the diaphragm in front of the 12th thoracic vertebra. It also transmits the thoracic duct and the azygos vein.

The left phrenic nerve pierces the top of the left dome of the diaphragm. The splanchnic nerves pierce the crus, the sympathetic trunk passes behind the medial arcuate ligament, the subcostal nerve and vessels pass behind the lateral arcuate ligament, and the superior epigastric vessels pass between sternal and costal attachments of the diaphragm.

Relations

The heart and lungs, within the pericardial and pleural sacs respectively, lie on the upper surface. The fibrous pericardium is firmly adherent to the central tendon. Inferiorly on the right side are the liver, right kidney and suprarenal gland, and on the left the fundus of the stomach, spleen, left kidney and suprarenal gland.

Development (see p. 30)

Defects in the development of the diaphragm can lead to protrusion of abdominal viscera into the thoracic cavity. At birth this causes severe respiratory embarrassment and requires urgent surgical treatment. The most common defect is deficiency or absence of the pleuroperitoneal membrane. A congenitally large oesophageal hiatus may produce similar symptoms to the acquired hiatus hernia (see p. 123).

RESPIRATION

The inspiratory and expiratory phases of respiration are produced by an alternating increase and decrease in the volume of the thoracic cavity. Respiration is usually described as quiet (e.g. at rest), deep (e.g. in exercise) and forced (e.g when the airway is obstructed). In quiet respiration, only the diaphragm is involved to any extent. Inspiration is aided by the weight of the liver attached to the undersurface of the diaphragm. This fact is made use of in the rocking method of aided respiration. Many of the details have been shown by X-ray examination.

Inspiration

The diaphragm contracts, descends, and increases the height of the cavity. The upper ribs are fixed by the scalene muscles and the remainder raised by the intercostal muscles. The manner in which the ribs move depends on the shape of their articulations with the transverse processes. Movement increases the anteroposterior and the transverse diameters of the chest. In forced inspiration fixation of the shoulder girdle enables serratus anterior and the pectoralis muscles to raise the ribs.

Expiration
This is produced mainly by elastic recoil of the lung tissue and costal cartilages. Simultaneous contraction of the abdominal wall muscles increases intra-abdominal pressure and forces the diaphragm upwards. The latter movement is exaggerated in forced expiration. (See also p. 109.)

THE BREAST (MAMMARY GLAND) Fig. 4.4

In the male and prepubertal female this is a rudimentary organ. The adult female breast is a soft, hemispherical structure, situated on the front of the thorax and consisting of glandular tissue and a varying amount of fat. The gland lies in the superficial fascia. The fascia invests it and forms radial septa which divide the

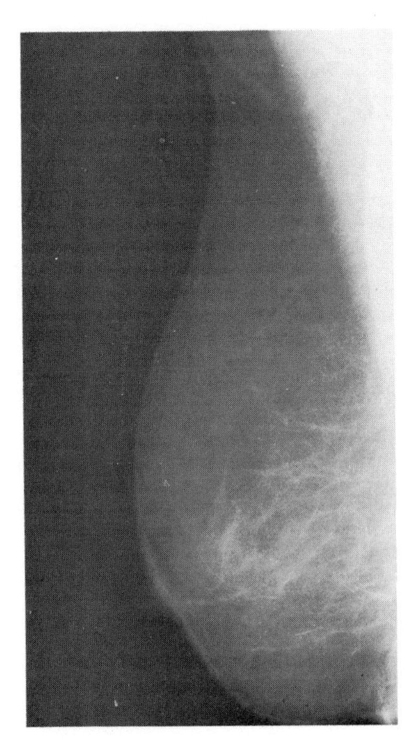

Fig. 4.4 Normal mammogram. Oblique view of the right breast with the nipple in profile using a soft tissue radiographic technique. Note the normal stromal pattern.

gland into lobules and attach it to the skin and the underlying deep fascia over pectoralis major. The base of the breast extends from the 2nd to the 6th rib and from the side of the sternum to near the midaxillary line. Its apex, the nipple, is surrounded by thin pigmented skin, the areola.

Blood supply

Branches of the internal thoracic, anterior intercostal and lateral thoracic arteries. The veins correspond to the arteries.

Lymph drainage

Lymph drains from the gland to a subcutaneous subareolar plexus and to a submammary plexus on pectoralis major. From these plexuses lymph drains laterally to the pectoral nodes in the axilla, superiorly to the infraclavicular and lower deep cervical nodes, inferiorly via the subcutaneous plexus to the anterior abdominal wall and to diaphragmatic nodes, and medially to the parasternal nodes and across the midline to communicate with plexuses of the breast of the opposite side. Complete removal of the possible affected nodes in a patient with breast cancer is thus impracticable. For this reason removal of the breast and pectoral nodes is frequently combined with radiotherapy to the regional nodes or with systemic chemotherapy.

Histology

The breast has a framework of fatty connective tissue and is divided by fascial septa into 15–20 lobes of glandular tissue. Each lobe comprises clusters of alveoli whose ducts unite to form a lactiferous duct which passes towards the nipple. Deep to the nipple each duct dilates to form a lactiferous sinus from which a short straight part of the duct opens on to the nipple. The nipple and areola contain some smooth muscle and modified sebaceous glands (the areolar glands) but little fat.

The mammary glands are modified skin glands and develop from an ectodermal downgrowth. Supernumerary mammary glands may be formed anywhere along the 'milk line' from the axilla to the groin. The resting (non-active) breast contains little glandular tissue. In pregnancy, however, the tissue undergoes marked proliferation and its blood supply increases prior to the onset of lactation.

— 5 —

The thoracic cavity and heart

The thoracic cavity is divided into right and left pleural cavities by a central partition, the **mediastinum**. The mediastinum is bounded behind by the vertebral column and in front by the sternum; inferiorly it is limited by the diaphragm and above is continuous with the structures in the root of the neck. It contains the heart, larger vessels, oesophagus, trachea and lymph nodes, all embedded in loose areolar tissue. By careful positioning of the person and the use of radio-opaque media, many of the organs can be investigated radiologically.

THE HEART

The heart is the muscular pump of the systemic and pulmonary circulations. There are four chambers (two atria and two ventricles), which are demarcated on its surface by coronary and interventricular sulci. The direction of flow of the blood is maintained by means of unidirectional valves placed between the atria and the ventricles, and between the ventricles and the large emerging aorta and pulmonary trunk. The heart lies within the pericardial sac, suspended by the large vessels. The organ is the size of a clenched fist, weighs about 300 g and is the shape of a flattened cone, possessing a base and an apex. It lies obliquely across the lower mediastinum behind the sternum.

The base faces posteriorly; it is square in outline and is formed mainly of the left atrium, receiving the four pulmonary veins. The apex is at the left inferior extremity of the heart and is formed by the tip of the left ventricle. The anterior (sternocostal) surface is formed mainly by the right ventricle which is separated by the vertical coronary sulcus from the right atrium above and by the anterior interventricular sulcus from the left ventricle. The surface merges superiorly with the beginning of the aorta and pulmonary artery.

The inferior (diaphragmatic) surface is formed mainly of the right and left ventricles anteriorly and part of the right atrium posteriorly. The inferior vena cava enters the right atrium at the right posterior corner of this surface which rests on the central tendon of the diaphragm. The left surface is formed by the left ventricle and a small part of the left atrium. It is in contact with the left lung. The right surface is formed by the right atrium which receives the superior vena cava above and the inferior vena cava below. It is in contact with the right lung.

The chambers of the heart

The right chambers of the heart pump blood through the lungs, and the left through the systemic circulation.

The **right atrium** (Fig. 5.1) is a thin-walled narrow chamber. Between the superior and inferior venae cavae, it forms the right border of the heart. Superomedially a small projection, the **right auricle**, overlaps the beginning of the ascending aorta. The posterior part of the wall behind the caval openings is smooth and separates the cavity from the left atrium. It is marked by a shallow oval depression, the **fossa ovalis**, bounded by a crescentric ridge, the **limbus** of the fossa ovalis. Anteriorly the wall is thicker and formed of parallel muscular ridges which pass transversely from the right auricle to a prominent vertical ridge, the **crista terminalis**, between the caval openings.

The superior caval orifice in the upper part of the cavity has no valve; to its left lies the cavity of the right auricle. The inferior caval orifice is guarded by a thin fold of endocardium anteriorly, the valve of the inferior caval opening and between them is a smaller orifice, the opening of the coronary sinus, also with an endocardial flap guarding it.

The **right ventricle** (Fig. 5.1) is a thick-walled elongated chamber projecting forwards and to the left of the right atrium and forming part of the anterior and inferior surfaces of the heart. The interventricular septum separates the right from the left ventricle and bulges into the right cavity, making it crescentic in cross section. The walls of the cavity are covered by interlacing muscle bands except superiorly where the smooth-walled wide infundibulum leads to the pulmonary orifice. Near the apex a distinct muscle bundle, the septomarginal

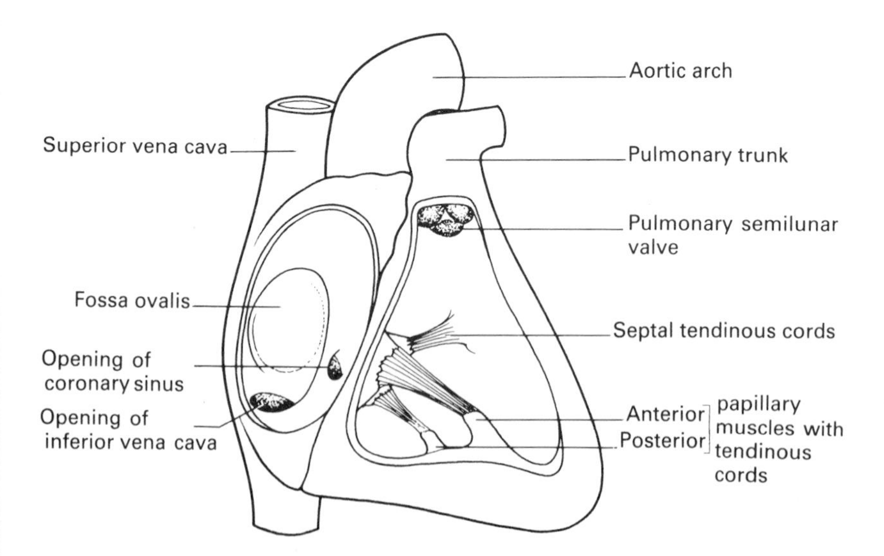

Fig. 5.1 Diagram showing the interior of the right atrium and the right ventricle. The irregularly-shaped right auricle is shown in front of the ascending aorta.

band, often crosses the cavity. The atrioventricular orifice lies postero-inferiorly. It is guarded by the **tricuspid valve** which possesses three thin cusps, each attached at its base to the fibrous ring surrounding the orifice. The cusps consist of a little fibrous tissue covered on both sides by endocardium. The atrial surfaces of the valve cusps are smooth. Their margins and ventricular surfaces are rough and anchored to the ventricular walls by fine tendinous cords which arise directly from the septum or from two small conical **papillary muscles** situated on the walls of the cavity. These muscles tighten the tendinous cords and prevent eversion of the cusps into the atrial cavity during ventricular contraction. The pulmonary orifice is a fibrous ring lying at the upper end of the smoother infundibulum. It is guarded by a valve possessing three semilunar valvules, one posterior and two anterior. In the middle of the free margin of each valvule is a thickened nodule.

The **left atrium**, a thin-walled, rectangular chamber, lies behind the right atrium and forms most of the base of the heart. Superiorly a small irregular

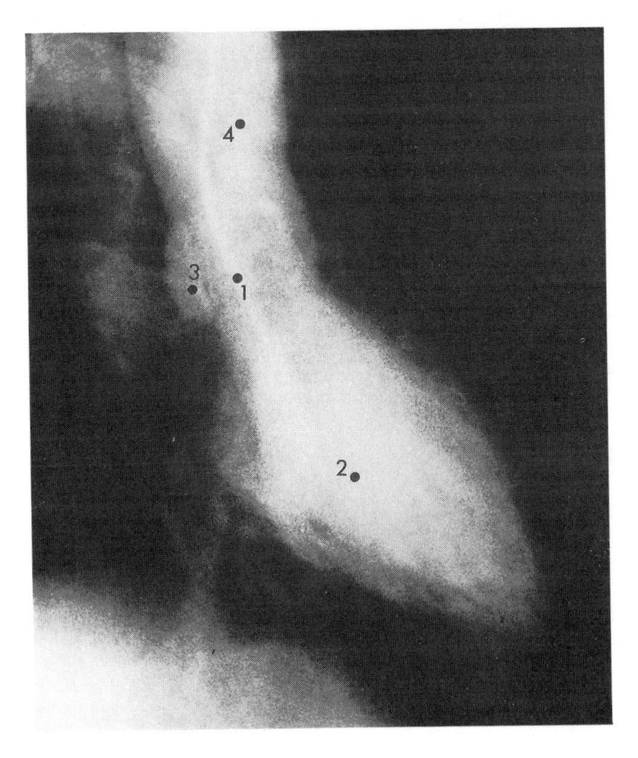

Fig. 5.2 Left ventriculogram. Right anterior oblique projection. The catheter (1) has been introduced into the right brachial artery and passed retrogradely into the left ventricle through the subclavian and brachiocephalic arteries, and the aorta and aortic valve. Contrast material has then been injected.

1. Catheter
2. Left ventricle
3. Right aortic sunus
4. Ascending aorta

projection, the left **auricle** overlies the left side of the pulmonary trunk. The four pulmonary veins enter symmetrically on the smooth posterior surface, two on each side; their orifices possess no valves. The left atrioventricular orifice lies in the anterior wall. The posterior wall of the atrium is separated by the cavity of the pericardium from the oesophagus, left bronchus and descending thoracic aorta (Fig. 6.11).

The **left ventricle** extends forwards and to the left from the left atrium and lies mainly behind the right ventricle. It forms the apex, the left border and surface, and part of the anterior and inferior surfaces of the heart (Fig. 5.2). Its walls are very thick and covered on the inside by muscular ridges except for a smooth area just below the aortic orifice, the vestibule. The conical cavity possesses two orifices, the left atrioventricular posteriorly and the aortic superiorly. The left atrioventricular orifice is guarded by the **mitral (bicuspid) valve**, the two cusps of which are attached to a fibrous ring surrounding the orifice. Their free margins and ventricular surfaces are anchored to papillary muscles on the ventricular wall by tendinous cords. The anterior cusp is the larger, lies between the aortic and mitral orifices, and blood flows over both its surfaces. The aortic orifice is a fibrous ring lying behind and to the right of the pulmonary orifice. It is guarded by a valve possessing three semilunar valvules (one anterior, two posterior) which are similar to those of the pulmonary valve.

The position of the interventricular septum is marked on the surface of the heart by the anterior and posterior interventricular sulci. The right ventricle, into which the septum bulges, lies anterior, the left ventricle posterior. The septum is very thick except for a thin membranous portion between the vestibule and the infundibulum.

BLOOD AND NERVE SUPPLY

ARTERIES (Fig. 5.3)

Right coronary artery

This arises from the anterior aortic sinus (see p. 78) and passes forwards between the right auricle and the pulmonary trunk to gain the right coronary sulcus in which it runs, firstly descending the anterior surface and then crossing the posterior surface of the heart where it anastomoses with the left coronary artery (Fig. 5.4).

Branches

(i) atrial and ventricular branches.

(ii) **marginal artery**—runs to the left towards the apex along the lower border of the heart.

(iii) **posterior interventricular artery**—runs along the posterior interventricular sulcus towards the apex and anastomoses with the anterior interventricular artery.

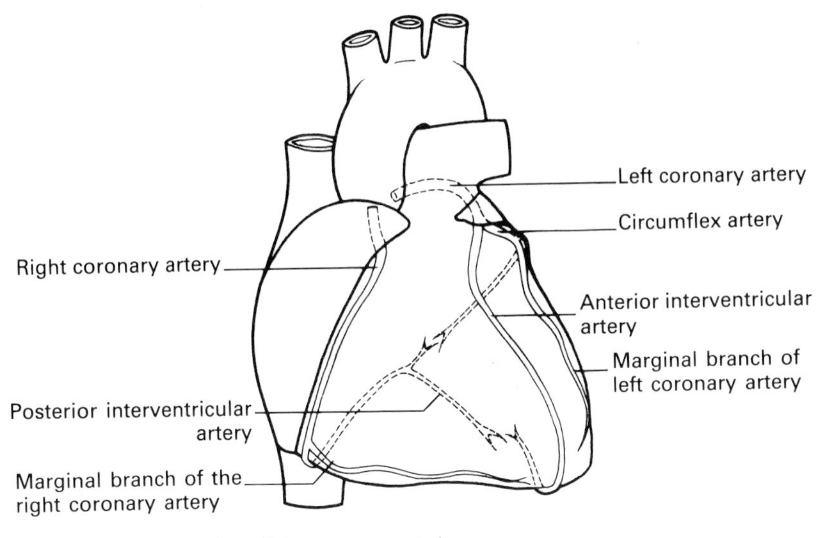

Fig. 5.3 The distribution of the coronary arteries.

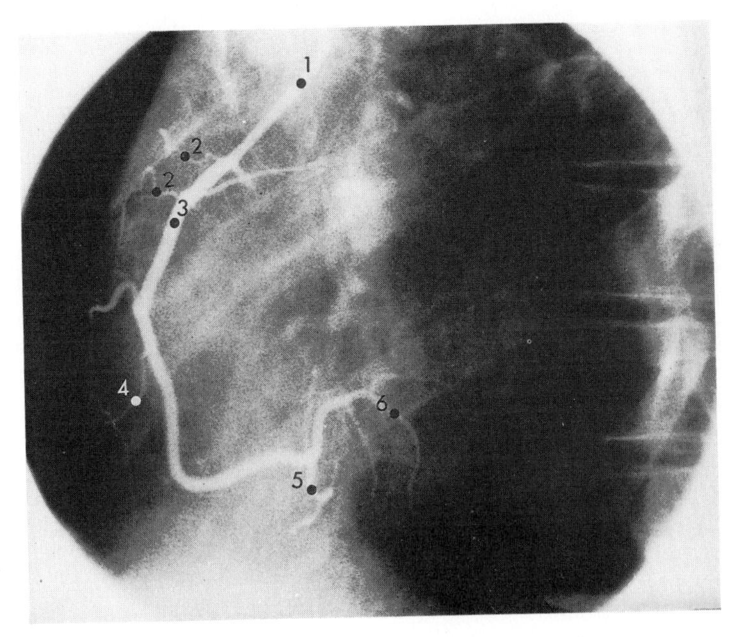

Fig. 5.4 Right coronary arteriogram. Left anterior oblique projection. The catheter (1) has been passed retrogradely as defined in 5.2 and selectively passed into the origin of the right coronary artery before the injection of contrast material.

1. Catheter
2. Atrial branches
3. Right main coronary artery
4. Small marginal branch
5. Posterior descending branch
6. Posterior interventricular artery

Left coronary artery

This arises from the left posterior aortic sinus (see p. 78). It passes forward between the left auricle and the pulmonary trunk to gain the left coronary sulcus (Fig. 5.5).

Branches

(i) atrial and ventricular branches.

(ii) **anterior interventricular artery**—descends in the anterior interventricular sulcus towards the apex, turns round the lower border and anastomoses with the posterior interventricular artery. It gives off a diagonal branch to the left ventricle.

(iii) **circumflex artery**—runs to the left in the coronary sulcus and then passes behind to anastomose with the terminal branch of the right coronary artery. It gives off the **marginal artery** which runs along the left border of the heart.

Generally the right ventricle is supplied by the right coronary artery, the left ventricle by the left, the interventricular septum by both, and the atria in a vari-

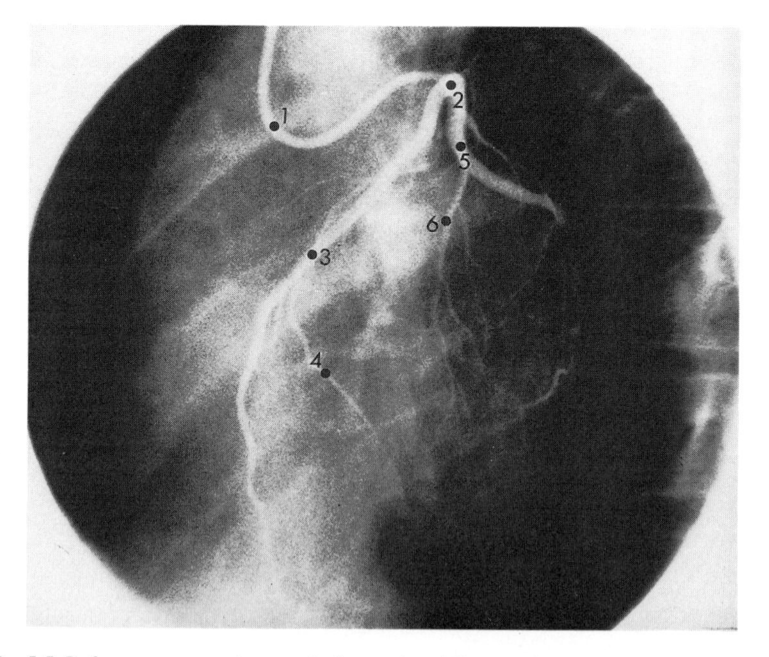

Fig. 5.5 Left coronary arteriogram. Left anterior oblique projection. The tip of the catheter (1) has been selectively introduced into the origin of the left coronary artery as described in 5.2 and contrast material injected.

1. Catheter
2. Stem of left coronary artery
3. Left anterior descending branch
4. Diagonal branch
5. Circumflex artery
6. Marginal artery

able manner. Although anastomoses between the two arteries are present in the coronary sulcus, near the apex and in the septum, sudden occlusion of a large branch may result in ischaemia of an area of heart muscle. If the area affected includes the conducting system or is relatively large, death of the patient may occur. Lesser degrees of damage decrease the work capacity of the heart and ischaemic pain (angina) may be experienced on exertion. Rupture of a papillary muscle may result in incompetance of the mitral or tricuspid valve. This, or other disease of the valve, may be surgically corrected by inserting an artificial valve in place of the defective one.

VEINS (Fig. 5.6)

These drain mainly via the coronary sinus. The anterior cardiac veins and the venae cordis minimae open into the right atrium.

The **coronary sinus** is formed at the left border of the heart as a continuation of the great cardiac vein. It passes to the right in the posterior coronary sulcus and enters the right atrium near the orifice of the inferior vena cava. It is about 3 cm long.

Tributaries

(i) **great cardiac vein**—is a large vessel which drains both ventricles. It ascends the anterior interventricular sulcus, turns around the left border of the heart and becomes the coronary sinus. A left marginal vein enters it.

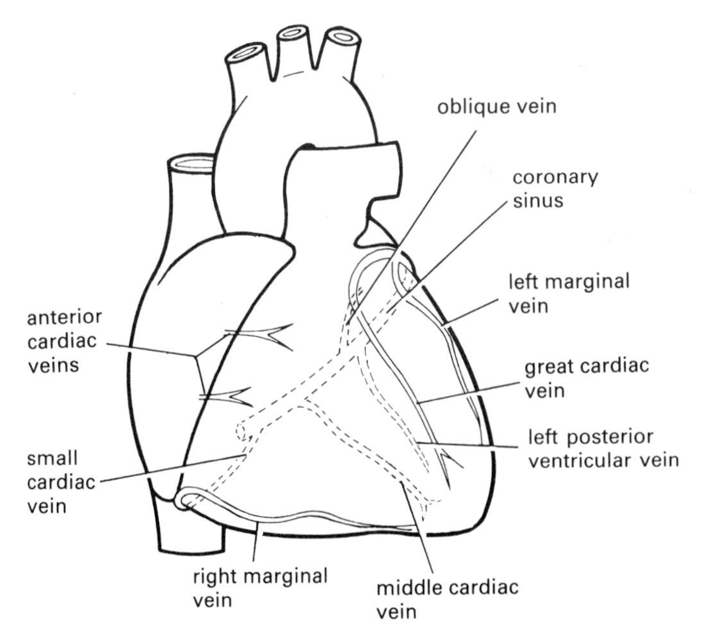

Fig. 5.6 The venous drainage of the heart.

(ii) **middle cardiac vein**—lies in the posterior interventricular sulcus.

(iii) **small cardiac vein**—runs to the left in the coronary sulcus and enters the right extremity of the coronary sinus. It is the continuation of the right marginal vein which runs along the lower border of the heart.

(iv) **left posterior ventricular vein**—runs up the back of the left ventricle and enters the coronary sinus.

(v) **oblique vein**—lies on the posterior wall of the left atrium. It is a remnant of the left superior vena cava of the embryo.

Anterior cardiac veins—several small veins which drain the anterior wall of the right ventricle and open into the right atirum.

Venae cordis minimae—are small veins draining much of the heart wall. They open directly into the cavities of the heart.

Lymph drainage

This is to the tracheobronchial lymph nodes.

Nerve supply

This is derived from the vagus and sympathetic fibres (upper thoracic segments) through the cardiac plexus (p. 101). The fibres are distributed with the coronary arteries. Parasympathetic ganglion cells are found on the heart walls. Sensory fibres subserving reflex activity pass in the vagus, and pain fibres pass in to the spinal nerves (T1–T3). Many pain fibres traverse the sympathetic ganglia before entering the nearby spinal nerves.

Histology

The heart wall is composed of three layers; a thin outer epicardium (serous pericardium), a very thick muscular (myocardial) layer and a thin inner endocardium of cubical endothelial cells. The myocardial cells are embedded in areolar tissue. Except for the AV bundle there is no myocardial continuity between the atria and the ventricles; they are separated by a pair of fibrous rings encircling both atrioventricular orifices in a figure-of-eight fashion. To these rings the myocardial muscle fibres are attached and they encircle the heart chambers in complex spirals and whorls. Cardiac muscle consists of branching striated fibres with a central nucleus and transverse intercalated discs at points of meeting of branches. The effectiveness of the heart depends largely on the state of cardiac muscle. Its efficiency as a pump also depends on the state of the valves.

THE CONDUCTING SYSTEM OF THE HEART

This system is formed of specialised cardiac muscle cells which in man are difficult to distinguish histologically from cardiac muscle. It consists of the sinuatrial (SA) node, the atrioventricular (AV) node, the AV bundle (of His), its right and left branches and a terminal subendocardial plexus (Purkinje fibres). This system initiates the complex cardiac muscle contractions comprising the cardiac cycle, and controls its regularity.

The **SA node** (pacemaker) is a small vascular area of conducting tissue lying in the wall of the right atrium at the upper end of the crista terminalis and to the right of the superior caval opening. Impulses are conducted from it through the atrial wall to the **AV node** which is a similar nodule lying in the septal wall of the right atrium above the opening of the coronary sinus. The **AV bundle** arises from the node, descends across the membranous part of the interventricular septum and divides into right and left branches which are distributed to their respective ventricles. Some of the right branch traverses the septomarginal band to reach the anterior ventricular wall. The branches ramify and form the subendocardial plexus which is distributed to the papillary muscles and ventricular walls.

THE PERICARDIUM (Fig. 5.7)

The pericardium is a fibroserous membrane surrounding the heart and the adjacent parts of the large vessels entering and leaving it. It consists of an outer fibrous pericardium and an inner serous pericardium.

The **serous pericardium** is a closed serous sac, invaginated by the heart, possessing visceral and parietal layers and enclosing a narrow pericardial cavity. Visceral pericardium covers the whole surface of the heart and is continuous with the parietal pericardium lining the inner surface of fibrous pericardium. The lines of reflection from the heart to the fibrous pericardium are shown in Figure 5.7. Its irregular form is related to the complex development of the heart. The serous pericardium is derived from the lining of the pericardial part of the coelomic cavity.

The **fibrous pericardium** forms a strong flask-shaped sac around the heart and serous pericardium. It blends inferiorly with the central tendon of the diaphragm and with the adventitia of the inferior vena cava, superiorly with the adventitia of the aorta, pulmonary trunk and superior vena cava, and posteriorly with that of the pulmonary veins.

DEVELOPMENT OF THE HEART AND PERICARDIUM

The embryonic heart develops from a vascular tube which hangs from the dorsal wall of the pericardial part of the coelomic sac. In its development two main processes are involved: (i) bending and differential growth, and (ii) division into right and left sides. (See p. 36).

BENDING AND DIFFERENTIAL GROWTH OF THE HEART TUBE

The heart is suspended from the dorsal wall of the embryo by a mesentery, the dorsal mesocardium. The caudal (venous) end of the heart tube receives

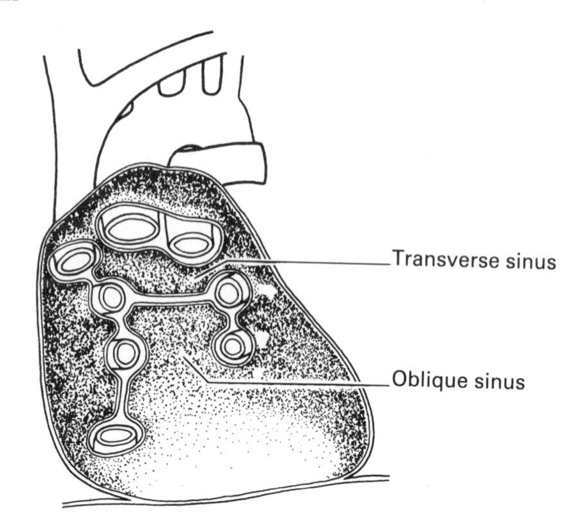

Fig. 5.7 Diagram of the interior of the pericardium, viewed from in front, after removal of the heart. The serous pericardium surrounding the vessels has been cut away.

vessels which develop into the superior and inferior venae cavae and the pulmonary veins. These vessels remain enclosed in a common sleeve of pericardium. The cephalic (arterial) end divides to form the aorta and pulmonary artery which similarly remain enclosed in a common sleeve of pericardium. The cephalic part of the tube elongates and descends, the arteries coming to lie in front of the veins. Simultaneously the middle of the dorsal mesocardium breaks down and forms a dorsal communication between the two sides of the pericardial cavity, the transverse sinus of the pericardium. This lies between the folded arterial and venous ends of the tube. In the adult the sinus lies behind the aorta and pulmonary artery, and in front of the superior vena cava and pulmonary veins.

The sleeve of pericardium surrounding the veins is stretched and forms a ∩-shaped cul-de-sac (the oblique sinus) bounded by two pulmonary veins on the left and two pulmonary veins and the venae cavae on the right (Fig. 5.7). It is separated from the transverse sinus superiorly by a double layer of pericardium.

Four dilatations develop in the primitive heart, the sinus venosus at the venous end, then the atrium, the ventricle and the bulbus cordis (the latter leading into the truncus arteriosus). The atrium is continuous with the ventricle through a narrow atrioventricular canal.

DIVISION INTO RIGHT AND LEFT SIDES

Longitudinal partitions divide the atrium, ventricle and bulbus cordis into right and left atria, right and left ventricles and the pulmonary and aortic trunks respectively.

DIVISION OF THE ATRIUM AND THE ATRIOVENTRICULAR CANAL

Two small projections, the endocardial cushions, unite across the narrow atrioventricular canal dividing it into right and left portions. In the atrial cavity a septum (the septum primum) descends and fuses with the endocardial cushions; it becomes perforated in its upper part. A septum secundum grows down on the right side of the septum primum, overlapping the perforation. The oblique communication which persists between the two cavities is known as the **foramen ovale**. The septum primum acts in a valve-like manner, opening the foramen ovale and allowing blood to flow from the right to the left atrium but not in the reverse direction. The lower edge of the septum secundum forms the limbus ovalis of the adult.

DIVISION OF THE VENTRICLE AND BULBUS CORDIS

The ventricle is divided by the development of a septum which fuses with the endocardial cushions above. The cavity of the bulbus cordis is divided by a spiral septum into the pulmonary and aortic trunks. The ventricular septum and the septum of the bulbus cordis unite with each other in such a way that the right ventricle leads into the pulmonary artery and the left ventricle into the aorta.

In their growth the ventricles incorporate part of the bulbus cordis thus forming the smooth-walled infundibulum and vestibule. The left atrium incorporates the pulmonary veins and the right atrium the sinus venosus and venae cavae, these forming the smooth-walled part of each cavity.

FETAL CIRCULATION (Fig. 2.3, 2.10, 2.11)

Oxygenated blood from the placenta passes through the liver in the ductus venosus and then into the inferior vena cava and right atrium. The angle between the superior and inferior venae cavae directs the placental blood through the foramen ovale into the left atrium. It then passes through the left ventricle into the aorta. Much of the blood goes to the head of the embryo to supply the developing brain. It returns to the heart in the superior vena cava and is directed into the right ventricle and then into the pulmonary artery. Blood in the pulmonary artery is shunted along the ductus arteriosus into the aorta, thus bypassing the lungs, and joins the aortic blood beyond the points of emergence of the cephalic blood flow. The blood in the thoracic and abdominal aorta passes to the rest of the body and much passes to the placenta in the umbilical arteries. The venous return from the body enters the inferior vena cava and mixes with the blood coming from the placenta through the umbilical vein and ductus venosus. At birth, the lungs take in air and there is a much greater flow of blood to the lungs with a subsequent increased venous return. This raises the pressure in the left atrium and so closes the foramen ovale. About the same time the ductus arteriosus and the ductus venosus also close, so establishing the adult pattern of circulation.

Cardiac abnormalities are among the most frequently encountered congenital defects and some are incompatible with independent life. One of the commonest abnormalities is a bicuspid aortic valve, this being usually without symptoms. Other abnormalities such as persistent patent ductus arteriosus, atrial septal defects (where the foramen ovale has failed to close) and ventricular defects may require surgical correction in childhood. Combined defects can also occur, the commonest consisting of a ventricular septal defect, pulmonary valvular stenosis (narrowing) and an aorta which communicates with both the right and left ventricular cavities. If the ductus fails to close, then pulmonary hypertension (high blood pressure in the lungs) results because of shunting of the systemic blood into the pulmonary circulation. Cardiac failure usually follows.

— 6 —

Other mediastinal structures

THE LARGE ARTERIES

THE AORTA (Figs. 6.3, 6.6, 30.1)

The aorta is the main arterial trunk of the systemic circulation and in the healthy state its walls contain a large amount of yellow elastic tissue. It arises from the left ventricle and ascends for a short distance before arching backwards over the root of the left lung and descending through the thorax and abdomen. It is divided into an ascending part, an arch, a descending thoracic part and an abdominal part.

Ascending aorta
This is a wide vessel about 5 cm long. It begins at the aortic orifice and ascends to the right around the pulmonary trunk as far as the level of the sternal angle.

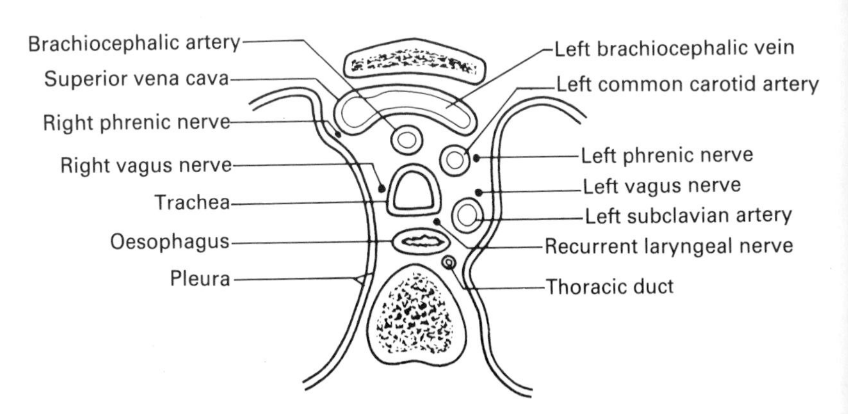

Fig. 6.1 Diagrammatic representation of the structures seen on a transverse section through the mediastinum at the level of the upper border of the body of the 3rd thoracic vertebra. The vertebra is shown at the bottom of the diagram and the manubrium at the top (see also caption to Fig. 6.2).

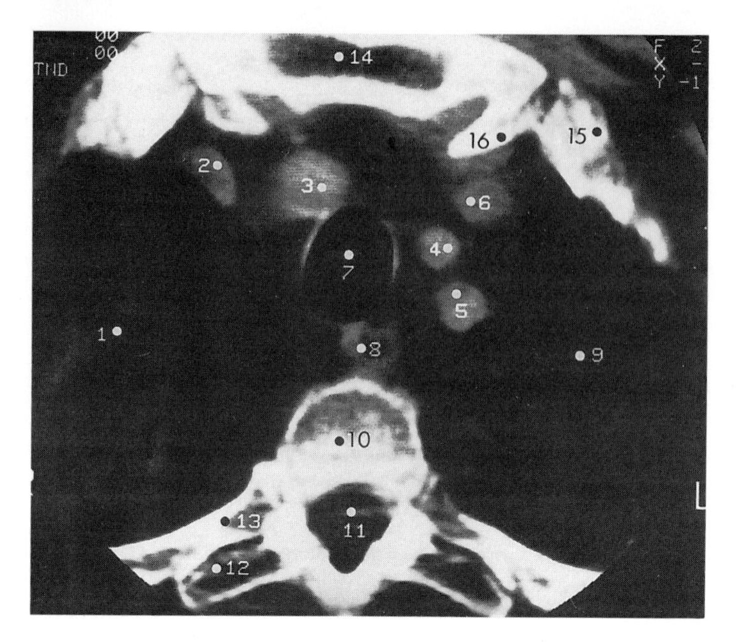

Fig. 6.2 CT transverse scan through the thorax at the level of the third thoracic vertebra.
The scan corresponds to Figure 6.1 and the level is indicated by line A in Figure 6.3. Computed tomography is an X-ray technique which uses a computer to reconstruct an image of a thin slice of the body. Differences in tissue density are made apparent and range from air, which is very black, to the dense white of compact bone. The intensity of the image can be varied to highlight an organ of specific density and a contrast material can be introduced into the gut, bladder or blood stream to increase the density of these structures and aid in their identification. Fat is black on the scan and serves to outline any organ it surrounds, this being particularly evident in abdominal scans. In the scans of the thorax, abdomen and pelvis the left side of the subject is on the right side of the picture which one has to envisage as if looking towards the head from the feet when the body has been cut through.

1. Right lung
2. Right brachiocephalic vein
3. Brachiocephalic artery
4. Left common carotid artery
5. Left subclavian artery
6. Left brachiocephalic vein
7. Trachea
8. Oesophagus
9. Left lung
10. Vertebral body
11. Vertebral canal
12. Transverse process
13. Rib
14. Manubrium sterni
15. Left clavicle
16. Left first rib

Above each of the semilunar folds (valvules) of the aortic valve is a dilatation, an **aortic sinus**.

Relations

It lies within the fibrous pericardium enclosed in a sheath of serous pericardium which is common to it and the pulmonary trunk. The lower part lies behind the

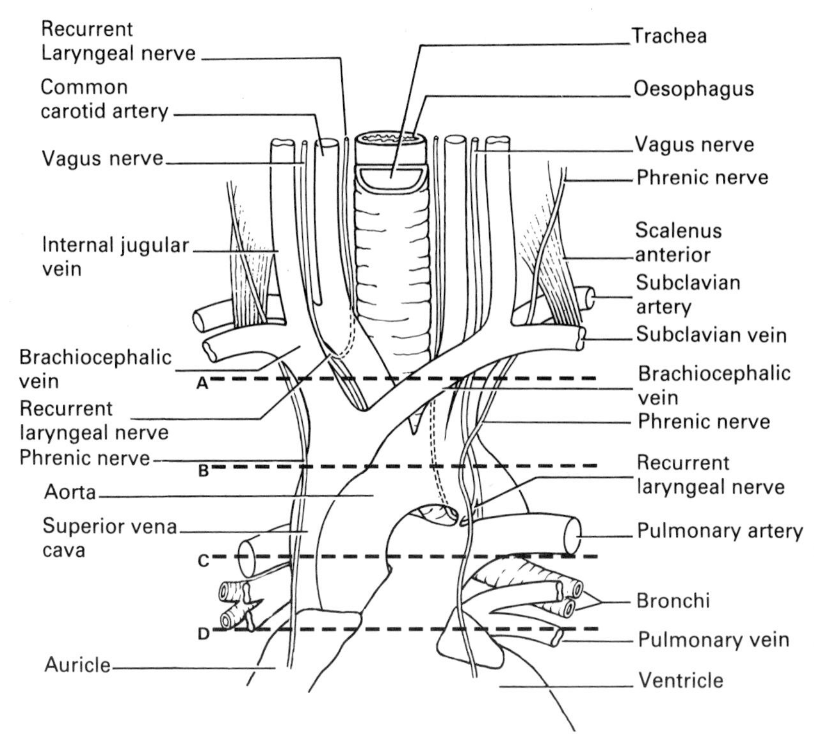

Fig. 6.3 Diagram of the principal relations of the arch of the aorta. The lines A, B, C and D correspond respectively to the CT scans shown in Figures 6.2, 6.7, 6.9 and 6.10

infundibulum of the right ventricle and the beginning of the pulmonary trunk and above this it is related to the sternum. Posteriorly from below upwards are the left atrium, right pulmonary artery and the right main bronchus. To the left lie the left auricle and the pulmonary trunk; to the right, the right auricle and the superior vena cava.

Branches

Right and left coronary arteries arise from the anterior and left posterior sinuses (p. 68).

Arch of the aorta

This part of the aorta passes upwards behind the manubrium from the level of the sternal angle. It arches backwards and to the left over the root of the left lung and descends to reach the left side of the 4th thoracic vertebra.

Relations

To its left lie the mediastinal pleura and lung, the left phrenic nerve in front and vagus behind. To its right lie the superior vena cava with the right phrenic nerve on its right side, the trachea and left recurrent laryngeal nerve, the oesophagus and thoracic duct and then the 4th thoracic vertebra, from before backwards. Inferiorly it crosses, from before backwards, the bifurcation of the pulmonary trunk, the ligamentum arteriosum, and the left main bronchus. Superiorly are its three branches, crossed anteriorly by the left brachiocephalic vein. The arch is connected inferiorly to the left pulmonary artery by the ligamentum arteriosum, a fibrous remnant of the ductus arteriosus. The left recurrent laryngeal nerve passes posteriorly around the ligamentum and the arch. The superficial part of the cardiac plexus lies anterior to the ligamentum and the deep part of the plexus inferior to the arch of the aorta. Remnants of the thymus gland may be found in front of the arch.

Branches

(i) the **brachiocephalic artery** is the first branch of the arch. It arises behind the manubrium and ascends, passing backwards and to the right as far as the right sternoclavicular joint where it divides into right subclavian and right common carotid arteries.

Relations: anteriorly the left brachiocephalic vein and thymus separate it from the manubrium. Posteriorly lies the trachea. To its right lies the right brachiocephalic vein and the superior vena cava and to its left the left common carotid artery.

It has only terminal branches, the right subclavian (p. 421) and the right common carotid arteries (p. 423).

(ii) the **left common carotid artery** (p. 423).

(iii) the **left subclavian artery** (p. 421).

Descending aorta

This descends from the left side of the 4th thoracic vertebra inclining medially to the front of the 12th vertebra where it passes through the diaphragm to become the abdominal aorta (p. 167).

Relations

Anteriorly, from above downwards, are the root of the left lung, the left atrium separated by pericardium, the oesophagus and the diaphragm. Posteriorly it lies on the vertebral column and hemiazygos veins. Its left side is in contact with the left pleura and lung and its right with the oesophagus above and the right lung and pleura below. The thoracic duct and azygos vein lie to its right side.

Branches

(i) 3rd–11th posterior intercostal arteries and the subcostal artery (p. 58).
(ii) two or three bronchial arteries.
(iii) four or five oesophageal arteries.
(iv) mediastinal and diaphragmatic arteries.

THE PULMONARY TRUNK (Fig. 6.4, 6.5, 6.8)

This wide vessel, about 5 cm long, begins at the pulmonary orifice, ascends posteriorly on the left side of the aortic trunk and ends by bifurcating into right and left pulmonary arteries in the concavity of the aortic arch. It lies in front of the transverse sinus and is contained within a common sleeve of serous pericardium with the ascending aorta. The right and left coronary arteries and the two auricles surround its base and it is covered anteriorly and on its left by the left pleura and lung.

Branches

(i) the **right pulmonary artery** passes to the hilus of the right lung behind the ascending aorta and superior vena cava, and in front of the oesophagus and right main bronchus.

(ii) the **left pulmonary artery** passes in front of the left bronchus and the descending aorta to the hilus of the lung. It is connected by a short fibrous band, the ligamentum arteriosum, to the lower surface of the aortic arch.

In the lungs these arteries and their branches accompany the bronchi and bronchioles.

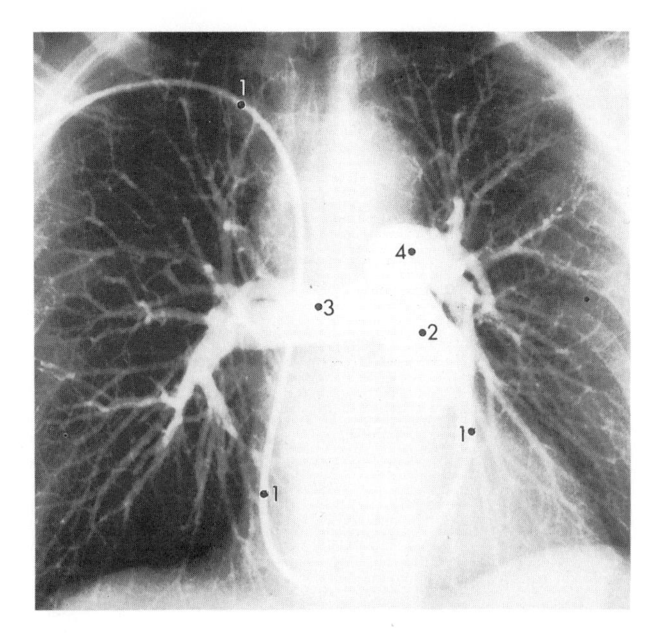

Fig. 6.4 Pulmonary arteriogram. The catheter (1) has been introduced percutaneously into the right basilic vein and passed through the right atrium and ventricle to enter the pulmonary trunk through the pulmonary valve. Contrast material has been injected.

1. Catheter
2. Pulmonary trunk
3. Right pulmonary artery
4. Left pulmonary artery

THE LARGE VEINS (Figs. 6.1, 6.3, 6.4, 6.6, 6.8)

The **pulmonary veins** are short wide vessels passing to the left atrium, usually two (upper and lower) from each lung. In the hilus of the lung they lie below and in front of the artery (see Figs. 7.1, 7.2). On the right the upper passes behind the superior vena cava, the lower behind the right atrium. On the left both veins pass in front of the aorta.

The **brachiocephalic veins** are formed behind the sternoclavicular joints by the union of the corresponding internal jugular and subclavian veins. After a short course they unite to form the superior vena cava about the middle of the right border of the manubrium.

The **right brachiocephalic vein** is about 3 cm long. It descends vertically behind the right border of the manubrium, anterolateral to the brachiocephalic artery. The right phrenic nerve descends along its lateral surface between it and the pleura.

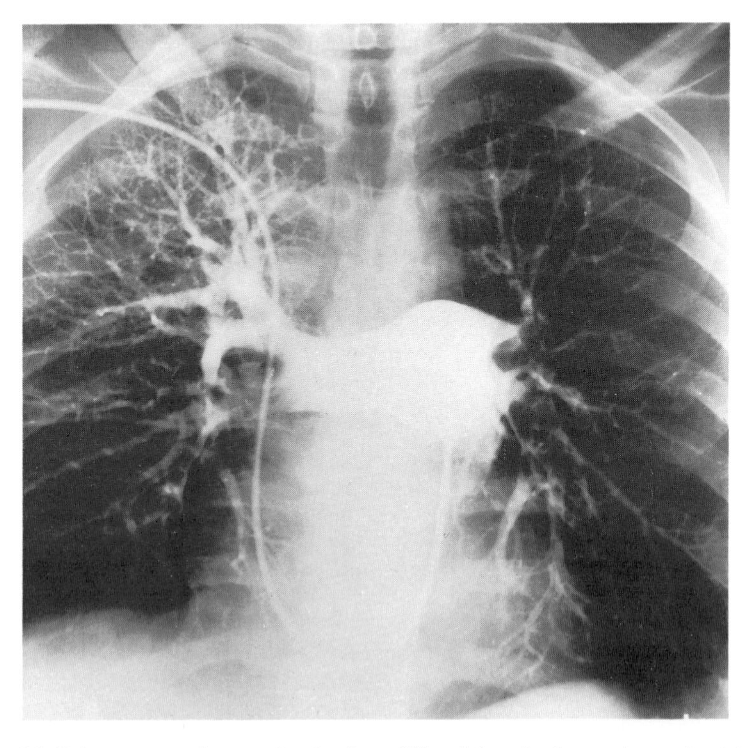

Fig. 6.5 Pulmonary arteriogram showing large filling defects in the arteries to the right lower, left upper and left lower lobes resulting in decreased blood flow to the lungs: only the right upper lobe is normal (compare with Fig. 6.4). The appearances are those of massive pulmonary embolism, blood clots having passed from leg and pelvic veins into the lungs.

The **left brachiocephalic vein** is about 6 cm long. It descends obliquely behind the manubrium crossing above the arch of the aorta and in front of its three large branches and the trachea. Thymic remnants may lie anteriorly.

Tributaries

Corresponding to some branches of the subclavian artery, viz:

(i) vertebral vein—drains neck muscles at the back and vertebral column.
(ii) inferior thyroid veins—usually unite in front of the trachea, and their common trunk enters the left brachiocephalic vein.
(iii) internal thoracic vein—drains anterior chest wall and diaphragm.
(iv) additionally, the thoracic duct enters the beginning of the left vein; the right lymph duct enters the beginning of the right vein. The left superior intercostal vein runs across the aortic arch to enter the left brachiocephalic vein.

The **superior vena cava** is a wide vessel about 7 cm long. It is formed by the union of the two brachiocephalic veins at the middle of the right border of the manubrium and descends vertically behind the manubrium and body of the sternum to enter the right atrium at the level of the 3rd right costal cartilage. It possesses no valves.

Relations

In its lower half it is covered by the fibrous and serous pericardium. Anterior are the right lung and pleura, and the manubrium. Posterior is the right lung root. Medial are the ascending aorta and the brachiocephalic artery. Lateral are the right phrenic nerve and the pleura and lung. The azygos vein is its only tributary, entering it posteriorly about its middle.

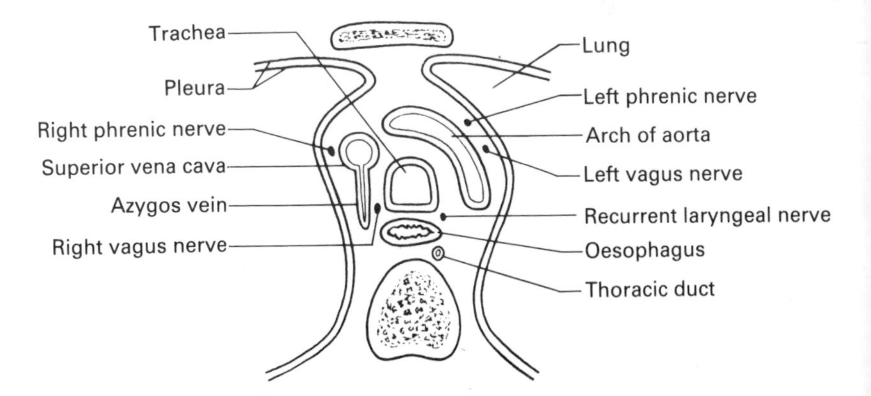

Fig. 6.6 Diagrammatic representation of the structures seen on a transverse section through the mediastinum at the level of the upper part of the body of the 4th thoracic vertebra (bottom), the arch of the aorta and the manubrium (top).

Determination of the right atrial pressure is important in the management of shocked patients' fluid needs. A catheter can be passed into the internal jugular or subclavian vein and thence into the superior vena cava and the right atrium.

The **azygos vein** is formed in the abdomen in front of the 2nd lumbar vertebra either as a branch of the inferior vena cava or by the union of the right subcostal and ascending lumbar veins. It leaves the abdomen through the aortic opening of the diaphragm. In the thorax it lies to the right of the aorta and thoracic duct and ascends on the right of the vertebral bodies covered by pleura. At the level of the 4th thoracic vertebra, it arches forwards over the hilus of the right lung and enters the superior vena cava. Above the hilus it lies between the lung and pleural sac laterally and the oesophagus and trachea medially.

Tributaries

(i) right ascending lumbar vein.
(ii) 4th–11th right posterior intercostal veins and subcostal vein.
(iii) right superior intercostal vein.
(iv) right bronchial veins.
(v) accessory hemiazygos and hemiazygos veins—these lie on the left side of the vertebral column and receive the 4th–11th left intercostal veins. They pass to the right in front of the mid-thoracic vertebrae to open separately into the azygos vein. The left ascending lumbar vein continues as the hemiazygos vein.

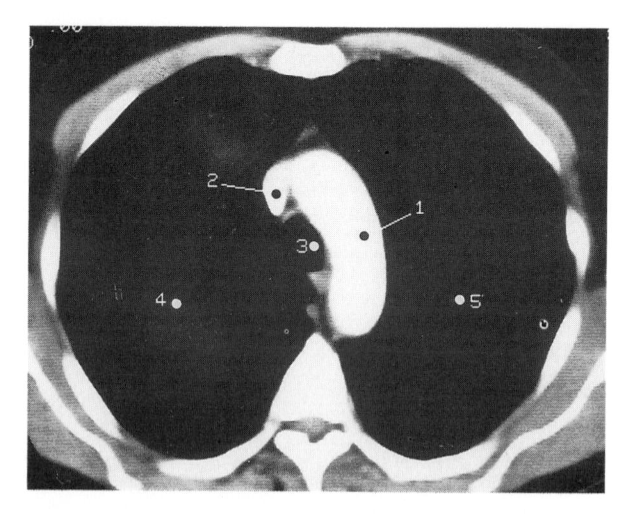

Fig. 6.7 CT transverse scan through the thorax at the level of the upper part of the body of the 4th thoracic vertebra, corresponding to Figure 6.6, and line B in Figure 6.3. (See also caption to Figure 6.2).

1. Aortic arch
2. Superior vena cava
3. Trachea
4. Right lung
5. Left lung.

The **inferior vena cava** is a wide vessel which pierces the central tendon of the diaphragm and, after a short intrathoracic course, enters the lower part of the right atrium. (See also pp. 61 and 169).

OTHER STRUCTURES

THE PHRENIC NERVES (Figs. 6.1, 6.3, 6.6, 6.8. See also p. 429)

The right and left phrenic nerves descend through the thorax, in front of each lung root.

The **right phrenic nerve** enters the thorax on the lateral side of the right brachiocephalic vein and in its descent it lies on the pericardium covering the superior vena cava, the right atrium and the inferior vena cava. It passes through the caval opening in the diaphragm. Throughout its course it is covered laterally by mediastinal pleura.

The **left phrenic nerve** enters the thorax in front of the left subclavian artery, behind the left brachiocephalic vein. In its descent it crosses the aortic arch and the pericardium covering the left ventricle and reaches and pierces the diaphragm. Throughout its course it is covered laterally by mediastinal pleura.

Both nerves supply the diaphragm and also give sensory branches to the mediastinal and diaphragmatic pleura, to the pericardium, and to the peritoneum. Pain from inflammation of the diaphragmatic pleura or peritoneum may be referred to the deltoid region of the shoulder which is innervated by the supraclavicular nerves, mainly derived like the phrenic nerve, from the C4 segment of the spinal cord.

THE VAGUS NERVES (Figs. 6.1, 6.3, 6.6, 6.8. See also p. 409)

Right vagus nerve

This enters the thorax posterolateral to the right brachiocephalic artery. It descends between the trachea medially, and the mediastinal pleura, the lung and the azygos vein laterally, to the back of the right main bronchus where it gives numerous branches to the pulmonary plexus before proceeding to the oesophageal plexus.

Left vagus nerve

This enters the thorax between the left common carotid and subclavian arteries, behind the left brachiocephalic vein. It descends inclining posteriorly across the left side of the aortic arch, giving off its recurrent laryngeal branch before passing medially behind the root of the lung to the oesophagus. The two vagi form the oesophageal plexus. From this plexus emerge anterior and posterior gastric nerves containing fibres of both vagi and sympathetic nerves. They descend through the oesophageal opening of the diaphragm. The anterior nerve supplies

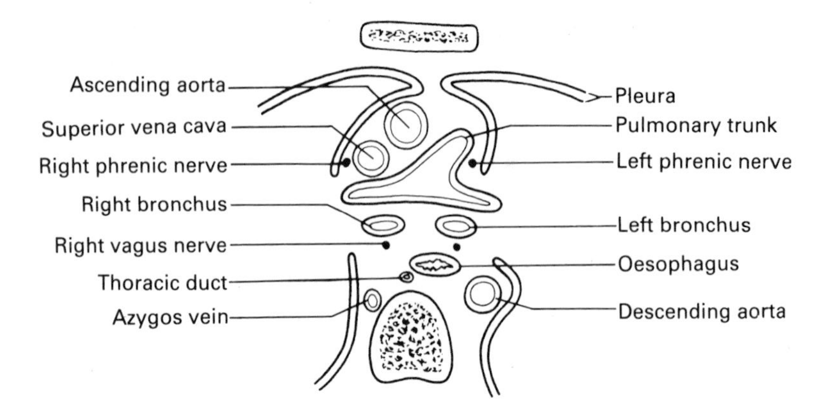

Fig. 6.8 Diagrammatic representation of the structures seen on a transverse section through the mediastinum at the level of the lower border of the 5th thoracic vertebra (bottom) and the sternum (top).

the front of the stomach, the duodenum, pancreas and liver; the posterior supplies the back of the stomach and gives branches which pass through the coeliac plexus to other abdominal viscera.

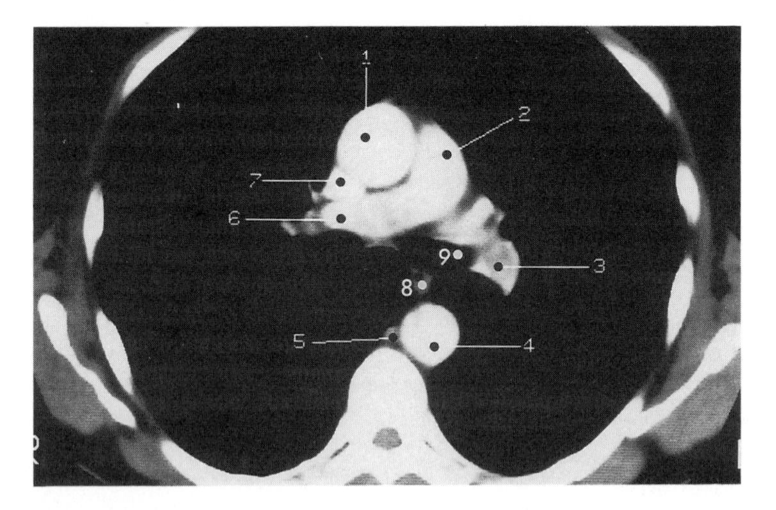

Fig. 6.9 CT transverse scan through the thorax at the level of the lower border of the 5th thoracic vertebra, corresponding to Figure 6.8 and line C in Figure 6.3. (See also caption to Figure 6.2).

1. Ascending aorta
2. Left main pulmonary trunk
3. Left main pulmonary artery
4. Descending thoracic aorta
5. Azygos vein
6. Right main pulmonary artery
7. Superior vena cava
8. Oesophagus
9. Left main bronchus

Branches

(i) left recurrent laryngeal nerve—winds around the ligamentum arteriosum and aortic arch and then ascends in the groove between the trachea and oesophagus into the neck (p. 410).

(ii) cardiac branches—to the cardiac plexus (p. 101).

(iii) branches to the pulmonary and oesophageal plexuses (p. 101).

THE THYMUS

This bilobed mass of lymphoid tissues lies in front of the trachea in the root of the neck and upper thorax. It is large at birth but atrophies after puberty. Variable in size, it may extend down beyond the aortic arch lying in front of the brachiocephalic veins and the left common carotid artery in the mediastinum. It is supplied by branches of the inferior thyroid and internal thoracic arteries and its veins drain to the internal thoracic and left brachiocephalic veins.

Histology and embryology

It is lobulated and formed of numerous cells embedded in a reticular network of fine collagen fibres. In the outer denser cortex, lymphocytes are predominant: the inner medulla contains paler round cells, thymocytes, and nests of concentrically arranged cells—Hassall's corpuscles. It is derived from the 3rd pair of pharyngeal pouches, and parathyroid tissue may be embedded in it.

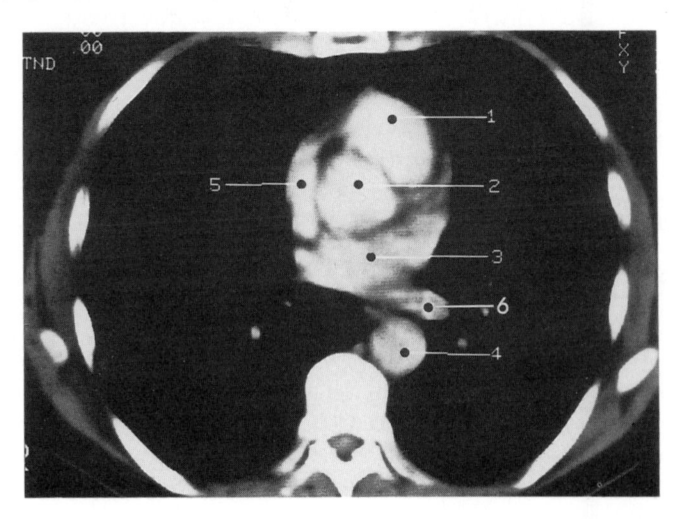

Fig. 6.10 CT transverse scan through the thorax at the level of the 6th thoracic vertebra, corresponding to line D in Figure 6.3. (See also caption to Figure 6.2)

1. Right ventricular outflow tract
2. Root of the aorta
3. Left atrium
4. Descending thoracic aorta
5. Right atrium
6. Left inferior pulmonary vein

THE OESOPHAGUS (Figs. 6.1, 6.3, 6.6, 6.11. See also pp. 123, 420)

The oesophagus descends through the thorax mainly to the left of the midline, curves forwards in its lower part away from the vertebral column and pierces the diaphragm at the level of the 10th thoracic vertebra.

Relations
In the upper part of the mediastinum it lies between the vertebral column posteriorly and the trachea anteriorly; the left recurrent laryngeal nerve lies in the groove between it and the trachea. Below, it is separated from the vertebral column, from above downwards, by the thoracic duct, the hemiazygos veins, the right posterior intercostal arteries and the descending thoracic aorta. Anteriorly below the trachea, it is crossed by the left bronchus, and below this it is separated by pericardium from the left atrium. On the right it is covered with mediastinal pleura and the azygos vein arches forwards above the lung root. On the left it is separated from the mediastinal pleura above by the left subclavian artery, the thoracic duct, the aortic arch and descending aorta. Its lower part is in contact with the pleura.

THE TRACHEA AND BRONCHI (Figs. 6.1, 6.3, 6.6, 6.8, 7.6, 7.7. See also p. 419)

The trachea lies in the midline anterior to the oesophagus, the left recurrent laryngeal nerve ascending in the groove between them. Anteriorly it is crossed by the brachiocephalic artery and then by the left brachiocephalic vein. On its left lie the common carotid and subclavian arteries above and the aortic arch below. On its right are the mediastinal pleura, the right vagus nerve and the azygos vein.

The tracheal bifurcation, the carina, is at the level of the sternal angle and the lower border of the 4th thoracic vertebra. It lies anterior to the oesophagus, behind and to the right of the bifurcation of the pulmonary trunk. It is separated from the right pulmonary artery by the deep part of the cardiac plexus and tracheobronchial lymph nodes. The trachea lengthens slightly on inspiration and recovers on expiration because of the contained elastic tissue.

The **extrapulmonary bronchi**: the main bronchus on each side arises at the bifurcation and descends passing laterally, enters the hilus of the lung where it divides, forming the intrapulmonary bronchial tree.

The **right bronchus** is about 3 cm long. It is wider and somewhat more vertical than the left and foreign bodies therefore tend to enter it more frequently. The right upper lobe bronchus arises from it just before it enters the hilus.

Relations: anteriorly the right pulmonary artery separates it from the pericardium and the superior vena cava. The arch of the azygos vein is above it and posteriorly lies the pulmonary plexus and the bronchial vessels.

The **left bronchus** is about 5 cm long.

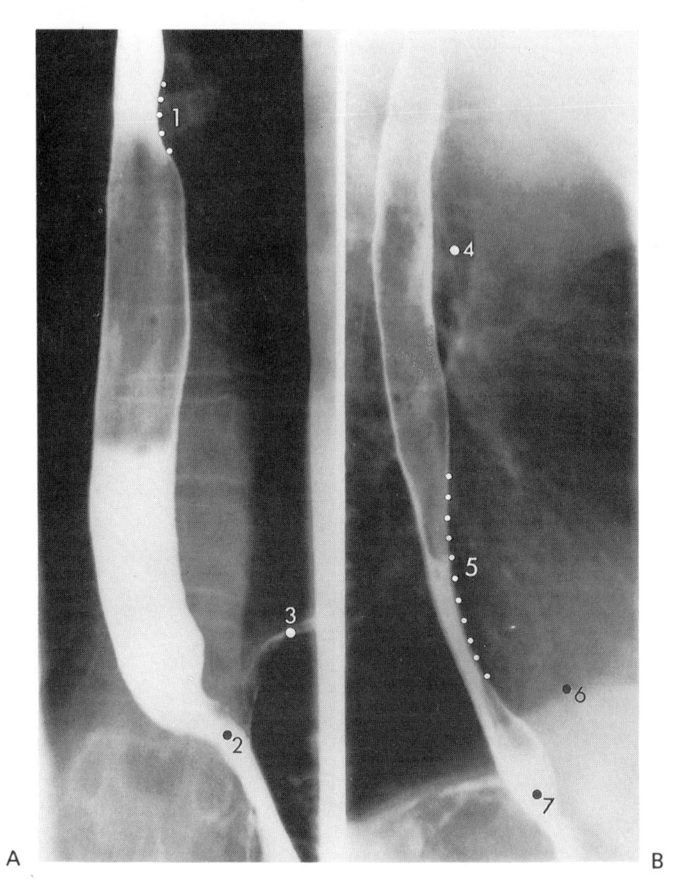

Fig. 6.11 Barium swallow. The patient has swallowed a mouthful of barium sulphate suspension just before these radiographs were taken. **A**. Anteroposterior view. **B**. Right lateral view.

1. Impression of aortic arch
2. Gastro-oesophageal junction
3. Fundus of stomach
4. Trachea
5. Cardiac impression (left atrium)
6. Diaphragm
7. Gastro-oesophageal junction.

Relations: anteriorly the left pulmonary artery separates it from the left atrium. The arch of the aorta is above it, and posteriorly lies the pulmonary plexus and bronchial vessels which separate it from the oesophagus and descending thoracic aorta.

The structure of the extrapulmonary bronchi is similar to that of the trachea (p. 419).

Direct viewing of the trachea and proximal main bronchi is possible with a bronchoscope. Instillation of a radio-opaque contrast medium, Lipiodol, into the bronchi, permits X-rays to show the arrangement of the distal smaller bronchi

(Figs. 7.6, 7.7). Bronchial cancer is one of the commonest malignant diseases and readily spreads to the tracheobronchial lymph nodes. It can produce early symptoms of bronchial obstruction.

— 7 —

The pleura and lungs

THE PLEURA

Each pleural sac is a closed cavity lined by a serous membrane which is invaginated by a lung. Parietal pleura lines the outer walls of the cavity and visceral (pulmonary) pleura covers the lungs. The layers of pleura are continuous around the root of the lung and are separated only by a thin layer of serous fluid. The two layers of pleura can glide easily on each other but are prevented from separating by the **surface tension** of the serous fluid and the **negative pressure** in the closed cavity. As a result, when the diaphragm descends and the thoracic cage expands, the lung also has to expand and so air is drawn in from the trachea.

The pleura is a fibro-elastic membrane lined by squamous mesothelial cells. The parietal pleura lines the ribs, costal cartilages and the intercostal spaces, the

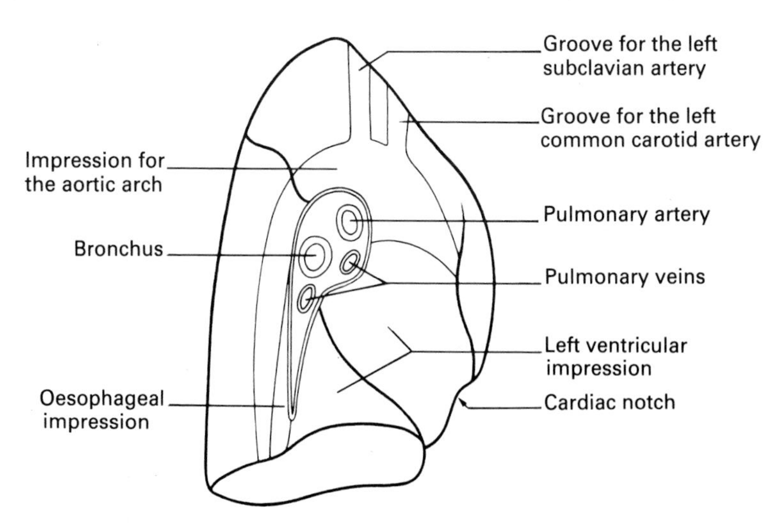

Fig. 7.1 The medial surface of the left lung. Some of the principal structures in contact with the lung are indicated. The hilus is surrounded by the cut edges of the pleura and pulmonary ligament (inferiorly).

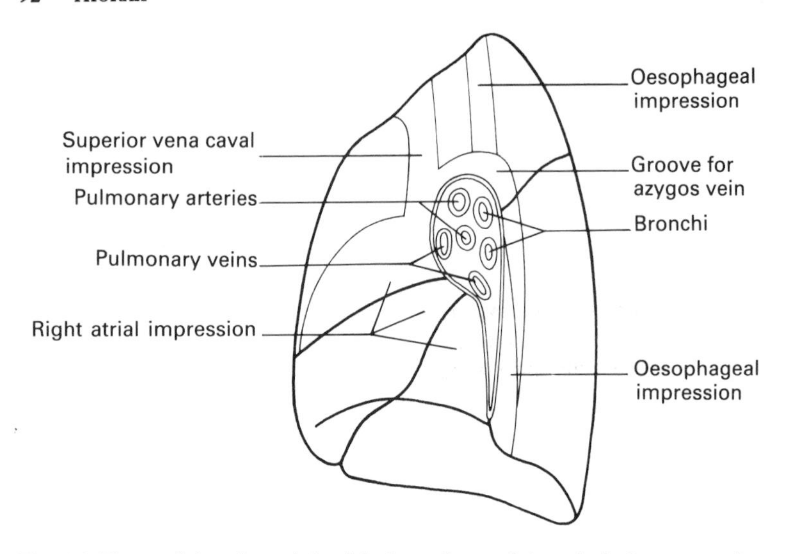

Fig. 7.2 The medial surface of the right lung. Some of the principal structures in contact with the lung are indicated. The hilus is surrounded by the cut edges of the pleura and pulmonary ligament (inferiorly).

lateral surface of the mediastinum and the upper surface of the diaphragm. Superiorly it extends beyond the thoracic inlet into the neck to form the cervical **dome** of pleura. Inferiorly, around the margin of the diaphragm, it forms a narrow gutter, the **costodiaphragmatic recess**. Anteriorly, in front of the heart, the left costal and mediastinal surfaces are in contact forming the **costomediastinal recess**. The mediastinal pleura invests the main bronchi and pulmonary vessels as an overlarge collar and passes on to the surface of the lung to become pulmonary pleura. This completely covers the lung and extends into its interlobar fissures.

The extent of the pleural sacs on each side should be noted. On both sides the upper limit of the cervical pleura lies about 3 cm above the medial third of the clavicle. From here the lines of pleural reflections descend behind the sternoclavicular joints to meet in the midline at the level of the 2nd costal cartilages. At the 4th cartilage, the left pleura deviates laterally and descends along the lateral border of the sternum as far as the 6th cartilage, whereas the right pleura continues vertically downwards near the midline to the 6th cartilage. At the 6th costal cartilage the lower border of both pleural reflections passes laterally just behind the costal margin to reach the 8th rib in the midclavicular line, the 10th rib in the midaxillary line and the 12th rib in the paravertebral line. The paravertebral line lies about 3 cm from the midline and corresponds to the tips of the transverse processes.

The pleura is supplied by blood from the tissues it covers. Pulmonary pleura has no pain fibres but the parietal pleura is richly supplied by the nerves in the subjacent tissues. Lymph from the pulmonary pleura passes to a superficial plexus in the lung and then to the hilar nodes. The parietal pleura drains to parasternal, diaphragmatic and posterior mediastinal nodes.

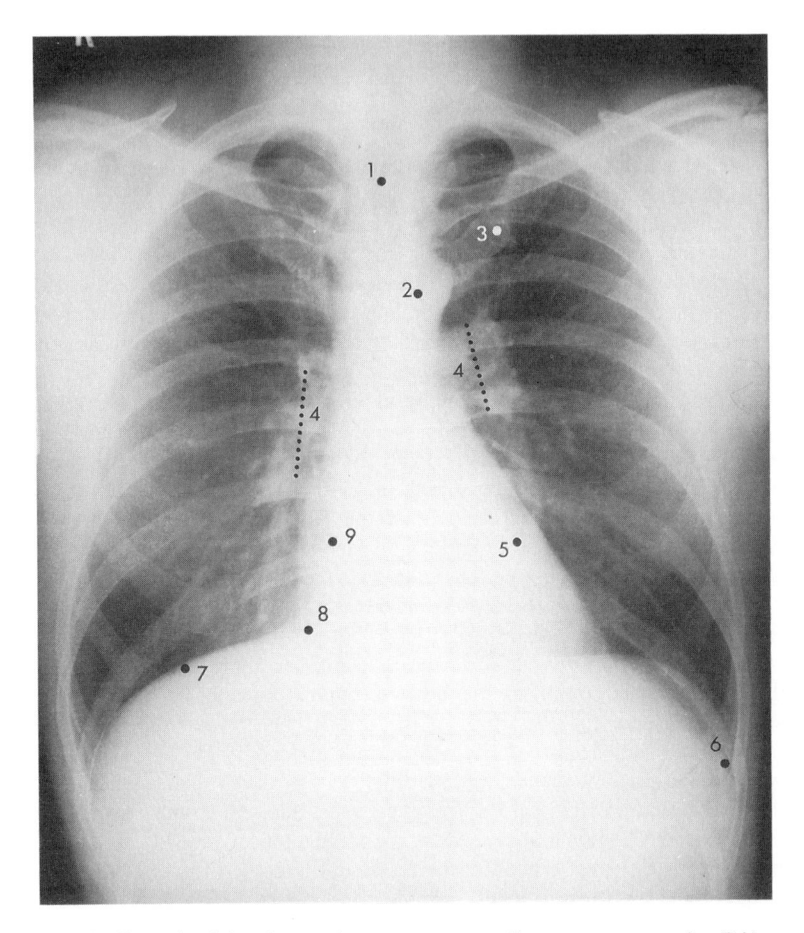

Fig. 7.3 Radiograph of the chest, using no contrast medium—postero-anterior (PA) view. The dense shadow occupying most of the upper abdomen is produced by the liver, which contains much blood. Many of the radiating streaks from the hilus of the lung are produced by the large blood vessels. Some hilar shadows may be caused by calcified lymph nodes. Shadows of the heart and large vessels are seen centrally. The term postero-anterior (PA) view refers to the direction of the X-rays. When examining a chest, the X-rays are usually directed from behind the patient who leans forward against the photographic film.

1. Trachea
2. Aortic knuckle
3. 1st costochondral junction
4. Hilar shadows
5. Left ventricle
6. Left costodiaphragmatic recess
7. Right dome of diaphragm
8. Inferior vena cava
9. Right atrium

THE LUNGS (Figs. 7.1, 7.2, 7.3)

The lungs are the paired organs of respiration. Each lies in its pleural sac attached to the mediastinum at the hilus. The lung is spongy and elastic in texture and cone-shaped to conform to the contours of the thoracic cavity. The right lung weighs about 620 g and the left about 560 g. The lungs have a characteristic mottled appearance on X-ray films. The clear areas are lung tissue and the dense shadows at the hilus and radiating outwards are caused by hilar tissues (lymph nodes) and by blood vessels (Fig. 7.3).

Each lung has an **apex** in the root of the neck and a **base** resting on the diaphragm. The base is separated by a sharp inferior border from a lateral convex costal surface and a medial concave (mediastinal) surface. In the centre of this latter surface, the structures forming the root of the lung are seen to be surrounded by a collar of pleura. The concavity of the medial surface is accentuated on the left to accommodate the left ventricle of the heart. The anterior border on the left side is deeply indented by the heart to form the **cardiac notch**. The posterior border is rounded and lies in the paravertebral sulcus.

The lungs are divided into lobes by fissures which extend deeply into their substance. An **oblique fissure** divides the left lung into an upper and a lower lobe; oblique and horizontal fissures divide the right lung into upper, middle and lower lobes. The oblique fissure of both lungs may be marked by a line curving around the chest wall from the spine of the 3rd thoracic vertebra to the 6th costochondral junction. The **horizontal fissure** is marked by a horizontal line passing laterally to the oblique line from the 4th right costal cartilage.

The **lower lobes** of both lungs lie below and behind the oblique fissure and comprise most of the posterior and inferior borders and parts of the medial and costal surfaces. The **upper lobe** of the left lung lies above and in front of the oblique fissure and comprises the apex, substantial portions of the mediastinal and costal surfaces and the whole of the anterior border including the cardiac notch. The equivalent part of the right lung is divided by the horizontal fissure into a large upper lobe and wedge-shaped anteriorly placed smaller **middle lobe**. A thin antero-inferior part of the left upper lobe, adjacent to the cardiac notch, is known as the lingula and represents the middle lobe. Variation exists in this lobar pattern. Fissures, especially the horizontal, may be incomplete or absent, and occasionally additional lobes are present.

The **hilus** of each lung contains a main bronchus, pulmonary artery, two pulmonary veins, the pulmonary nerve plexus, and lymph nodes, all surrounded by the collar of pleura whose narrow inferior extension is known as the pulmonary ligament. On both sides the bronchus lies behind the pulmonary artery, and the two pulmonary veins lie antero-inferior to both other structures.

THE INTRAPULMONARY BRONCHI AND BRONCHOPULMONARY SEGMENTS

The right upper lobe bronchus arises from the right main bronchus just before the hilus, and soon after entering the lung substance it divides into three segmental bronchi. The middle lobe bronchus arises from the right bronchus just

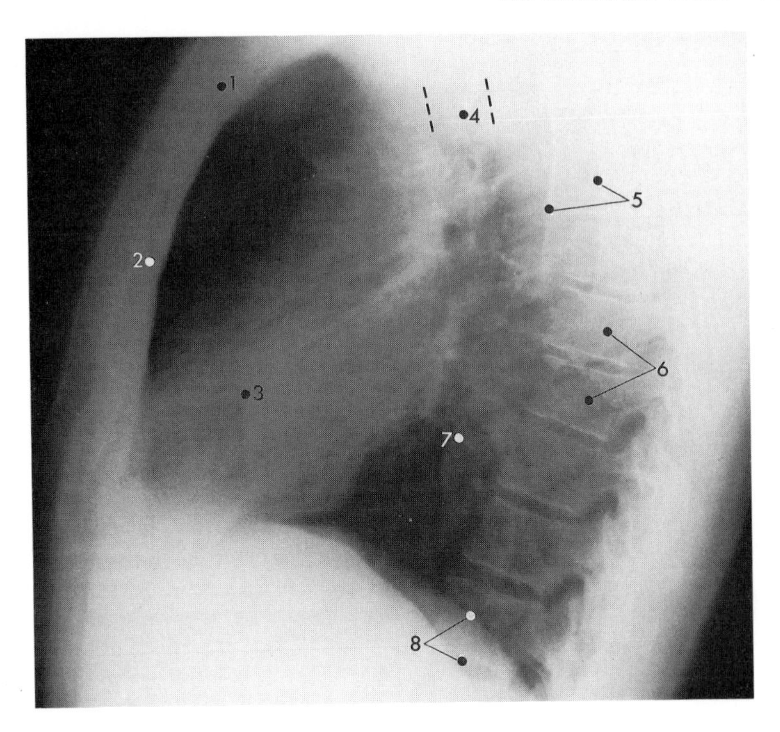

Fig. 7.4 Radiograph of the chest—left lateral view. Note the large space between the heart and the sternum in front, and between the heart and the vertebral bodies behind.

1. Manubriosternal joint
2. Sternum
3. Cardiac shadow
4. Trachea
5. Borders of scapulae
6. Vertebral bodies
7. Retrocardiac space
8. Domes of diaphragm

below the upper lobe bronchus and divides into two segmental bronchi. The continuation of the right bronchus passes to the lower lobe and divides into five segmental bronchi.

The left upper lobe bronchus arises from the main bronchus within the lung and divides into five segmental bronchi, two passing to the lingula. The continuation of the left bronchus passes to the lower lobe and divides into five segmental branches. Each segmental bronchus is distributed to a functionally independent unit of lung tissue—a **bronchopulmonary segment**. There are three segments in each upper lobe and two in the right middle lobe which are equivalent to the two in the lingula of the left lung. The five segments in the right lower lobe are more or less equivalent to those in the left lower lobe.

Relations

PARIETAL RELATIONS:

The borders of the lungs closely follow the lines of pleural reflection on the chest wall except below, where the inferior border of the lung lies approximately two

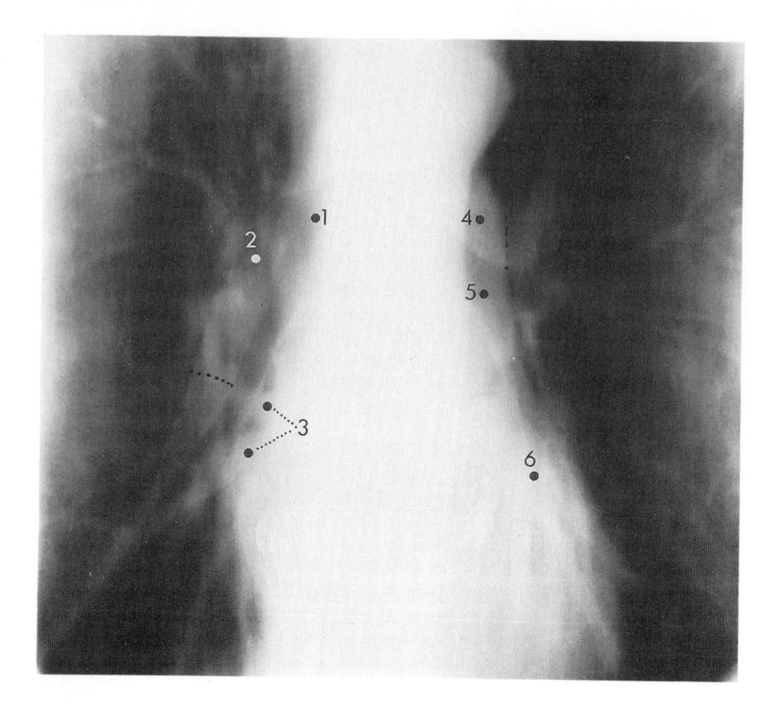

Fig. 7.5 Tomogram of chest through hilar region showing soft tissue shadows, anteroposterior view. By the technique of tomography, it is possible to obtain a radiograph of an 'isolated layer' of the body.
1. Right main bronchus
2. Right pulmonary artery
3. Right pulmonary veins
4. Left pulmonary artery
5. Left main bronchus
6. Left pulmonary vein

intercostal spaces above the inferior limit of the pleura, and in front, where the anterior border of the left lung in the region of the cardiac notch lies about 3 cm lateral to the pleural reflection in this region. The costal surface is related to the thoracic wall. The base is separated by the diaphragm from the right lobe of the liver on the right side and the left lobe of the liver, stomach and spleen on the left. The apex is covered by the dome of pleura which lines the suprapleural membrane, a fibrous sheet extending from the transverse process of the 7th cervical vertebra to the inner border of the 1st rib. The subclavian vessels arch over the apex above the pleura and membrane. Posteriorly the branch of the 1st thoracic nerve to the brachial plexus, the superior intercostal artery, the sympathetic trunk and the pleura separate the apex of the lung from the neck of the 1st rib.

The **medial relations** of the two lungs differ (Figs. 7.1, 7.2).

On the left: there is a large concavity antero-inferiorly for the left ventricle of the heart. It is continuous superiorly with a groove for the aortic arch which passes in front of and above the hilus. Above the groove the surface is in contact

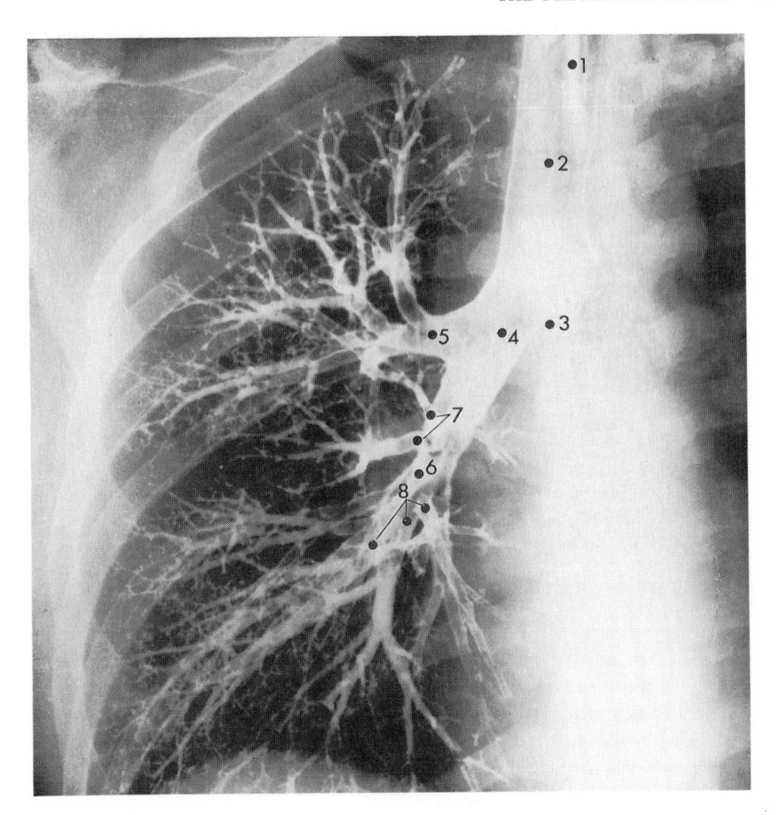

Fig. 7.6 Bronchogram of right lung, anteroposterior view. Contrast material is introduced into the lung along a catheter inserted into the trachea.

1. Intubation catheter
2. Trachea
3. Carina
4. Right main bronchus
5. Right upper lobe bronchus
6. Right middle lobe bronchus
7. Right apical lower lobe bronchi
8. Right basal lower lobe bronchi

with the left brachiocephalic vein, left common carotid and left subclavian arteries, and the oesophagus.

On the right: there is a shallow concavity in front of the hilus for the right atrium. This is continuous above with a groove for the superior vena cava and with a short one below for the inferior vena cava. The azygos vein grooves the lung as it arches forwards above the hilus. The oesophagus, which lies near the posterior border, is in contact throughout its length save where the azygos vein separates it from the lung. Between the superior vena cava and the oesophagus superiorly is the trachea.

Blood supply

Lung tissue is supplied by the bronchial branches of the descending thoracic aorta. Some of this blood returns to the heart by the pulmonary veins. The bron-

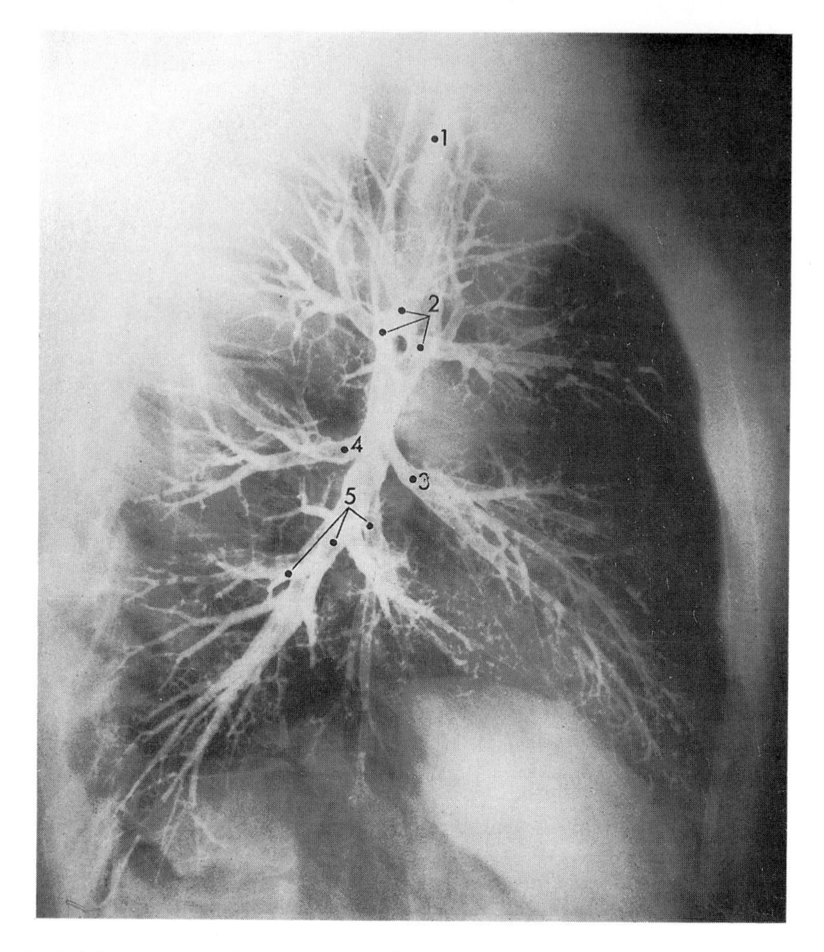

Fig. 7.7 Bronchogram of right lung, lateral view.
1. Trachea
2. Right upper lobe bronchi
3. Right middle lobe bronchus
4. Right apical lower lobe bronchus
5. Right basal lower lobe bronchi

chial veins drain to the azygos or accessory hemiazygos veins. Poorly oxygenated blood is conveyed to the lung alveoli by the pulmonary artery whose branches correspond to and accompany the bronchial tree. The terminal branches end in a capillary network within the alveolar walls. Venous tributaries from the network return with the bronchi to upper and lower pulmonary veins.

Lymph drainage

This is via a superficial subpleural lymph plexus and a deep plexus of vessels accompanying the bronchi. Both groups drain through hilar (bronchopulmonary)

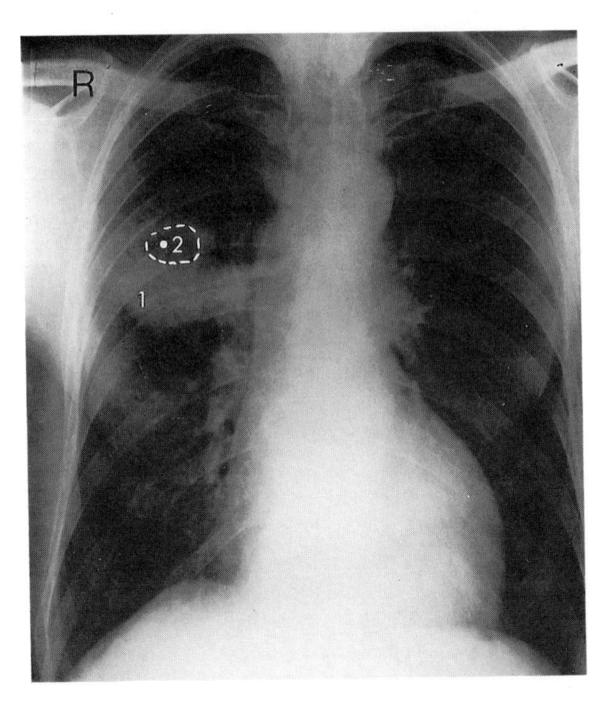

Fig. 7.8 Postero-anterior radiograph of chest showing on the right side a round white shadow (1) with a dark upper area (2) within it. The latter is air within an abscess cavity. In this case the lesion proved to be cancer of the lung with necrosis of its upper part.

lymph nodes to tracheobronchial nodes around the bifurcation of the trachea and thence to mediastinal lymph trunks.

Nerve supply
Sympathetic (bronchodilator) fibres from the upper thoracic segments and parasympathetic (bronchoconstrictor) fibres from the vagus, both pass through the pulmonary plexus. Sensory fibres pass mainly in the vagus.

Histology
The walls of intrapulmonary bronchi, lined by respiratory epithelium, are formed of fibrous tissue and smooth muscle, and contain incomplete rings of hyaline cartilage. The bronchi divide and redivide until their diameter is less than 1 mm after which they are known as bronchioles, the walls of which lack mucous glands and cartilaginous plates. The bronchioles subdivide and form respiratory bronchioles whose walls, lined by nonciliated columnar epithelium, contain smooth muscle. Each respiratory bronchiole opens into a number of alveolar ducts from which arise alveolar sacs whose walls possess the terminal respiratory saccules, the alveoli. Alveolar walls are lined by simple squamous epithelium.

Embryology
The lungs develop from a single ventral pharyngeal outgrowth which bifurcates. The lung buds enlarge into the coelomic pericardioperitoneal canals (see p. 28 and 39).

X-rays of the chest allow accurate localisation of disease of the lung, in addition to defining abnormal collections of air or fluid in the pleural space (see p. 91). The use of radio-opaque contrast substances greatly facilitates localisation of the lesion (Figs. 7.6, 7.7). Excision of a segment, a lobe or a whole lung can be undertaken. A knowledge of the pattern of division of the bronchi helps when a part of the lung has to be drained of pus (Fig. 7.8). The mechanisms of respiration are outlined on page 62.

— 8 —

The autonomic nervous system and lymph drainage

AUTONOMIC NERVOUS SYSTEM

The autonomic nerve supply of the thoracic viscera is distributed through visceral plexuses, viz. the cardiac, pulmonary and the oesophageal plexuses. These plexuses receive sympathetic contributions from the cervical and thoracic sympathetic trunks and parasympathetic contributions from the vagi (p. 85).

THE VISCERAL PLEXUSES

The **cardiac plexus**: the superficial (ventral) part lies in front of the ligamentum arteriosum; the deep (dorsal) part in front of the bifurcation of the trachea. They communicate with each other and functionally are a single unit, receiving branches from each of the cervical sympathetic ganglia, the upper thoracic ganglia, and parasympathetic cardiac branches from both vagi. Branches pass with the coronary arteries to supply heart muscle, coronary arteries, the conducting system of the heart, and some branches pass to the pulmonary plexus.

The **pulmonary plexus** lies in front of and behind the lung roots. It receives branches from the upper four thoracic sympathetic ganglia and from both vagi. It supplies the lung substance and visceral pleura.

The **oesophageal plexus** is a network surrounding the lower oesophagus. It receives branches from the upper thoracic sympathetic ganglia and from both vagi. It supplies the oesophagus and forms anterior and posterior vagal trunks which pass with the oesophagus into the abdomen as anterior and posterior gastric nerves (p. 85).

THE THORACIC SYMPATHETIC TRUNK

Each ganglionated trunk lies alongside the vertebral column in front of the necks of the ribs. It continues superiorly into the cervical sympathetic trunk and inferiorly into the lumbar. It usually possesses 12 ganglia, one corresponding to each thoracic nerve, but often the 1st is fused with the inferior cervical ganglion to form a large **stellate ganglion** above the neck of the 1st rib. Each ganglion is attached to its spinal nerve by a white (preganglionic) and a grey (postganglionic) ramus.

Branches

(i) rami communicantes pass to the spinal nerves.

(ii) the upper four ganglia give branches to the cardiac, pulmonary and oesophageal plexuses.

(iii) the greater, lesser and least **splanchnic nerves:** the greater arises from the 5th–9th ganglia; the lesser from the 10th–11th and the least (renal nerve) from the 12th. They descend over the posterior thoracic wall medial to the sympathetic chain and enter the abdomen by piercing the corresponding crus of the diaphragm to reach the abdominal plexuses. Their fibres synapse in the coeliac and other pre-aortic ganglia.

The **vagi** are described on page 85.

LYMPH DRAINAGE

The vessels may be divided into those draining the chest wall and those draining the thoracic viscera.

The **chest wall** drains by superficial and deep vessels. The superficial vessels drain, together with those of the breasts, to the pectoral and central groups of axillary lymph nodes. Some superficial vessels cross the midline and communicate with vessels of the opposite side. The vessels draining the deep tissues of the wall pass to three groups of nodes:

(i) **Parasternal nodes**—lie alongside the internal thoracic artery. Their efferents pass to the bronchomediastinal lymph trunk.

(ii) **Intercostal nodes**—lie at the posterior end of the intercostal spaces. Their efferents pass to the right lymph duct and the thoracic duct.

(iii) **Diaphragmatic nodes**—lie on the upper surface of the diaphragm. Their efferents drain to the parasternal and posterior mediastinal nodes and communicate through the diaphragm with the vessels draining the upper surface of the liver.

The **thoracic viscera** drain to three main groups of nodes:

(i) **Anterior mediastinal nodes**—lie in front of the brachiocephalic veins and drain the thymus and anterior pericardium. Their efferents join those of the tracheobronchial nodes.

(ii) **Tracheobronchial nodes**—lie alongside the trachea and bronchi and drain the lungs, bronchi, trachea and heart. Their efferents join with vessels from the anterior mediastinal nodes and form the bronchomediastinal lymph trunk. On the right side the trunk ascends alongside the trachea to join the right lymph duct, and on the left to join the thoracic duct.

(iii) **The posterior mediastinal nodes**—lie alongside the oesophagus behind the pericardium. They drain the oesophagus and posterior pericardium. Their efferents pass to the thoracic duct.

The thoracic duct and the smaller right lymph duct are the final vessels along which lymph passes before entering the bloodstream.

THE THORACIC DUCT

This is about 45 cm long. It arises in the abdomen from the cisterna chyli (p. 209) and enters the thorax through the aortic opening of the diaphragm to the right of the aorta. It ascends in front of the vertebral column on the left of the azygos vein and behind the oesophagus. In front of the 5th thoracic vertebra it passes to the left behind the oesophagus and continues its ascent to the level of the 7th cervical vertebra where it arches laterally behind the carotid sheath and then forwards in front of the subclavian artery to enter the beginning of the left brachiocephalic vein.

TRIBUTARIES: drain all the body below the diaphragm and the left half above the diaphragm and, in addition, the posterior part of the right chest wall. It usually receives the united left jugular and subclavian lymph trunks and the left bronchomediastinal lymph trunk.

THE RIGHT LYMPH DUCT

This is a short vessel which is often absent. It is formed in the neck by the union of right jugular, subclavian and bronchomediastinal lymph trunks. It enters the beginning of the right brachiocephalic vein and drains lymph from the right upper limb, the right side of the head, neck and the thorax, with the exception noted above.

PART THREE

Abdomen and pelvis

— 9 —

The abdomen and peritoneum

The abdomen is that part of the trunk between the diaphragm and pelvis. For descriptive purposes the anterior abdominal wall is divided into nine regions by two horizontal and two vertical planes. The horizontal planes are the **transpyloric** (midway between the jugular notch and the crest of the pubis), and the **trans-tubercular** (through the tubercles of the iliac crests). The vertical planes are the right and left **lateral** which pass through the midinguinal points (between anterior superior iliac spine and symphysis pubis). The regions so formed are—centrally, from above downwards, the epigastric, the umbilical and the pubic. On each side of these there are the hypochondriac, lateral and inguinal regions. The transpyloric plane is coincident with the level of the tip of the 9th costal cartilage anteriorly and the body of the 1st lumbar vertebra posteriorly. The subcostal plane (through the lowest part of the costal margin) is sometimes used instead of the transpyloric plane.

Skeleton: vertebrae—see page 43; sacrum—see page 174; ribs—see page 54; pelvis—see pages 171 and 175.

ANTERIOR ABDOMINAL WALL

The skin of the anterior abdominal wall is supplied by the 6th thoracic to 1st lumbar spinal nerves. T6 supplies the epigastric region, T10 the umbilical region, and L1 the groin. The superficial fascia is in two layers. The more superficial is fatty and continuous with the superficial fascia of the thorax and thigh. The deeper is membranous and best developed in the lower abdomen. It is attached inferiorly to the iliac crest, the fascia lata of the thigh and the pubic tubercle. It is continued in front of the symphysis pubis, between and below the two pubic tubercles, and is attached inferiorly to the ischiopubic rami and the posterior border of the perineal membrane. (See p. 197 for relations of this fascia in the perineum.) Fractures of the pelvis may be accompanied by a torn urethra (p. 198). If such a patient micturated, urine would escape and spread over the lower abdomen deep to this membranous layer.

Laterally, the anterior abdominal wall consists of three sheet-like muscles in separate layers, an outer external oblique, a middle internal oblique and an inner transversus abdominis. Anteriorly they become aponeurotic, fuse and form the sheath around rectus abdominis.

RECTUS ABDOMINIS

Each muscle lies ventrally alongside the midline enclosed in a fibrous sheath. It is formed of vertically running fibres and is wider above than below.

Attachments

SUPERIOR—the anterior surface of the 5th, 6th and 7th costal cartilages and the xiphoid process.

INFERIOR—the pubic crest and the front of the pubic symphysis.

There are two or three horizontal tendinous intersections in the muscle which are attached to the anterior layer of the sheath.

The **rectus sheath** is formed by the aponeuroses of the three flat abdominal muscles which join in the midline to form a strong fibrous raphé, the **linea alba**. Above the costal margin the posterior wall of the sheath is absent and the anterior wall is formed by the external oblique aponeurosis. Between the costal margin and a point midway between the umbilicus and the pubic symphysis, the internal oblique aponeurosis splits to enclose the muscle and is reinforced anteriorly by the external oblique and posteriorly by the transversus abdominis. The lower free edge of the posterior wall forms the **arcuate line**. Inferior to it the rectus muscle lies on the transversalis fascia and peritoneum, and the anterior wall is formed by all three aponeuroses. The sheath contains (i) rectus abdominis, (ii) superior and inferior epigastric vessels (which anastomose in the sheath), (iii) the 7th–11th intercostal and subcostal nerves, (iv) lymph vessels.

EXTERNAL OBLIQUE

Attachments

SUPERIOR—from the outer surfaces at the lower eight ribs, interdigitating with serratus anterior and latissimus dorsi.

INFERIOR—the fleshy posterior fibres pass downwards to the anterior half of the iliac crest. The muscle has a free posterior border. Anteriorly, the fibres become aponeurotic, pass medially and downwards, in front of the rectus abdominis to be attached to the xiphoid process, the linea alba and the front of the body of the pubis. The pubic crest forms the base of a triangular deficiency in the aponeurosis, the **superficial inguinal ring**. The apex of the deficiency points laterally and its thickened borders form the lateral and medial crura. Between the pubic tubercle and the anterior superior iliac spine the thickened lower border of the aponeurosis is known as the **inguinal ligament**. This lower border turns backwards and inwards on itself and gives attachment laterally to internal oblique and transversus abdominis. Its medial 2 cm is expanded and attached to the pectineal line on the pubis. It forms the pectineal part (lacunar ligament) which, with the inguinal ligament proper, forms the gutter-like floor of the inguinal canal. Below the inguinal ligament, the aponeurosis is continuous with the fascia lata of the thigh.

INTERNAL OBLIQUE

Attachments

LATERAL—in a continuous line from the thoracolumbar fascia, the anterior two-thirds of the iliac crest and the lateral two-thirds of the inguinal ligament.

MEDIAL—the vertical posterior fibres ascend to the costal margin. Most of the fibres pass upwards and medially and form an aponeurosis which splits around the upper part of rectus abdominis and reaches the linea alba. In the lower abdomen the fibres pass in front of rectus abdominis but leave a free edge posteriorly known as the arcuate line. The fibres from the inguinal ligament arch medially over the spermatic cord, unite with the transversus abdominis aponeurosis and form the **conjoint tendon** (inguinal falx) which passes anterior to the lower part of the rectus abdominis and reaches the crest and medial part of the pectineal line of the pubis.

TRANSVERSUS ABDOMINIS

Attachments

LATERAL—in a continuous line from the inner surfaces of the lower six costal cartilages (interdigitating with the diaphragm), the thoracolumbar fascia, the anterior two-thirds of the iliac crest and the lateral one-third of the inguinal ligament.

MEDIAL—the horizontal upper fibres become aponeurotic lateral to the rectus abdominis muscle and reinforce the posterior wall of its sheath from the xiphoid to the arcuate line. Below this the aponeurosis blends with the internal oblique aponeurosis and forms the conjoint tendon.

Deep to transversus abdominis lie the transversalis fascia and peritoneum.

Actions

The muscles form an elastic wall that both supports the abdominal viscera and yet allows changes in their size and position during diaphragmatic movements. They assist in expiration and in expulsive efforts such as defaecation, micturition, parturition (childbirth) and vomiting. The muscles of both sides act together in flexing the trunk; their unilateral action produces lateral flexion of the trunk and the oblique muscles can produce trunk rotation.

Nerve supply

The lower five intercostal nerves, the subcostal nerve and the 1st lumbar nerve.

Blood supply

The anterior abdominal wall receives its blood supply from the superior and inferior epigastric, the intercostal and subcostal arteries, and branches of the

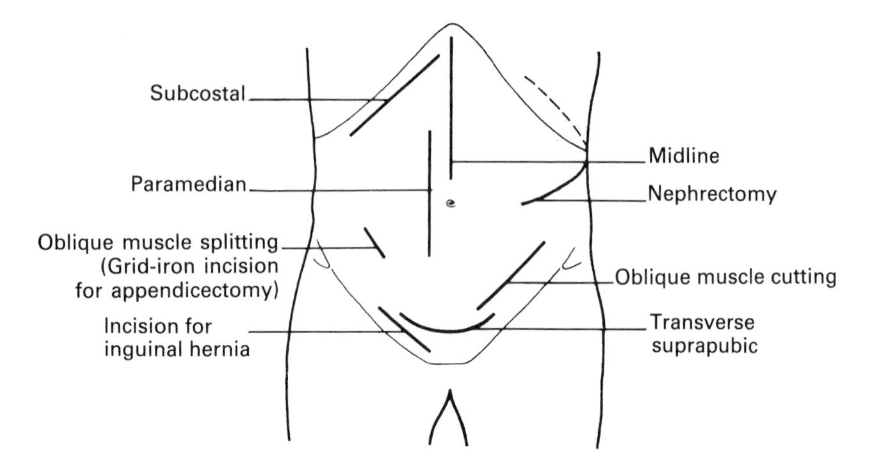

Fig. 9.1 Diagram showing the position of some of the more common surgical incisions through the anterior abdominal wall.

femoral artery. The veins mainly correspond to the arteries.

Surgical access to the abdomen is gained through a variety of incisions. Surgeons operating on the gall bladder, stomach, duodenum and spleen frequently use either midline (through the linea alba) or paramedian incisions (see Fig. 9.1). The latter are made vertically about 2 cm from the midline through the anterior rectus sheath and, after the rectus abdominis muscle has been retracted laterally, through the posterior rectus sheath and peritoneum. Oblique muscle splitting incisions in the right iliac regions through external and internal obliques and transversus abdominis muscles provide access to the appendix. Longer oblique muscle cutting incisions allow access to the right colon or, on the left side, to the sigmoid colon and rectum. Lower paramedian incisions provide similar access to the latter structures. Pelvic organs can be approached through a low transverse incision through the anterior rectus sheath and peritoneum. The rectus muscles may be divided transversely but it is usually sufficient to retract them laterally.

A hernia is a protrusion of an organ or tissue through its containing wall (see Fig. 9.2).

Midline hernias—these are not uncommon, particularly just above the umbilicus. They occur through small defects in the linea alba. A hernia may also be associated with a scar from a surgical incision.

THE INGUINAL CANAL (Fig. 9.2)

This is an oblique path through the anterior abdominal wall. It extends from the **deep inguinal ring**, a deficiency in the transversalis fascia just above the midpoint of the inguinal ligament, to the **superficial inguinal ring**, a deficiency in the external oblique aponeurosis, lying just above the pubic tubercle. The canal is about 4 cm long and possesses anterior and posterior walls and a floor.

Anterior wall

Formed throughout by the external oblique aponeurosis and reinforced laterally by the internal oblique.

Posterior wall

Formed by the transversalis fascia throughout and reinforced by the conjoint tendon medially. The inferior epigastric artery ascends in this wall, medial to the deep ring.

Floor

The grooved upper surface of the inguinal ligament and its medial pectineal (lacunar) part.

The spermatic cord in the male and the round ligament of the uterus in the female traverse the canal.

THE SPERMATIC CORD

The spermatic cord is formed when the testis descends through the inguinal canal into the scrotum, carrying with it its duct, vessels and nerves. A prolongation of peritoneum, the **processus vaginalis**, precedes the testis down the fetal inguinal canal and is later largely obliterated. The spermatic cord gains three fascial coverings from the layers through which it passes:

(i) internal spermatic fascia—from the transversalis fascia,
(ii) cremasteric fascia and muscle—from the internal oblique,
(iii) external spermatic fascia—from the external oblique.

It contains three nerves, three arteries, lymph vessels and three other structures, namely:

(i) the genital branch of the genitofemoral nerve, the ilio-inguinal nerve and autonomic nerves,
(ii) the testicular artery (from the aorta), the cremasteric artery (from the inferior epigastric artery), and the artery to the ductus deferens (from the inferior vesical artery),
(iii) lymph vessels draining to the para-aortic lymph nodes round the renal vessels,
(iv) the pampiniform plexus of veins, the ductus (vas) deferens, and the remains of the processus vaginalis.

Fig. 9.2 The inguinal region in the male. **A.** Structures lying deep to the inguinal ligament. **B.** Position of the incision parallel to the inguinal ligament, to open up the inguinal canal. **C.** The spermatic cord lying in the canal. The external oblique aponeurosis has been divided and the heavy dotted line indicates an incision in the internal oblique muscle to expose deeper structures. **D.** The deep inguinal ring with the spermatic cord lies medial to the edge of transversus abdominis muscle but lateral to the inferior epigastric vessels. The cord has been displaced outwards. **E.** Diagram showing an indirect inguinal hernia within the coverings of the spermatic cord and medial to it, a direct inguinal hernia. The spermatic cord has been displaced outwards. **F.** Diagram of a section along the length of the inguinal canal. The arrow coming through the deep ring and the superficial ring indicates an indirect inguinal hernia. The other arrow indicates a direct inguinal hernia. Both hernias emerge through the superficial inguinal ring.

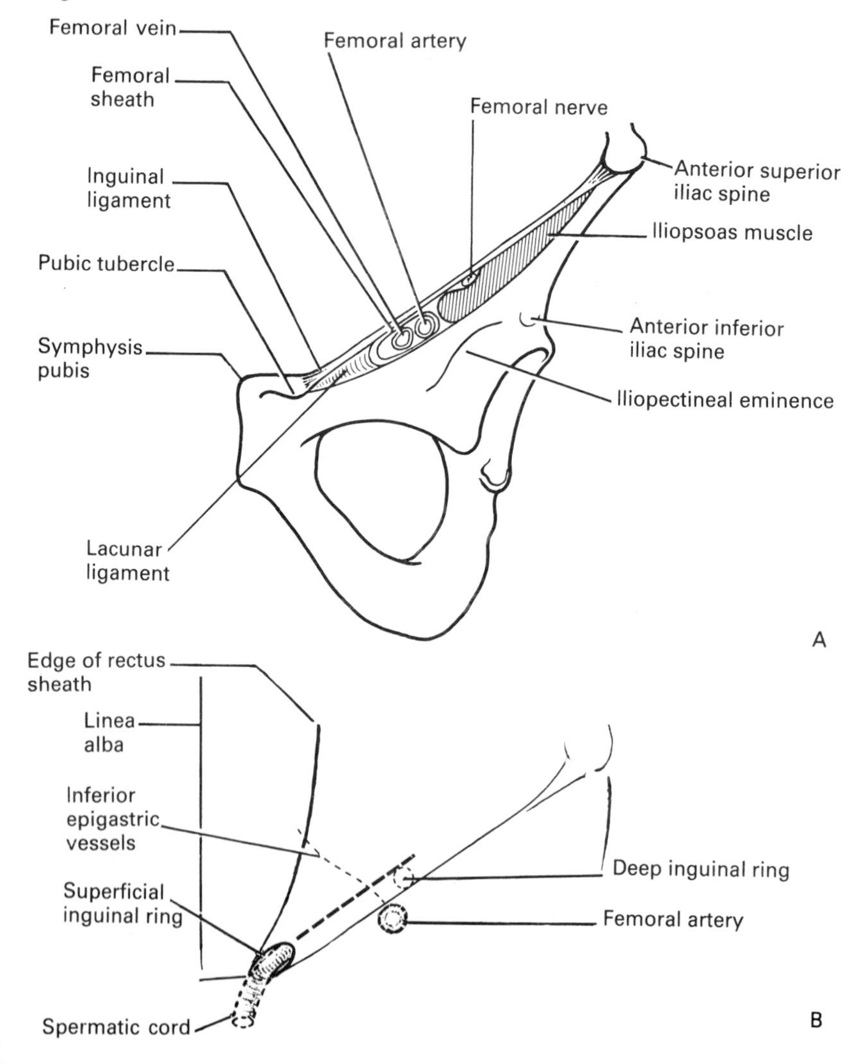

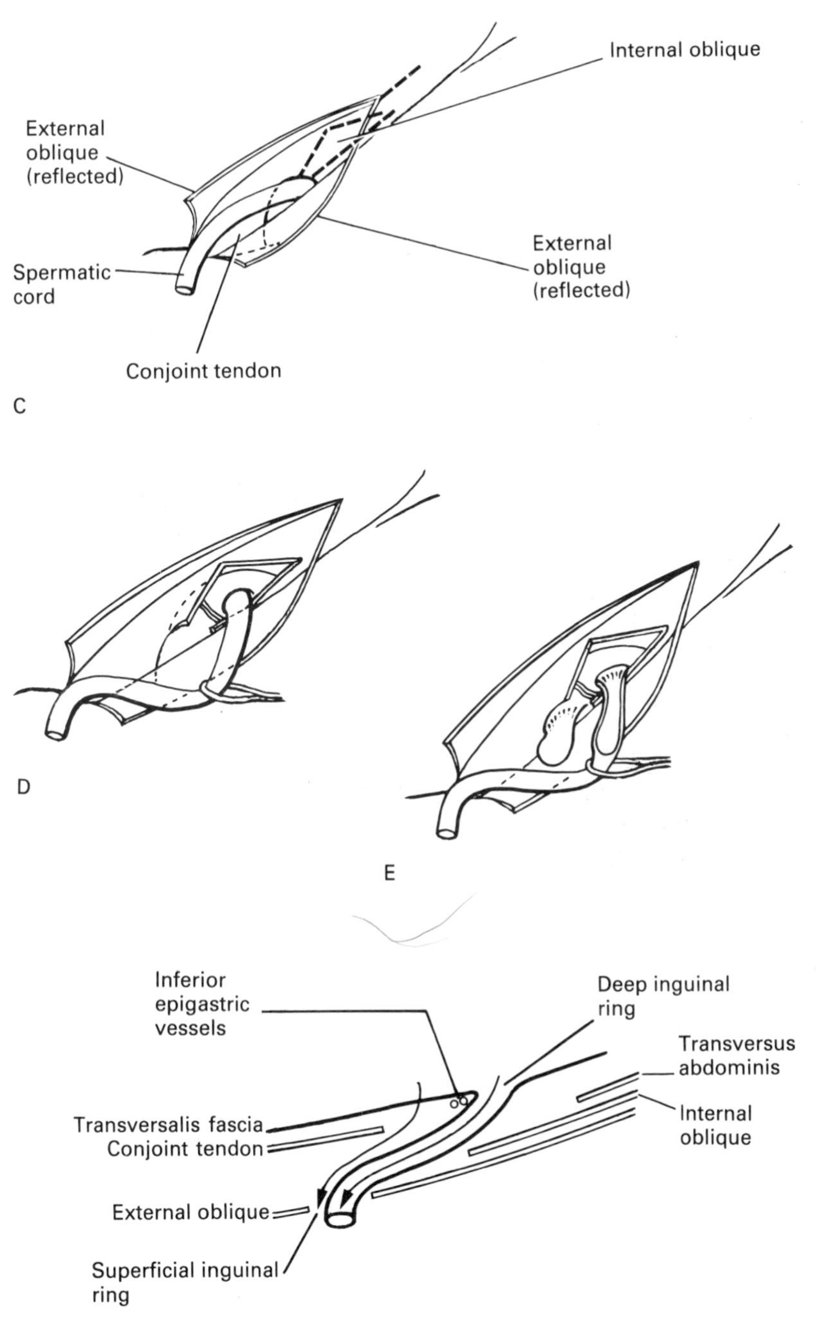
Internal oblique
External
oblique
(reflected)
Spermatic
cord
External
oblique
(reflected)
Conjoint tendon
C
D
E
Inferior
epigastric
vessels
Deep inguinal
ring
Transversus
abdominis
Transversalis fascia
Conjoint tendon
Internal
oblique
External oblique
Superficial inguinal
ring
F

Inguinal hernia—an indirect inguinal hernia passes through the deep ring and the whole length of the canal, usually in a patent processus vaginalis. This defect may not become apparent until young adult or middle life. A direct inguinal hernia enters only the lower part of the canal by pushing through a weakened posterior wall.

THE TESTIS AND EPIDIDYMIS

The two testes are oval glandular organs. Each is suspended in the scrotum by the spermatic cord and is about 4 cm long and 2.5 cm in diameter. It has upper and lower poles. The epididymis, with the ductus deferens on the medial side, is applied to its posterolateral aspect. The other aspects of the gland and the sides of the epididymis are covered by the visceral layer of a closed serous sac, the **tunica vaginalis**, the lower end of the processus vaginalis. The tunica vaginalis, testis, epididymis and ductus deferens are surrounded by extensions of the spermatic cord coverings, namely: (i) internal spermatic fascia; (ii) cremaster muscle and cremasteric fascia; (iii) external spermatic fascia; (iv) superficial fascia which contains some smooth muscle (the dartos); and (v) scrotal skin. Each testis is separated from its fellow by the median scrotal septum. Vessels and nerves enter the gland at its lower pole posteriorly.

The epididymis

This is a tightly coiled tube about 6 cm long applied to the posterolateral surface of the testis. It consists of an enlarged upper extremity (the head), a body and a tapering lower extremity (the tail). The head is attached to the upper pole of the testis by the efferent ducts of the testis. The lumen of the tube is continuous with these efferent ducts above and with the ductus deferens below (see p. 194).

Blood supply

This is through the testicular artery, a branch of the aorta, and to a lesser extent, the artery to the ductus, a branch from a vesical artery. Venous blood drains via the pampiniform plexus to the right and left testicular veins. The right drains into the inferior vena cava and the left into the left renal vein.

Nerve supply

Sympathetic fibres, which originate in the 10th thoracic segment pass to the gland along the testicular artery. Afferent fibres enter the spinal cord in spinal nerves, especially the genitofemoral and ilio-inguinal nerves.

Lymph drainage

Lymph vessels pass with the arteries to para-aortic nodes in the region of the renal arteries.

Histology

The testis is covered by thin mesothelium (the tunica vaginalis), continuous with the peritoneum in the embryo. Under this is a dense fibrous coat, the tunica albuginea which is thickened posteriorly to form the mediastinum testis. From the mediastinum incomplete septa radiate into the gland and divide it into 200–400 compartments, each of which contains two to four convoluted seminiferous tubules. These tubules open into the plexiform rete testis in the mediastinum and about 20 efferent ducts pass from the rete to the head of the epididymis where they become convoluted prior to forming the duct of the epididymis. The seminiferous tubules consist of a thin basement membrane and, after puberty, contain developing spermatozoa. Scattered groups of interstitial cells are present between the tubules. The efferent ducts and the duct of the epididymis have muscular walls and are lined by ciliated columnar and pseudostratified epithelium respectively. The ductus deferens has muscular walls and the lining is much folded and mostly nonciliated.

Embryology

The testis develops from the coelomic mesothelium of the posterior abdominal wall between the mesonephros on each side of the dorsal mesentery. The tubules of the mesonephros become the efferent ducts and the head of the epididymis. The mesonephric duct becomes the ductus deferens (see Figs. 2.8, 2.9). The testis is attached by its lower pole to a mesodermal mass, the gubernaculum. During development, the testis, preceded by the gubernaculum, descends and passes through the inguinal canal. At birth it lies in the scrotum. A tubular process of peritoneum, the processus vaginalis, passes into the scrotum with the descending testis. Its neck is usually obliterated at the time of birth but the distal part remains as the tunica vaginalis of the testis.

Undescended testis: usually the testis has descended through the inguinal canal into the scrotum by the time of birth, but it may be retarded or impeded along the line of descent and so remain undescended. It may pass through the external inguinal ring but fail to enter the scrotum and remain in an ectopic position. It often lies in a subcutaneous fascial pouch just lateral to the external ring.

Hydrocele of the testis: is an accumulation of liquid in the tunica vaginalis. Hydrocele of the cord also occurs.

Torsion (twisting) of the cord: results in impairment of the blood supply to the testis and rarely occurs outside the 10 to 20-year-old age group.

PERITONEAL CAVITY

PERITONEUM

Peritoneum is a thin serous membrane (mesothelium) enclosing the peritoneal cavity. The part lining the abdominal walls is known as parietal and that covering the viscera as visceral peritoneum. The visceral and parietal layers are separated by a small amount of serous fluid. Viscera invaginate the sac to a varying extent. Some are almost completely invested in peritoneum and carry with them double layers known as mesenteries, ligaments, folds or omenta. Others are incompletely invested and have bare areas in contact with the abdominal walls. This invagination of the viscera decreases the size and increases the complexity of the peritoneal cavity.

Arrangement (Figs. 9.3, 9.4)
Near the midline and between the umbilicus and the oesophagus, a double layer of peritoneum passes back from the anterior abdominal wall and the diaphragm, and separates to enclose the liver. Anteriorly these two layers are in apposition and form the **falciform ligament**. Superiorly the right layer is pulled from the diaphragm to form the coronary and right triangular ligaments and it encloses much of the bare area of the liver. The left layer forms the left triangular ligament. On the visceral surface of the liver (Fig. 11.1), the two layers reunite along the fissure for the ligamentum venosum and the porta hepatis to form the **lesser omentum**. The two layers of the lesser omentum pass back to the oesophagus, the stomach and the beginning of the duodenum, and separate to enclose these three structures. The layers come together again along the greater curvature of the stomach. The peritoneum is attached to the posterior abdominal wall from the descending (2nd) part of the duodenum, along the anterior border of the pancreas, the anterior surface of the left kidney and the under surface of the diaphragm as far as the oesophageal opening. Between these attachments and the greater curvature of the stomach, the **greater omentum** hangs down in a lax double fold in front of the transverse colon and mesocolon, and small intestine.

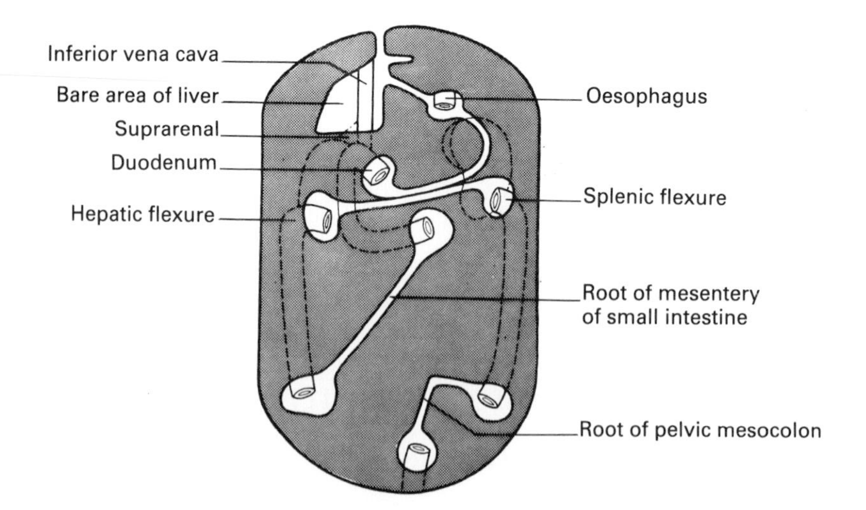

Fig. 9.3 Diagrammatic lay-out of the lines of attachment to the posterior abdominal wall of the peritoneal mesenteries of the alimentary tract. The positions of the two kidneys are indicated behind the hepatic and splenic flexures.

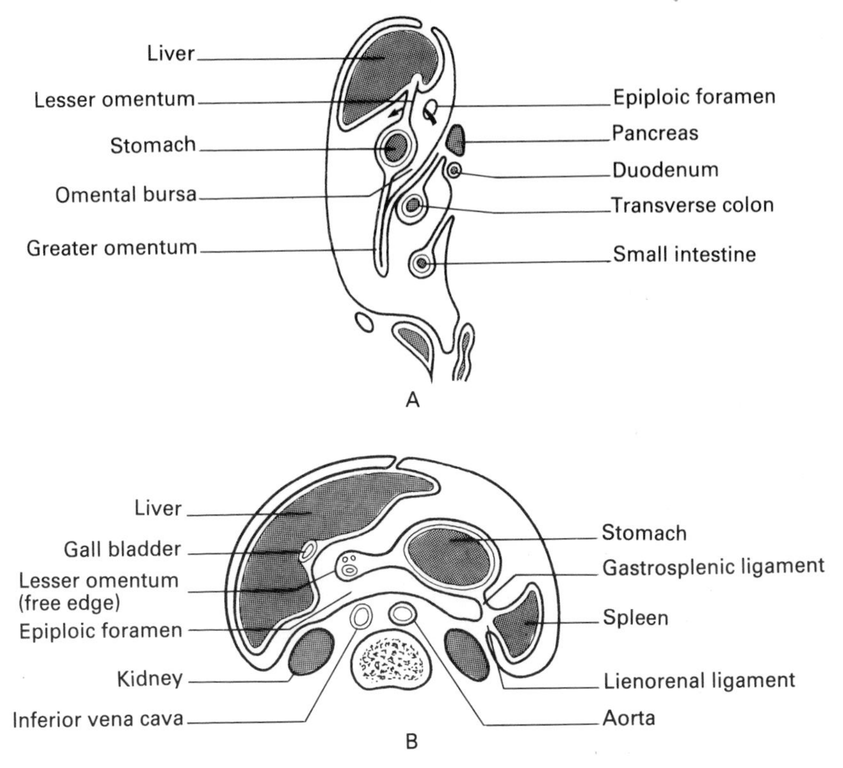

Fig. 9.4 Diagrammatic lay-out of the peritoneum; A, in a sagittal section to the left of the epiploic foramen. The symphysis pubis, urinary bladder and rectum are shown, from front (on left) to back; B, in a horizontal section through the epiploic foramen.

Between the stomach and the anterior surface of the left kidney, the peritoneum encloses the spleen and so is divided into gastrosplenic (gastrolienal) and lienorenal ligaments. The four peritoneal layers of the greater omental fold fuse with each other and become adherent to the upper surface of the transverse mesocolon, so that the transverse colon appears to be enclosed within the layers of the greater omentum (Fig. 9.4A).

The transverse mesocolon is attached to the posterior abdominal wall horizontally across the descending (2nd) part of the duodenum, the anterior border of the pancreas and the anterior surface of the left kidney. The parietal peritoneum covering the posterior abdominal wall below the transverse mesocolon also covers the horizontal (3rd) part of the duodenum and the anterior surfaces of the ascending and descending colons, as these structures lie on the posterior abdominal wall. Peritoneum forms a mesentery for the sigmoid colon. This is attached along a ∧-shaped line whose apex lies over the left sacro-iliac joint. This same parietal peritoneum forms a mesentery for the jejunum and ileum which is attached to the posterior abdominal wall along a line joining the duodenojejunal flexure and the ileocolic junction.

The **omental bursa** (lesser sac) of peritoneum is a diverticulum of the peritoneal cavity lying behind the stomach and in front of the posterior abdominal wall. Its opening, the epiploic foramen, is directed to the right, and is bounded by the liver superiorly, the inferior vena cava posteriorly, the duodenum inferiorly and the free edge of the lesser omentum anteriorly (Fig. 9.4B). The omental bursa extends downwards into the greater omentum, to the left as far as the hilus of the spleen and superiorly into a recess between the liver (caudate lobe) and the diaphragm. Its anterior and posterior relations are shown in Figure 9.4.

The abdominal peritoneum extends down into the pelvis to cover the pelvic walls and the upper surfaces of the viscera. It covers the superior surface of the bladder. In the male it covers the upper end of the seminal vesicles and dips down into the rectovesical fossa, whence it ascends to the front and sides of the rectum. In the female, peritoneum also covers the anterior and posterior surfaces of the uterus and uterine tubes. The peritoneal fold over each tube extends to the pelvic walls and is known as the **broad ligament** of the uterus. It separates a shallow uterovesical pouch in front (which is present between the uterus and bladder) from a deeper recto-uterine posteriorly (pouch of Douglas). This latter pouch has the rectum behind and the uterus and upper part of the vagina in front.

Embryology (see also pp. 30 and 39)
The development of the abdominal alimentary tract can be described in five stages (Fig. 9.5).

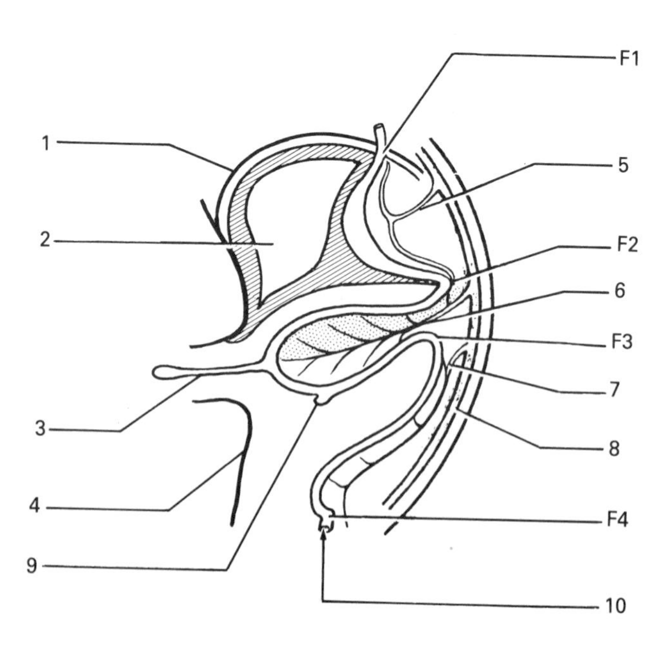

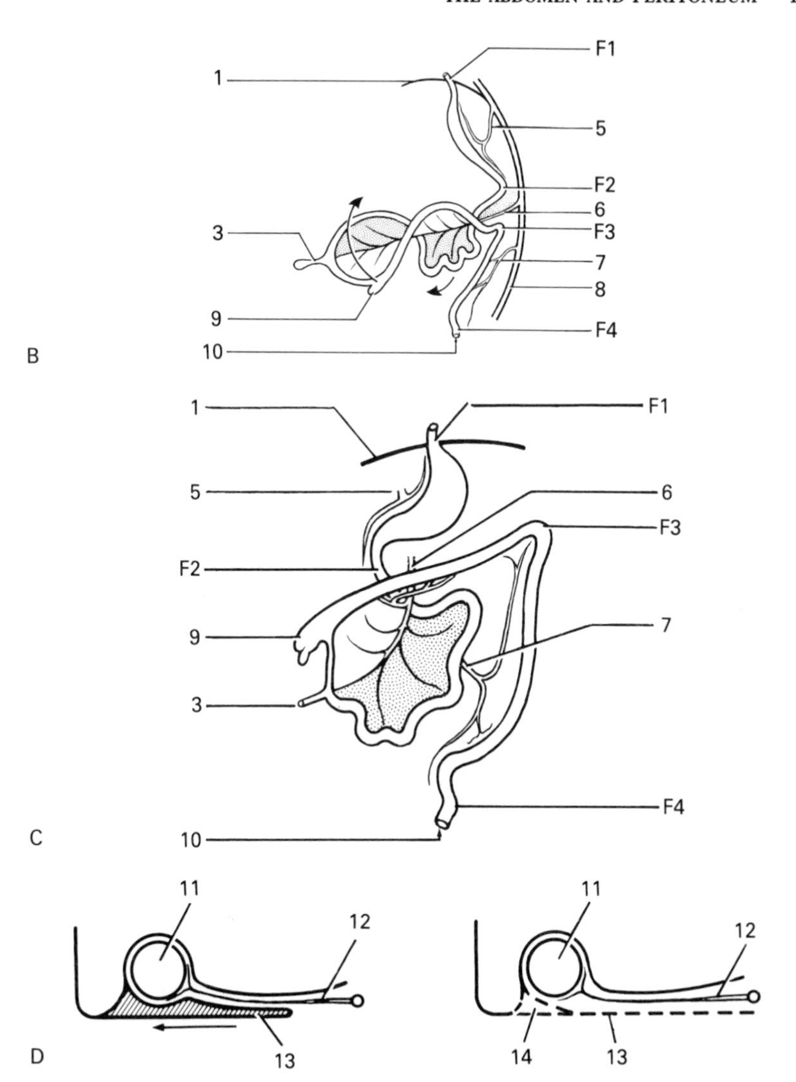

Fig. 9.5 Diagrams illustrating stages in the development of the gut. The ventral mesentery is cross-hatched in **A**, and the pre-arterial dorsal mesentery of the midgut is stippled. **A**. Stages of elongation and herniation, lateral view. **B**. Stage of rotation, lateral view. **C**. Stage of retraction, front view. **D**. Stage of fixation. Relatively fixed points—F1. Oesophagus passing through diaphragm. F2. Duodenum on posterior abdominal wall. F3. Splenic flexure on posterior abdominal wall. F4. Recto-anal junction passing through pelvic floor.
1. Diaphragm 2. Liver 3. Vitello-intestinal duct in the body stalk 4. Anterior abdominal wall 5. Coeliac trunk to foregut 6. Superior mesenteric artery to midgut 7. Inferior mesenteric artery to hindgut 8. Aorta 9. Caecum 10. Anus 11. Ascending or descending colon 12. Colic artery 13. Fold of peritoneum being 'absorbed' 14. Retrocolic pouch which may remain

Stage 1, elongation: most parts increase in length especially between four relatively fixed regions, (a) the oesophagus where it passes through the developing diaphragm, (b) the proximal part of the duodenum lying on the posterior abdominal wall, (c) the region where the midgut becomes the hindgut (future splenic flexure), and (d) the rectum where it passes through the pelvic diaphragm.

Stages 2, herniation: the midgut becomes too large for the abdominal cavity, and a loop passes into the body stalk. The proximal limb of the loop is supplied by pre-arterial branches of the superior mesenteric artery.

Stage 3, rotation: in the foregut, this is through about 90° in an anticlockwise direction, so that the right side of the upper coelomic cavity comes to lie behind the stomach, and the left side extends in front. The ventral mesentery lies obliquely across the upper abdomen and is much distorted by the enlarging liver (Fig. 9.4B). The midgut loop rotates anticlockwise through about 270° round the axis of the superior mesenteric artery so that the pre-arterial limb of the loop comes to lie below and to the left. The postarterial limb comes to lie above and to the right.

Stage 4, retraction: on its return, the midgut loop fills much of the abdomen; the pre-arterial loop (future small intestine) lies below and to the left of the postarterial loop (future ascending and transverse colons). The caecum and appendix are at first high in the abdomen, under the developing liver, but later elongation of the ascending colon and the fixation of the hepatic flexure of the colon brings the appendix and caecum into the region of the right iliac fossa.

Stage 5, fixation: in some regions the gut loses its dorsal mesentery, is covered in front only by peritoneum and has a relatively large bare area against the posterior abdominal wall.

These changes in the size and disposition of the abdominal alimentary tract mainly take place during the second and third months of fetal life. The stomach, jejunum, ileum, transverse and sigmoid colons retain their mesenteries; the duodenum, ascending and descending colons lose their mesenteries and lie on the posterior abdominal wall covered only anteriorly by peritoneum.

The primitive stomach rotates, its left surface becoming anterior. The spleen develops in the left side of its dorsal mesentery and in so doing divides this mesentery into gastrosplenic and lienorenal ligaments. Below the greater curvature of the stomach the dorsal mesentery increases in size and forms a double fold—the greater omentum—in front of the transverse colon and mesocolon.

The liver develops in the ventral mesentery which provides its peritoneal covering and grows towards the right causing the lesser omentum and stomach to trap behind them a pouch of the peritoneum called the omental bursa (lesser sac).

Minor anomalies are common and are due to defects in the timing of the stages and the direction and degree of rotation. Major anomalies, e.g. situs inversus, also occur and may involve reversal of all processes (clockwise, and no rotation). Occasionally incomplete retraction into the abdomen occurs and a part of the cavity or of the gut may be in the umbilical cord at birth. This is usually associated with maldevelopment of the anterior abdominal wall.

Infection of the peritoneal cavity may lead to abscess formation in one of its recesses. In the lower abdomen infection may localise in the rectovesical or recto-uterine pouch and so cause a pelvic abscess. In the upper abdomen abscesses may develop in the right and left subphrenic spaces, (on each side of the falciform ligament), the right subhepatic space (between the liver and the right kidney), and less commonly in the omental bursa of the peritoneum. (See p. 146.)

— 10 —

The abdominal alimentary tract

The alimentary tract, from lips to anus, has differing structure, shape and size according to the main local function. The basic tube has a wall consisting of four layers:

(a) a mucous membrane lining which has a surface epithelium, a lamina propria (mainly loose areolar tissue with glands and lymph tissues) and a thin sheet of smooth muscle, the muscularis mucosae. There is a change in the type of epithelium at the gastro-oesophageal junction from stratified squamous in the oesophagus to columnar in the stomach. A change in the reverse direction occurs at the recto-anal junction from columnar in the rectum to stratified squamous in the anal canal.

(b) a submucous layer with vascular areolar tissue and some glands.

(c) smooth muscular layers arranged as inner circular and outer longitudinal. At each end of the tube, including the oesophagus, striated muscle is present.

(d) the outermost covering is usually a serous layer (peritoneum in the abdomen) but where no peritoneum is present it is formed of areolar tissue (see also p. 143).

THE ABDOMINAL OESOPHAGUS

This is about 3 cm long. It enters the abdomen on the left of the midline through fibres of the right crus of the diaphragm posterior to the central tendon and becomes continuous with the stomach at the **cardiac orifice**. It lies between the diaphragm and the left lobe of the liver and is covered anterolaterally by peritoneum. Gastric vessels and vagal nerves lie in its walls. (See also pp. 88, 213 and 420). Reflux of gastric contents is normally prevented by the lower oesophageal sphincter and the oblique angle of entry of the oesophagus into the stomach. If reflux does occur, inflammation of the lower oesophagus and subsequent stricture formation may result. Not infrequently, the oesophageal opening in the diaphragm becomes incompetent and a part of the stomach is allowed to slide into the mediastinum. This condition, known as a hiatus hernia, also predisposes to the reflux of gastric contents.

THE STOMACH (Figs. 10.1, 10.2)

This is a dilatation of the alimentary canal between the oesophagus and the duodenum. Although variable in size, shape and position, it is usually J-shaped and situated in the left hypochondrium and epigastrium, its lower part extending to the level of the umbilicus. But it may be shorter and more horizontal (steer-horn stomach) or longer, drooping into the pelvis (fishhook stomach). Differ-

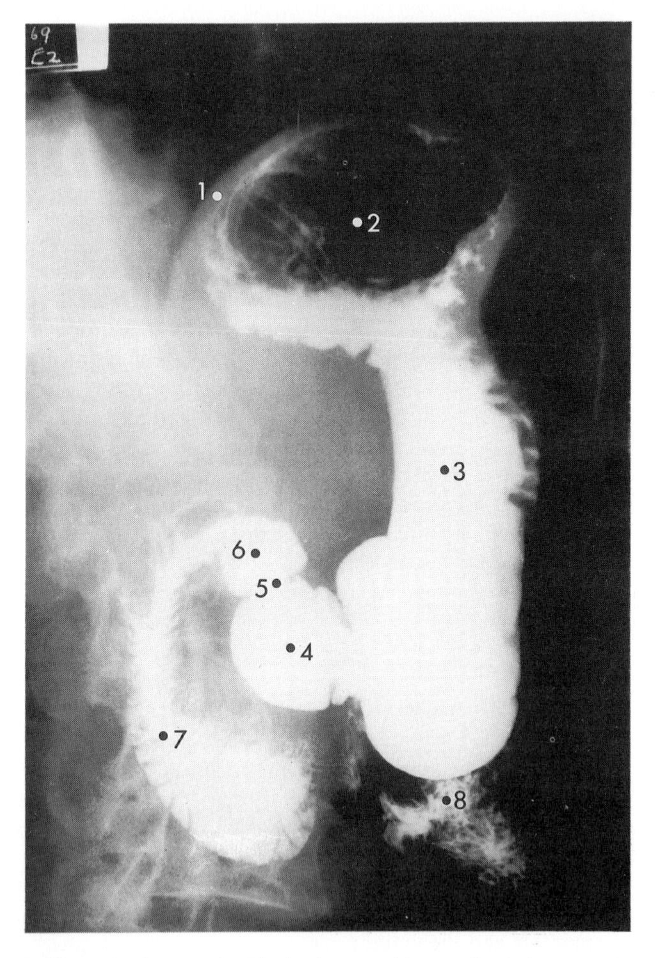

Fig. 10.1 Radiograph of stomach. The barium meal has outlined the stomach and some has entered the duodenum and jejunum.

1. Left dome of diaphragm
2. Gas in fundus
3. Body
4. Pyloric antrum
5. Pylorus
6. Duodenal cap
7. Descending part of duodenum
8. Jejunum

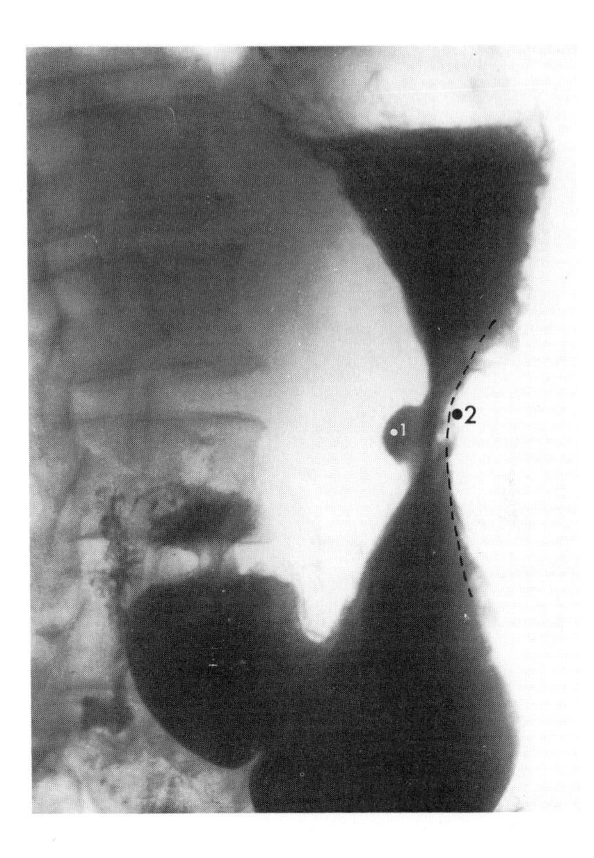

Fig. 10.2 Barium meal examination. Erect oblique view of stomach showing a large pouch (1) of barium projecting from the lesser curvature of the stomach. This represents a benign gastric ulcer. There is some narrowing of the body of the stomach (2).

ences in the size, shape, motility and rate of emptying of the stomach can be demonstrated by radiographs after a barium meal. The pattern of the mucosal folds can also be seen and irregularities detected. The air in the fundus can often be seen on a plain radiograph. Changes in position and posture of the subject and states of nervous tension will alter markedly the appearance of the stomach. The activity of the pyloric sphincter is also very obvious.

The stomach is divided into a fundus, a body and a pyloric portion. It possesses anterior and posterior surfaces, both covered by peritoneum, and lesser (right) and greater (left) curvatures. The **fundus** is that part above the level of the oesophageal opening (**cardiac orifice**). The **body** extends from the fundus to the **angular notch** (the most dependent part of the lesser curvature) and the **pyloric portion** from the notch to the pyloric sphincter (pylorus) which separates the stomach from the duodenum. The pyloric part is subdivided into a proximal

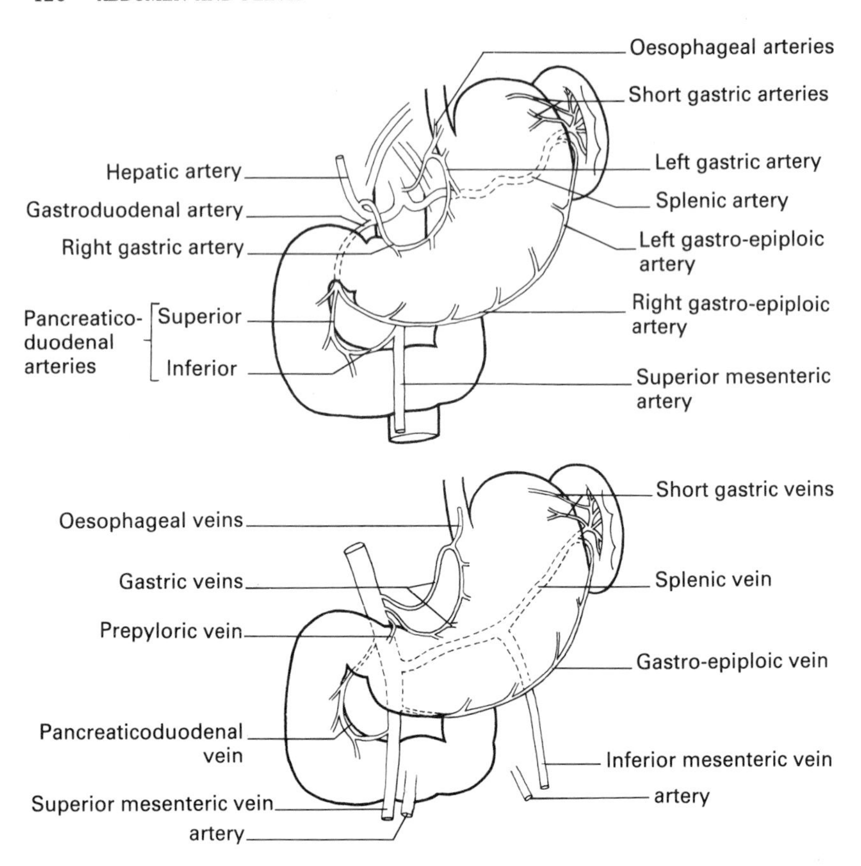

Fig. 10.3 The arterial supply from the coeliac trunk (above) and the venous drainage to the portal vein (below) of the stomach and the duodenum.

dilated **pyloric antrum** and a distal tubular **pyloric canal**. The **lesser curvature** is the upper right border and extends from the right of the oesophagus to the pylorus; to it is attached the lesser omentum. The angular notch is near its lower end. The **greater curvature** extends from the left of the oesophagus over the fundus to the pylorus, forming the left border of the stomach. It gives attachment to the gastrosplenic ligament and the greater omentum.

Relations

ANTERIOR—the left lobe of the liver, diaphragm and anterior abdominal wall.

POSTERIOR (stomach bed)—the omental bursa separates it from the diaphragm above; the pancreas, splenic artery, the left kidney and suprarenal gland, and the spleen from right to left across the middle; and the transverse mesocolon and colon below.

SUPERIOR—the lesser omentum and the gastric vessels.
INFERIOR—the greater omentum and the gastro-epiploic vessels.

Blood supply
The left and right gastric arteries supply the lesser curvature and adjacent surfaces; the left and right gastro-epiploic arteries supply the greater curvature, and the short gastric branches from the splenic artery supply the fundus. All these arteries are derived from the coeliac trunk (p. 136). All the blood from the stomach passes to the portal vein. The right and left gastric veins enter it directly, the right gastro-epiploic vein drains to the superior mesenteric vein and the left gastro-epiploic vein and the short gastric veins drain to the splenic vein (Fig. 10.3).

Lymph drainage
All lymph passes to the coeliac group of pre-aortic nodes. It flows from the lesser curvature via nodes along the left gastric artery which communicate with nodes at the porta hepatis; from the fundus, lymph drains to nodes along the splenic artery; from the rest of the greater curvature lymph passes to nodes around the gastro-epiploic, splenic and hepatic arteries.

Histology
The stomach wall has four coats:

(*a*) *mucous*—this is lined with columnar epithelium and limited externally by the muscularis mucosae. Three types of glands are found:

(i) cardiac glands—long ducts with mucus secreting cells found around the cardiac orifice.
(ii) gastric glands—short ducts and long straight alveoli lined with peptic (enzyme secreting) and oxyntic (acid secreting) cells; some mucous cells are found in the fundus and body. Endocrine cells are also found in the body.
(iii) pyloric glands—long ducts and short coiled alveoli with mucus (mainly) and gastrin secreting cells in the pyloric part.

(*b*) *submucous*.
(*c*) *muscular*—three layers of fibres: innermost oblique, middle circular (which is thickest at the pyloric sphincter) and outer longitudinal.
(*d*) *serous*—peritoneal coat.

THE PYLORUS

This sphincter is a thickening of the circular muscle coat and is found usually at the level of the upper border of the 1st lumbar vertebra. The sphincter regulates the flow of material between the stomach and the duodenum. Its position is marked on the anterior abdominal wall about 3 cm to the right of the midline on the transpyloric plane.

The diagnosis of all gastric and duodenal disease is aided by a barium meal (Fig. 10.2). The oesophagus, stomach and duodenum may also be directly visualised by a gastroscope.

Gastric ulcers are usually situated along the lesser curve, particularly in the region of the angular notch.

THE SMALL INTESTINE

This narrow tube extends from the stomach to the colon. Its length (about 5 m) depends on the state of tone of its muscle wall. It consists of the duodenum, jejunum and ileum.

THE DUODENUM (Fig. 10.4)

This lies on the posterior abdominal wall in the epigastric and umbilical regions. It is about 25 cm long, extends from the pylorus to the duodenojejunal flexure and embraces the head of the pancreas in a C-shaped curve which is open to the left. Near the pylorus and the duodenojejunal flexure, the surface is completely covered by peritoneum; the remainder of the tube is retroperitoneal and relatively immobile. For descriptive purposes it is divided into four parts which can

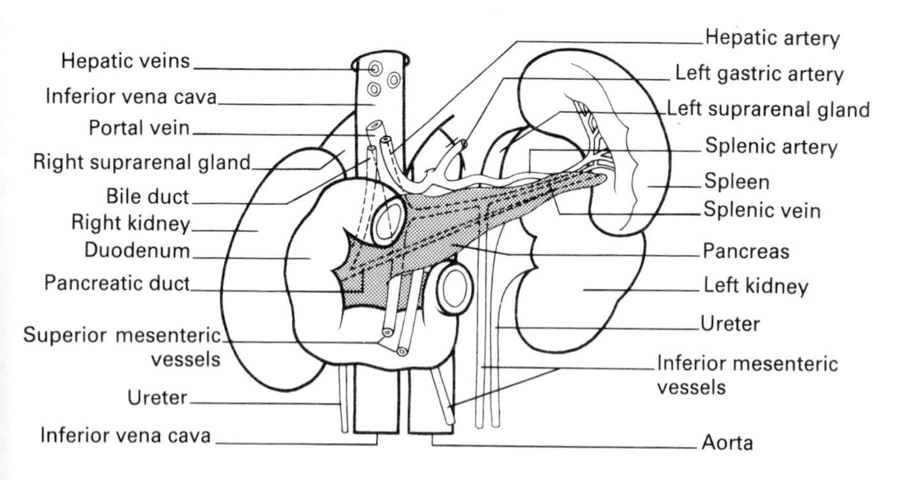

Fig. 10.4 Diagram indicating some of the principal structures in the upper part of the posterior abdominal wall. (The pancreas is stippled.) The inferior mesenteric vein is seen to be a tributary of the splenic vein which joins with the superior mesenteric vein and forms the portal vein.

be visualised with X-rays. The part seen adjacent to the pylorus is called the duodenal cap.

The 1st part (5 cm long) passes from the pylorus upwards, backwards and to the right, and then turns downwards to join the 2nd part (8 cm long) which descends to the right side of the 3rd lumbar vertebra. The 3rd part (10 cm long) passes to the left across the posterior abdominal wall and turns upwards to become the 4th part (3 cm long) which ascends to the duodenojejunal flexure on the left side of the 2nd lumbar vertebra.

Relations

Superior (1st) part, the first 2 cm, has similar peritoneal relations to the stomach. The lesser omentum passes from the upper border and the greater omentum from the lower. Posteriorly the omental bursa separates it from the pancreas, and anteriorly it is separated from the fundus of the gall bladder and the liver by the greater sac. The rest is retroperitoneal. Superiorly is the opening into the omental bursa and the free border of the lesser omentum; posteriorly is the bile duct, portal vein and gastroduodenal artery. Inferiorly it rests on the pancreas, and anteriorly the greater sac separates it from the liver and the neck of the gall bladder.

Descending (2nd) part is crossed anteriorly by the attachment of the transverse mesocolon, and posteriorly it is in contact with the right suprarenal gland, the hilus of the right kidney and the right psoas. Above the transverse colon and mesocolon it is in contact with the liver, and below with coils of small intestine. On its left lies the head of the pancreas, the bile duct and the pancreaticoduodenal vessels. The duodenal papilla, a common opening for the pancreatic and bile ducts, opens about halfway down its posteromedial wall.

Horizontal (3rd) part lies below the head and uncinate process of the pancreas and crosses, from right to left, the right ureter and right psoas, the gonadal vessels, the inferior vena cava, the inferior mesenteric artery and the aorta. It is crossed anteriorly by the superior mesenteric vessels and the root of the mesentery of the small intestine, and covered anteriorly and inferiorly by coils of small intestine.

Ascending (4th) part lies to the left of the vertebral column on the left psoas. The inferior mesenteric vein, the left ureter and the lower pole of the left kidney lie on its left side. The pancreas is medial to it and it is related anteriorly to the root of the mesentery and the coils of small intestine.

Histology

The duodenum has a similar basic structure to that of the stomach but the mucous membrane shows numerous villi and some of the glands (duodenal mucous glands) extend through the muscularis mucosae into the submucous layer.

Duodenal ulcers are a common affliction and the vast majority occur in the superior part of the duodenum. When posteriorly situated, penetration by the

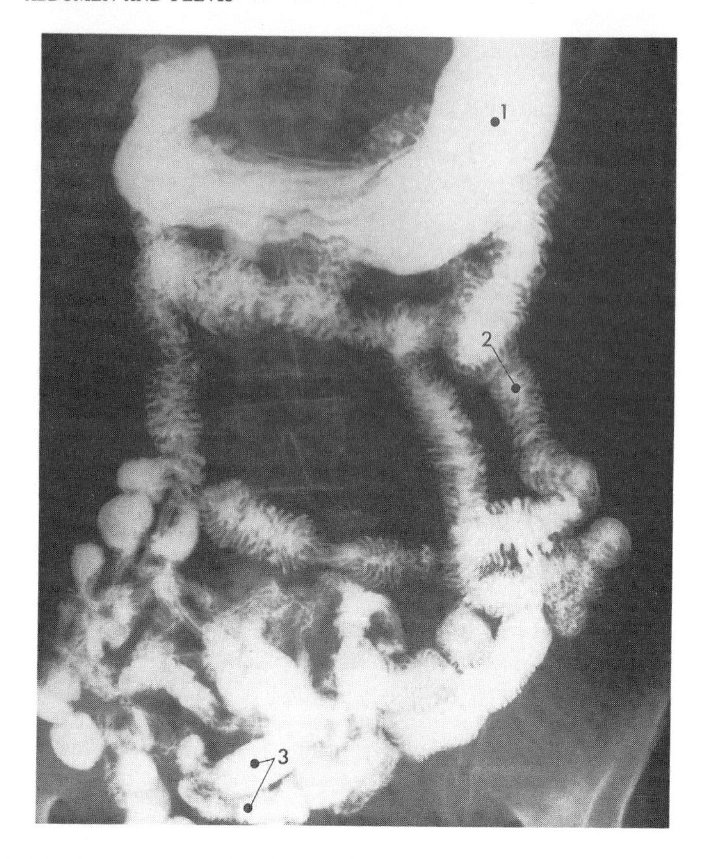

Fig. 10.5 Barium meal and follow-through. Barium has passed through the pylorus and now fills the duodenum and small intestine.
1. Stomach 2. Jejunum 3. Loops of ileum

ulcer may cause bleeding from the gastroduodenal artery or erosion into the head of the pancreas. Anterior ulcers are more frequently complicated by perforation into the general peritoneal cavity. Treatment of ulcers aims at reducing the amount of acid secreted by the stomach by removal of part of the stomach (partial gastrectomy), by denervation of the stomach (selective vagotomy), or by pharmacological means.

THE DUODENOJEJUNAL FLEXURE

This lies on the left of the 2nd lumbar vertebra and faces downwards and to the left. Small peritoneal recesses are sometimes found behind the left side of the 4th part of the duodenum and behind the adjacent part of the inferior mesenteric vein.

THE JEJUNUM AND ILEUM (Fig. 10.5)

These are the proximal and distal parts of the coiled small intestine, extending from the duodenum to the colon. They are suspended from the posterior abdominal wall by a mesentery and are situated below the transverse colon mainly in the central abdomen but parts may extend into the pelvis. They are covered with peritoneum except at their mesenteric attachment. The jejunum is wider and thicker-walled than the ileum; its mucous membrane is thrown into circular folds with many villi but it has fewer aggregations of lymphoid tissue in its wall and less fat in its mesentery than the ileum. The arrangement of blood vessels varies between the jejunum and ileum.

Relations
The shorter jejunum lies above and to the left of the ileum. The root of the mesentery passes downwards from the left side of the 2nd lumbar vertebra to the right sacro-iliac joint crossing in turn the left psoas, aorta, inferior vena cava, right gonadal vessels, right psoas and ureter (Figs. 9.3, 10.4).

THE LARGE INTESTINE (Fig. 10.6)

This extends from the ileocaecal junction to the anus and is about 1.5 m long. It consists of the caecum, the appendix, the ascending, transverse, descending and sigmoid colon, the rectum and the anal canal. The longitudinal muscle of the caecum and colon lies outside a continuous circular coat and is restricted to three bands, **taeniae coli**, which are shorter than the rest of the wall and cause it to be sacculated. Fatty tags, **appendices epiploicae**, project outwards from the wall of the large intestine and are covered by peritoneum. The mucous membrane has no villi but mucous cells are numerous.

THE CAECUM

This blind sac (about 8 cm wide and 8 cm long) is continuous with the ascending colon. It is situated in the right iliac fossa, completely invested in peritoneum. The taeniae coli converge on the appendix which is attached to the posteromedial wall. The **ileocaecal orifice**, an oval slit, opens on its medial wall and is surrounded by a thickening of the ileal circular muscle coat, the ileocaecal sphincter.

Relations
It lies on iliacus and psoas and the lateral femoral cutaneous nerve; anteriorly are the small intestine and the anterior abdominal wall. The femoral nerve and external iliac vessels lie on its left side.

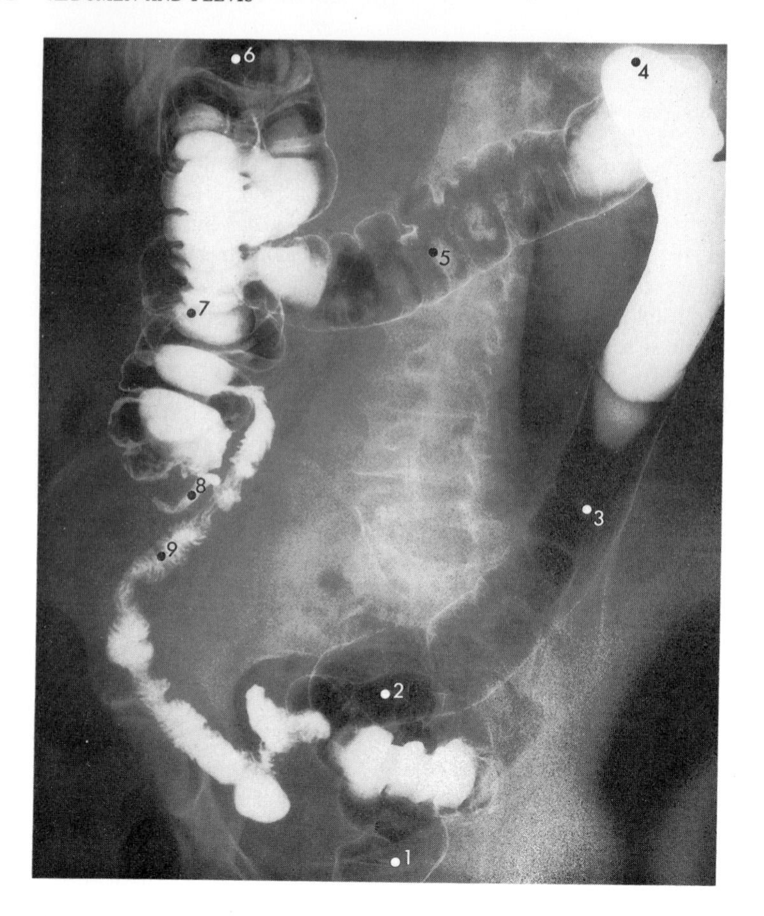

Fig. 10.6 Radiograph after partial evacuation following a barium enema. The contrast material outlines first the rectum and colon and passes into the terminal ileum. Note the relatively high position of the hepatic flexure.

1. Rectum
2. Sigmoid colon
3. Descending colon
4. Splenic flexure
5. Transverse colon
6. Hepatic flexure
7. Ascending colon
8. Appendix
9. Reflux into terminal ileum.

THE VERMIFORM APPENDIX

This narrow diverticulum (about 8 cm long) arises from the posteromedial aspect of the caecum about 3 cm below the ileocaecal orifice. It is clothed in peritoneum which connects it by a mesentery, the mesoappendix, to the terminal ileum. The appendicular artery lies within this fold. Aggregations of lymphoid tissue are found in the wall and may replace the muscle coat in places. The appendix lies most commonly behind the caecum or within the pelvis. It is very mobile and

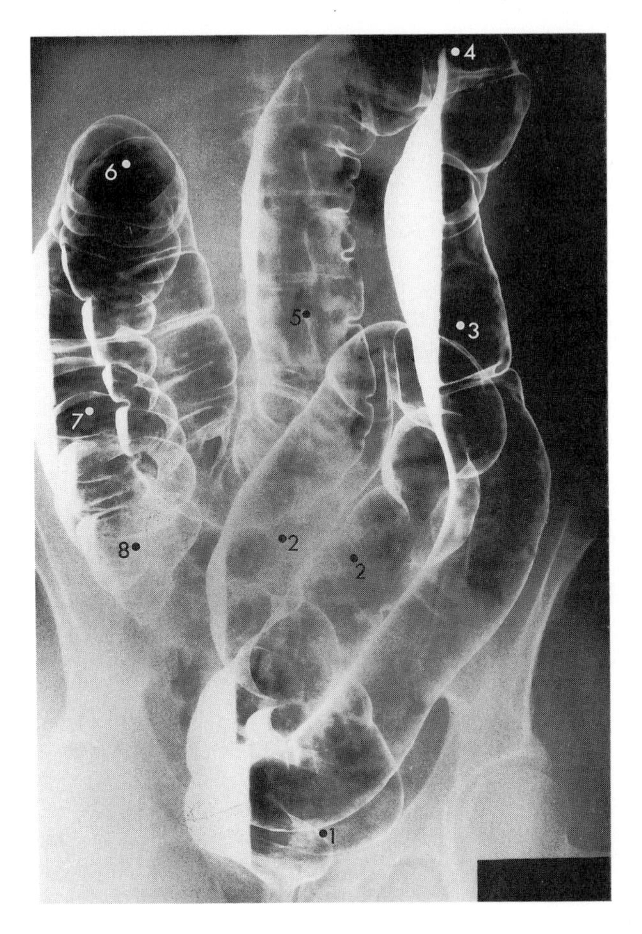

Fig. 10.7 Radiograph taken with patient lying on the right side after barium and air have been introduced into the colon. Note the position of the heavier barium and the horizontal fluid lines.

1. Rectum
2. Sigmoid colon
3. Descending colon
4. Splenic flexure
5. Transverse colon
6. Hepatic flexure
7. Ascending colon
8. Caecum

its relations are variable. Inflammation of the appendix, usually following obstruction of its lumen by faeces (large gut contents), is a common condition and it may lead to peritonitis.

THE ASCENDING COLON (Fig. 10.6)

This lies in the right lateral region and extends from the ileocaecal orifice to the under surface of the right lobe of the liver where it turns to the left, forming the **right colic (hepatic) flexure**. It is about 15 cm long; peritoneum covers it anteriorly and on both sides and fixes it to the posterior abdominal wall, thus forming a **paracolic sulcus** along its right side. At the upper end, this sulcus leads into a subphrenic space.

Relations

Posteriorly it lies on iliacus, quadratus lumborum and the lower pole of the right kidney. Its peritoneal surfaces are in contact with coils of small intestine.

THE TRANSVERSE COLON (Fig. 10.6)

This extends upwards from the right to the **left colic (splenic) flexure** across the abdomen suspended by the transverse mesocolon. It is about 50 cm long.

Relations

Initially it lies directly on the descending part of the duodenum and the head of the pancreas, but in its subsequent course it is attached by its mesentery to the body of the pancreas and is related anterosuperiorly to the liver, gall bladder, stomach, greater omentum and spleen. Posteriorly it lies on the descending part of the duodenum, head of the pancreas, small intestine and the left kidney.

THE DESCENDING COLON (Fig. 10.6)

This is the narrowest part of the colon. It lies in the left lateral region and extends from the left colic flexure to the pelvic brim where it continues as the sigmoid colon. It is about 30 cm long. Peritoneum covers it anteriorly and on both sides, fixing it to the posterior abdominal wall and forming a paracolic sulcus on its left side. The left colic flexure lies higher than the right and is attached to the diaphragm by a fold of peritoneum, the phrenicocolic ligament.

Relations

Posteriorly it lies on the lower pole of the left kidney and the diaphragm, quadratus lumborum, iliacus and psoas. Its peritoneal surfaces are in contact with coils of small intestine.

THE SIGMOID COLON

This lies in the left iliac region and extends from the pelvic brim to the front of the 3rd sacral segment where it becomes continuous with the rectum. Its length (usually about 40 cm) and position are variable. It is attached to a ∧-shaped mesentery, the sigmoid mesocolon (Fig. 9.3), to the pelvic wall. The apex

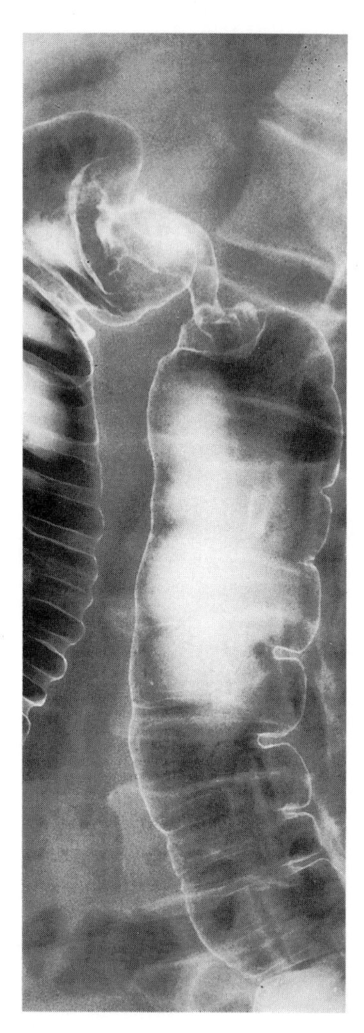

Fig. 10.8 Barium enema with air insufflation. The segment of gut shown is the splenic flexure of the colon viewed at an oblique angle. Note the segment of narrowing caused by a cancer of the colon surrounding the wall and cutting down the size of the lumen, giving a typical 'apple core' appearance.

of the ∧ overlies the left ureter anterior to the bifurcation of the left common iliac artery and the left sacroiliac joint.

Relations

Posteriorly it lies on the left ureter and common iliac vessels; superiorly it is covered by coils of small intestine and inferiorly it lies on the bladder in the male or the uterus and bladder in the female.

The rectum and anal canal are described on page 180.

Cancer of the large intestine arises from mucosal epithelial cells and invades the deeper layers of the wall, eventually reaching the peritoneum. Spread occurs to lymph vessels and nodes, and via the portal venous system to the liver.

ARTERIAL SUPPLY OF THE GASTRO-INTESTINAL TRACT

This is from the coeliac, the superior mesenteric and the inferior mesenteric arteries, each arising from the front of the aorta and supplying respectively the abdominal foregut, the midgut and the hindgut and their derivatives.

THE COELIAC TRUNK (Figs. 10.3, 10.9)

This short trunk arises just above the pancreas. It passes anteriorly for 2 cm and divides into left gastric, common hepatic and splenic arteries. It is surrounded by the coeliac plexus of nerves and the coeliac group of lymph nodes.

LEFT GASTRIC ARTERY

This passes upwards and to the left behind the omental bursa to the oesophagus where it turns into the lesser omentum. It anastomoses with the right gastric artery on the lesser curvature of the stomach and supplies the lower third of the oesophagus and some of the stomach.

COMMON HEPATIC ARTERY

This passes to the right towards the superior part of the duodenum and then turns into the lesser omentum in which it ascends in front of the portal vein. It divides into right and left terminal branches at the porta hepatis.

Branches

(i) **right gastric artery**—runs along the lesser curvature of the stomach to anastomose with the left gastric artery in the lesser omentum.

(ii) **gastroduodenal artery**—descends behind the duodenum and divides into the **right gastro-epiploic** and **superior pancreaticoduodenal arteries**. The former passes along the greater curvature to anastomose with the left gastro-epiploic; the latter descends in the groove between the duodenum and the head of the pancreas, and supplies both these structures.

(iii) **right** and **left hepatic arteries**—ramify in the liver; the right gives a small **cystic artery** to the gall bladder. Variations in the pattern of these vessels are important in surgery of the gall bladder.

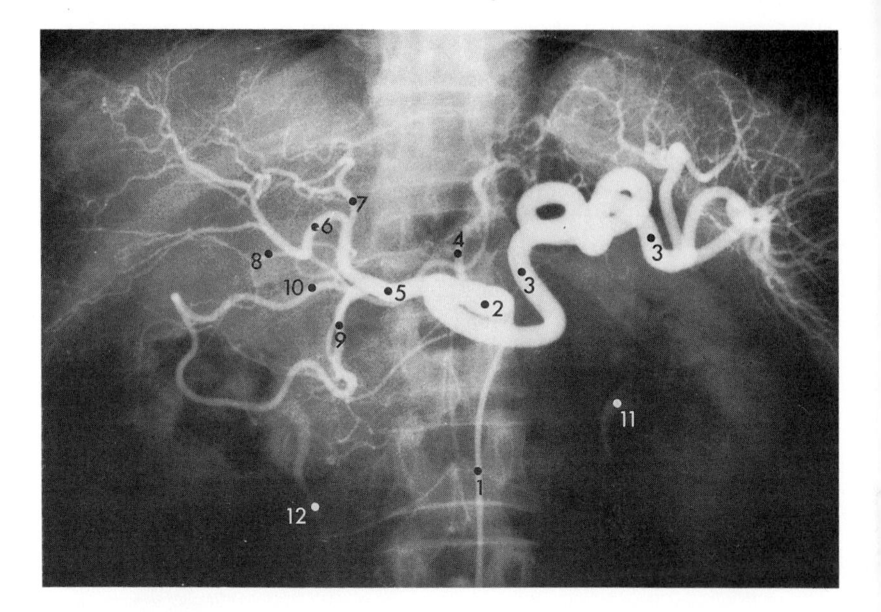

Fig. 10.9 Selective coeliac axis arteriogram. The catheter (1) has been introduced percutaneously into the right femoral artery and passed through the iliac artery and aorta into the origin of the coeliac trunk. The renal pelves and proximal ureter are faintly seen because contrast material from a previous injection has been excreted by the kidneys.

1. Catheter
2. Coeliac trunk
3. Splenic artery
4. Left gastric artery
5. Common hepatic artery
6. Right hepatic artery
7. Left hepatic artery
8. Cystic artery
9. Gastroduodenal artery
10. Right gastro-epiploic artery
11. Left renal pelvis
12. Right ureter

SPLENIC ARTERY

This runs a tortuous course to the left along the upper border of the pancreas and passes in the lienorenal ligament to the hilus of the spleen where it divides into terminal branches. It lies behind the omental bursa and crosses the left crus of the diaphragm, left psoas and the left suprarenal gland and kidney.

BRANCHES

(i) **pancreatic arteries**—several.

(ii) **short gastric arteries**—several, pass in the gastrosplenic ligament to the fundus of the stomach.

(iii) **left gastro-epiploic artery**—runs in the gastrosplenic ligament to the greater curvature of the stomach and anastomoses with the right gastro-epiploic artery.

THE SUPERIOR MESENTERIC ARTERY (Figs. 10.4, 10.10)

This arises just below the coeliac trunk, descends to the right behind the body of the pancreas and over its uncinate process. It crosses the horizontal part of the duodenum and enters the root of the mesentery of the small intestine. In the mesentery it supplies the small intestine and ends in the right iliac fossa by dividing into ileocolic and right colic arteries.

Relations

Its origin is surrounded by the superior mesenteric plexus of nerves and pre-aortic lymph nodes, and lies behind the pancreas and splenic vein. It passes downwards on the left renal vein, the uncinate process, and the horizontal part of the duodenum, covered above by the body of the pancreas and below by the peritoneum. Its vein lies on its right side. Within the mesentery, accompanied by veins, lymph vessels and nerves, it crosses over in turn the inferior vena cava, the right psoas, the right ureter and the right gonadal vessels.

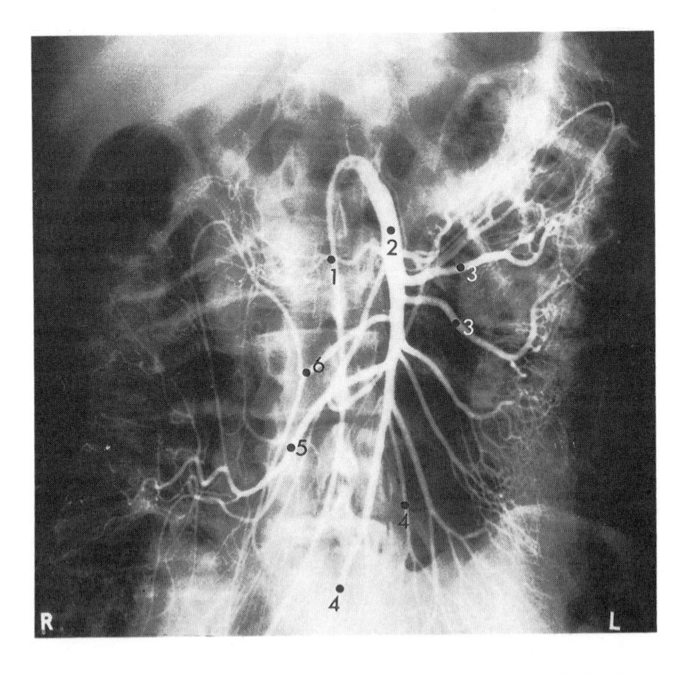

Fig. 10.10 Selective superior mesenteric arteriogram. The catheter (1) has been introduced percutaneously into the right femoral artery and passed through the iliac artery and aorta into the origin of the superior mesenteric artery. The middle colic branch has not been clearly demonstrated in this picture.

1. Catheter
2. Superior mesenteric artery
3. Jejunal branches
4. Ileal branches
5. Ileocolic artery
6. Right colic artery

Branches

(i) **inferior pancreaticoduodenal artery**—runs to the right between the pancreas and the duodenum to anastomose with the superior pancreaticoduodenal artery. It supplies the pancreas and duodenum.

(ii) **jejunal and ileal arteries**—15–20 in number, arise from the left side of the artery within the mesentery. By repeated divisions and side-to-side anastomoses two to five tiers of arterial arcades are formed whose terminal branches enter the small intestine.

(iii) **ileocolic artery**—descends to the right and divides into ascending and descending branches; the former supplies the lower part of the ascending colon and anastomoses with the right colic artery; the latter gives anterior and posterior caecal branches. The appendicular artery is a branch of the posterior caecal artery.

(iv) **right colic artery**—descends to the right behind the peritoneum of the posterior abdominal wall, across the right psoas and ureter, and divides into ascending and descending branches. It supplies the ascending colon and anastomoses with the ileocolic and middle colic arteries.

(v) **middle colic artery**—passes upwards on the body of the pancreas to reach the mesocolon within which its branches supply the right two-thirds of the transverse colon.

THE INFERIOR MESENTERIC ARTERY (Figs. 10.4, 10.11)

This arises behind the duodenum and descends to the left across the posterior abdominal wall. It continues beyond the brim of the pelvis as the superior rectal artery.

Relations

It is covered by the peritoneum of the posterior abdominal wall and crosses the left psoas and the left common iliac vessels. Its vein and the left ureter are on its left side.

Branches

(i) **left colic artery**—ascends to the left across the left psoas and ureter as far as the left colic flexure. Its branches supply the transverse and descending colons and anastomose with the middle colic and sigmoid arteries.

(ii) **sigmoid arteries**—enter the sigmoid mesocolon and supply the sigmoid colon and the lower descending colon.

(iii) **superior rectal artery**—is the continuation of the inferior mesenteric artery in the pelvis. It descends in the lower part of the sigmoid mesocolon and reaches the back of the rectum. It supplies the rectum and proximal two-thirds of the anal canal and anastomoses with the inferior rectal branch of the internal pudendal artery.

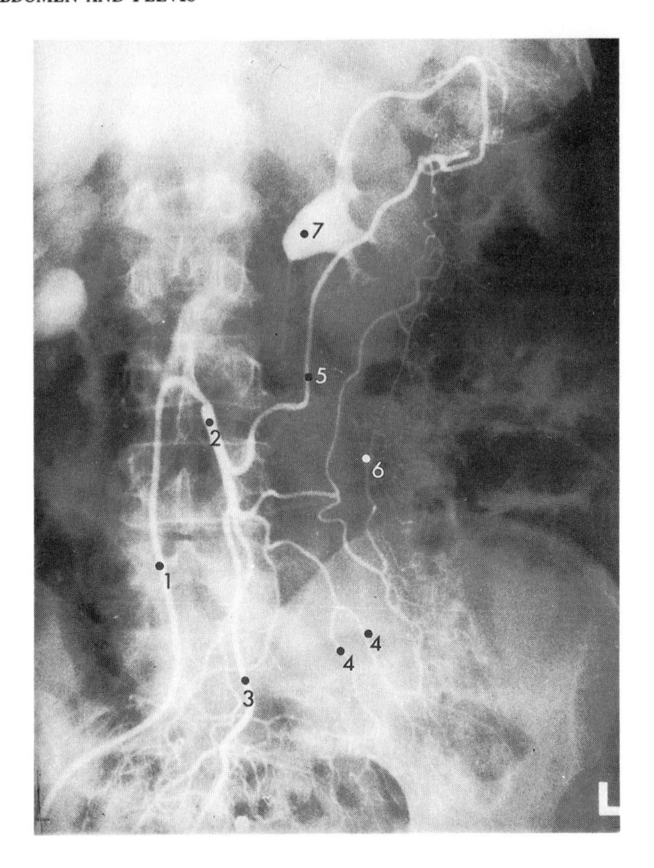

Fig. 10.11 Selective inferior mesenteric arteriogram. The catheter (1) has been introduced percutaneously through the right femoral artery and passed up the iliac artery and aorta into the origin of the inferior mesenteric artery. Contrast material from a previous injection is being excreted by the kidneys.

1. Catheter
2. Inferior mesenteric artery
3. Superior rectal artery
4. Sigmoid arteries
5. Left colic artery
6. Marginal artery
7. Left renal pelvis

The terminal arteries supplying the gut form a continuous anastomotic arcade, the marginal artery, along the mesenteric border of the small intestine and the concavity of the large intestine. The anastomosis is effective and resection of a part of the gut is made easier. The ileal and colic vessels are usually embedded in fat.

VENOUS DRAINAGE OF THE GASTRO-INTESTINAL TRACT (THE HEPATIC PORTAL SYSTEM)

THE PORTAL VEIN (Figs. 2.11D, E, 10.3, 10.4)

The gastro-intestinal tract from the lower end of the oesophagus to the upper part of the anal canal, the spleen, pancreas and gall bladder drain into the liver via the portal vein which is formed by the union of the splenic and superior mesenteric veins behind the neck of the pancreas. It ascends in the free edge of the lesser omentum to the porta hepatis where it divides into right and left branches.

Relations

At its origin it lies in front of the inferior vena cava, behind the neck of the pancreas, and below and then behind the superior part of the duodenum. In the free edge of the lesser omentum it lies behind the common hepatic artery on the left, and the bile duct on the right, and is separated from the inferior vena cava by the opening into the omental bursa.

Tributaries

(i) **right** and **left gastric veins**—draining the lesser curvature of the stomach.

(ii) **superior pancreaticoduodenal vein**.

(iii) **para-umbilical veins**—pass in the falciform ligament to the left branch of the portal vein.

THE SUPERIOR MESENTERIC VEIN (Figs. 10.3, 10.4)

This is formed in the mesentery in the right iliac fossa. It ascends on the right of its artery to the root of the mesentery, crosses anterior to the horizontal part of the duodenum and uncinate process of the pancreas and joins the splenic vein behind the neck of the pancreas. Its tributaries correspond to the branches of its artery and it also receives the right gastro-epiploic vein.

THE SPLENIC VEIN (Figs. 10.3, 10.4)

This is formed in the hilus of the spleen by the union of tributaries from the organ, the left gastro-epiploic and short gastric veins. It passes to the right behind the tail and body of the pancreas to join the superior mesenteric vein behind the neck of the pancreas.

Relations

Within the lienorenal ligament it is accompanied by its artery and the tail of the pancreas, and it crosses the left kidney, the left psoas and the aorta. The splenic artery lies above it.

Tributaries

(i) **short gastric veins**—draining the fundus of the stomach.

(ii) **left gastro-epiploic vein**—draining the left side of the greater curvature. Both these veins lie in the gastrosplenic ligament.

(iii) **inferior mesenteric vein**—the continuation of the superior rectal vein above the pelvic brim. It ascends on the posterior abdominal wall, to the left of its artery and of the duodenojejunal flexure. It passes behind the body of the pancreas and enters the splenic vein.

(iv) numerous **pancreatic veins** entering along its course.

PORTAL-SYSTEMIC ANASTOMOSES

The portal system may anastomose with the systemic venous system in the following situations:

(i) the lower end of the oesophagus—between the left gastric and the azygos veins.
(ii) the lower part of the anal canal—between the superior and inferior rectal veins.
(iii) the umbilical region of the anterior abdominal wall—between the epigastric veins and the para-umbilical veins in the falciform ligament of the liver.
(iv) the bare areas of the gastro-intestinal tract and its related organs, e.g. veins between the bare area of the liver and the diaphragm.

When the portal vein is obstructed by a thrombus, or the venous flow through the liver is impeded by fibrosis (cirrhosis), these portal-systemic communications enlarge. Severe haemorrhage may then occur especially from the dilated lower oesophageal veins (varices).

LYMPH DRAINAGE OF THE INTESTINAL TRACT

The intestinal mucosa is richly supplied with lymph vessels which drain via submucosal and subserous plexuses to nodes on the surface of the viscus. These drain into intermediate groups of nodes arranged along arteries (in the mesentery or on the abdominal wall) which in turn drain to one of three groups of **pre-aortic nodes** arranged about the origins of the coeliac, superior and inferior mesenteric arteries.

The **inferior mesenteric group** drains the distal one-third of the transverse colon, the descending and the sigmoid colon and the upper part of the rectum.

The **superior mesenteric group** drains the distal half of the duodenum, the jejunum, ileum, caecum and appendix, and the ascending and proximal transverse colon.

The **coeliac group** drains the stomach and proximal duodenum, the liver, spleen and pancreas. It also drains the superior and inferior mesenteric groups of nodes. Efferent vessels pass to the cisterna chyli (p. 209).

HISTOLOGY OF THE INTESTINAL TRACT

The walls of the intestinal tract possess three coats: an inner mucous membrane, a muscular layer and an outer serous (peritoneal) layer.

The mucous membrane contains numerous lymph nodules. It is divided by the muscularis mucosae, a layer of smooth muscle, into the mucosa and the submucosa. The mucosa is lined by columnar epithelium with numerous goblet cells. It is invaginated to form simple intestinal glands and in the small intestine forms numerous finger like projections, the villi, which contain blood and lymph vessels. As well as digestive enzymes, the cells of the mucosa produce mucus, and certain hormones such as gastrin, secretin and CCK-PZ from endocrine cells. The submucosa consists of loose areolar tissue containing blood vessels and the submucous (Meissner's) nerve plexus.

The muscular coat is in two layers, an inner circular and an outer longitudinal one between which lies the myenteric (Auerbach's) nerve plexus.

The serous (peritoneal) layer is lined by flattened mesothelium derived from the lining of the coelomic cavity of the embryo.

Regional variations

The duodenum has mucous glands (Brunner's) whose coiled pits extend into the submucosa.

The jejunum and ileum: the villi of the jejunum are longer and more numerous than those of the ileum. The lymph tissue of the ileum is aggregated to form well-marked nodules (Peyer's patches).

The appendix has a mucous membrane containing many large lymphoid follicles.

The large intestine has no villi. Intestinal glands are well developed and contain many goblet cells. The longitudinal muscle layer as far as the rectum forms three discrete bundles of fibres, the taeniae coli, and the remaining intervening wall is sacculated.

Embryology

Some aspects of the development of the gut and its rotation are described on page 118.

— 11 —

The liver, spleen and pancreas

THE LIVER (Figs. 11.1, 11.2, 11.3)

This is the largest organ in the body and weighs about 1.5 kg. It lies under the diaphragm mainly in the right hypochondrium and is wedge-shaped, possessing rounded diaphragmatic, and visceral (inferior) surfaces. A normal liver may occasionally be palpated just below the right costal margin on deep inspiration.

The irregular **visceral surface** is set obliquely and faces downwards, backwards and to the left. It is divided into a right and a smaller left lobe by a continuous **interlobar fissure** which contains the ligamentum venosum above and the ligamentum teres below. The ligamentum venosum is the remnant of the ductus venosus which, in the fetus, joins the umbilical vein to the inferior vena cava and by-passes the liver. The ligamentum teres is the obliterated umbilical vein. In the adult, both ligaments are attached to the left branch of the portal vein. The right lobe contains the small **quadrate lobe** in front and the **caudate lobe** behind. The **porta hepatis** is a transverse sulcus through which pass the portal vein, hepatic artery, bile duct and lymph vessels. The caudate lobe lies

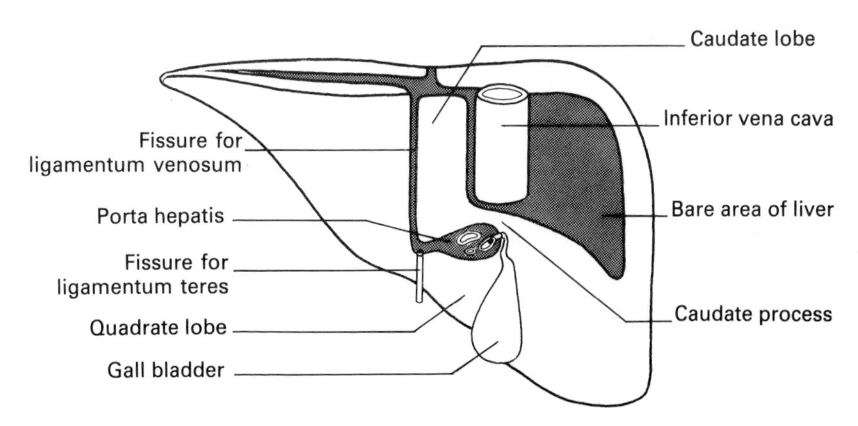

Fig. 11.1 The liver, posterior and visceral aspects. The bare area extends forwards and to the left between the layers of the left triangular ligament. The right triangular ligament is at the right extremity of the bare area.

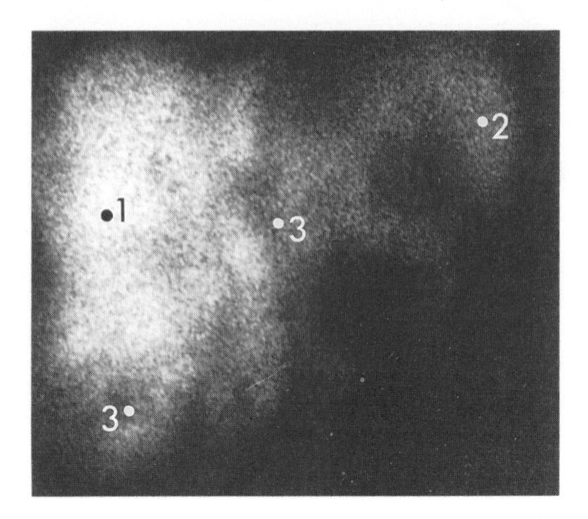

Fig.11.2 Colloid scan of the liver and spleen. Technetium-99m tin colloid has been injected intravenously and has been taken up by the Kupffer cells of the liver and the spleen outlining these organs. The liver is markedly enlarged and the lucent defects (3) are secondary tumour deposits within its substance which have not taken up the colloid. Dynamic scans using other labelled radiopharmaceuticals can be used to measure blood flow through regions and renal filtration or excretion providing indices of physiological function.
1. Liver
2. Spleen
3. Liver metastases

between the interlobar fissure and the inferior vena cava, above the porta hepatis, and communicates with the right lobe proper by the **caudate process**, a narrow strip of tissue between the inferior vena cava and the porta hepatis. The quadrate lobe lies between the interlobar fissure and the gall bladder, and below the porta hepatis.

The liver is largely covered by peritoneum. It is connected to the stomach by the lesser omentum, the two layers of which are attached to the liver around the porta hepatis and along the fissure for the ligamentum venosum. The liver is attached to the anterior abdominal wall and the diaphragm by four peritoneal folds:

(i) **The falciform ligament** is sickle-shaped and extends from the diaphragm and anterior abdominal wall above the umbilicus to the anterior and superior surfaces of the liver where its attachments demarcate the right and left lobes. Its inferior free border contains the ligamentum teres which can be traced across the visceral surface to the porta hepatis. The ligament separates the right and left subphrenic spaces (peritoneal pouches) and, with the lesser omentum, forms the remains of the ventral mesentery of the abdominal foregut.

(ii) **The coronary ligament** has upper and lower layers and connects the posterior surface of the right lobe to the diaphragm. Between these two layers the liver is not covered with peritoneum and is in direct contact with the diaphragm (the bare area).

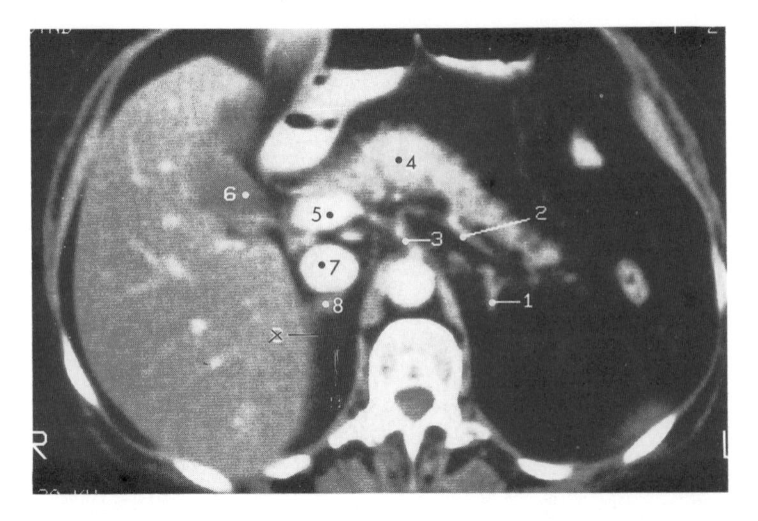

Fig. 11.3 CT transverse abdominal scan through the body of the first lumbar vertebra (see also caption to Figure 6.2).

1. Medial part of the left suprarenal gland
2. Splenic artery
3. Coeliac axis
4. Pancreas
5. Portal vein
6. Gall bladder
7. Inferior vena cava
8. Body of right adrenal gland

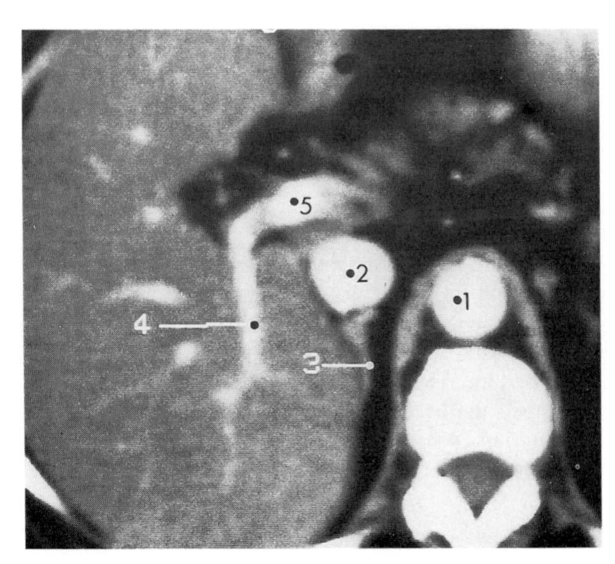

Fig. 11.4 CT transverse abdominal scan at a slightly higher level than Figure 11.3.

1. Aorta
2. Inferior vena cava
3. Medial part of the right suprarenal gland
4. Right branch of the portal vein
5. Portal vein

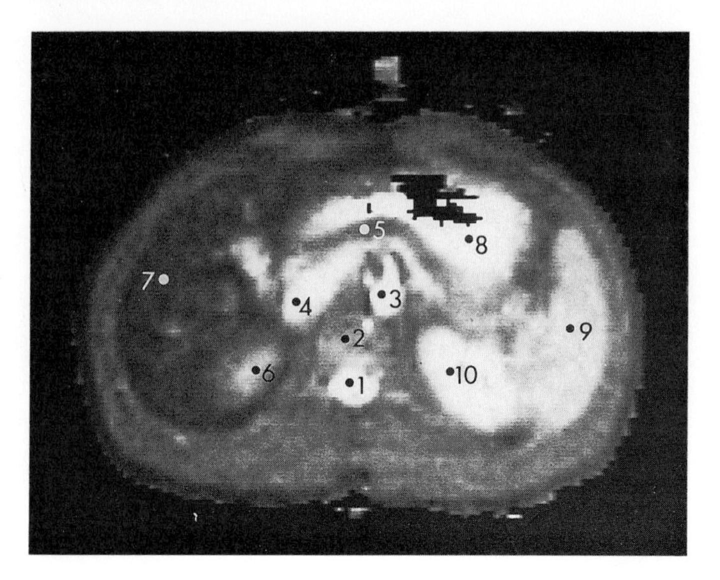

Fig. 11.5 Magnetic resonance image (MRI) of transverse section of the abdomen at the level of the first lumbar vertebra. MRI is a new imaging modality capable of producing cross-sectional images like its predecessor, X-ray computed tomography (CT). Unlike CT it does not use ionising radiation but rather utilises an apparently safe interaction between static magnetic fields, radiowaves and atomic nuclei. Rather than depict tissue electron density like CT it shows hydrogen density in thin slices. Compare with CT scan 11.3.

1. Vertebral canal
2. Body of the first lumbar vertebra
3. Aorta
4. Inferior vena cava
5. Pancreas
6. Upper pole of the right kidney
7. Liver
8. Stomach
9. Spleen
10. Left kidney

(iii) **The right triangular ligament** is the right extremity of the coronary ligament where its upper and lower layers are in continuity.

(iv) **The left triangular ligament** similarly is at the left extremity and connects the posterior surface of the left lobe to the diaphragm. Because of the surface tension between the visceral and parietal layers of peritoneum, and of the vessels passing from the liver to the diaphragm, the liver moves in respiration. This is utilised in certain forms of assisted respiration (the rocking method) where the weight of the liver is allowed to pull the diaphragm downwards and thus assist inspiration. When the bed is rocked in the opposite direction, the weight of the liver pushes the diaphragm up and assists expiration.

Relations

The liver is separated by the diaphragm from the heart and pericardium, lungs, pleura, and chest wall. The bare area of the posterior part of the diaphragmatic surface is also related to the inferior vena cava and the right suprarenal gland.

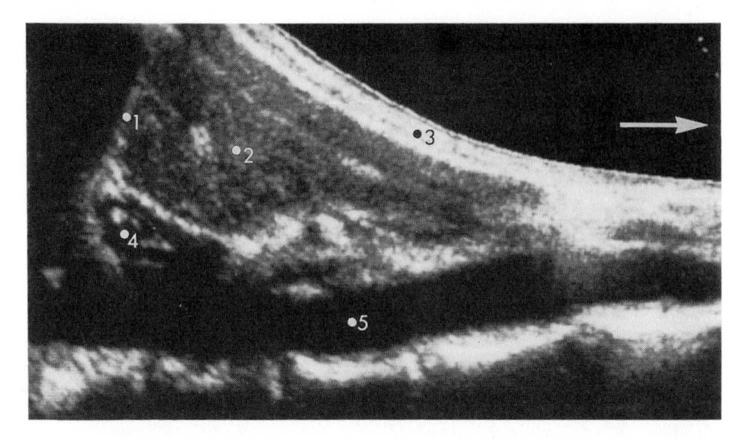

Fig. 11.6 Longitudinal ultrasound scan of the upper abdomen 1 cm to the left of the midline. In an ultrasound scan, high frequency sound waves are passed through a thin slice of the body and an image is made up of their reflections from the constituents of the slice. Dense structures produce a bright (white) image. The arrow is pointing to the feet of the subject.

1. Diaphragm
2. Left lobe of liver
3. Abdominal wall
4. Gastro-oesophageal junction
5. Aorta

Visceral surface: the left lobe is in contact with the abdominal oesophagus, the fundus and body of the stomach, and the lesser omentum, the latter separating it from the pancreas. The gall bladder lies in a shallow fossa along the right margin of the quadrate lobe: it is usually in direct contact with the liver and its inferior surface is covered by peritoneum. The quadrate lobe and gall bladder are related to the superior part of the duodenum superiorly and to the transverse colon inferiorly. The caudate lobe is separated by the upper recess of the omental bursa from the diaphragm, and the caudate process forms the upper boundary of the opening into the omental bursa. The right lobe proper is separated from the right colic flexure inferiorly and the right kidney superiorly by the subhepatic (hepatorenal) peritoneal space (pouch) which communicates with the right medial paracolic sulcus and with the omental bursa (lesser sac).

Blood supply

The hepatic artery and portal vein divide at the porta hepatis into right and left branches from which interlobular vessels ramify through the liver tissue. Blood drains by several hepatic veins to the inferior vena cava.

Nerve supply

From both vagi via the anterior gastric nerve, and sympathetic fibres via the coeliac plexus.

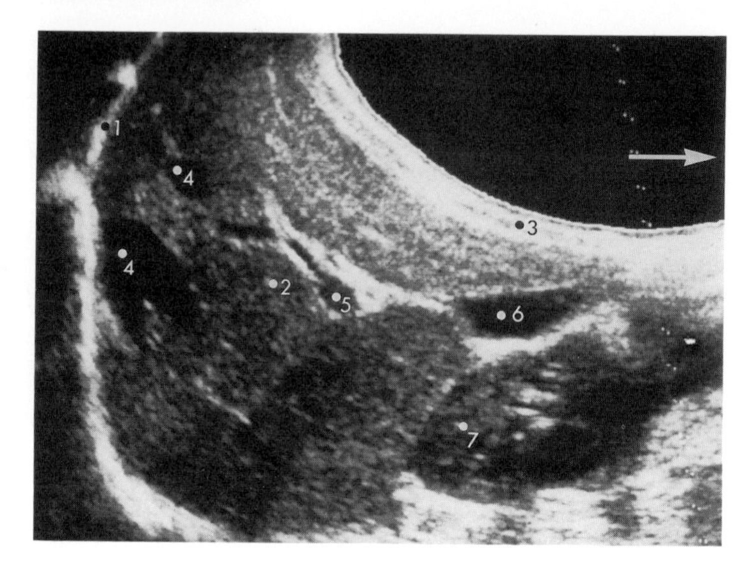

Fig. 11.7 Longitudinal ultrasound scan of the upper abdomen 6 cm to the right of the midline. (See also caption to Figure 11.6). The arrow is pointing to the feet of the subject.

1. Diaphragm
2. Right lobe of the liver
3. Abdominal wall
4. Hepatic veins
5. Right branch of the portal vein
6. Gall bladder
7. Upper pole of the right kidney

Lymph drainage

By superficial and deep plexuses. The superficial plexus on the diaphragmatic surface drains through the diaphragm to the anterior mediastinal nodes and on the visceral surface drains to nodes around the porta hepatis. The deep plexus drains either with the hepatic veins and the inferior vena cava to the posterior mediastinal nodes or with the portal vein to nodes around the porta hepatis.

Histology

The liver has a fibrous capsule which is invaginated at the porta by the portal vein, hepatic artery and hepatic ducts to form sheaths around these structures in their intrahepatic course. The liver parenchyma is arranged as closely packed lobules about 1 mm in diameter. Each is composed of radiating double plates of liver cells separated by a vascular sinusoidal network. The sinusoids have an incomplete endothelial lining of which some cells are macrophagic (reticulo-endothelial) in function. This network receives blood from interlobular branches of the hepatic artery and portal vein and drains to a central (intralobular) vein from which sublobular collecting vessels convey blood to the hepatic veins. Bile formed by the liver cells is discharged into the bile canaliculi within the layers of the cell plates. From the canaliculi, interlobular ducts pass to the right and left hepatic ducts.

Embryology
The liver arises as a ventral outgrowth of the foregut into the ventral mesentery. The stalk remains as the bile duct. As the liver bud enlarges it comes into contact with the mesenchyme of the septum transversum, and cords and plates of liver cells penetrate into its substance. Between the plates of liver cells endothelial sinusoids develop which communicate with vitelline veins from the gut. The hepatic artery, a branch of the coeliac trunk, takes oxygenated blood into the sinusoids. Hepatic veins drain blood from the sinusoids into the future inferior vena cava. The umbilical veins (later only the left one persists) pass through the developing liver as the ductus venosus, taking blood from the placenta straight to the inferior vena cava and the heart. The gall bladder and cystic duct appear early as a diverticulum from the bile duct in the ventral mesentery. (See also p. 39.)

THE EXTRAHEPATIC BILIARY SYSTEM (Figs. 11.8,11.9,11.10)

This conveys bile from the liver to the duodenum. It consists of the right and left hepatic ducts, the common hepatic duct, the gall bladder and cystic duct, and the bile duct.

The **common hepatic duct** is formed by the union of the right and left **hepatic ducts** in the porta hepatis. It descends in the free edge of the lesser omentum and is soon joined by the **cystic duct** from the gall bladder and forms the bile duct.

The **bile duct** is 8 cm long and extends from the junction of the cystic and common hepatic ducts to the duodenal papilla. It descends in the free edge of the lesser omentum, then behind the superior part of the duodenum and the head of the pancreas.

Relations: In the lesser omentum the portal vein is posterior to the bile duct, and separates it from the opening of the omental bursa. The hepatic artery lies on its left side. Behind the duodenum it is anterior to the portal vein and accompanied by the gastroduodenal artery on its left side. Behind the head of the pancreas it lies on the inferior vena cava and, turning to the right opens, in common with the main pancreatic duct, into the **ampulla**. It enters the descending part of the duodenum at the apex of the **duodenal papilla** 10 cm beyond the pylorus. The opening into the ampulla is surrounded by a sphincter of smooth muscle.

The **gall bladder** is a pear-shaped sac lying in the right hypochondrium on the visceral surface of the right lobe of the liver. It has a fundus, body and neck. Its fundus and inferior surfaces are usually covered by peritoneum.

Relations: The fundus extends beyond the inferior border of the liver, is in contact with the anterior abdominal wall deep to the tip of the 9th right costal cartilage and lies on the beginning of the transverse colon. The body passes

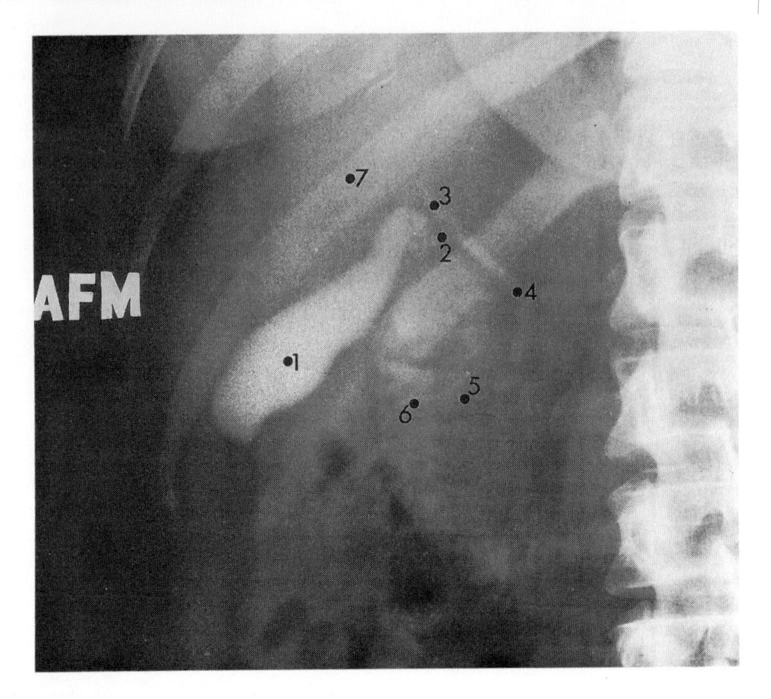

Fig. 11.8 Cholecystogram. Contrast material is absorbed by the duodenum, excreted by the liver and concentrated by the gall bladder. The extrahepatic biliary system is outlined (AFM: photograph was taken after fatty meal).

1. Gall bladder
2. Cystic duct
3. Common hepatic duct
4. Bile duct
5. Site of biliary ampulla
6. Duodenum (descending part)
7. 11th rib

upwards towards the porta hepatis and overlies the descending part of the duodenum. The neck is sinuous and is continuous with the cystic duct which descends in the lesser omentum. It overlies the superior part of the duodenum.

Vessels: The cystic artery, most commonly a branch of the right hepatic artery, supplies the gall bladder, and an accompanying vein drains into the portal vein. Blood from most of the extrahepatic biliary system drains directly into the liver. Lymph drains via nodes in the porta hepatis to the coeliac group of pre-aortic nodes.

The gall bladder, in common with the rest of the biliary tract, is lined by columnar epithelium with numerous mucus secreting cells. Macroscopically, its mucous membrane shows a honeycombed appearance and is thrown into numerous folds (rugae). There is a constant spiral fold, the spiral valve, in the neck which probably prevents kinking of the duct.

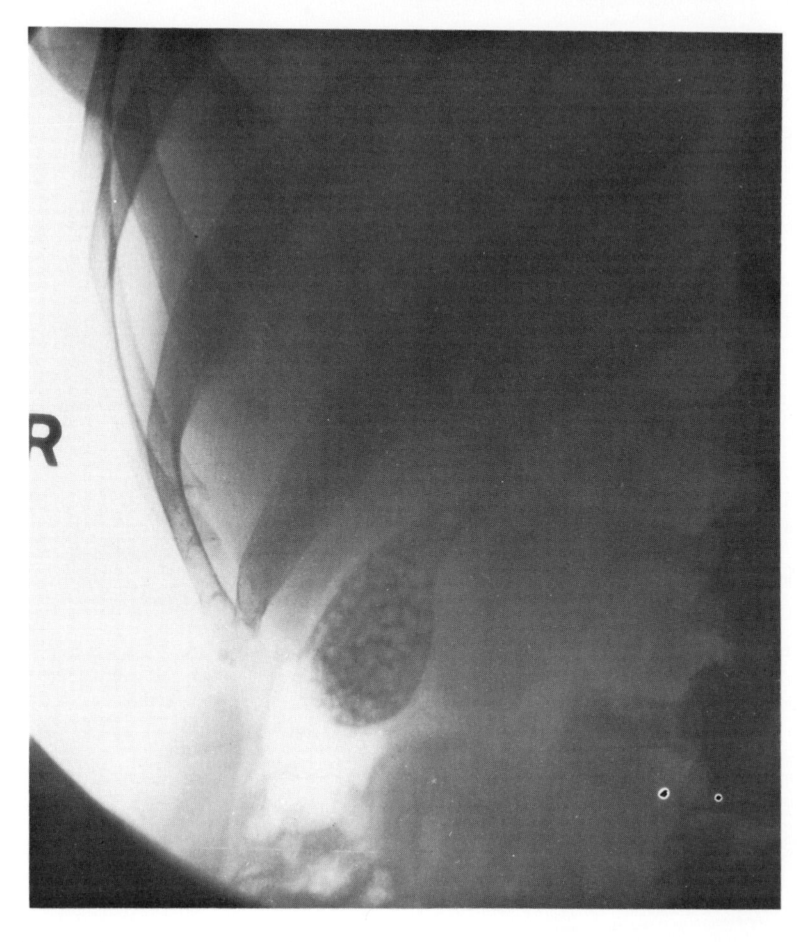

Fig. 11.9 Cholecystogram showing numerous round grey translucent filling defects within the gall bladder. These represent multiple nonopaque gall stones. Compare this with the normal gall bladder shown in Fig. 11.8

The gall bladder concentrates and stores bile and adds mucus to it. Some radio-opaque dyes are excreted by the liver and concentrated by the normal gall bladder. Irregularities of function and anomalous positions of the organ may be detected by radiological examination (Fig. 11.8). Stones form in the gall bladder or in the ducts and may pass to the duodenum (Fig. 11.9). The pattern of symptoms and signs will be influenced by the position of the stones and the degree of blockage to the flow of bile. Pain from the gall bladder and the diaphragm may be referred to the right shoulder region (see p. 430). Stones in the bile duct may cause obstructive jaundice. Postoperatively the anatomy of the biliary tree is revealed by a cholangiogram (Fig. 11.10).

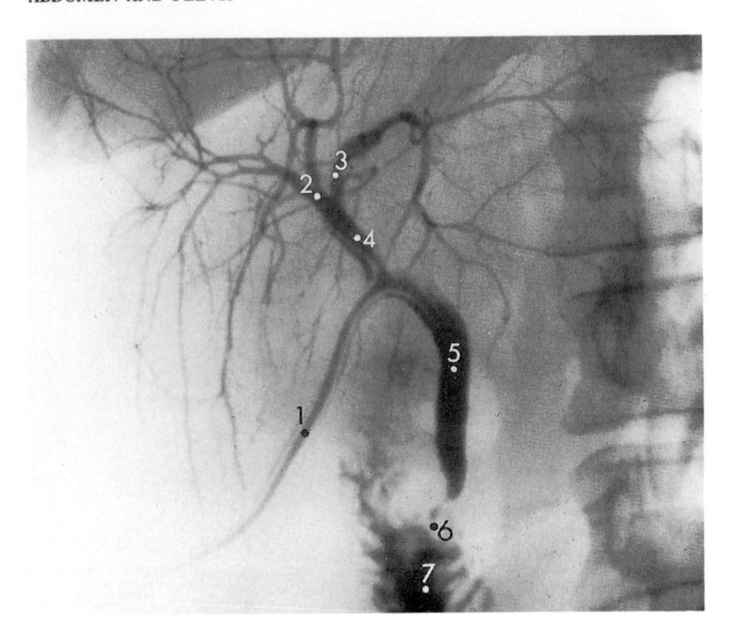

Fig. 11.10 Biliary tree. Contrast material has been introduced by the T-shaped tube which had been left in place temporarily after an operation at which the gall bladder was removed. The cystic duct was ligated close to the common hepatic duct.

1. T-tube
2. Right hepatic duct
3. Left hepatic duct
4. Common hepatic duct
5. Bile duct
6. Site of biliary ampulla
7. Duodenum

THE SPLEEN (Fig. 10.4)

This large lymphoid organ is situated deeply in the left hypochondrium where it lies obliquely beneath the 9th, 10th and 11th ribs. It is about 3 cm thick, 8 cm wide and 13 cm long and weighs about 200 g. The spleen has posterior and anterior ends, inferior and superior (notched) borders, diaphragmatic and visceral surfaces, the latter bearing the hilus.

Relations

The diaphragm separates the diaphragmatic surface from the left pleural sac, the lung and the 9th, 10th and 11th ribs. The visceral surface is related to the stomach anteriorly, the left kidney and suprarenal gland posteriorly, and the splenic flexure of the colon inferiorly. The tail of the pancreas, which lies in the lienorenal ligament, reaches as far as the hilus. The anterior end reaches the midaxillary line and lies on the phrenicocolic ligament, a fold of peritoneum passing from the diaphragm to the left colic flexure.

The organ is invested in peritoneum and suspended at its hilus by two peritoneal folds, the lienorenal and the gastrosplenic ligaments. These folds contain the splenic vessels and their branches and form the lateral limit of the omental bursa.

Blood and nerve supply
The splenic artery and vein, and branches of the coeliac plexus.

Lymph drainage
This occurs along the splenic artery via the suprapancreatic nodes to the coeliac group of pre-aortic nodes.

Histology
It is largely covered by peritoneum and has a fibrous capsule from which trabeculae extend into the organ to form a framework around the lymphoid tissue. The larger vessels run along the trabeculae. The arteries then pass through lymph nodules (white pulp) before dilating to form sinusoids in the diffuse internodular tissue (red pulp). The sinusoids drain via collecting veins into larger trabecular veins.

Embryology
The spleen develops on the left side of the dorsal mesentery of the stomach. It is mesodermal in origin, and later incorporates vascular and lymphatic elements. The spleen functions as part of the macrophage (reticulo-endothelial) and haemopoietic systems and also possibly as a vascular storage organ.

In its deep situation the spleen is usually well protected from injury but severe blows over the left lower chest may rupture the organ, particularly if the overlying ribs are fractured. Severe intraperitoneal haemorrhage then results and the spleen will need to be removed. An enlarged spleen is more easily damaged because it is more fragile.

THE PANCREAS (Fig. 10.4)

This is both an endocrine and an exocrine gland. It is about 15 cm long, lobulated and weighs about 80 g. It is retroperitoneal and extends across the posterior abdominal wall from the curve of the duodenum to the hilus of the spleen. The pancreas has a head with an uncinate process, a neck, a body and a tail.

The **head** is the expanded right extremity of the gland bearing inferiorly an **uncinate process** which passes to the left. The **body** is triangular in section and has anterior, inferior and posterior surfaces; the **tail** is the narrowed left extremity and lies in the lienorenal ligament.

Relations

(i) the head is embraced by the curve of the duodenum. Anteriorly from above downwards it is covered by the pylorus, the transverse colon and coils of small intestine. Posteriorly, the head lies on the inferior vena cava, the right renal vessels and the bile duct. The uncinate process lies on the left renal vein and the aorta, and is crossed anteriorly by the superior mesenteric vessels.

(ii) the neck lies between the pylorus and gastroduodenal artery anteriorly, and the beginning of the portal vein posteriorly.

(iii) the body—the coeliac artery lies above it and the common hepatic and splenic arteries run along its upper border. Its anterior surface, covered by peritoneum of the omental bursa, is related to stomach and lesser omentum. Its inferior surface, covered by peritoneum of the greater sac, is related to the duodenojejunal flexure and coils of jejunum. The transverse mesocolon is attached to the border between the anterior and inferior surfaces. From right to left, the body lies on the aorta and superior mesenteric artery, the left crus of the diaphragm, the left renal vessels, the left suprarenal gland and the hilus and anterior surface of the left kidney. The splenic vein lies behind it throughout its length and is joined by the inferior mesenteric vein.

(iv) the tail lies with the splenic vessels between the two layers of the lienorenal ligament and reaches the hilus of the spleen.

The **pancreatic duct** traverses the length of the gland from left to right, draining the smaller interlobular ducts. In the head it descends to join the bile duct in the ampulla and opens into the descending part of the duodenum. An accessory pancreatic duct may drain the lower part of the head and the uncinate process and may open separately into the duodenum about 3 cm proximal to the main duct. These ducts usually communicate.

Blood supply

Branches of the splenic and superior and inferior pancreaticoduodenal arteries. The veins from the body and tail join the splenic vein, and the pancreaticoduodenal veins pass to the superior mesenteric vein or portal vein.

Nerve supply

This is from the splanchnic nerves and the vagi via the coeliac plexus.

Lymph drainage

This occurs via the suprapancreatic nodes alongside the splenic artery to the coeliac group of pre-aortic nodes.

Histology

Serous alveoli and their associated ducts give the gland its lobulated appearance. The ducts are lined with columnar epithelium. Scattered among the alveoli are clumps of closely packed, paler-staining cells, the islets (of Langerhans), which produce insulin and glucagon. The islets contain a mixture of acidophilic and basophilic cells.

The pancreatic enzymes produced by the alveoli play an important role in digestion, and the hormones produced by the islet tissue regulate the level of the blood sugar.

Embryology

The pancreas develops from the foregut. A ventral bud arises in common with the liver outgrowth from the duodenum and a dorsal bud arises proximal to it. Differential growth and rotation of the duodenum bring the buds together dorsally and fusion occurs. The pancreatic duct is formed from the ducts of both buds; the accessory from the dorsal duct only.

Occasionally the pancreas is found encircling the duodenum and this may cause obstruction of the gut. Pancreatic tissue may be found in other parts of the gut. Inflammation of the gland may result in the escape of secretions into the surrounding tissues or into the peritoneal cavity causing severe abdominal pain and peritonitis. Cancer of the head of the pancreas usually constricts the bile duct and causes progressive jaundice.

— 12 —

The kidneys, suprarenal glands and posterior abdominal wall

THE KIDNEYS (Figs. 10.4, 12.1, 12.2 and 12.3)

The two kidneys lie on the posterior abdominal wall on either side of the vertebral column. Their hila are about the level of the transpyloric plane. Each is about 10 cm long, 5 cm wide, 3 cm thick and weighs about 100 g. Each has upper and lower poles, anterior and posterior surfaces, a lateral convex border and a medial border whose concavity with the hilus, gives the kidney its characteristic shape.

Each is obliquely set so that its upper pole is nearer the midline than its lower. The right kidney is slightly lower than the left because of the bulk of the liver on the right side. The kidneys are embedded in a mass of perirenal fat which is enclosed by part of the fascia transversalis (perirenal fascia). Superiorly the kidney is separated from the suprarenal gland by a layer of this fascia. Inferiorly the two layers remain separate as far as the iliac fossa; laterally they fuse but medially remain separate. The deeper layer passes behind the large vessels and fuses with the anterior longitudinal ligament in front of the vertebral column. The more superficial (anterior) layer passes medially and fuses with its fellow of the opposite side in front of the aorta.

Relations

The posterior relations are similar on the two sides—the diaphragm separates the upper pole from the costodiaphragmatic recess of pleura and the 11th and 12th ribs. Below this the kidney lies on psoas, quadratus lumborum and transversus abdominis from medial to lateral. Quadratus lumborum is crossed by the subcostal nerve and vessels and the iliohypogastric and ilioinguinal nerves.

Anterior relations of the **left kidney**—the suprarenal gland lies on its upper medial aspect, the spleen on its upper lateral aspect and the stomach between these two areas. The body of the pancreas and the splenic vessels cross the hilus and middle part, and the lower pole is in contact with coils of jejunum medially and the left colic flexure laterally.

Anterior relations of the **right kidney**—its upper pole is capped by the suprarenal gland. The hilus is covered by the descending part of the duodenum and

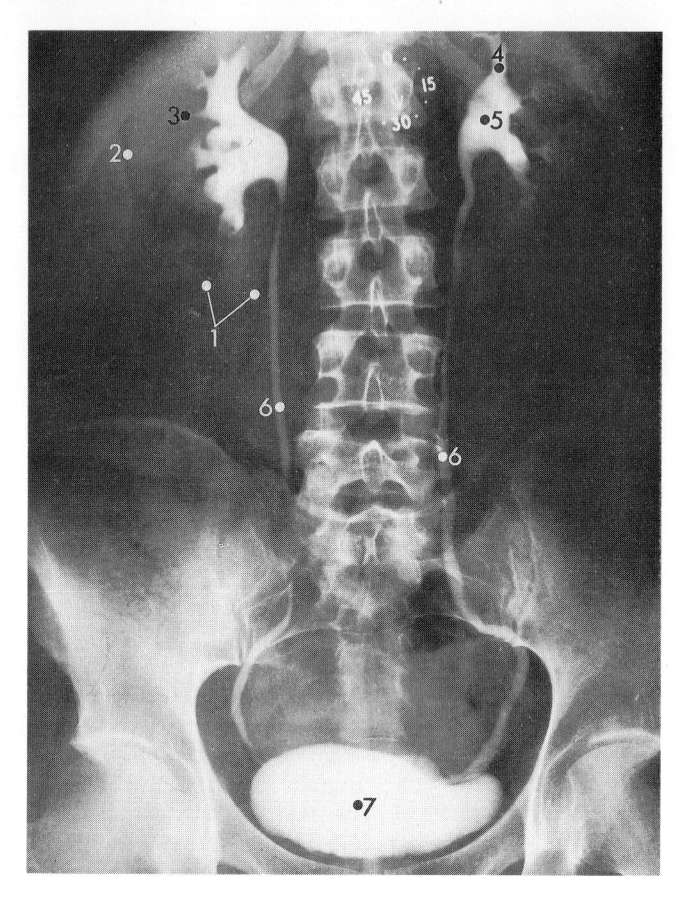

Fig. 12.1 Intravenous urogram (IVU). Radio-opaque material is injected intravenously and is excreted by the kidneys. The urinary tracts are outlined, including the substance of the kidneys. The 'clock' indicates the time which has elapsed (25 minutes) since the injection was made.

1. Lower pole of right kidney
2. Lateral margin of kidney
3. Minor calyx
4. Major calyx
5. Pelvis of left ureter
6. Left and right ureters
7. Urinary bladder

the lower pole by the right colic flexure laterally and the jejunum medially. The remainder of the surface is related to the visceral surface of the liver. At the **hilus** of each kidney the renal vein, artery and the pelvis of the ureter (its dilated upper end) lie in that order, from before backwards. A branch of the renal artery often passes posterior to the pelvis of the ureter. The pelvis of the ureter passes through the hilus into a narrow space, the renal **sinus**, where it divides into two large branches, the **major calyces**. Each of these divides into three or four **minor**

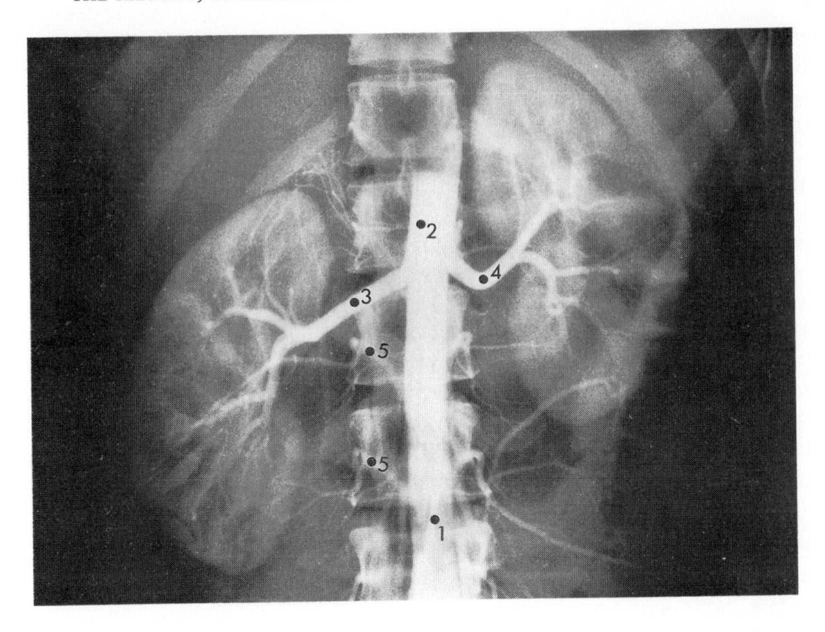

Fig. 12.2 Midstream aortogram showing both renal arteries. The catheter (1) has been introduced percutaneously into the right femoral artery and advanced through the external and common iliac arteries into the aorta to the level of the renal arteries.
1. Catheter
2. Aorta
3. Right renal artery
4. Left renal artery
5. Lumbar arteries

calyces, into which project **pyramids** of renal tissue, the **papillae**. The calyces and renal vessels within the renal sinus are embedded in fat.

Blood supply

Each renal artery rises from the aorta and divides into branches near the hilus (Fig. 12.2). Venous tributaries unite to form the single wide renal veins which drain into the inferior vena cava.

Nerve supply

Sympathetic nerves originate in the 12th thoracic and 1st lumbar segments, pass through the coeliac ganglia and along the renal artery to the organ. The parasympathetic nerves are derived from the vagi. (See also p. 213).

Lymph drainage

Into the para-aortic nodes around the origin of the renal arteries.

Histology
Each kidney consists of a **medulla** (formed of the pyramids) and a **cortex** which is both outside and between the pyramids. In the latter situation the cortex forms the renal **columns** through which pass the renal vessels. The histological and functional unit of the kidney is the nephron (renal tubule) composed of a glomerular (Bowman's) capsule, a proximal convoluted tubule, a loop of Henle, a distal convoluted tubule and a collecting duct. The cup-like glomerular capsule lies in the cortex around a tuft of capillaries, the glomerulus, and leads into the proximal convoluted tubule. The loop of Henle descends into the medulla nearly as far as the apex of the pyramid and then returns to the cortex and continues as the distal convoluted tubule. The distal tubules join the collecting ducts which pass through the medulla and open on the surface of a renal papilla into a minor calyx. The glomerular capsule is lined by thin specialised epithelium; cubical epithelium lines the proximal and distal convoluted tubules (the former has a brush border) and adjacent parts of the loop of Henle. The rest of the loop is lined by simple squamous and the collecting ducts by columnar epithelium.

Embryology
See pages 31 and below.

Pus resulting from inflammation of the kidney is at first retained within the perirenal fascia but may rupture into the surrounding space in the fascia transversalis and spread over the posterior abdominal wall.

THE URETERS

These are narrow muscular tubes 25 cm long which convey urine from the kidneys to the bladder. The dilated upper end, the **pelvis**, holds about 8 ml. The upper half of the ureter lies on the posterior abdominal wall and the remainder is within the pelvic cavity.

Relations
The abdominal course of the ureters is similar in male and female but the relations of the two sides differ. In the pelvis the course is different in each sex but similar on each side.

ABDOMINAL PART. The **right** ureter descends on psoas and crosses over the genitofemoral nerve. Anteriorly, it is crossed by the descending part of the duodenum, the right colic, the ileocolic and gonadal vessels, and the root of the mesentery. Elsewhere it is covered by and adherent to peritoneum. The **left** ureter descends on psoas and crosses over the genitofemoral nerve. It is crossed by the left colic and gonadal vessels, the root of the sigmoid mesocolon. Elsewhere it is covered by and adherent to peritoneum.

PELVIC PART. In the **male** it crosses the common iliac vessels in front of the sacroiliac joint and descends on the pelvic wall to the ischial spine before turning

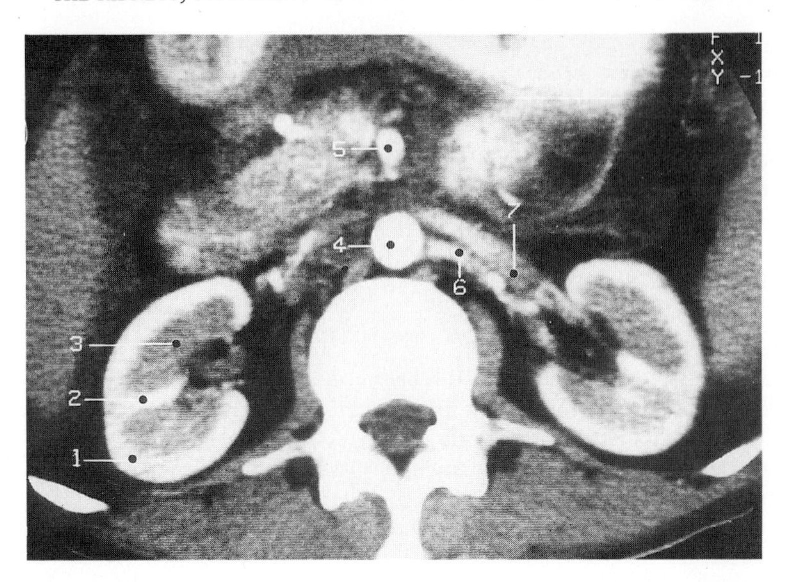

Fig. 12.3 CT transverse scan through the abdomen at the level of the 2nd lumbar vertebra. (See also caption to Figure 6.2).

1. Cortex of the right kidney
2. Cortical column
3. Medulla of the right kidney
4. Aorta
5. Superior mesenteric artery with superior mesenteric vein lying alongside it, slightly anteriorly and to the right
6. Left renal artery
7. Left renal vein

forwards and medially above levator ani. The ductus deferens crosses it superiorly from lateral to medial near the ischial spine. In the **female**, the ureter descends in a similar manner as far as the ischial spine where it turns forwards and medially under the root of the broad ligament. Here it is crossed superiorly by the uterine artery and lies in close relation with the lateral fornix of the vagina just before entering the bladder.

Blood supply

This is by a longitudinal anastomosis between renal, gonadal and inferior vesical arteries, supplemented by vessels from the posterior abdominal wall. The veins drain into the renal, gonadal and internal iliac veins.

Nerve supply

From autonomic fibres originating in the upper lumbar and sacral segments. Sensory nerves enter the lower thoracic and upper lumbar segments of the cord. (See also p. 211).

Lymph drainage

To para-aortic and internal iliac groups of nodes.

Embryology (of kidney and ureter)
In the fetus excretory tissue is bilaterally represented by two mesodermal masses, the mesonephros and a more caudal metanephros. The ureter, calyces and collecting ducts develop from a bud of the mesonephric duct. The remainder of the kidney is derived from the metanephric cap which surmounts the ureteric bud. (See also p. 31). Sometimes a double kidney or double ureter is found on one or both sides. Occasionally a 'horse-shoe' kidney is present, the caudal ends of the two kidneys being fused together in front of the aorta and inferior vena cava.

Radiology
Injection of contrast media, intravenously or through a ureteric catheter, shows up the outline of the calyces, pelvis, ureter and bladder (see Fig. 12.1). Stones, distortions and dilatations may be seen. Injections into the renal arteries may reveal abnormalities which are characteristic of malignant tumours.

Pain from an obstructed ureter is referred to dermatomes T12 and L1 and is felt in the lower abdomen or groin.

THE SUPRARENAL (ADRENAL) GLANDS (Fig. 10.4)

These two endocrine glands are situated on the upper poles of the kidneys. Each is enclosed in a sheath continuous with the perirenal fascia and weighs about 5 g. They are asymmetrical. The right gland is pyramidal in shape and lies in front of the right crus of the diaphragm behind the inferior vena cava and the bare area of the liver. The left gland is crescentic and its lower (medial) end reaches the hilus of the kidney. It lies in front of the left crus of the diaphragm, and behind the splenic vessels and the pancreas. It forms part of the stomach bed. The two glands are separated by the coeliac plexus lying on the front of the aorta.

Blood supply
By suprarenal arteries from the phrenic and renal arteries and the aorta. A large vein drains into the inferior vena cava on the right side and to the renal vein on the left side.

Nerve supply
Preganglionic fibres from the splanchnic nerves end in the suprarenal medulla. (See also p. 211).

Histology
The gland has an outer cortex and an inner medulla. The cortex is yellow and convoluted, and consists of polyhedral lipoid-filled cells arranged in three layers:

(i) outer, zona glomerulosa—cells arranged in clusters.
(ii) middle, zona fasciculata—cells in radially arranged columns.
(iii) inner, zone reticulata—cells irregularly arranged.

The medulla is more vascular and contains perivascular collections of large ovoid chromaffin cells.

Embryology
The cortex develops from coelomic mesothelium and the medulla from the ectodermal cells of the neural crest.

Function
The cortex and medulla are independent endocrine glands. Cortical hormones are concerned with salt transport, electrolyte balance and blood pressure regulation as well as general metabolism and sexual development. The anterior lobe of the pituitary gland plays a large part in the control of cortical hormones. The medulla secretes adrenaline and noradrenaline and is controlled by the sympathetic nerves.

THE POSTERIOR ABDOMINAL WALL

This consists of the five lumbar vertebrae (see p. 46, muscles, fascia and their related structures).

MUSCLES AND FASCIA

PSOAS MAJOR

Attachments

VERTEBRAL
(i) from the five intervertebral discs above each of the lumbar vertebrae and the adjacent vertebral bodies.
(ii) the lumbar transverse processes.
(iii) fibrous arches which cross the concave sides of the lumbar vertebral bodies.

FEMORAL
The lesser trochanter.

Action
(i) flexion and medial rotation of the femur at the hip joint.
(ii) flexion of the trunk.

Nerve supply
Branches from 2nd and 3rd lumbar nerves.

QUADRATUS LUMBORUM

Attachments

INFERIOR—the iliolumbar ligament and the adjacent part of the iliac crest.

SUPERIOR—inferior border of the 12th rib.

Along its medial edge, fibres are attached to the transverse processes of the lumbar vertebrae.

Action
Fixation of the 12th rib thus facilitating contraction of the diaphragm. It is also a lateral flexor of the trunk.

Nerve supply
Branches from the 12th thoracic and the upper four lumbar nerves.

ILIACUS—see page 178.

THE THORACOLUMBAR FASCIA

This encloses the muscles of the posterior abdominal wall and consists of three layers which fuse at the lateral border of these muscles to be continuous with the internal oblique and the transversus abdominis. The stronger posterior layer covers erector spinae and is reinforced by the aponeurosis of latissimus dorsi. Medially it is attached to the lumbar and sacral spines. Superiorly it is continuous over the postvertebral muscles and becomes the deep fascia of the back and neck.

The middle layer lies between erector spinae and quadratus lumborum. It is attached medially to the tips of the lumbar transverse processes and extends vertically between the iliac crest and the 12th rib. The thin anterior layer covers quadratus lumborum, separating it from psoas and peritoneum. It is attached medially to the front of the lumbar transverse processes and passes from the iliac crest to the 12th rib. Superiorly it is thickened to form the lateral arcuate ligament.

THE PSOAS FASCIA

This part of the fascia transversalis forms a thick sheath over psoas and extends behind the inguinal ligament into the thigh. It is thickened superiorly and forms the medial arcuate ligament.

NERVES AND VESSELS OF THE ABDOMEN

NERVES

These comprise the subcostal nerve and the upper branches of the lumbosacral plexus.

The **subcostal nerve** is the ventral ramus of the 12th thoracic nerve. It enters the abdomen under the lateral arcuate ligament and lies on quadratus lumborum behind the kidney. It supplies the muscles of the posterior abdominal wall, and the skin and peritoneum in the suprapubic region.

The lumbar part of the **lumbosacral plexus** is formed in the substance of psoas by the ventral rami of the first four lumbar nerves.

Branches

(i) and (ii) **iliohypogastric** and **ilio-inguinal nerves**—pass laterally around the posterior abdominal wall. The former supplies the skin around the superficial inguinal ring and the latter passes through the inguinal canal. They supply the skin and the lateral muscles of the anterior abdominal wall, and the skin over the pubis and external genitalia.

(iii) **genitofemoral nerve**—descends on the surface of psoas; the genital branch traverses the inguinal canal and the femoral branch passes behind the inguinal ligament and enters the thigh. The nerve supplies the cremasteric and dartos muscles and the skin covering the genitalia and the femoral triangle.

(iv) **lateral femoral cutaneous nerve**—this descends on iliacus and enters the thigh deep to the lateral end of the inguinal ligament. It supplies the parietal peritoneum of the iliac fossa and the skin over the lateral aspect of the thigh.

(v) **femoral nerve** (from the posterior divisions of the 2nd, 3rd and 4th lumbar nerves)—descends in the groove between iliacus and psoas, passes deep to the inguinal ligament and enters the thigh on the lateral side of the femoral artery (see also p. 296). In the abdomen it supplies iliacus.

(vi) **obturator nerve** (from the anterior divisions of the 2nd, 3rd and 4th lumbar nerves)—descends medial to psoas and runs along the lateral pelvic wall to the obturator groove through which it passes and reaches the thigh. In the pelvis it passes between the internal iliac vessels and the lateral pelvic wall above the obturator artery. In the female it lies lateral to the ovary. It has no branches in the pelvis; its branches in the thigh are described on page 291.

(vii) **lumbosacral trunk** (4th and 5th lumbar nerves)—descends medial to psoas over the lateral part of the sacrum and joins the sacral part of the lumbosacral plexus (see p. 211).

THE ABDOMINAL AORTA (Fig. 10.4)

The abdominal aorta is the continuation of the thoracic aorta and enters the abdomen in front of the 12th thoracic vertebra and between the two crura of the diaphragm. It descends on the posterior abdominal wall slightly to the left of the midline and ends on the body of the 4th lumbar vertebra by dividing into the two common iliac arteries.

Relations

The aorta is surrounded by networks of autonomic nerves and ganglia, and lymph vessels and nodes. The cisterna chyli lies between it and the right crus of the diaphragm. Posteriorly, it lies on the bodies of the upper four lumbar vertebrae, intervertebral discs and the left lumbar veins. On its right side lies the inferior vena cava, separated from it superiorly by the right crus of the diaphragm and the cisterna chyli. On its left side lie the left sympathetic trunk and from above downwards the left crus of the diaphragm, the pancreas, the ascending part of the duodenum and coils of small intestine. Anteriorly it is covered from above downwards by the peritoneum of the omental bursa, the pancreas and splenic vein, the left renal vein, the horizontal part of the duodenum and coils of small intestine. The stomach, transverse colon, greater and lesser omenta lie more superficially, in front of the omental bursa.

Branches

(*a*) *midline* (*unpaired*)—to the alimentary tract, i.e. coeliac, superior and inferior mesenteric arteries.

(*b*) *intermediate* (*paired*)—to the gonads, kidneys and suprarenal glands.

(*c*) *parietal* (*paired*)—to the abdominal walls, i.e. phrenic and lumbar arteries.

(*d*) *caudal* (*unpaired*)—median sacral.

UNPAIRED

(i) those arteries supplying the abdominal alimentary tract and its derivatives; they arise from the front of the aorta (see p. 136), coeliac trunk, superior mesenteric artery and inferior mesenteric artery.

(ii) **median sacral artery**—arises from the back of the aorta at the bifurcation and descends into the pelvis, supplying the lower part of the rectum and the front of the sacrum.

PAIRED

(i) **inferior phrenic arteries**—are the 1st branches of the aorta, they supply the crura of the diaphragm and the suprarenal glands.

(ii) **suprarenal arteries**

(iii) **renal arteries**—arise at the level of the 2nd lumbar vertebra and pass laterally and divide into terminal branches at the hilus of the kidney. The shorter left artery crosses the left crus of the diaphragm and psoas behind the body of the pancreas. The right artery crosses the right crus and psoas behind the head of the pancreas and the inferior vena cava. Each artery supplies the neighbouring part of the ureter and the suprarenal gland.

(iv) **testicular arteries**—arise at the level of the 3rd lumbar vertebra and descend obliquely across the posterior abdominal wall, around the pelvic brim to the deep inguinal rings. Each passes through the inguinal canal in the spermatic cord, reaches and supplies a testis.

Relations: in the abdomen the arteries cross psoas and the ureter and the right artery first crosses the inferior vena cava. They are covered by peritoneum and crossed by colic vessels. Along the pelvic brim each crosses over the external iliac artery. It also supplies part of the ureter.

The **ovarian arteries** have a similar course in the abdomen but at the pelvic brim descend across the external iliac vessels, pass into the suspensory ligaments and reach and supply the ovaries. Each artery also supplies the ureter and anastomoses with the uterine artery in the broad ligament.

(v) **lumbar arteries**—the four pairs pass backwards around the upper four vertebral bodies deep to psoas and supply the vertebral column, the spinal cord and the posterior abdominal wall.

Embryology

The main embryonic vessels comprise a single ventral aorta which arises from the truncus arteriosus and communicates through pharyngeal arch arteries with paired dorsal aortae descending on each side of the midline. The 4th left pharyngeal arch artery forms the arch of the aorta, and the fused dorsal aortae form the descending thoracic and then the abdominal aorta (Fig. 2.10).

Degenerative arterial disease produces some loss of elasticity of the wall of the aorta or its branches. A common consequence is thinning of the wall and dilatation of the vessel. The swelling thus produced is known as an aneurysm and is liable to rupture.

THE INFERIOR VENA CAVA (Fig. 10.4)

This is formed in front of the body of the 5th lumbar vertebra by the union of the two common iliac veins. It ascends on the posterior abdominal wall to the right of the midline as far as the caval opening in the central tendon of the diaphragm, through which it passes at the level of the 8th thoracic vertebra. After a short intrathoracic course, it opens into the right atrium.

Relations

Inferiorly the inferior vena cava lies on the right psoas, sympathetic trunk and the lower two lumbar vertebrae, and superiorly on the right crus of the diaphragm, separated from it by the right renal, suprarenal and inferior phrenic arteries and the right suprarenal gland. Anteriorly, it is covered from below upwards by the right common iliac artery, the posterior parietal peritoneum and the root of the mesentery of the small intestine, the right gonadal artery, the horizontal part of the duodenum, the head of the pancreas and the bile duct, the superior part of the duodenum, the opening into the omental bursa and the posterior surface of the liver in which it is embedded. The aorta is in contact with its left side, the right ureter lies on its right and it is intimately related to the neighbouring lymph and autonomic plexuses.

Tributaries

(i) lower two lumbar veins.
(ii) right gonadal vein.
(iii) renal veins—are formed in the hilus of the kidney and they lie anterior to their arteries. The right vein, about 3 cm long, lies behind the descending part of the duodenum; the left, about 8 cm long, lies behind the body of the pancreas, crosses the aorta and receives the left suprarenal and left gonadal veins.
(iv) right suprarenal vein.
(v) hepatic veins—convey portal and systemic blood from the liver and comprise two or three large trunks and a variable number of small vessels. They have no extrahepatic course.
(vi) phrenic veins.

Embryology

The inferior vena cava has a composite origin (see Fig. 2.11 and p. 37).

(*a*) The part below the right renal vein develops from the right supracardinal vein.

(*b*) The part between the renal vein and the liver develops from the right subcardinal vein and its anastomosis with the right supracardinal vein.

(*c*) The remaining part develops from an anastomosis between the right subcardinal vein and the hepatic veins.

In the region of the kidneys, the right and left subcardinal veins anastomose and this communication remains as the left renal vein. The right gonadal vein enters the inferior vena cava directly and the left is a tributary of the left renal vein.

— 13 —

The pelvis

The bony pelvis comprises the two hip bones, the sacrum and the coccyx. Anteriorly the hip bones articulate with each other at the symphysis pubis; posteriorly the sacrum articulates with both hip bones at the sacro-iliac joints. Through this bony girdle the weight of the trunk is transmitted to the lower limbs (when standing) or to the seat (when sitting). The caudal parts of the alimentary and urogenital systems are contained in the pelvis and in parturition (birth) the child passes through the pelvis.

HIP (INNOMINATE) BONE (Fig. 13.1)

This large, irregularly shaped bone consists of three parts, the ilium above, ischium below and pubis in front. In the child, they meet in a Y-shaped epiphyseal cartilage at the **acetabulum**, a cup-shaped fossa which articulates with the head of the femur. The fossa is prominently situated on the lateral aspect of the bone. By late puberty the three parts of the hip bone have fused together. The ilium expands upwards in a fan-like manner whilst below, the pubis and ischium unite to enclose the obturator foramen and form part of the wall of the lesser pelvis.

The **ilium** forms about two-fifths of the acetabulum and expands superiorly into a fluted, flattened plate, the **ala**, for muscular attachments. It has lateral (gluteal) and medial (pelvic) surfaces which are separated by the **iliac crest** superiorly and anterior and posterior borders. The gluteal surface gives attachment to muscles of the lower limb (p. 285).

The medial surface is divided into a posterior sacropelvic surface and an anterior iliac fossa. On the **sacropelvic surface** is the **auricular area** for the sacrum and behind this is a roughened area, the **iliac tuberosity** for the interosseous sacro-iliac ligaments. Between the **iliopubic eminence**, which marks the junction of the ilium and pubis, and the articular surface the bone is thickened to form the iliac part of the **arcuate line**.

The short anterior border of the ilium bears two small projections; the subcutaneous **anterior superior iliac spine** giving attachment to sartorius and the lateral end of the inguinal ligament, and the **anterior inferior iliac spine** to

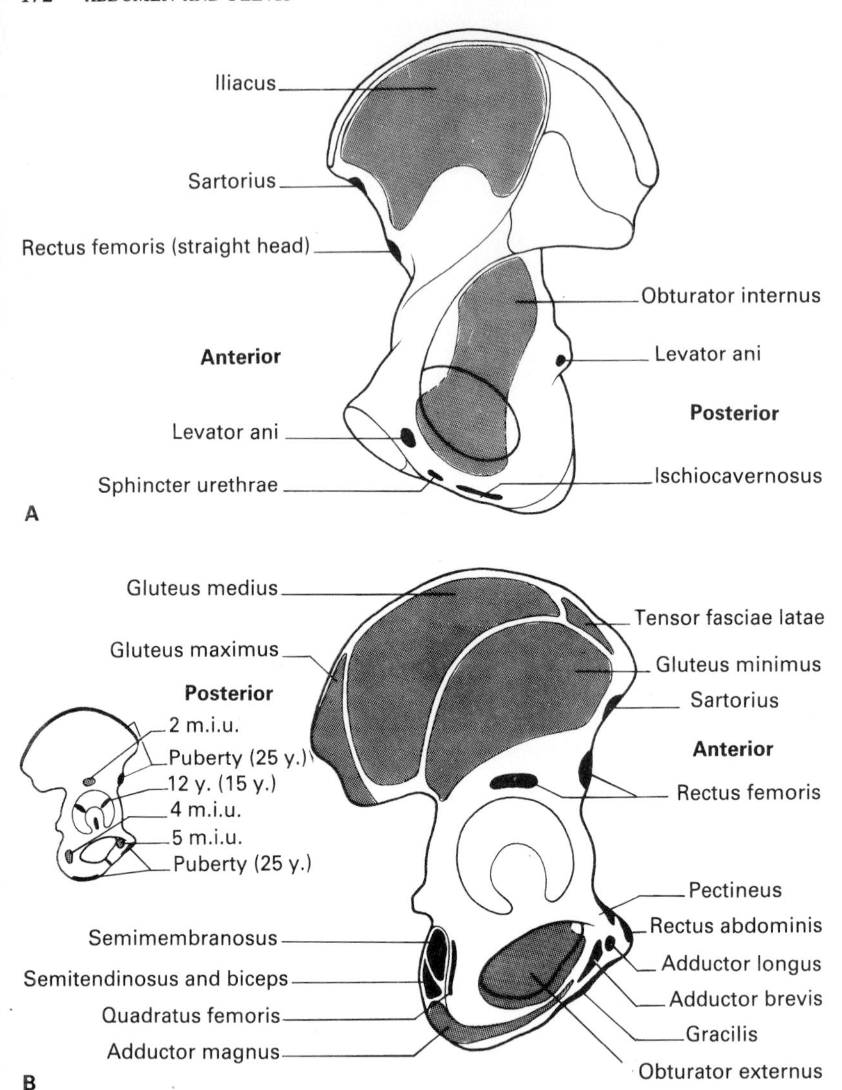

Fig. 13.1 The right hip bone; **A**, medial (internal) and **B**, lateral (external) surfaces. Inset: ossification (m.i.u., months in utero; y., years). The ischial and pubic rami fuse together at about 15 years.

which are attached rectus femoris and the iliofemoral ligament. The posterior border also bears two projections, the subcutaneous **posterior superior iliac spine** and the **posterior inferior iliac spine**. The border gives attachment to the upper part of the sacrotuberous ligament. The inferior spine lies just above the

greater sciatic notch, a deep indentation lying below and behind the ilium. It is largely filled by piriformis.

The **iliac crest**, the curved subcutaneous upper margin of the ilium, extends from the anterior to the posterior superior iliac spine. Its outward convexity is accentuated by a lateral prominence, the **iliac tubercle**, 5 cm from the anterior spine. The crest is strong and gives attachment to muscles of the lateral and posterior abdominal walls, and to gluteus maximus of the buttock at the back. The highest points of the iliac crests (intercristal plane) are at the level of the spine of the fourth lumbar vertebra. A lumbar puncture is performed at this level for the L4–L5 intervertebral space is wide and allows access to the subarachnoid space below the end of the spinal cord.

The **ischium** forms the postero-inferior part of the hip bone. It consists of a body and a ramus. The **body** forms two-fifths of the acetabulum and inferiorly expands to form the **ischial tuberosity**. The roughened tuberosity gives attachment to the hamstring and adductor magnus muscles and to the sacrotuberous ligament. The femoral surface of the body of the ischium faces downwards and gives attachment to obturator externus. The dorsal surface lies above the ischial tuberosity and is crossed by the sciatic nerve, piriformis and obturator internus. The smooth pelvic surface gives attachment to obturator internus. The anterior border gives attachment to the obturator membrane. The posterior border is continuous with that of the ilium and bears a conical **ischial spine** which separates the greater sciatic notch superiorly from the **lesser sciatic notch** inferiorly. The spine gives attachment to levator ani and the sacrospinous ligament, and is crossed by the internal pudendal vessels, the pudendal nerve and the nerve to obturator internus. The **ischial ramus** passes forwards from the ischial tuberosity to join the inferior pubic ramus and form the **ischiopubic ramus**. It bounds the obturator foramen inferiorly and gives attachment to the adductor muscles of the thigh.

The **pubis** has a body and superior and inferior pubic rami, and helps to bound the obturator foramen. Its **body** has outer and inner (pelvic) surfaces and a superior border, the pubic crest. The outer surface gives attachment to adductor longus. The smooth pelvic surface is related to the bladder and gives attachment to levator ani. The oval **symphyseal surface** is covered with hyaline cartilage and joined to its fellow by a fibrocartilaginous disc and ligaments.

The thick palpable **pubic crest** gives attachment from within outwards to rectus abdominis, the conjoint tendon and the aponeurosis of external oblique. Laterally it projects to form the **pubic tubercle** to which is attached the medial end of the inguinal ligament. The **superior ramus** fuses with the ilium at the **iliopubic eminence** and forms about one-fifth of the acetabulum. It bears two ridges, which diverge from the pubic tubercle. The sharp upper ridge, the **pectineal line**, passes laterally to become continuous with the iliac part of the **arcuate line**; it forms part of the pelvic brim and gives attachment to part of the inguinal ligament and pectineus. The lower and anterior ridge, the **obturator crest**, passes to the acetabulum and overlies the obturator groove on the inferior surface of the ramus. Between the two ridges pectineus is attached. The pelvic surface of the ramus is smooth and related to the bladder. The **inferior pubic ramus** descends to fuse with the ischial ramus and form the narrow ischiopubic ramus. Its inferior border meets its fellow of the opposite side to form the **pubic arch**. Its femoral surface is roughened and gives attachment to the adductor muscles of the thigh; its pelvic surface in the male has an everted area for the attachment of a crus of the penis.

SACRUM (Fig. 13.2)

This is formed of five fused sacral vertebrae. It is triangular in shape and possesses a base superiorly, an apex inferiorly, and dorsal, pelvic, and two lateral surfaces. It is divided by paired rows of foramina on the dorsal and pelvic surfaces into a median portion which comprises the fused bodies of the vertebrae and a pair of lateral masses, formed of the fused costal elements and transverse processes. The pelvic and dorsal **sacral foramina** communicate by corresponding intervertebral foramina with the central sacral canal. They convey the ventral and dorsal rami of the sacral nerves. The **base** is directed upwards and forwards, and is formed of the upper surface of the 1st vertebra. Centrally, there is an oval articular facet on the upper surface of the body for the disc separating it from the 5th lumbar vertebra. The projecting anterior margin is known as the **promontory**. Behind the body is the upper opening of the sacral canal. On each side of the opening project the **superior articular processes**. Lateral to the body are the fan-shaped lateral parts. The small apex articulates with the coccyx. The pelvic surface is concave and its median part is marked by four transverse ridges which end at the pelvic sacral foramina. The smooth lateral mass here gives attachment to piriformis. The upper part of the pelvic surface is in contact with peritoneum and the lower part with rectum. The dorsal surface is convex and irregular. In the midline where it roofs the sacral canal it bears a **median sacral crest** and laterally two less prominent intermediate sacral crests: the latter are continuous inferiorly with the **sacral cornua** which are joined to the coccyx. Each row of dorsal sacral foramina lies between the intermediate crest and an irregular lateral sacral crest with its transverse tubercles. The posterior wall of

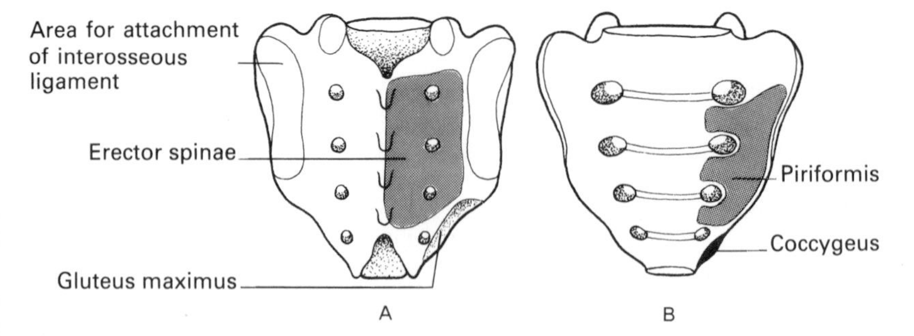

Fig. 13.2 The sacrum: **A.** dorsal (posterior) and **B.** pelvic (anterior) surfaces. In **A.**, the sacral hiatus (inferiorly) is bounded by the sacral cornua.

the sacral canal is deficient inferiorly between the cornua, forming the **sacral hiatus** which is closed by fibrous tissue. The **sacral canal** is triangular in section and contains the cauda equina and cerebrospinal fluid within the meninges, and extradurally the internal vertebral venous plexus, spinal nerves and fat. The dorsal surface gives attachment to parts of erector spinae, gluteus maximus and to the thoracolumbar fascia.

The lateral surface bears a roughened **auricular surface** for the ilium, above and behind which is a pitted area for the attachment of the strong interosseous sacro-iliac ligaments.

The **coccyx** is small, triangular and is formed by the fusion of four coccygeal vertebrae. The 1st may be separate and articulates superiorly with the apex of the sacrum. It gives attachment to the sacrotuberous and sacrospinous ligaments and coccygeus, levator ani and gluteus maximus muscles.

The sacral hiatus is used to introduce anaesthetic solutions into the sacral canal so as to block the sacral nerves (usually in childbirth). This epidural anaesthetic affects mainly the pudendal nerves and other cutaneous nerves to the perineum (p. 204).

JOINTS AND LIGAMENTS OF THE PELVIC GIRDLE

THE SACRO-ILIAC JOINT

This is a synovial joint of the plane variety, between the irregular auricular surfaces of the sacrum and ilium.

Ligaments

(a) CAPSULE
This is attached to the articular margins.

(b) CAPSULAR THICKENINGS

(i) **ventral sacro-iliac ligament**—a short, flat band strengthening the anterior part of the capsule.

(ii) **interosseous sacro-iliac ligament**—is very strong and consists of stout fibres uniting the pitted areas above and behind the auricular surfaces of both bones.

(iii) **dorsal sacro-iliac ligament**—lies superficial to the interosseous ligament and extends from the posterior superior iliac spine to the lateral tubercles of the sacrum. It is very strong.

(c) ACCESSORY LIGAMENTS

(i) **sacrotuberous ligament**—is a strong thick band, narrower in the middle than at each end, which extends from the posterior iliac spines, the transverse tubercles and the lateral crest of the sacrum, and the coccyx to the ischial tuberosity.

(ii) **sacrospinous ligament**—converges from the lateral border of the sacrum and coccyx to the ischial spine. Coccygeus lies on its pelvic surface.

These two ligaments form the posterior boundaries of the greater and lesser sciatic foramina.

(iii) **iliolumbar ligament**—extends from the transverse process of the 5th lumbar vertebra to the posterior part of the iliac crest.

Intracapsular structures

Synovial membrane lines the nonarticular surfaces of the joint. In adults the cavity may be partly obliterated by fibrous adhesions between the joint surfaces.

Functional aspects

Movement is limited to slight gliding and rotation by the reciprocal irregularities of the articular surfaces. Stability is maintained almost entirely by ligaments. The weight of the trunk transmitted through the vertebral column tends to rotate the sacral promontory towards the symphysis. This forward rotation is checked by the sacrospinous and sacrotuberous ligaments. Forward gliding movements are opposed by the interosseous and dorsal sacro-iliac ligaments. Hormonal influences in late pregnancy produce softening of the ligaments and increased mobility. The joint is supplied by the lateral sacral vessels and the 4th and 5th lumbar nerves.

Relations

Dorsal—erector spinae and gluteus maximus; *ventral*—lumbosacral trunk, obturator nerve, psoas, the bifurcation of the common iliac vessels, and the ureter; *inferior*—piriformis and superior gluteal vessels and nerve.

THE SACROCOCCYGEAL JOINT

This is a secondary cartilaginous joint, the two bones being connected by a fibrocartilaginous intervertebral disc and by some small sacrococcygeal ligaments. There is considerable mobility which is increased in late pregnancy and the joint may be dislocated.

THE SYMPHYSIS PUBIS

This is a secondary cartilaginous joint between the symphyseal surfaces of the two pubic bones. The surfaces are covered by hyaline cartilage and are united by a fibrocartilaginous disc. The joint is strengthened on all surfaces by interpubic ligaments. The inferior ligament is thick and forms the anterior border of the pubic arch. Little separation of the pubic bones is possible except during late pregnancy.

THE ARTICULATED PELVIS

In the erect posture, the pelvis lies obliquely so that the anterior superior iliac spines and the top of the pubic symphysis lie in the same vertical plane.

The pelvis is divided into greater and lesser parts by an imaginary plane through the pelvic brim which is bounded by the arcuate lines anteriorly and laterally, and the promontory of the sacrum posteriorly. The **greater (false) pelvis** is the expanded part above the pelvic brim which forms the lower part of the abdominal cavity. The **lesser (true) pelvis** lies below the pelvic brim. In the female it forms the birth canal and, since its dimensions are not much greater than those of a fetal head, it is a space of much obstetric importance. It has an inlet, cavity, and outlet.

The **inlet** is the pelvic brim. In the male it is heart-shaped but in the female it is round or oval and its transverse diameter (13 cm) is greater than its antero-posterior diameter (11 cm). The **cavity** is a short curved canal, the posterior wall being about three times as long as the anterior wall. It is bounded by the lesser pelvic surfaces of the hip bones, the sacrum and the sacrospinous and sacrotuberous ligaments. The **outlet** is diamond-shaped, extending from the lower border of the pubic symphysis to the coccyx and bounded laterally by the ischiopubic rami and the sacrotuberous ligaments. In the female the anteroposterior diameter of the outlet (the conjugate, 13 cm) is greater than the transverse (12 cm).

It should be noted that the maximum diameter of the inlet is transverse and of the outlet, anteroposterior. The large diameters of the fetal head (anteroposterior) normally occupy the widest diameter of the part of the pelvis within which the head is lying. Therefore, in normal labour the head undergoes partial rotation during descent.

Sexual differences are more evident in the pelvis than in other bones. In the **male** the pelvis is rougher, thicker and heavier; the brim is heart-shaped rather than round or oval; the cavity is longer and the outlet narrower. In addition, the ischiopubic ramus is everted (due to attachment of the crus), the acetabular fossa is larger, the greater sciatic notch forms a more acute angle and the subpubic angle is less than 90° (in the female it is greater than 90°). Finally, the auricular surface of the sacrum extends over three vertebrae in the male, and in the female only over two vertebrae.

Fractures of the pelvis are commonly the result of crush injuries. Their most serious consequences are the associated soft tissue injuries. Bladder and urethral injuries are associated with pubic fractures, and major vessel (common and internal iliac) damage with sacro-iliac fractures and dislocations.

MUSCLES AND FASCIA OF THE PELVIS (Figs. 13.1–13.3)

The **pelvic walls** are lined by a layer of muscles which, in turn, is covered by pelvic fascia. The muscles are iliacus lying in the iliac fossa, and obturator internus and piriformis on the walls of the lesser pelvis.

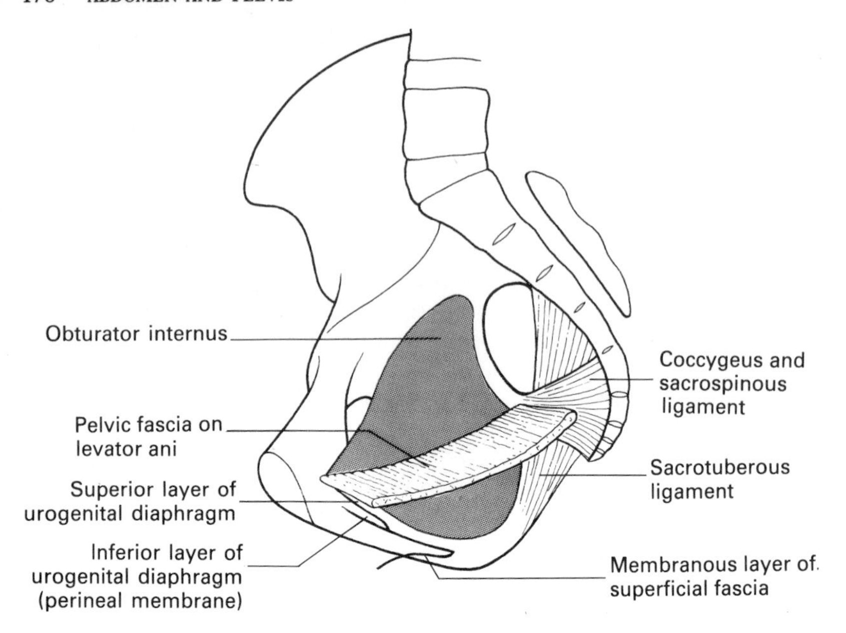

Fig. 13.3 Diagram of the right lateral pelvic wall showing some of the muscular, ligamentous and fascial attachments. Levator ani has been cut anteroposteriorly. Below levator ani is the ischiorectal fossa.

Iliacus

Attachments

PELVIC—the upper two-thirds of the iliac fossa.

FEMORAL—the fibres converge and pass under the inguinal ligament lateral to psoas with which it forms a common tendon. The tendon is attached to the lesser trochanter and to the area just below.

Action

(i) flexion of the trunk on the thigh.
(ii) flexion and medial rotation of the femur on the trunk.

Nerve supply: femoral nerve.

The **fascia iliacus** covers the iliacus muscle and is continued over psoas (p. 166) and, below the inguinal ligament, into the thigh.

Obturator internus

Attachments

PELVIC—pelvic surface of the obturator membrane and the adjacent bone.

FEMORAL—the fibres converge on the lesser sciatic foramen and groove the ischium as they turn laterally and leave the pelvis. Its tendon passes to the medial surface of the greater trochanter.

Action
Lateral rotation of the femur.

Nerve supply
Nerve to obturator internus, a branch of the sacral plexus.

Piriformis

Attachments
SACRAL—the lateral part of the pelvic surface of the sacrum.
FEMORAL—it passes laterally through the greater sciatic foramen to the upper border of the greater trochanter.

Action
Lateral rotation of the femur.

Nerve supply
Direct branches from the sacral plexus.

Levator ani
This is a broad thin muscle.

Attachments

LATERAL
(i) the back of the body of the pubis.
(ii) the pelvic surface of the ischial spine.
(iii) the fascia covering obturator internus between these two points.

MEDIAL
The fibres descend to the midline and there meet fibres of the opposite side in a raphé. The anterior fibres pass backwards around the prostate or vagina to the fibrous perineal body. The middle fibres pass backwards and downwards around the rectum to the fibrous anococcygeal body. Some of these fibres blend with the anal sphincter muscles. The posterior fibres pass to the coccyx and a midline raphé between it and the anococcygeal body.

Action
It provides muscular support for the pelvic viscera, particularly when intra-

abdominal pressure is raised as in defaecation, micturition and parturition, and it reinforces the rectal and urethral sphincters.

The gutter-like arrangement of the two muscles rotates the fetal head into the anterposterior plane as it descends through the pelvis.

Nerve supply
Branches of the 3rd and 4th sacral nerves and of the pudendal nerve.

Coccygeus
This triangular muscle passes from the ischial spine to the sides of the lower sacrum and coccyx. Its actions are complementary to those of levator ani.

The **pelvic floor** is a fibromuscular diaphragm separating the pelvic cavity above from the perineum and ischiorectal fossae below. It is formed by the two levator ani muscles anteriorly and the two coccygei muscles posteriorly.

The **pelvic fascia** covers all the walls and the surfaces of all viscera lying in the pelvic cavity. It is divided into parietal and visceral parts. The parietal fascia covers the walls and floor of the pelvic cavity. It is thick over obturator internus where it gives attachment to levator ani and is continuous superiorly with the fascia iliacus and the transversalis fascia. On levator ani it varies in thickness and is expansile. Note that spinal nerves lie external to it and the pelvic vessels internal.

The visceral fascia is continuous with the parietal and covers the bladder, uterus and rectum. Thickenings around the lower parts of these viscera form ligaments which pass to the pelvic walls; they are considered with their respective organs.

THE RECTUM AND ANAL CANAL (Figs. 10.6, 10.7, 13.4, 13.6)

The **rectum** is the part of the alimentary tract extending from the sigmoid colon to the anal canal. It lies in the posterior part of the pelvis and is about 12 cm long. It begins in front of the 3rd sacral segment and curves forwards, usually making a prominent loop to the left, as far as the tip of the coccyx. Inferiorly it widens and forms the rectal **ampulla**. It has no mesentery—its upper one-third is covered by peritoneum on its front and sides, the middle one-third on its front only and its lower one-third is embedded in pelvic fascia.

Relations
Laterally in its upper part it is in contact with coils of intestines; both are covered by peritoneum. Below this it is related to levator ani and coccygeus, and to the rectal vessels. Posteriorly the superior rectal artery, the 3rd, 4th and 5th sacral

nerves, the sympathetic trunk, and lateral and median sacral vessels separate it from the lower part of the sacrum, the coccyx and the anococcygeal body in the midline, and from the lower part of piriformis, coccygeus and levator ani on each side. Anteriorly in both sexes the upper two-thirds forms the posterior wall of a peritoneal pouch, the rectovesical in the male and the recto-uterine in the female. Deep to the pouch in the male are the seminal vesicles, ducti deferentia, the bladder and the prostate gland. In the female in its lower one-third it is in contact with the posterior wall of the vagina and uterus.

The **anal canal** is the terminal part of the alimentary tract. It descends as the continuation of the rectum but turns posteriorly through the pelvic floor and opens externally at the anus. It is about 4 cm long. Posteriorly it is related to the anococcygeal body; laterally, levator ani partly separates it from the ischio-rectal fossa; and anteriorly the perineal body separates it in the female, from the lower end of the vagina, and in the male, from the bulb of the penis and the prostate gland. The anal canal is surrounded by internal and external anal sphincter muscles.

The **internal sphincter**, an involuntary muscle, is a continuation of the circular muscle coat of the rectum. It surrounds the upper two-thirds of the anal canal and is innervated by the pelvic plexus. Sympathetic stimulation contracts the muscle.

The **external sphincter** encircles the lower two-thirds of the anal canal and is arranged in deep, superficial and subcutaneous parts (from above downwards). It is supplied by the inferior rectal nerves.

(i) The deep part surrounds the middle part of the anal canal. It has no bony attachment but is reinforced by the fibres of levator ani. Functionally this is an essential part of the anal sphincter.

(ii) The superficial part surrounds the lower part of the anal canal. It is attached to the anococcygeal body posteriorly and the perineal body anteriorly.

(iii) The subcutaneous part is a thick ring of muscle surrounding the anal orifice.

Blood supply

The superior, middle and inferior rectal arteries. The veins form communicating plexuses in the submucous coat and in the outer subserous coat of the rectum. These plexuses drain via the superior rectal vein to the portal vein and via the middle and inferior rectal veins to the internal iliac vein. The plexuses form potential portal-systemic anastomoses (p. 142).

Lymph drainage

The rectum and upper two-thirds of the anal canal—to the internal and common iliac nodes, the sacral nodes and along the superior rectal arteries to the pre-aortic nodes; the lower one-third of the anal canal—to upper superficial inguinal nodes. Cancer of the rectum commonly spreads via the lymph vessels; thus surgical attempts at cure must include removal of the appropriate lymph nodes. For similar reasons the surgical treatment of the less common cancer of the anal canal may demand removal of the inguinal nodes (see p. 334).

Nerve supply

The rectum and upper part of the anal canal—sympathetic nerves derived from aortic and pelvic plexuses contract the circular muscle and the internal sphincter; parasympathetic nerves from the pelvic splanchnic nerves 'empty' the viscus. The lower one-third of the anal canal and the external sphincter are under voluntary control and are supplied by the inferior rectal nerves and the perineal branch of the 4th sacral nerves.

Histology

Like the rest of the large intestine, the rectum and upper part of the anal canal have serous, muscular, submucous and mucous coats. The serous (peritoneal) coat is absent inferiorly; the muscular coat comprises a complete outer longitudinal and an inner circular layer; the submucous coat contains many veins embedded in fine areolar tissue; the mucous membrane is thick, lined with columnar epithelium and contains numerous mucous glands and aggregations of lymphoid tissue. The mucous membrane of the upper two-thirds of the anal canal forms marked longitudinal folds, known as anal columns, which in the newborn are connected inferiorly by horizontal crescentic folds, the anal valvules. The lower one-third of the anal canal is lined by stratified squamous epithelium. The muscle of the external anal sphincter is striated. Anal continence is aided by the disposition of the mucous membrane into three anal cushions. Congestion of these and prolapse of these results in the common condition of haemorrhoids (piles).

Embryology

The rectum and upper two-thirds of the anal canal are derived from the cloaca: the lower one-third of the canal and the external anal sphincter from the wall of an ectodermal pouch, the proctodeum. The separating membrane usually breaks down before birth. This dual development accounts for the differences in the blood supply, lymph drainage, nerve supply and histology between these two parts (see p. 40).

Defaecation

Faeces entering the rectum give rise to a desire to defaecate. Defaecation begins with massive peristaltic waves passing through the colon which advance more faeces into the rectum. The rectum then contracts and expels faeces through a relaxed anal canal. Simultaneous contraction of the diaphragm and anterior abdominal wall helps this expulsion. The anal canal is emptied by the contraction of levator ani pulling forwards the anorectal angle and closing the lumen of the canal. Finally, contraction of the voluntary external anal sphincter occurs. Reflexes producing the desire to defaecate can be voluntarily inhibited as long

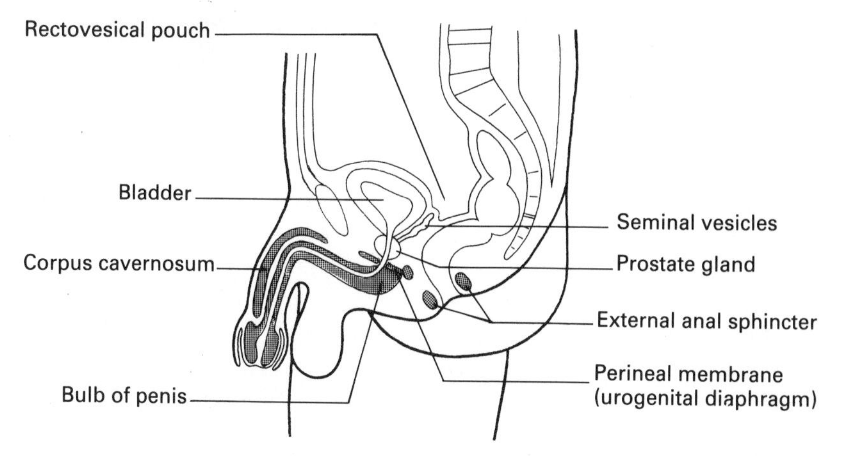

Fig. 13.4 Diagrammatic lay-out of a median section of a male pelvis.

as the anal sphincter is contracted. The anal canal is opened only during the passage of faeces or flatus, continence being maintained by the sphincter muscles.

On **digital examination** of the rectum, the tip of the coccyx and the sacrum can be felt posteriorly. Anteriorly, in the male, the firm prostate and above it the seminal vesicles and the ducti deferentia are palpable. In the female the cervix can be felt through the rectal and vaginal walls.

THE BLADDER (Figs. 13.4, 13.5, 13.6, 14.1)

This reservoir receives urine from the kidneys and subsequently expels it via the urethra. It is a hollow muscular organ lying in the anterior part of the pelvis, outside the peritoneum, and surrounded by extraperitoneal fibrous tissue. It is dilatable and its size and shape vary with the amount of urine it contains. When empty it is pyramidal and presents an apex behind the pubic symphysis, a base posteriorly and a superior and two inferolateral surfaces. When full it is ovoid and bulges up behind the anterior abdominal wall, especially in the child.

From the apex the median umbilical ligament, a remnant of the urachus, ascends to the umbilicus. The ureters enter the posterolateral angles of the base, and the urethra leaves inferiorly at the narrow neck, surrounded by the vesical sphincter.

Peritoneum covers its superior surface and, in the female, passes on to the body of the uterus thus forming the uterovesical pouch. In the male it descends over the upper part of the posterior surface before passing on to the rectum and forms the rectovesical pouch.

Pelvic fascia surrounds the bladder and is thickened around its neck and forms ligaments which attach it to the back of the pubis (pubovesical and puboprostatic), the lateral walls of the pelvis (lateral ligaments of the bladder) and the rectum (posterior ligament). In the male, the bladder is also supported

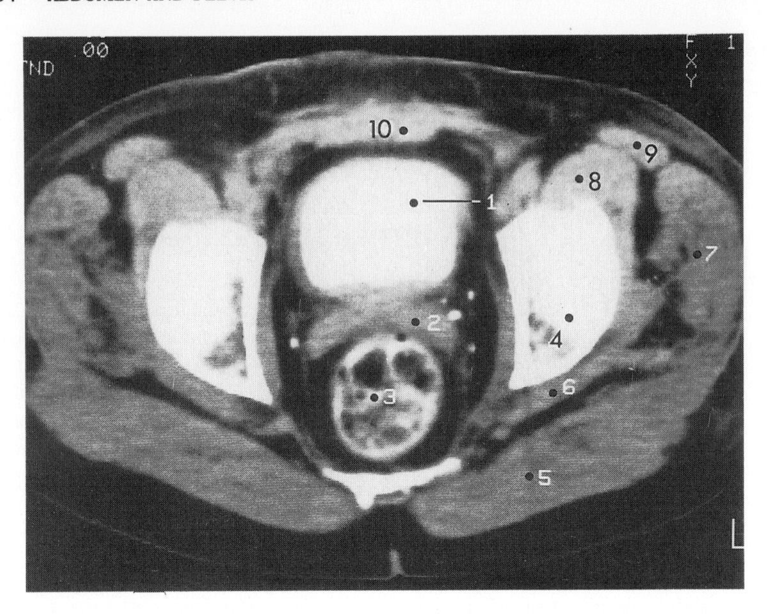

Fig. 13.5 CT transverse scan of the male pelvis. (See also caption to Figure 6.2.)

1. Bladder containing contrast material
2. Left seminal vesicle
3. Rectum
4. Roof of acetabulum
5. Gluteus maximus muscle
6. Obturator internus muscle
7. Gluteus medius and minimus muscles
8. Ilio-psoas muscle
9. Sartorius muscle
10. Rectus abdominis muscle

by its continuity with the prostate. In the female the bladder, though fixed by pelvic fascia to the cervix and the anterior vaginal fornix, is more mobile.

The interior of the bladder is covered by mucous membrane which is thrown into many folds except over a smooth triangular area, the **trigone**, lying between the two ureteric orifices above and the internal urethral orifice below.

Relations

The inferolateral surfaces are separated by a fat-filled retropubic space from the pubic bones, levator ani and obturator internus muscles. The superior surface is in contact with coils of intestine and, in the female, the uterus. The base lies in front of the rectum separated by the vagina in the female and the seminal vesicles and ducti deferentia in the male. Inferiorly the neck overlies the prostate in the male and the urogenital diaphragm in the female (see Figs. 13.4, 13.6).

Blood supply

Superior and inferior vesical branches of the internal iliac arteries; the veins drain via the vesical plexus into the internal iliac veins.

Lymph drainage
To the internal and external iliac nodes.

Nerve supply
Sympathetic fibres, from the 1st and 2nd lumbar segments via the pelvic plexuses—they may be motor to the vesical sphincter and inhibitory to the muscular wall. Parasympathetic fibres, from the pelvic splanchnic nerves (S2, 3, 4)—they are motor to the muscular wall and inhibitory to the vesical sphincter muscle. Sensory fibres are found in both the sympathetic and parasympathetic supplies.

Histology
The bladder has mucous, submucous, muscular, subserous and serous (peritoneal) coats. The mucous coat is lined with transitional epithelium; the submucosa contains some elastic tissue; the muscular coat is composed of whorls of interlacing smooth muscle fibres (**detrusor** muscle) which are thickest around the internal urethral orifice and form the **vesical sphincter**. Thickenings around the ureteric orifices maintain the obliquity of the ureter in the bladder wall and prevent reflux of urine during micturition. The subserous coat contains much fibrous tissue and the serous coat covers only the superior surface and the upper part of the base of the organ.

Embryology
The bladder develops mainly from the ventral part of the cloaca (the urogenital sinus), though the trigone develops from the incorporated caudal ends of the mesonephric (Wolffian) ducts. The median umbilical ligament is a remnant of the urachus. Rarely its cavity persists in the umbilical cord and urine may then drain out at the umbilicus after birth. (See also p. 42.)

Micturition
A urinary volume of more than about 300 ml in the adult bladder usually provokes the desire to micturate. Micturition begins with the slight relaxation of the vesical sphincter allowing some urine to enter the prostatic urethra. This initiates a reflex contraction of the bladder wall (detrusor muscles) and further relaxation of the vesical and urethral sphincter muscles. Reflexes producing the desire to micturate can be voluntarily inhibited; the urethral sphincter is contracted and this is accompanied by relaxation of the detrusor muscle and an increase in tone of the vesical sphincter. However, once urine enters the urethra, micturition usually becomes inevitable.

FEMALE INTERNAL GENITAL ORGANS (Figs. 13.6, 13.7)

These comprise the uterus and uterine tubes, the ovaries and the vagina. They are all related to the broad ligament of peritoneum.

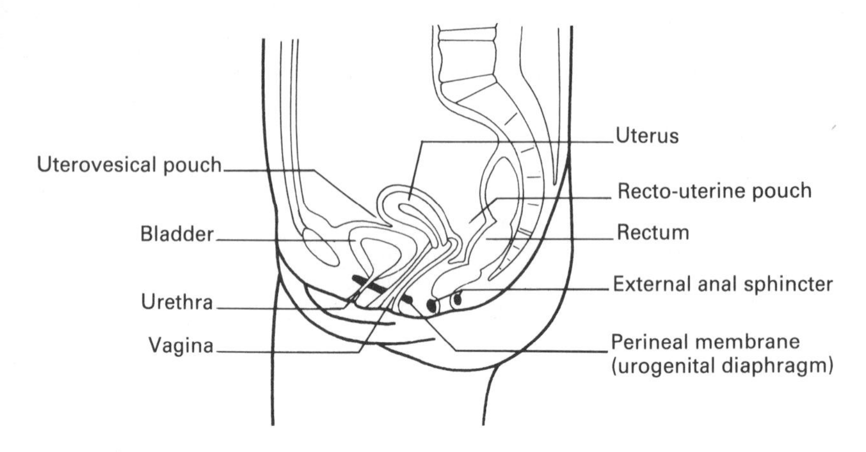

Fig. 13.6 Diagrammatic lay-out of a median section of a female pelvis.

The **broad ligament** is a bilateral double fold of peritoneum extending from the lateral pelvic wall to the lateral margin of the uterus where it is continuous with the serous coat. The layers are separated by a variable amount of fat and fibrous tissue. The uterine tubes lie in the medial two-thirds of its free upper border; the remaining one-third, the **suspensory ligament** of the ovary, contains the ovarian vessels. The ovary is attached to the posterior surface by a peritoneal fold, the **mesovarium**, and the **ovarian ligament** passes between the two layers from the ovary to the body of the uterus where it is continuous with the **round ligament of the uterus**. Within the base of the broad ligament lies fibrous tissue, the **parametrium**, and its thickenings. The uterine artery crosses the ureter just lateral to the cervix of the uterus in the parametrium.

THE UTERUS (Figs. 13.6–13.9)

The adult uterus is a pear-shaped, thick walled, hollow muscular organ lying between the bladder and rectum. It is about 8 cm long, 5 cm wide and 3 cm thick. The uterus opens into the vagina inferiorly and its long axis is usually directed forwards at right angles to the vagina. In this position it is said to be anteverted. It consists of two parts. The **body**, which forms the upper two-thirds, meets the cervix at an angle which faces forwards. This position of the uterus is known as anteflexion. The body is flattened anteroposteriorly and overhangs the bladder. It is pierced at its upper lateral angles by the uterine tubes. The part above the tubes is the **fundus**.

The cylindrical **cervix** projects through the upper part of the anterior vaginal wall. The larger supravaginal part of the cervix is surrounded by fibrous tissue, the parametrium, which is thickened and forms three pairs of ligaments (see below). The conical vaginal part is surrounded by a sulcus, the **vaginal fornix**, which is deepest posteriorly. The cervix bears on its apex the smooth round external opening of the fusiform cervical canal. After childbirth, the opening becomes oval and more irregular in shape.

Peritoneum covers the larger part of the uterus. Posteriorly it covers the fundus, body and supravaginal parts of the cervix and passes on to the posterior wall of the vagina. Anteriorly it covers the fundus and the body only before passing forwards on to the superior surface of the bladder (Fig. 13.6). From the anterior and posterior surfaces the peritoneum passes laterally to the lateral pelvic wall and forms the broad ligaments of the uterus.

Ligaments

The **round ligament** is a fibromuscular band which passes from the upper lateral angle of the uterus through the broad ligament to the deep inguinal ring and then through the inguinal canal to end in the labium majus. It is continuous with the ovarian ligament. Both are derived from the gubernaculum (p. 192).

The **parametrium** surrounds the supravaginal cervix. It is thickened forming three paired ligaments on the upper surface of levator ani:

(i) uterosacral ligaments—pass posteriorly around the rectum to the sacrum.
(ii) lateral cervical (cardinal) ligaments—lie in the base of the broad ligament and pass to the lateral pelvic wall.
(iii) pubocervical ligaments—pass forwards to the body of the pubis.

The uterus is supported and stabilised by the parametrial ligaments. The round ligaments pull the fundus forwards and help to maintain the anteverted position. Further support is afforded by the adjacent viscera and levator ani.

Relations

Posterosuperiorly, coils of small intestine and sigmoid colon separate it from the rectum. Antero-inferiorly is the bladder and on each side are the broad ligaments and their contents. The vaginal part of the cervix is surrounded by the fornix and vaginal walls which separate it from the rectum posteriorly and the ureter and uterine artery laterally.

Blood supply

This is from the uterine artery, a branch of the internal iliac artery. The artery anastomoses superiorly with the ovarian artery, and inferiorly it supplies the vagina. The veins drain via the uterine venous plexus in the base of the broad ligament to the uterine and internal iliac veins.

Lymph drainage

This is mainly to the external and common iliac nodes. Some vessels pass along the round ligament to the superficial inguinal nodes, and others with the ovarian vessels to para-aortic nodes.

Nerve supply
Sympathetic and parasympathetic fibres from the pelvic plexus.

Histology
The uterus has serous (peritoneal), muscular and mucous coats. The muscular coat (myometrium) is composed of bundles of smooth muscle fibres which interlace and are intermingled with collagen fibres and blood vessels. It is about 2 cm thick. The mucous coat (endometrium) consists of a vascular fibrous stroma, containing many leucocytes, covered by a single layer of columnar ciliated epithelium which is invaginated to form many simple glands which may extend into the myometrium. During menstruation the superficial endometrium disintegrates and is shed following haemorrhage from its blood vessels. After the menopause, the endometrium becomes thinner and less vascular. In the cervix the mucous membrane is less vascular, its branched glands are shorter. The earliest sign of cancer of the uterus or cervix may be intermenstrual or postmenopausal bleeding. These symptoms must therefore be fully investigated.

Embryology
The paramesonephric (Müllerian) ducts fuse caudally and form the uterus and upper vagina. Their cranial ends become the uterine tubes. (See also p. 33).

Age changes
Until puberty, the cervix and body are of equal size. After puberty the body enlarges. In pregnancy there is a 30-fold increase in size produced by hypertrophy of the muscle fibres. Atrophy of muscle and endometrium occurs after the menopause.

The uterus provides protection and nourishment for the fetus. Contraction of the muscle propels the fetus through the birth canal.

THE UTERINE TUBES (Fig. 13.7)

These lie in the upper borders of the broad ligament. Each is about 10 cm long and opens medially into the superolateral angle of the uterus. It passes laterally to curl around and overlap the lateral surface of the ovary and opens into the peritoneal cavity. It is usually divided, from medial to lateral, into:

(i) the uterine part, within the wall of the uterus,
(ii) the narrow isthmus,
(iii) the dilated and convoluted ampulla,
(iv) the funnel-shaped infundibulum,
(v) the finger-like fimbriae, one of which is attached to the ovary.

Throughout its course the tube is related to coils of small intestine.

Blood supply: uterine and ovarian vessels.

Nerve supply: sympathetic fibres from the pelvic plexus.

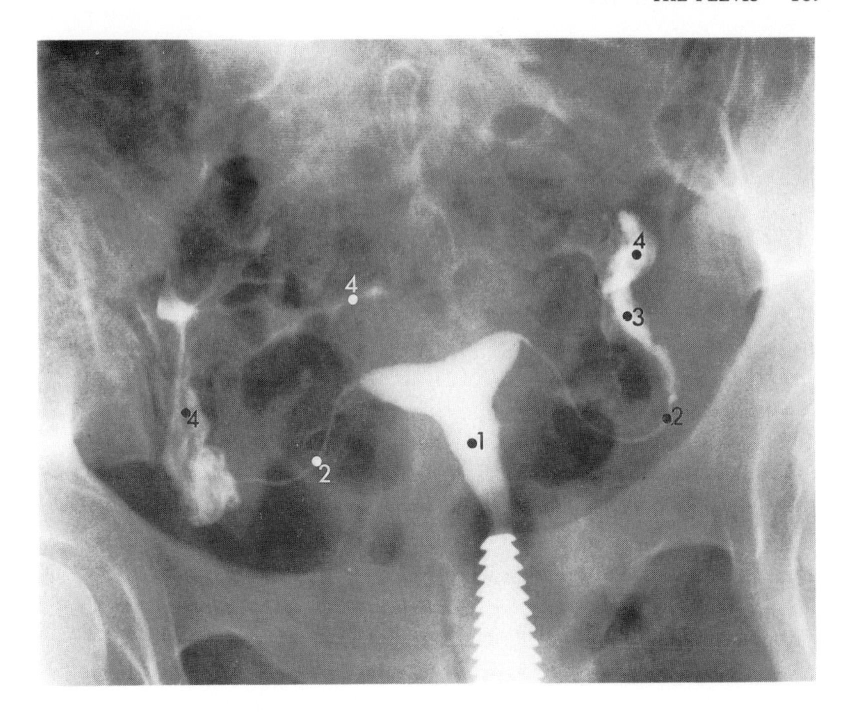

Fig. 13.7 Hysterosalpingogram. Contrast material has been injected through the cervix and fills the uterus and uterine tubes. Some has spilled into the peritoneal cavity.

1. Uterus
2. Uterine tube
3. Ampulla of tube
4. Contrast material in peritoneal cavity

Lymph drainage: along the ovarian artery to the para-aortic nodes.

Histology and embryology: the tube possesses three coats, a serous coat, a muscular coat of smooth muscle fibres and a thick mucous coat possessing numerous longitudinal folds of mucous membrane. It is lined by ciliated columnar epithelium with many mucous cells. The tube develops from the cranial part of the paramesonephric duct (see p. 33). Its function is to convey ova to the uterus.

The patency of the tubes can be demonstrated radiologically by using a contrast medium (see Fig. 13.7).

THE OVARY

This is an almond-shaped organ about 3 cm long, having medial and lateral surfaces, and tubal and uterine ends. It lies on the back of the broad ligament attached by a double fold of peritoneum, the mesovarium, which is continuous with the germinal epithelium that surrounds the ovary.

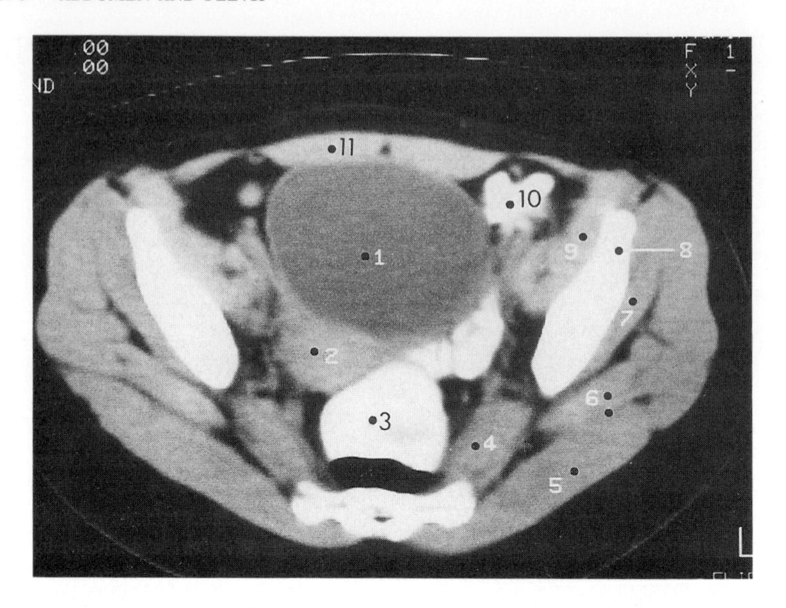

Fig. 13.8 CT transverse scan of the female pelvis (see also caption to Figure 6.2).
1. Bladder (full)
2. Uterus
3. Rectum filled with contrast material
4. Piriformis muscle
5. Gluteus maximus muscle
6. Gluteus medius muscle
7. Gluteus minimus muscle
8. Ilium
9. Iliopsoas muscle
10. Bowel containing contrast material
11. Rectus abdominis muscle

Relations

The lateral surface is adjacent to the pelvic wall. Laterally the ovary is clasped by the infundibulum and fimbriae of the uterine tube. Its tubal end is attached to the tube by one of the fimbriae and its uterine end to the uterus by the fibrous ovarian ligament.

Blood supply

Ovarian artery and branches of the uterine artery. Its veins drain via a pampiniform plexus to the ovarian vein and then to the inferior vena cava on the right side and the renal vein on the left.

Lymph drainage

Along the ovarian artery to para-aortic nodes.

Nerve supply

Sympathetic fibres derived from the 10th thoracic segment via the aortic plexuses.

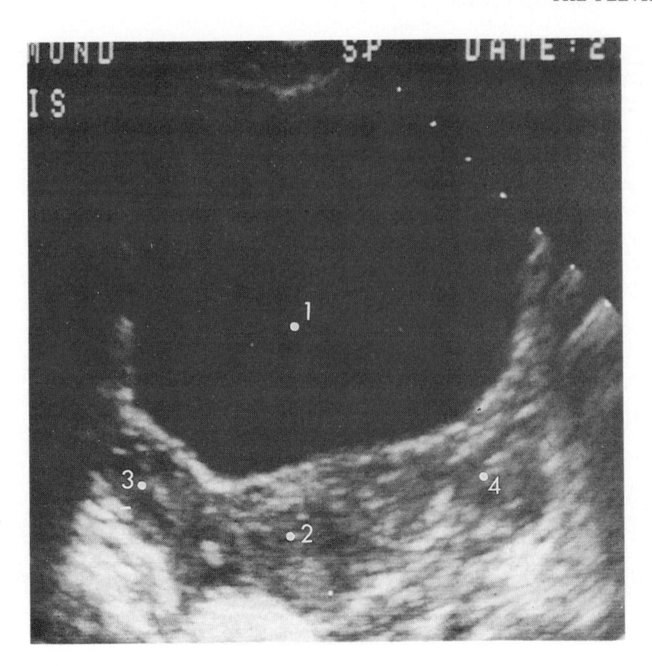

Fig. 13.9 Ultrasound transverse scan of the female pelvis. (See also caption to Figure 11.6).

1. Bladder
2. Uterus
3. Right ovary containing follicles
4. Left ovary

Histology

The ovary is composed of a fibrous stroma surrounded by a layer of cubical cells, the germinal epithelium. The stroma contains a large number of developing follicles and small groups of interstitial cells. Its outer layer is condensed and forms the tunica albuginea.

The immature follicle consists of an **ovum** surrounded by a single layer of cubical cells. After puberty, maturation of some follicles occurs; the follicular wall proliferates and a fluid-filled cavity incompletely divides it into a layer of cells surrounding the ovum, the cumulus oophorus, and an outer layer, the stratum granulosum.

At ovulation, the follicle ruptures and the ovum is expelled into the uterine tube. Its follicular remnant undergoes changes to form a lipoid-filled cyst, the **corpus luteum**. Fibrous remnants of old corpora lutea are seen in adult ovaries. The interstitial cells and the cells of the corpus luteum (under the influence of the pituitary gland) produce oestrogenic and progestational hormones.

Embryology

The ovary develops from the coelomic mesothelium adjacent to the paramesonephric ridge on the posterior abdominal wall. Like the testis, it descends,

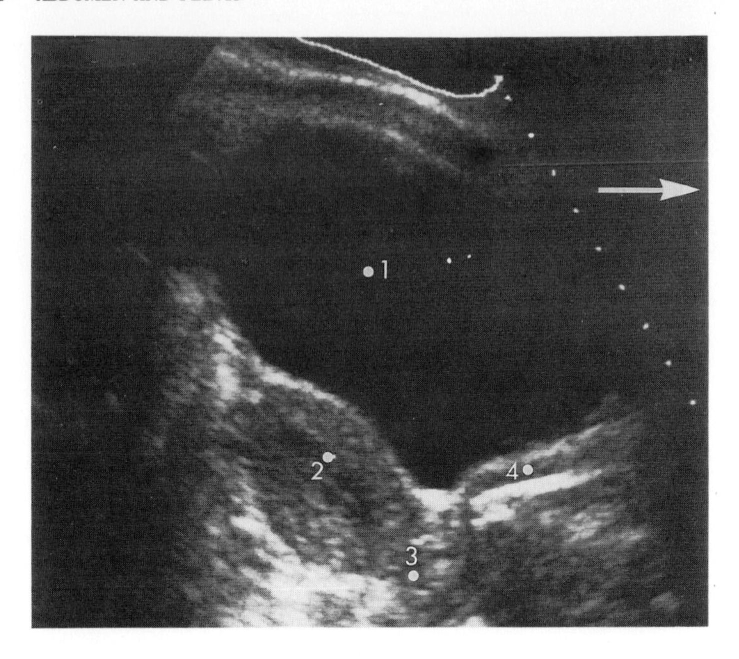

Fig. 13.10 Longitudinal ultrasound scan of the female pelvis (see also caption to Figure 11.6). The arrow is pointing to the feet of the subject.

1. Bladder
2. Body of uterus
3. Cervix of uterus
4. Vagina

preceded by the gubernaculum which persists as the fibrous ligament of the ovary and the round ligament of the uterus. Occasionally remnants of the mesonephric ducts are found in the broad ligament near the ovary.

THE VAGINA (Fig. 13.6)

This is a canal about 8 cm long, extending from the uterus to the pudendal cleft. It lies behind the bladder and urethra and in front of the rectum, its axis forming an angle of 90° or so with the uterus. Its anterior and posterior walls are normally in contact except at its upper end into which projects the cervix surrounded by a deep sulcus, the fornix. This is deepest posteriorly. The lower end is partially occluded by a perforated membrane, the hymen.

Relations

Anterior—from above downwards, the uterus, the bladder and the urethra; *posterior*—from above downwards, the peritoneum of the uterorectal pouch (pouch of Douglas), separating the vagina from the rectum, then the anal canal and perineal body; *lateral*—the base of the broad ligament containing the ureter and the uterine vessels. Below this is levator ani, here forming the vaginal sphincter. The lowest part of the vagina lies in the perineum (p. 200).

Blood supply

The vaginal branches of the internal iliac and the uterine arteries. The veins drain via plexuses on its lateral walls to the internal iliac veins.

Lymph drainage

The upper two-thirds, to the internal and external iliac nodes, the lower one-third, to the upper superficial inguinal nodes.

Histology and embryology

The vagina has muscular and mucous coats. The muscular coat consists of outer longitudinal and inner circular layers of smooth muscle. It is lined by a corrugated mucous membrane covered by stratified squamous epithelium. There are no glands. The vagina is supported by the surrounding fascia continuous above with the pelvic fascia and below with the perineal body. Its upper part is derived from the lower ends of the fused paramesonephric ducts; the lower part by canalisation of an ectodermal thickening.

MALE INTERNAL GENITAL ORGANS (Fig. 13.4)

These comprise the prostate, ducti deferentia, seminal vesicles and ejaculatory ducts.

THE PROSTATE

This is a fibromuscular organ containing glandular tissue lying between the neck of the bladder and the urogenital diaphragm. It resembles a truncated cone about 3 cm in diameter and possesses a base superiorly, an apex inferiorly, and anterior, posterior and two inferolateral surfaces. It is traversed by the urethra and more posteriorly, in its upper half, by the two ejaculatory ducts. These three structures and fibrous septa divide it into a median and two lateral lobes. The **median lobe** lies posteriorly between the urethra and ejaculatory ducts. The **lateral lobes** are below and lateral to the median lobe; they are continuous anteriorly but separated posteriorly by a midline sulcus. (See also p. 200.)

Pelvic fascia invests the organ and its surrounding venous plexus and forms the prostatic sheath. The sheath is continuous above with the fascia around the bladder neck; in front with the puboprostatic ligaments, and behind with the rectovesical fascia covering the seminal vesicles and the front of the rectum.

Relations
Superior—the bladder neck and seminal vesicles; *inferior*—urethral sphincter; *anterolateral*—the levator ani muscles; and *posterior*—the rectum.

Blood supply
The inferior vesical and middle rectal branches of the internal iliac artery. The veins drain via the prostatic and the vesical venous plexuses to the internal iliac veins and to the internal vertebral venous plexus.

Lymph drainage
To internal iliac nodes.

Nerve supply
Sympathetic fibres are derived from the pelvic plexus, parasympathetic from the pelvic splanchnics (S2, 3, 4).

Histology and embryology
The glandular alveoli are lined by columnar epithelium and surrounded by a fibromuscular stroma. Their ducts open into the prostatic urethra. The gland is surrounded by a fibrous capsule and the prostatic venous plexus. The gland develops as outgrowths from that part of the urogenital sinus which forms the prostatic urethra. Its thin alkaline secretion is added to the seminal fluid. The prostatic urethra is described on page 200.

In middle age there is often hypertrophy of the glandular tissue particularly in the median lobe which enlarges and projects into the bladder distorting the vesical sphincter and interfering with micturition. Cancer of the prostate may produce the same symptoms but can usually be distinguished by the hard irregular surface palpable on rectal examination. Secondary deposits may be carried via the internal vertebral veins to the bodies of vertebrae, which may later collapse and cause severe back pain.

THE DUCTI DEFERENTIA, SEMINAL VESICLES AND EJACULATORY DUCTS

Ducti (vasa) deferentia
Each is a narrow muscular tube. It is a continuation of the canal of the epididymis and extends from the tail of the epididymis through the scrotum, inguinal canal and pelvis to the ejaculatory duct. It is about 45 cm long.

Relations
In the scrotum and the inguinal canal, where it is palpable, it forms part of the spermatic cord. It enters the pelvis at the deep inguinal ring, lateral to the

inferior epigastric vessels, and then runs along the lateral wall of the pelvis, covered only by peritoneum, crossing in turn the external iliac vessels, the obliterated umbilical artery, obturator vessels and nerve, vesical vessels and the ureter. At the ischial spine it turns medially above the ureter and descends posterior to the bladder and medial to the seminal vesicles. Here it is dilated and known as the ampulla. Its terminal portion joins the duct of the seminal vesicle and forms an ejaculatory duct.

Seminal vesicles

Each is a coiled, blind, sacculated diverticulum from the ductus deferens, about 5 cm long. It lies obliquely, above the prostate, behind the base of the bladder and lateral to the ampulla of the ductus. Posteriorly it is separated from the rectum by the rectovesical fascia below and the rectovesical pouch of peritoneum above. The narrow lower end joins the ductus just above the prostate and is here palpable on rectal examination.

Ejaculatory ducts

Each is formed behind the neck of the bladder by the union of the duct of the seminal vesicle and the terminal part of the ductus. It traverses the upper half of the prostate to open into the prostatic urethra alongside the prostatic utricle (p. 200). It is 2 cm long.

Blood supply: all these structures are supplied by the inferior vesical and middle rectal arteries. Their veins drain through the vesical venous plexus to the internal iliac veins.

Lymph vessels: pass to the internal iliac nodes.

Nerve supply: sympathetic fibres are derived from the pelvic plexuses; they are motor to the walls of the ductus deferens and seminal vesicles, and their section may produce sterility because ejaculation is not then possible.

Embryology: each ductus and ejaculatory duct develops from the mesonephric duct; and the seminal vesicle is a diverticulum from it. (See also p. 32.)

Function: on ejaculation the ductus carries sperm to the ampulla where they are mixed with ampullary and vesicular secretions, before continuing along the ejaculatory duct and following the prostatic secretion down the urethra. After cutting and tying the ducts (vasectomy), there is atrophy of the seminiferous tubules of the testis. Sperm, however, may remain viable in the tract for several months.

— 14 —

The perineum

The perineum is the part of the pelvic cavity below the pelvic diaphragm. It is a diamond-shaped space between the pubic symphysis and the coccyx. It is bounded anterolaterally by the ischiopubic rami, laterally by the ischial tuberosities and posterolaterally by the lower borders of the sacrotuberous ligaments. It is divided by a line joining the ischial tuberosities into an anal triangle posteriorly and a urogenital triangle anteriorly.

The **anal triangle** contains the anal canal in the midline (see p. 180) and the ischiorectal fossae laterally.

The **ischiorectal fossa** (Fig. 13.3) is a wedge-shaped space, lying with its base inferiorly between the ischium and the anal canal. It is bounded superomedially by levator ani attached to the obturator fascia above, and blending with the external anal sphincter below. Laterally it is bounded by fascia on obturator internus, and inferiorly by perineal skin. Posteriorly it extends between the sacrotuberous and sacrospinous ligaments deep to gluteus maximus; anteriorly it extends between levator ani and the superior fascia of the urogenital diaphragm.

The two fossae communicate with each other round the anal canal and are separated by the anococcygeal body, the anal canal and the perineal body. Each fossa contains:

(i) a mass of loose fatty tissue
(ii) the internal pudendal vessels and the pudendal nerve which lie in the lateral wall within a sheath, the pudendal canal, formed by a split in the fascia covering obturator internus. The canal lies about 3 cm above the ischial tuberosity
(iii) the inferior rectal nerve and vessels which pass medially across its roof
(iv) scrotal nerves and vessels
(v) perineal branch of the 4th sacral nerve which crosses the fossa and supplies the peri-anal skin and the external sphincter muscle.

The **urogenital triangle** (Fig. 13.3) is divided by the inferior layer (perineal membrane) of the urogenital diaphragm into superficial and deep compartments (pouches).

The **urogenital diaphragm** is a triangular double layer of fascia (stronger in the male than the female) stretching across the pubic arch between its attachments to the ischiopubic rami. It has a free posterior border, the centre of which is attached to a fibrous mass, the **perineal body**. It is pierced by the urethra, vessels and nerves passing to structures in the superficial perineal pouch and, in the female, by the vagina also. Its inferior layer (perineal membrane) gives attachment to the bulb and crura of the penis or clitoris. The superior (deep) layer is continuous with the pelvic fascia where the viscera pass through between the levatores ani. It fuses posteriorly with the posterior border of the perineal membrane. The superficial and deep perineal pouches differ markedly in the two sexes and are considered separately.

THE MALE UROGENITAL TRIANGLE

The superficial perineal pouch (Fig. 13.3)
This lies between the membranous layer of the superficial fascia and the inferior layer of the urogenital diaphragm. The superficial fascia is attached posteriorly to the free border of the diaphragm, laterally to the ischiopubic rami and anteriorly it is continuous between the pubic tubercles with the membranous layer of superficial fascia of the anterior abdominal wall. It provides inferiorly a fascial sheath around the penis and scrotum. The pouch contains the root and body of the penis, the superficial perineal muscles, the scrotal vessels and nerves. The testis and spermatic cord lying within the scrotum are in a space continuous with the superficial pouch (Fig. 13.3).

Leakage of urine into the superficial perineal pouch spreads firstly into the subfascial tissues of the scrotum and penis and then ascends the anterior abdominal wall deep to the superficial fascia. This may occur in cases of rupture of the spongy urethra after direct trauma or after pelvic fractures.

The penis
The penis comprises three longitudinal cylinders of erectile tissue, the central corpus spongiosum and the two lateral corpora cavernosa, all covered by fascia and skin. It has a root attaching it to the perineal membrane, and a pendulous cylindrical body. The **corpus spongiosum** lies inferiorly (ventral) and is expanded to form the bulb of the penis posteriorly and the glans penis anteriorly. The bulb is attached to the perineal membrane and is covered by the bulbospongiosus muscle. The corpus spongiosum is traversed by the urethra which opens at the external urethral orifice on the apex of the glans. The two **corpora cavernosa** are united dorsally and their anterior extremities are embedded in the glans. Posteriorly, beneath the symphysis pubis they diverge and form the **crura** of the penis. The crura are covered by the ischiocavernosus muscles. The membranous layer of superficial fascia surrounds the body of the penis and fuses with the corpora just behind the glans. It gives attachment to the short suspensory ligament of the penis which descends from the front of the symphysis pubis.

This layer of superficial fascia is devoid of fat, consists of loose areolar tissue and contains superficial vessels and nerves.

The skin over the body of the penis is thin and hairless and is prolonged forwards as a fold, the **prepuce**, over the glans. The glans is covered with thin skin. Ventrally there is a narrow skin fold, the frenulum, between the glans and the prepuce. This contains a small artery which has to be ligated in excision of the prepuce (circumcision).

Blood supply

By branches of the internal pudendal artery. The artery to the bulb supplies the bulb, corpus spongiosum and the glans; the deep artery of the penis, the corpora cavernosa; and the dorsal artery of the penis, the skin. The veins correspond to the arteries; the corpora drain to the internal pudendal vein and also through the deep dorsal vein of the penis to the prostatic plexus.

Lymph drainage

To the upper superficial inguinal nodes.

Nerve supply

The skin is supplied by the dorsal nerve of the penis and the ilio-inguinal nerve. The skin over the glans is very sensitive. Parasympathetic vasodilator nerves from the pelvic splanchnics supply the erectile tissue.

Histology

Each of the three corpora is surrounded by a thick fibrous coat (the tunica albuginea) and the two corpora cavernosa are separated by an incomplete fibrous septum. From the tunica and the septum numerous fibro-elastic trabeculae arise, subdividing the cavities of the corpora into blood-filled cavernous spaces lined by flattened endothelium. The tortuous terminal branches of the penile arteries open directly into these spaces. Erection of the penis is caused by an increased blood flow in the penile arteries distending the cavernous tissue co-incident with a reduced venous drainage.

Embryology

The penis forms from a swelling, the genital tubercle, lying cranial to the urogenital opening. The penile urethra develops from a ventral groove on the penis (see Fig. 2.9 and p. 35).

The scrotum

This is a pouch of thin, rugose skin enclosing the two testes and spermatic cords, and separating them is a midline septum. Its walls contain a layer of smooth muscle (the dartos). It develops from the fusion of two lateral scrotal swellings.

Superficial muscles of the perineum

The superficial muscles lie in the superficial perineal pouch. They are supplied by the perineal branch of the pudendal nerve.

(i) **bulbospongiosus**—covers the bulb of the penis. The two halves of the muscle arise from the perineal body and from a common midline raphé on the under surface of the bulb. The fibres wind around the penis to be attached to the dorsal surface.

(ii) **ischiocavernosus**—these two muscles cover the crura of the penis. Posteriorly each is attached to the ischiopubic ramus, anterior to the corpus cavernosum. Bulbospongiosus and ischiocavernosus constrict the corpora, playing a subsidiary role in erection and micturition but a major role in ejaculation.

(iii) **superficial transverse perineal muscle** is a scarcely definable transverse bundle running along the posterior edge of the perineal membrane.

The deep perineal pouch

This lies between the superior and the inferior layers of the urogenital diaphragm and below levator ani. It is a closed space, filled mainly by the sphincter urethrae muscle surrounding the membranous urethra. The bulbo-urethral glands and ducts, the internal pudendal vessels and pudendal nerves and their branches, lie in the pouch and leave it by piercing the perineal membrane.

The sphincter urethrae muscle

This is a broad sheet of striated muscle surrounding the membranous urethra, attached on each side to the ischiopubic ramus. Urinary continence is largely dependent on the action of this muscle which is supplied by the perineal branch of the pudendal nerve.

THE FEMALE UROGENITAL TRIANGLE

As in the male, the perineal membrane divides the triangle into superficial and deep pouches but these are almost completely divided again by the passage of the vaginal canal (Fig. 13.6).

The superficial pouch contains the crura of the clitoris, the bulb of the vestibule, the greater vestibular glands, superficial perineal muscles and the terminal parts of the urethra and vagina. The sphincter urethrae which lies in the deep pouch also gains attachment to the vaginal wall. All these tissues may be temporarily distorted during the passage of the child in parturition.

Female external genitalia comprise the **mons pubis**, a fatty elevation over the symphysis, and two pairs of skin folds, an outer **labia majora** and an inner **labia minora** surrounding a median cleft, the **vestibule**, into which opens the urethra and posteriorly the vagina. Anterior to the urethral opening is the **clitoris**, a small sensitive mass of erectile tissue attached to the ischiopubic rami.

It is homologous with the penis and has a very similar structure, possessing two corpora cavernosa capped by a diminutive glans. A fibrous mass, the **perineal body**, separates the vagina and anal canal and is commonly damaged during childbirth. In the superficial pouch and alongside the vagina are the vestibular bulb and the greater vestibular glands.

(i) the **vestibular bulb** comprises two elongated masses of erectile tissue lying on the perineal membrane, one on each side of the vaginal opening. They are united anteriorly by a median commissure and are covered by the bulbospongiosus muscle.

(ii) the two **greater vestibular glands** lie behind the bulb, one on each side of the vaginal opening.

THE URETHRA (Figs. 13.4, 13.6, 14.1)

The male urethra

This is an S-shaped tube about 20 cm long extending from the internal (vesical) urethral orifice through the prostate, deep perineal pouch and corpus spongiosum to the external urethral orifice. It is divided into prostatic, membranous and spongy parts.

(i) the **prostatic urethra** is the widest and most dilatable part. It is about 3 cm long and descends through the gland in a slightly forward direction. The upper part of its posterior wall is marked by a fusiform elevation, the **urethral crest** which makes the urethral cavity ∩-shaped on transverse section. On the summit of the crest is a small pit, the **prostatic utricle**, on each side of which open the ejaculatory ducts. Into the groove on the sides of the crest open the 20–30 prostatic ducts.

(ii) the **membranous urethra** is narrow and more rigid. About 1 cm in length, it descends through the deep perineal pouch, about 3 cm behind the symphysis pubis, surrounded by the **sphincter urethrae** muscle. Two bulbo-urethral glands lie posterolateral to it embedded in the sphincter muscle. They secrete a lubricant fluid along ducts which pierce the perineal membrane and enter the spongy urethra.

(iii) the **spongy urethra** is about 16 cm long. It traverses the whole length of the corpus spongiosum. Its slit-like lumen is narrowest at the external urethral orifice and is dilated posteriorly in the bulb forming the intrabulbar fossa and anteriorly in the glans forming the navicular fossa.

Blood and nerve supply

This comes from the inferior vesical and branches of the internal pudendal artery. Its veins drain to the prostatic venous plexus and the internal pudendal vein. The mucous membrane is supplied mainly by the pudendal nerve.

Lymph drainage

This is from the prostatic and membranous parts to the internal iliac nodes, and from the spongy part to the superficial inguinal nodes.

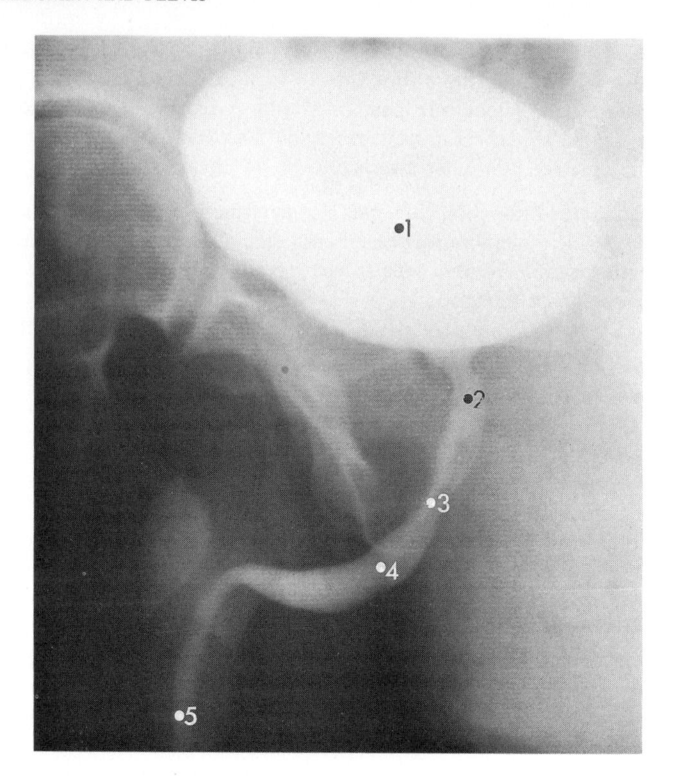

Fig. 14.1 Urethrogram. Contrast material, introduced into the bladder, outlines the bladder and urethra.
1. Urinary bladder
2. Prostatic urethra
3. Membranous urethra
4. Bulbous urethra
5. Spongy urethra

Histology

It possesses muscular, submucous and mucous coats. The mucous membrane is very vascular. In the prostatic part it is lined with transitional squamous epithelium. It possesses numerous mucous glands.

Embryology

The prostatic and membranous parts of the urethra are developed from the urogenital sinus; the spongy part by the fusion of the lips of the urethral groove on the ventral surface of the penis. (See p. 35.)

The female urethra

This is about 3 cm long. It descends from the neck of the bladder through the deep perineal pouch where it is surrounded by the sphincter muscle, the superficial pouch and opens into the vestibule. The external urethral orifice lies between the clitoris and the vaginal opening. It is similar to the proximal parts of the male urethra in structure and develops from the urogenital sinus. (See p. 35.) Stress incontinence occurs when the sphincters are incompetent.

VESSELS AND NERVES OF THE PERINEUM

The pudendal nerve

The pudendal nerve is a branch of the sacral plexus. It leaves the pelvis through the lower part of the greater sciatic foramen, crosses the ischial spine and enters the perineum through the lesser sciatic foramen. In the perineum it passes forwards within the pudendal canal where it ends by dividing into the perineal nerve and the dorsal nerve of the penis (clitoris).

Branches

(i) **inferior rectal nerve**—arises in the posterior part of the pudendal canal. It passes medially across the roof of the ischiorectal fossa to reach and supply levator ani, the external anal sphincter and the peri-anal skin.

(ii) **dorsal nerve of the penis**—runs forwards on the medial side of the ischiopubic ramus into the deep perineal pouch and pierces the perineal membrane to gain the dorsum of the penis. It supplies the glans and skin of the penis and carries parasympathetic fibres to the corpora.

(iii) **perineal nerve**—runs forwards into the superficial pouch. It supplies the perineal muscles and, via its scrotal (labial) branch, the perineal and scrotal (labial) skin.

The internal pudendal artery

This is a branch of the internal iliac artery. It leaves the pelvis and enters the perineum alongside the pudendal nerve, and its branches correspond to those of the nerve to a large extent. It ends in the deep perineal pouch by dividing into deep and dorsal arteries of the penis (clitoris) and the artery to the bulb.

Branches

(i) **inferior rectal artery**—accompanies the inferior rectal nerve and supplies the lower part of the anal canal.

(ii) **scrotal (labial) branches.**

(iii) **artery to the bulb**—arises in the deep perineal pouch. It pierces the perineal membrane supplying the corpus spongiosum.

(iv) **deep artery of the penis**—supplies the corpus cavernosum.

(v) **dorsal artery of the penis**—pierces the perineal membrane and passes along the dorsum of the penis supplying the prepuce and glans.

Lymph drainage
From the perineal skin to the superficial and deep inguinal nodes.

Cutaneous nerves
Perineal skin is supplied by the perineal branch of the posterior femoral cutaneous nerve and the scrotal (labial) branches of the perineal nerve. Injection of local anaesthetic into the pudendal nerves in the pudendal canals will anaesthetise the skin and deeper tissues of the perineum, and is performed as an aid to obstetric procedures.

— 15 —

The blood vessels, lymph drainage and nerves

ARTERIES

THE COMMON ILIAC ARTERIES

These arise at the aortic bifurcation on the front of the 4th lumbar vertebra slightly to the left of the midline. Each descends laterally to the front of the sacro-iliac joint where it ends by dividing into internal and external iliac arteries (Fig. 15.1).

Relations
The common iliac veins lie behind and to the right of the arteries. Each artery is crossed anteriorly by sympathetic nerve fibres and, at its bifurcation, by the ureter. It is covered by peritoneum which separates it from coils of intestine. On the left it is also crossed by the superior rectal artery and the root of the sigmoid mesocolon. Posteriorly it crosses the bodies of the 4th and 5th lumbar vertebra, sympathetic trunk, psoas muscle, obturator nerve and the lumbosacral trunk. The origin of the inferior vena cava lies behind the right artery.

Branches
None, other than its terminal branches.

THE INTERNAL ILIAC ARTERY

This is a terminal branch of the common iliac artery. It arises in front of the sacro-iliac joint and descends on the posterior pelvic wall to the greater sciatic notch where it ends by dividing in a variable manner into parietal and visceral groups of branches.

Relations
Anterior—the ureter which separates it from the ovary; *posterior*—the lumbosacral trunk and the sacro-iliac joint; *lateral*—the external iliac vein and obturator

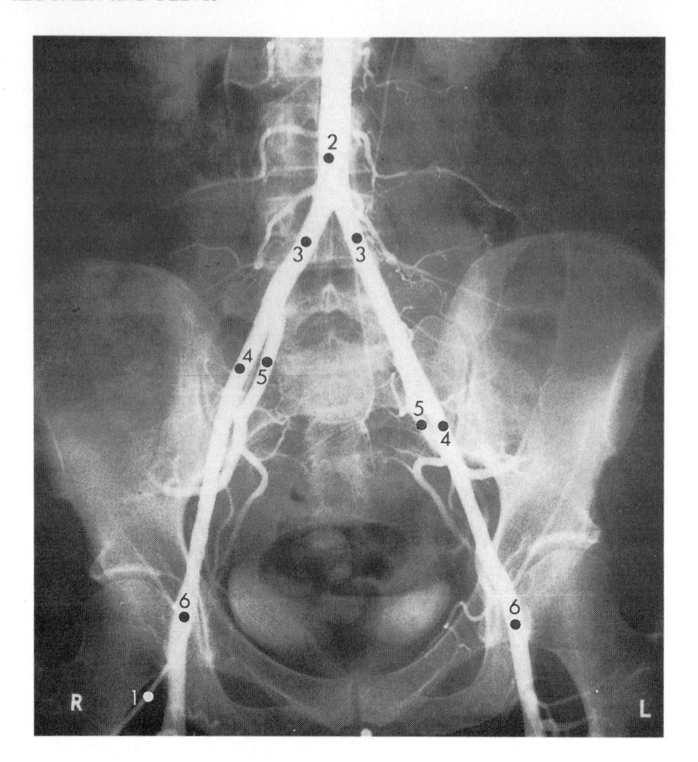

Fig. 15.1 Arteriogram of the aortic bifurcation and the main branches of the iliac arteries. (See also Figures 21.4, 22.5, 22.6).

1. Fine plastic catheter which has been inserted percutaneously into the femoral artery and along which contrast material has been injected
2. Aorta
3. Common iliac arteries
4. External iliac arteries
5. Internal iliac arteries
6. Femoral arteries

nerve separate it from psoas and obturator internus; and *medial*—coils of intestines covered by peritoneum.

Branches

VISCERAL GROUP

(i) **superior vesical artery**—is a branch of the umbilical artery (see below) and supplies the upper part of the bladder.

(ii) **uterine artery**—passes medially on levator ani and crosses over the ureter just above the lateral fornix of the vagina, gains the broad ligament in which it ascends tortuously anastomosing with the ovarian artery. It supplies the vagina, uterus and uterine tubes.

(iii) **middle rectal artery**—when present supplies the seminal vesicles, prostate and the muscular wall of the lower rectum.

(iv) **vaginal artery**—passes anteriorly and supplies the vagina, bladder and ureter. (This corresponds to the inferior vesical artery in the male which supplies the lower end of the ureter, the bladder, the ductus deferens, the seminal vesicles and the prostate.)

PARIETAL GROUP

(i) **umbilical artery**—is obliterated soon after birth but can be seen passing forwards between the lateral pelvic walls and the bladder beyond its superior vesical branch.

(ii) **obturator artery**—runs forwards on the obturator internus below its nerve, leaves the pelvis by passing above the obturator membrane on whose outer surface it ramifies to supply adjacent muscles. It also gives a branch to the hip joint.

(iii) **internal pudendal artery**—passes backwards between piriformis and coccygeus, and leaves the pelvis through the greater sciatic foramen. Its subsequent course is described on page 203.

(iv) **inferior gluteal artery**—accompanies the pudendal artery and leaves the pelvis by the greater sciatic foramen. In the gluteal region it supplies the gluteal muscles, takes part in an anastomosis around the hip joint and gives a branch to the sciatic nerve.

(v) **superior gluteal artery**—runs backwards between the lumbosacral trunk and the sacral plexus and passes through the greater sciatic foramen above piriformis. In the gluteal region it is distributed to the gluteal muscles and takes part in an anastomosis around the hip joint.

(vi) **lateral sacral artery**—this descends anterior to the sacral plexus and anastomoses over the coccyx with the median sacral artery and its fellow of the opposite side. It supplies the contents of the sacral canal.

(vii) **iliolumbar artery**—ascends on the pelvic surface of the sacrum to the medial border of psoas. It supplies iliacus and the muscles of the posterior abdominal wall.

THE EXTERNAL ILIAC ARTERY

This is a terminal branch of the common iliac artery. It arises over the sacro-iliac joint and passes forwards around the pelvic brim to a point below the inguinal ligament, midway between the anterior superior iliac spine and the pubic symphysis (the midinguinal point) where it leaves the abdominal cavity and becomes the femoral artery.

Relations

Posterior—psoas; *medial*—the external iliac vein. In the male it is crossed by the testicular vessels and the ductus deferens; in the female by ovarian vessels and the round ligament. It is covered anteriorly by peritoneum.

Branches

(i) **inferior epigastric artery**—ascends medial to the deep inguinal ring, behind the anterior abdominal wall entering the rectus sheath whose contents it supplies and within which it anastomoses with the superior epigastric artery.

(ii) **deep circumflex iliac artery**—passes laterally towards the iliac crest and supplies the muscles in that region.

Both of these arteries arise just above the inguinal ligament.

VEINS

These drain mainly to the internal iliac vein, whose tributaries correspond to the branches of its artery. The veins from the viscera drain via venous plexuses surrounding the viscera. These plexuses, namely, the prostatic, vesical, uterine, vaginal and rectal, communicate with one another and with the external and internal vertebral venous plexuses. (These communications may account for the frequent spread of cancer from pelvic organs to the vertebrae and bones of the skull.)

The internal and external iliac veins unite at the pelvic brim and form the common iliac vein. They lie posteromedial to the accompanying arteries and have a similar course and relations.

The common iliac vein is formed anterior to the sacro-iliac joint by the union of the internal and external iliac veins. It ascends obliquely, meets its fellow on the right side of the 5th lumbar vertebra and forms the inferior vena cava. Both veins lie behind and on the right of their arteries and their relations correspond. The left vein is longer than the right and is crossed by the right common iliac artery. They also receive the iliolumbar veins.

The gonadal veins—pampiniform plexuses form veins which drain to the inferior vena cava on the right and the renal vein on the left side.

LYMPH DRAINAGE OF THE ABDOMEN AND PELVIS
(Figs. 15.2, 15.3)

Abdomen

The lymph vessels of the anterior abdominal wall pass to axillary, anterior mediastinal and superficial inguinal nodes (see pp. 102, 273, 334). The abdominal viscera and posterior abdominal wall drain to aortic nodes which are subdivided into pre-aortic groups.

Pre-aortic nodes are arranged around the origins of the three arteries supplying the alimentary tract and its derivatives, and drain the corresponding structures. Their efferents unite and form the intestinal lymph trunk which enters the cisterna chyli.

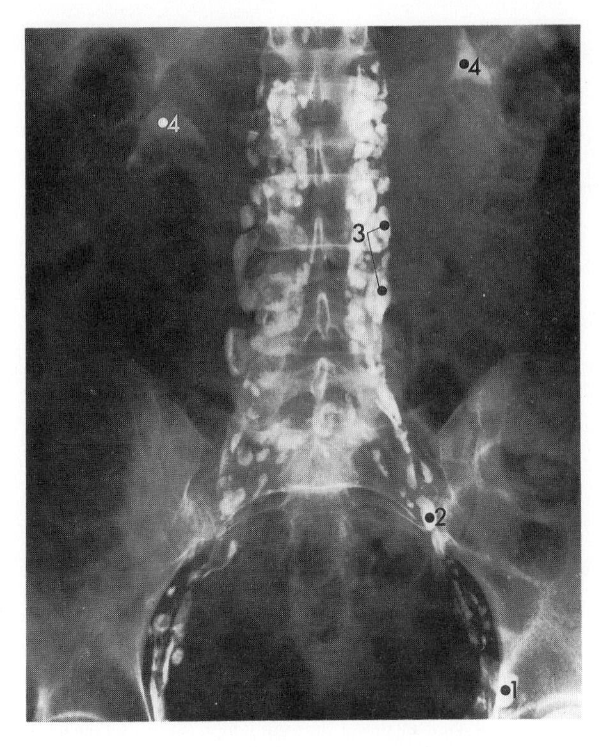

Fig. 15.2 Lymphangiogram. Contrast material is introduced into a lymph vessel in each foot and is then carried up to the external iliac nodes and thence to the para-aortic nodes (see also Figs. 15.3 and 24.3). Some contrast material has been excreted by the kidneys.

1. External iliac nodes
2. Common iliac nodes
3. Para-aortic nodes
4. Pelvis of ureter

Para-aortic nodes lie alongside the aorta around the origins of the paired lateral arteries. They drain the posterior abdominal wall, the kidneys, the supra-renals and the gonads and, through the common iliac nodes, the pelvic viscera and lower limbs. Their efferents unite and form the right and left lumbar lymph trunks.

The cisterna chyli is a thin walled slender sac 5–7 cm long lying between the aorta and the right crus of the diaphragm in front of the upper two lumbar vertebrae. It leads directly into the thoracic duct (p. 103) and receives the right and left lumbar lymph trunks, the intestinal lymph trunk and efferents from nodes in the lower posterior intercostal spaces.

Pelvis

The lymph nodes draining the pelvis and perineum are arranged around the iliac and sacral vessels and are named accordingly. The **external iliac nodes** mainly

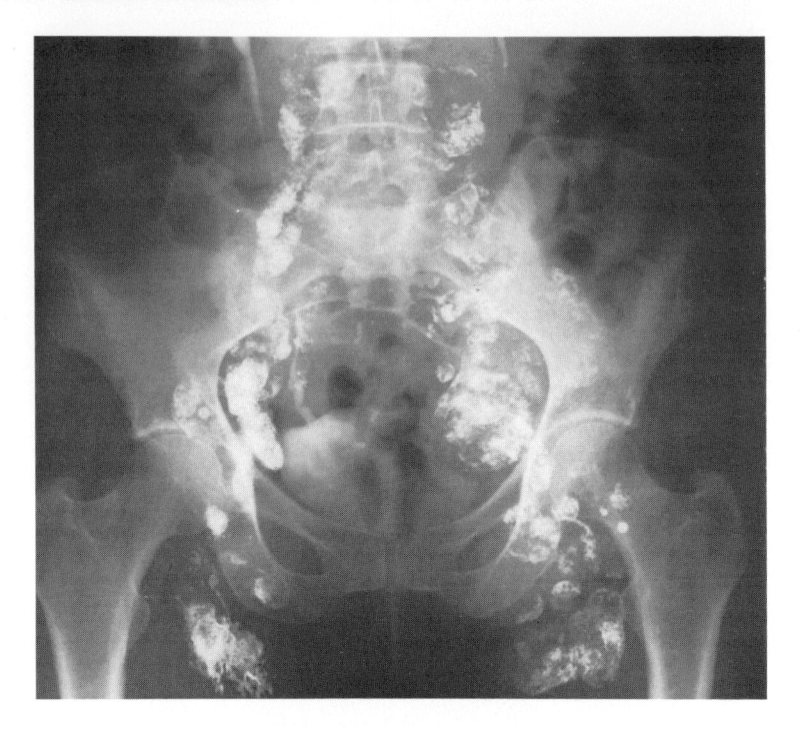

Fig. 15.3 Anteroposterior view of pelvis showing lymph nodes, following lymph angiography. Note the gross enlargement and deformity of the lymph nodes (compare with Figs. 15.2 and 24.3) due to replacement by malignant tissue.

receive vessels from the lower limb, lower abdominal wall and the superficial perineum through the inguinal nodes. It also drains the bladder, prostate, lower uterus and cervix. The **internal iliac nodes** drain all the pelvic viscera except the ovary, and in addition, the deep gluteal region and the deep perineum. Lymph vessels from the testis and ovary drain directly to the para-aortic nodes round the renal vessels. A small group of sacral nodes drains the back of the rectum. Efferent vessels from the internal and external nodes pass to the **common iliac nodes** and thence to the **para-aortic nodes**.

SPINAL NERVES

These comprise the lumbosacral and coccygeal plexuses. The obturator nerve (p. 291), the femoral nerve (p. 296) and the sciatic nerve (p. 289) are the largest branches. (See also p. 167.)

Lumbosacral plexus

The sacral part of the lumbosacral plexus is formed by the ventral rami of the 4th and 5th lumbar nerves and the upper four sacral nerves. It lies in front of the sacrum on the surface of piriformis deep to the pelvic fascia. Its branches can be conveniently divided into pelvic branches and those leaving the pelvis. (See below.)

Branches

PELVIC

(i) *muscular*—to piriformis, levator ani, coccygeus and the external anal sphincter.

(ii) **pelvic splanchnics**—these form the pelvic parasympathetic outflow (p. 213) and are derived from the 2nd, 3rd and 4th sacral nerves.

LEAVING THE PELVIS

(Most of these pass backwards through the greater sciatic foramen above or below piriformis muscle.)

(i) above piriformis, the **superior gluteal nerve**—enters the gluteal region where it lies between gluteus medius and gluteus minimus. (It supplies both these muscles and tensor fasciae latae.)

(ii) below piriformis, are the **sciatic nerve** (p. 289), the **pudendal nerve** (p. 203), the **inferior gluteal nerve** (supplying gluteus maximus), the nerve to obturator internus, the nerve to quadratus femoris, the posterior femoral cutaneous (to the back of the thigh). The perforating cutaneous nerve pierces coccygeus muscle and supplies the gluteal region.

Coccygeal plexus

The coccygeal plexus is formed of the ventral rami of the 4th and 5th sacral and the coccygeal nerves. It lies on coccygeus and its branches supply the skin over the coccyx.

AUTONOMIC NERVOUS SYSTEM IN ABDOMEN AND PELVIS

As elsewhere it is formed of sympathetic and parasympathetic parts and supplies the smooth muscle of viscera and vessels. Pain afferent fibres from the viscera pass into the nearest spinal nerves.

SYMPATHETIC PART

This is derived from the thoracic, lumbar and sacral sympathetic trunks, the fibres passing mainly through the coeliac, aortic, hypogastric and pelvic plexuses.

(a) Lumbar sympathetic trunk

This enters the abdomen by passing under the medial arcuate ligament and descends under the medial margin of psoas to the pelvic brim. It usually has four ganglia.

Branches

(i) spinal branches—pass to the lumbar nerves (grey rami communicantes).
(ii) visceral branches—pass to the aortic plexus and also descend anterior to the common iliac vessels to the hypogastric plexus.
(iii) vascular branches—pass to the common iliac artery.

(b) Sacral sympathetic trunk

This descends on the pelvic surface of the sacrum medial to the pelvic sacral foramina and joins its fellow in front of the coccyx at the ganglion impar. It is continuous under the iliac vessels with the abdominal sympathetic trunk above. It has four or five small ganglia.

Branches

(i) spinal branches—pass to the sacral nerves (grey rami communicantes).
(ii) visceral branches—pass to the pelvic plexuses.
(iii) vascular branches—pass to the branches of the internal iliac artery.

(c) Coeliac plexus

This is formed of the two intercommunicating coeliac ganglia which lie on each side of the origin of the coeliac artery. Each ganglion receives greater, lesser and least (renal) splanchnic nerves from the thoracic sympathetic trunk (p. 101) and branches pass down on the front of the aorta and form aortic plexuses. From the coeliac and aortic plexuses, postsynaptic fibres pass with branches of the aorta and supply the upper part of the alimentary tract and its derivatives, also the kidneys, and the gonads. Presynaptic fibres pass through the plexus and end in the suprarenal glands (medullary part).

(d) Aortic plexus

This lies on the front of the lower abdominal aorta and receives branches from the coeliac plexus and the lumbar sympathetic trunk. Postsynaptic fibres pass along the inferior mesenteric artery to the lower abdominal viscera. It gives branches to the hypogastric plexus.

(e) Hypogastric plexus

This lies in the midline below the aortic bifurcation. It is formed of branches from each lumbar sympathetic trunk and the aortic plexus. It divides into right and left branches which descend to the pelvic plexuses.

(f) Pelvic plexuses
Each of these lies on the side wall of the rectum and is formed by branches from the hypogastric plexus, and the sacral sympathetic trunk. Postsynaptic fibres supply the pelvic viscera.

PARASYMPATHETIC PART

This is derived from the vagi (anterior and posterior vagal trunks which become the anterior and posterior gastric nerves) and the pelvic splanchnics (S2, 3, 4).

(a) Anterior and posterior vagal trunks
These are a mixture of both vagi. They are distributed through the coeliac plexus and along the branches of the aorta to the alimentary tract and its derivatives as far as the splenic flexure.

(b) Pelvic splanchnic nerves
These pass to the hypogastric and pelvic plexuses. From the hypogastric plexus some fibres ascend to the inferior mesenteric artery and are distributed along it, supplying the gut distal to the splenic flexure. From the pelvic plexuses fibres are distributed to the pelvic viscera and are mainly responsible for evacuating the contents. Some branches may produce vasodilatation of erectile tissue.

Surgical removal or chemical blockage of the sympathetic trunks or plexuses will produce vasodilatation and relaxation of the sphincters of the gut. Duodenal ulcers usually heal if the gastric acid output is reduced by surgically dividing branches of the anterior and posterior vagal trunks. Such a vagotomy, however, reduces the motility of the stomach and in order to prevent gastric stasis and distension, the 'drainage' of the stomach has to be improved by anastomosing the stomach to the jejunum or by widening the pyloric aperture (pyloroplasty). (See also p. 101.) Injury to the pelvic splanchnic nerves causes inability to empty the bladder voluntarily, and possibly impotence.

PART FOUR

Upper limb

— 16 —

The shoulder region

The human upper limbs are specialised organs for sensation and prehension (grasping) and show little of the primitive functions of locomotion and support. Without great loss of power there has developed the ability to perform finely controlled movements. The power of the limb is dependent on the strength of its muscles, tendons and bones and on its capacity to fix the joints in positions which allow the muscles to work at their maximum advantage. The ability of the limb to perform finely controlled movements is due to its well-developed sensory nerve supply, its large cerebral representation and its great mobility. This last factor includes the mobility of the shoulder girdle, the movements of pronation of the forearm and opposition of the thumb. Much information about the state of the skeleton (e.g. age) and of the soft tissues can be obtained in the living person by radiological examination.

THE SHOULDER GIRDLE

The shoulder girdle is the means by which the humerus of the upper limb is attached to the axial skeleton. The girdle consists of the scapula (with which the humerus articulates) joined to the clavicle which articulates with the sternum and 1st costal cartilage. Compression forces from the arm are transmitted from the scapula through the coracoclavicular and costoclavicular ligaments to the 1st rib. The whole system is very mobile and the joints are easily dislocated.

CLAVICLE (Fig. 16.1)

This is a long bone but atypical in that it lacks a medullary cavity. The expanded medial end articulates with the manubrium sterni and first costal cartilage. The body has two curvatures, being convex anteriorly in its medial two-thirds and concave anteriorly in its lateral third. The flattened lateral end articulates with the medial side of the acromion.

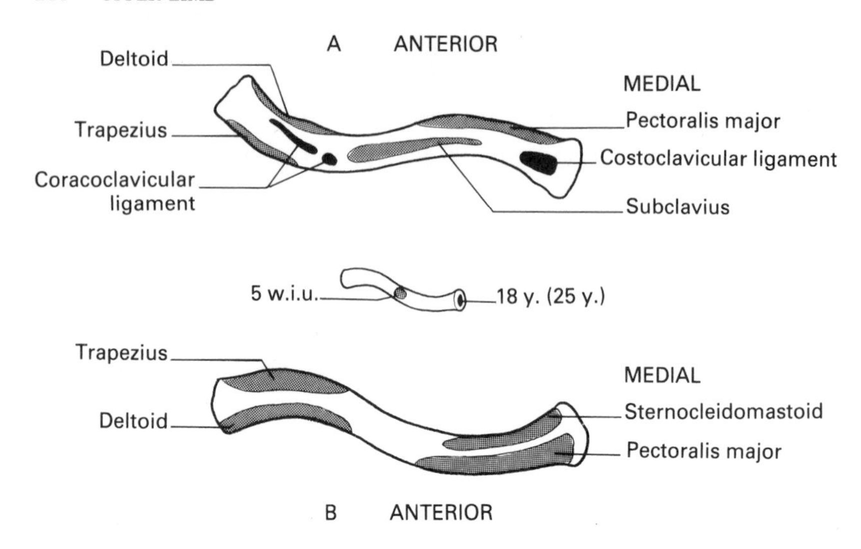

Fig. 16.1 The right clavicle. **A.** lower and **B.** upper surfaces. Inset: ossification (w.i.u., weeks in utero; y., years).

The inferior surface has a roughening near the medial end for the costoclavicular ligament and near the lateral end for the coracoclavicular ligament. Pectoralis major medially and deltoid laterally are attached to the anterior surface, sternocleidomastoid is attached to the superior surface medially and trapezius to the posterior surface laterally.

The bone transmits part of the weight of the limb to the trunk and also allows the limb to swing clear of the trunk. In fractures of the clavicle, the limb falls forwards and downwards on to the chest. The clavicle is the first bone in the body to ossify (5th week) and does so mainly in mesenchyme. The medial end becomes fibrocartilaginous; a secondary centre of ossification is associated with it and appears about the 18th year and joins the body of the bone about the 25th year. It is one of the last bones to stop growing. The bone is subcutaneous throughout its length and is probably the most frequently fractured part of the skeleton.

SCAPULA (Fig. 16.2)

This flat triangular bone is situated on the posterolateral aspect of the chest wall over the 2nd to 7th ribs. It possesses costal and dorsal surfaces; superior, medial and lateral borders; lateral, superior and inferior angles.

The ridged **costal surface** is slightly concave and gives attachment to subscapularis. The **dorsal surface** is divided by a shelf-like projecting **spine** into a smaller supraspinous and larger infraspinous fossa giving attachment to supraspinatus and infraspinatus muscles respectively. The two areas communicate

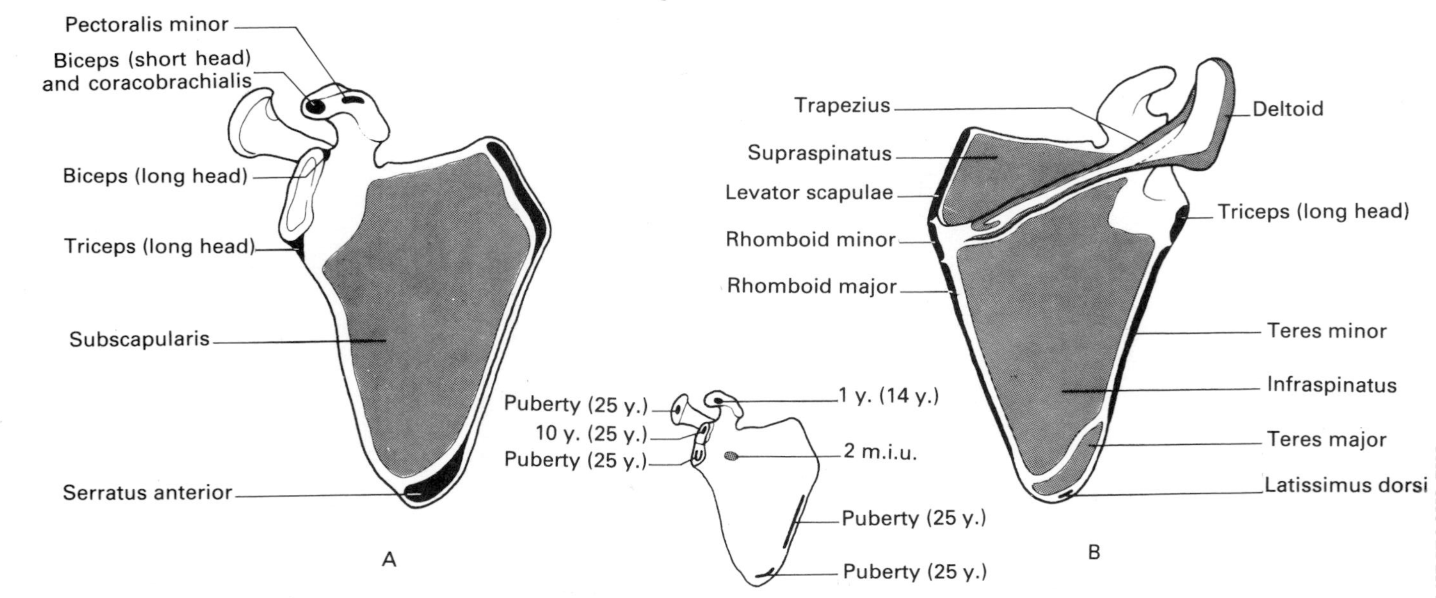

Fig. 16.2 The right scapula. **A.** costal and **B.** dorsal surfaces. Inset: ossification (m.i.u., months in utero; y., years).

laterally around the base of the spine. The spine is expanded laterally into a flattened **acromion** which projects laterally and then anteriorly and articulates with the clavicle on its medial side. The acromion and spine give attachment to trapezius and deltoid.

The superior border passes downwards and laterally. Near the lateral end the beak-like **coracoid process** projects upwards and then forwards. It gives attachment to the coracobrachialis, short head of biceps and pectoralis minor muscles and the coracoclavicular ligament. The border is notched near the base of the coracoid process. A ligament turns the notch into a foramen for the passage of the suprascapular nerve and gives attachment to omohyoid muscle. The medial border gives attachment to the levator scapulae and the two rhomboid muscles dorsally, and the serratus anterior ventrally. The two teres muscles are attached to the thickened lateral border.

The lateral angle is expanded to form the flat **glenoid cavity** for articulation with the humerus. Tubercles above and below the glenoid cavity give attachment to the long heads of biceps and triceps respectively. The latissimus dorsi, teres major and serratus anterior are attached to the stout inferior angle.

The spine and acromion, the coracoid process and inferior angle are palpable. The inferior angle is an important surface landmark and overlies the 7th rib in the resting position.

HUMERUS (Fig. 16.3)

This is a long bone possessing an upper end, body and lower end. The upper **end** includes the head and greater and lesser tuberosities. The **head** is hemispherical and articulates with the glenoid cavity of the scapula; it is bounded by the **anatomical neck** of the bone. The **greater tuberosity** lies laterally behind the **lesser tuberosity** and is separated from it by the **intertubercular groove**. The greater tuberosity has three facets—a superior for the attachment of the supraspinatus, a middle for the infraspinatus and a lower for teres minor. Subscapularis is attached to the lesser tuberosity and the tendon of the long head of biceps lies in the intertubercular groove. The junction between the upper end and body is called the **surgical neck** because fractures occur in this region.

The **body** is stout, cylindrical superiorly and flattened anteroposteriorly below. Half way down its lateral surface is the V-shaped deltoid tuberosity. On the posterior aspect of the bone the **radial groove** runs downwards and laterally. Above it, is attached the lateral head of triceps, and below is the medial head of triceps. Coracobrachialis is attached to the medial side. Pectoralis major is attached to the lateral lip of the lower part of the intertubercular groove, latissimus dorsi to its floor and teres major to its medial lip. At the lower end of the shaft there are well-marked lateral and medial **supracondylar ridges**. To the former are attached brachioradialis and extensor carpi radialis longus. The lower anterior surface of the bone gives attachment to brachialis.

The **lower end** is expanded and forms lateral and medial **epicondyles**. A

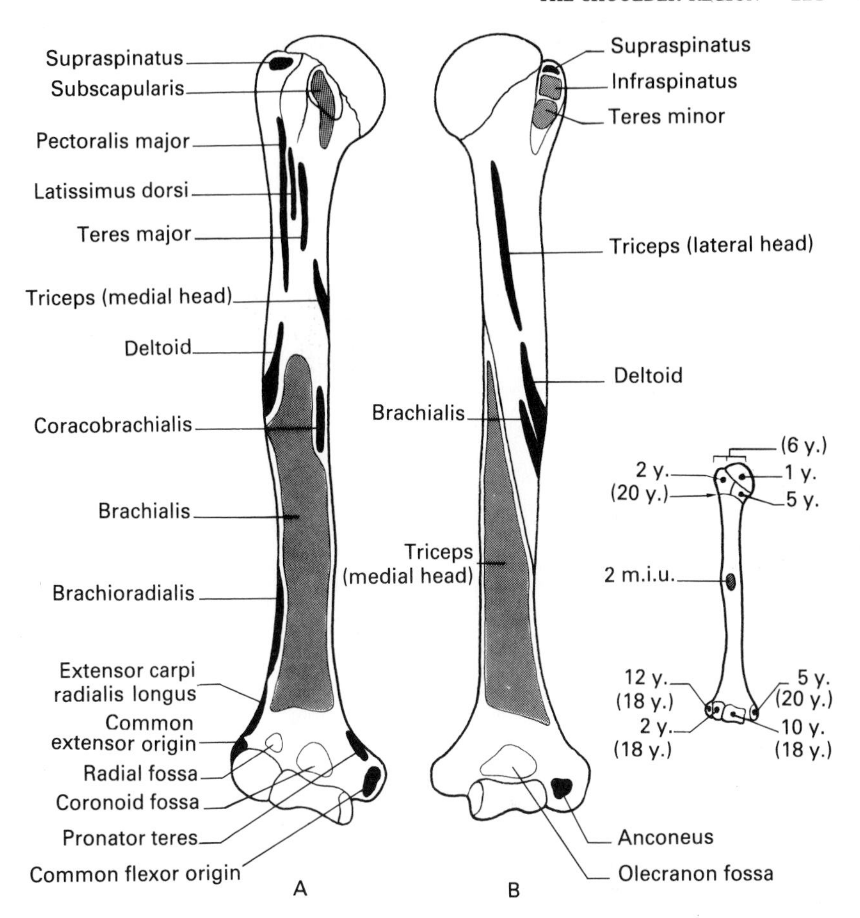

Fig. 16.3 The right humerus. **A.** anterior and **B.** posterior surfaces. Inset: ossification (m.i.u., months in utero; y., years).

rounded **capitulum** laterally and a pulley-shaped **trochlea** medially articulate with the radius and ulna respectively. Fossae are present in front for the head of the radius laterally and for the coronoid process of the ulna medially. Behind is a large fossa for the olecranon. The common extensor and flexor muscle origins are attached to the front of the lateral and medial epicondyles respectively.

The epicondyles are palpable subcutaneously and the greater tuberosity is palpable through the lateral fibres of deltoid. Much of the body is also palpable.

Fractures of the humerus can be complicated by injury to closely related nerves and vessels. In particular, the axillary nerve may be injured in fractures of the surgical neck, the radial nerve in fractures of the body, the brachial artery in fractures of the supracondylar region, and the ulnar nerve in injuries to the medial epicondyle.

THE STERNOCLAVICULAR JOINT

This is a synovial joint with features of the ball and socket variety. The shallow concavity formed by the manubrium sterni and first costal cartilage articulates with the medial end of the clavicle. The joint surfaces are lined by fibrocartilage.

Ligaments

(a) CAPSULE
Attached to the articular margins of the joint.

(b) CAPSULAR THICKENINGS
Anterior and posterior sternoclavicular ligaments.

(c) ACCESSORY LIGAMENTS
(i) **costoclavicular ligament**—a strong band between the inferior aspect of the medial end of the clavicle and the upper surface of the first costal cartilage and the first rib.
(ii) **interclavicular ligament**—passes across the jugular notch and unites the medial ends of the two clavicles.

Intracapsular structure
A fibrocartilaginous disc completely divides the joint into medial and lateral compartments each having its own synovial membrane.

Functional aspects
MOVEMENT: the clavicular movements possible at this joint are elevation and depression, forward and backward movement in a horizontal plane, circumduction and axial rotation. These increase the range of movement of the upper limb and they are more fully discussed under scapular movements. The costoclavicular ligament acts as a fulcrum around which clavicular movements occur. Thus movement of the lateral end of the clavicle is accompanied by proportionately decreased movement of the medial end in the opposite direction.
STABILITY: this is maintained by the intra-articular disc and the joint ligaments, particularly the costoclavicular. The disc prevents the clavicle from overriding the sternum.

Relations
A brachiocephalic vein is formed posterior to each joint.

THE ACROMIOCLAVICULAR JOINT

This is a synovial joint of the plane variety, it lies subcutaneously. The small oval facet on the lateral aspect of the clavicle articulates with the medial side of the acromion. In dislocation of the joint, the clavicle will ride over the acromion.

Ligaments

(a) CAPSULES
Weak, attached to articular margins.

(b) CAPSULAR THICKENINGS
None.

(c) ACCESSORY LIGAMENTS
(i) **coracoclavicular ligament**—between the upper surface of the coracoid process and the inferior surface of the lateral end of the clavicle. This ligament is strong and transmits much of the weight of the upper limb to the clavicle. It provides a fulcrum for scapular movements.
(ii) **acromioclavicular ligament**—a quadrangular band covering the upper aspect of the joint and uniting adjacent bony margins.

Intracapsular structures
An articular disc is sometimes found but is rarely complete.

Functional aspects
MOVEMENT: slight gliding and rotation are possible and these are always associated with scapular movements about the clavicle.
STABILITY: this is maintained mainly by the coracoclavicular ligament.

THE SHOULDER JOINT

This is a synovial joint of the ball and socket variety, between the shallow glenoid cavity of the scapula and the hemispherical head of the humerus. The glenoid cavity is deepened by the **glenoidal labrum**, a ring of fibrocartilage attached to its boundaries; even so its articular area remains less than a third of that of the humeral head.

Ligaments

(a) CAPSULE
Strong but lax, especially inferiorly. Attached proximally to the glenoidal labrum and distally to the articular margin of the head of the humerus, except inferiorly where it encroaches for 1–2 cm on to the neck of the bone.

(b) CAPSULAR THICKENINGS
Anteriorly, the **glenohumeral ligaments**.

(c) ACCESSORY LIGAMENTS
(i) **coracohumeral ligament**—a broad band passing from the base of the coracoid process to the greater tuberosity.

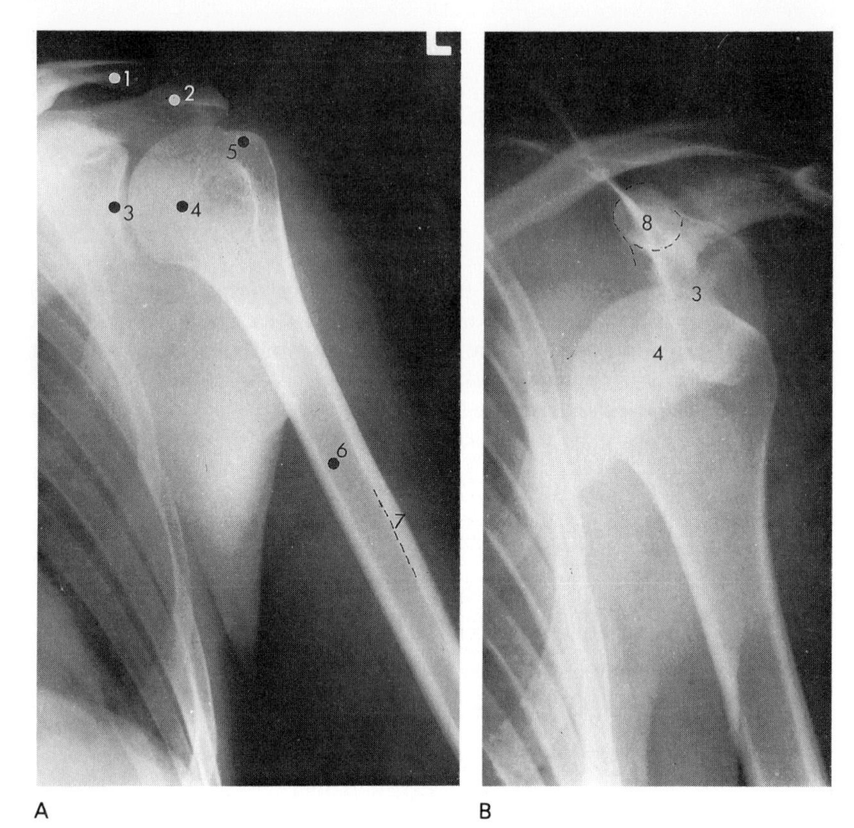

Fig. 16.4 Anteroposterior views of left shoulder. **A.** Normal. **B.** Anterior dislocation of head of humerus which lies in the subcoracoid position.
1. Clavicle 2. Acromion 3. Glenoid process 4. Head of humerus 5. Greater tuberosity 6. Body of humerus 7. Dividing line between cortex and medulla of body 8. Coracoid process

(ii) **transverse humeral ligament**—unites the greater and lesser tuberosities, bridging the intertubercular groove and tying down the long tendon of biceps.
(iii) **glenoidal labrum.**

Intracapsular structures
The synovial membrane lines the nonarticular surfaces of the joint and communicates between the glenohumeral ligaments with the subscapular bursa. It encloses the tendon of the long head of biceps in a tubular sheath as this structure passes across the joint cavity.

Functional aspects

MOVEMENT
The shoulder joint has a wider range of movement than any other joint; it is

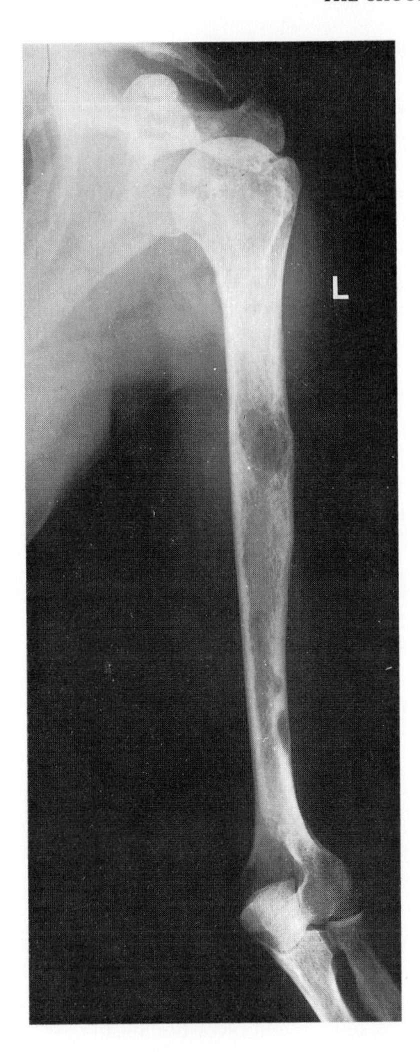

Fig. 16.5 Humerus showing areas of bone destruction in the body indicating bony metastases from a tumour.

capable of flexion, extension, abduction, adduction, circumduction and medial and lateral rotation.

However: (*a*) Associated movement of the shoulder girdle increases the range of these movements. (*b*) The plane of the joint is not sagittal, it passes forwards and medially. Thus anatomical flexion of the limb is accompanied by some abduction at the joint. (*c*) Abduction of the humerus beyond 90° is associated with lateral rotation, thus preventing the greater tuberosity from impinging on the acromion.

FLEXION—clavicular head of pectoralis major, anterior fibres of deltoid and coracobrachialis.

EXTENSION—posterior fibres of deltoid. Extension from a flexed position is reinforced by latissimus dorsi, the sternocostal head of pectoralis major, and teres major.

ABDUCTION—initiated by supraspinatus and continued by deltoid.

ADDUCTION—pectoralis major, latissimus dorsi, subscapularis and teres major.

MEDIAL ROTATION—pectoralis major, anterior fibres of deltoid, latissimus dorsi, teres major and subscapularis.

LATERAL ROTATION—posterior fibres of deltoid, teres minor and infraspinatus.

STABILITY

This joint is more frequently dislocated than any other. The short articular muscles, subscapularis, supraspinatus, infraspinatus, and teres minor, by their close proximity, are the most important factors in maintaining its stability. The absence of muscles inferiorly accounts for the frequency of downward dislocation. The lax capsule here is of little support and the shallow glenoid cavity affords little bony stability. The axillary nerve may be injured. Superiorly the coraco-acromial arch acts as a socket for the head of the humerus and prevents upward dislocation.

Blood supply

Circumflex humeral and suprascapular arteries.

Nerve supply

Axillary and suprascapular nerves.

Relations (Fig. 16.6)

Anterior—subscapularis muscle and bursa; *posterior*—infraspinatus and teres minor; *superior*—supraspinatus, subacromial bursa and coraco-acromial arch; inside the joint is the tendon of the long head of biceps; *inferior*—long head of triceps and axillary nerve. The nerve may be damaged in downward dislocation of the humerus. Deltoid muscle embraces the joint.

Bursae

(i) **subscapular**—lies deep to subscapularis, separating it from the neck of the scapula and shoulder joint. It communicates with the joint cavity.

(ii) **subacromial**—lies above supraspinatus and separates it from the coraco-acromial arch and the deep surface of deltoid.

Movements of the scapula and shoulder girdle

Scapular movements are always associated with movement at the sternoclavicular and acromioclavicular joints and usually with movement at the shoulder joint.

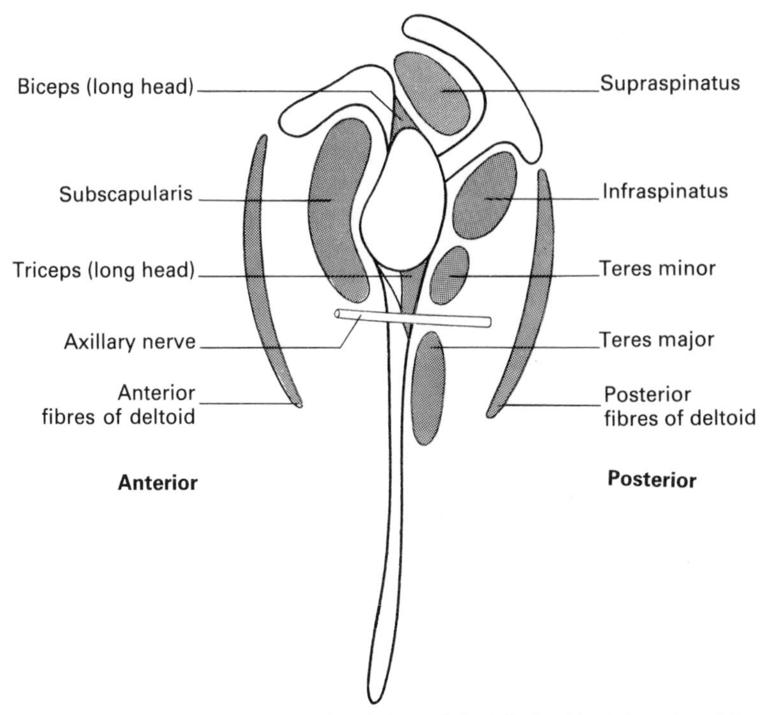

Fig. 16.6 Diagram indicating the main relations of the left shoulder joint, viewed from the lateral aspect.

The bone may be elevated, depressed, moved forwards or retracted on the chest wall and rotated laterally or medially.

(*a*) ELEVATION—upper fibres of trapezius and levator scapulae.

(*b*) DEPRESSION—serratus anterior and pectoralis minor aided by gravity.

During these movements the medial end of the clavicle moves in the opposite direction to the lateral end.

(*c*) FORWARD MOVEMENT ON THE CHEST WALL, as in pushing or punching—serratus anterior and pectoralis minor while latissimus dorsi holds the inferior angle of the bone on the chest wall.

(*d*) RETRACTION, as in bracing the shoulders—trapezius and the rhomboid muscles.

In forward movement and retraction of the scapula, the medial end of the clavicle swings in the opposite direction and there is slight gliding at the acromioclavicular joint.

(*e*) LATERAL ROTATION involves lateral movement of the inferior angle and upward rotation of the glenoid fossa. It is produced by serratus anterior and trapezius.

(*f*) MEDIAL ROTATION—levator scapulae, the rhomboid and pectoralis minor muscles.

Fixation of the humerus and shoulder girdle by placing the elbows on a table allows many of these muscles to act as accessory muscles of respiration during forced inspiration. Their trunk attachments (e.g. the ribs) are approximated to the more fixed limb attachments (e.g. the humerus or scapula) and the capacity of the chest is increased.

MUSCLES CONNECTING THE UPPER LIMB TO THE TRUNK

(Figs. 16.1, 16.2, 16.3, 16.6)

ANTERIOR GROUP

1. Pectoralis major

A powerful muscle arising from the anterior chest wall by two heads. It lies in the anterior axillary wall. Its upper border is separated from deltoid by the deltopectoral groove in which lies the cephalic vein.

Attachments

TRUNK

(i) *clavicular head*—medial half of the anterior surface of the clavicle.

(ii) *sternocostal head*—anterior surface of the sternum, the upper six costal cartilages and the upper part of the external oblique aponeurosis.

HUMERAL

Lateral lip of the intertubercular groove, the clavicular head lying anteriorly.

Action

(i) adduction and medial rotation of the humerus.
(ii) the clavicular fibres flex the humerus.
(iii) the sternocostal fibres extend the humerus from a flexed position, e.g. the follow-through of a tennis player's service. With the arm raised and fixed, this action pulls the trunk upwards.
(iv) an accessory muscle of respiration.

Nerve supply

Lateral and medial pectoral nerves (branches of the lateral and medial cords of the brachial plexus respectively).

2. Pectoralis minor

This muscle lies deep to pectoralis major in the anterior axillary wall and overlaps the middle part of the axillary artery and the brachial plexus.

Attachments

COSTAL—the 3rd, 4th and 5th ribs near their costochondral junctions.

SCAPULAR—the medial border of the coracoid process.

Action

(i) draws the scapula forwards and medially around the chest wall.
(ii) directs the glenoid cavity downwards.
(iii) accessory muscle of respiration.

Nerve supply

Medial pectoral branch of the medial cord of the brachial plexus.

3. Serratus anterior

This covers the medial axillary wall.

Attachments

COSTAL—lateral aspects of the upper eight ribs; the lower slips interdigitate with external oblique muscle.

SCAPULAR—costal surface of the medial border of the scapula; the lower four slips pass to the inferior angle.

Action

(i) pulls the scapula forwards around the chest.
(ii) laterally rotates the scapula, directing the glenoid cavity upwards.
(iii) an accessory muscle of respiration.

Nerve supply

Long thoracic nerve (from the 5th, 6th and 7th cervical nerves).

POSTERIOR GROUP

1. Trapezius

The trapezius is a broad flat triangular muscle lying superficially on the back of the neck and upper trunk.

Attachments

TRUNK—the superior nuchal line of the occipital bone, the ligamentum nuchae, all the thoracic spines and the supraspinous ligaments.

SCAPULAR

(i) the upper fibres run downwards and laterally to the posterior aspect of the lateral third of the clavicle.
(ii) the middle fibres run horizontally to the medial border of the acromion and the superior lip of the scapular spine.
(iii) the lower fibres ascend to the tubercle on the scapular spine.

Action

(i) rotation of the scapula in elevation and controlled descent of the arm.
(ii) retraction of the scapula (as in bracing the shoulders).
(iii) raising of the scapula.
(iv) the lower fibres may assist in depression of the scapula.
(v) one trapezius alone pulls the head backwards and laterally, both together extend the neck and head.

Nerve supply

This is by the spinal accessory nerve.

2. Latissimus dorsi

This is a powerful muscle with an extensive origin.

Attachments

TRUNK

(i) spines and supraspinous ligaments of the lower six thoracic vertebrae deep to trapezius.
(ii) by the thoracolumbar fascia to the lumbar and sacral vertebral spines.
(iii) posterior part of the iliac crest.
(iv) lower four ribs (interdigitating with external oblique).

Its fibres pass upwards and laterally, gain a small slip from the inferior angle of the scapula, and converge on a strap-like tendon which winds around the lower border of teres major to reach its humeral attachment.

HUMERAL

To the floor of the intertubercular groove.

Action

(i) a powerful extensor of the humerus particularly from the flexed position, e.g. the follow-through of a tennis player's service. With the arm fixed and raised above the head, this same movement will pull the trunk upwards.
(ii) adduction and medial rotation of the humerus.
(iii) holds the inferior angle of the scapula against the chest wall.
(iv) an accessory muscle of respiration (in coughing).

Nerve supply

This is by the thoracodorsal branch of the posterior cord of the brachial plexus.

3. Levator scapulae and the rhomboids (major and minor)

These muscles lie on the posterior thoracic wall deep to trapezius. They run downwards and laterally in that order from the vertebral column to the medial border of the scapula. Levator scapulae is attached opposite the supraspinous fossa, rhomboid minor opposite the spine and rhomboid major opposite the infraspinous fossa.

Action
These three muscles act together to elevate and medially rotate the scapula.

Nerve supply
Branches from 3rd to the 5th cervical nerves.

SCAPULOHUMERAL MUSCLES (Figs. 4.1, 4.2, 16.1, 16.3, 16.6).

1. Deltoid
This is a powerful multipennate muscle. Deep to it lie the subacrómial bursa, the short articular muscles around the shoulder joint, the upper end of the humerus and the shoulder joint.

Attachments
PROXIMAL—from the anterior border of the lateral third of the clavicle, the lateral border of the acromion and the lower lip of the scapular spine.
DISTAL—the V-shaped deltoid tuberosity on the lateral aspect of the body of the humerus.

Action
(i) *lateral fibres*—abduction of the humerus.
(ii) *anterior fibres*—medial rotation and flexion of the humerus.
(iii) *posterior fibres*—lateral rotation and extension of the humerus.

Nerve supply
Axillary nerve (from the posterior cord of the brachial plexus).

2. Teres major
This is a thick muscle forming part of the posterior axillary wall.

Attachments
SCAPULAR—lower part of the lateral border and the inferior angle.
HUMERAL—medial lip of the intertubercular groove.

Action
Adduction and medial rotation of the humerus.

Nerve supply
Branches from the posterior cord of the brachial plexus.

3. Short articular muscles (cuff-muscles)
(i) **Subscapularis**
Attachments:
SCAPULAR—the ridged costal surface (separated from the neck of the scapula by the subscapular bursa. The muscle forms part of the posterior axillary wall.
HUMERAL—lesser tuberosity.
It is supplied by nerves from the posterior cord of the brachial plexus.
(ii) **Supraspinatus**—extends from the supraspinous fossa of the scapula to the upper facet on the greater tuberosity of the humerus and
(iii) **Infraspinatus**—extends from the infraspinous fossa of the scapula to the middle facet on the greater tuberosity of the humerus.
Both muscles are supplied by the suprascapular branch of the upper trunk of the brachial plexus.
(iv) **Teres minor**—passes between the upper two-thirds of the lateral border of the scapula and the lower facet on the greater tuberosity of the humerus. It is supplied by the axillary nerve.
The tendons of these four muscles blend closely with the shoulder joint capsule and are important factors in maintaining the joint's stability.

Actions
Subscapularis—medial rotation of the humerus.
Supraspinatus—abduction of the humerus.
Infraspinatus and teres minor—lateral rotation of the humerus.

Injury to the shoulder joint is apt to give rise to spasm of these cuff-muscles, so producing a stiff (frozen) shoulder. The tendon of supraspinatus may degenerate and rupture, so making difficult the initial stages of abduction.

— 17 —

The axilla and upper arm

THE AXILLA

This is a fat-filled space between the lateral thoracic wall and the upper limb. Its shape is that of a truncated pyramid with apex, base and four walls.

Apex
This is bounded by the superior border of the scapula, the outer border of the 1st rib and the middle third of the clavicle. Through it the axilla communicates with the posterior triangle of the neck.

Base
This is formed by axillary fascia, subcutaneous tissue and skin.

Anterior wall
This contains superficially pectoralis major, and deep to this is pectoralis minor enclosed in the clavipectoral fascia.

Posterior wall
This extends lower than the anterior and is composed of subscapularis, latissimus dorsi and teres major, from above downwards.

Medial wall
This comprises the upper ribs and intercostal spaces, these structures being covered by slips of serratus anterior.

Lateral wall
The narrow intertubercular groove of the humerus, into the lips of which, muscles of the anterior and posterior walls are inserted.

Contents
(*a*) axillary artery and vein
(*b*) cords and terminal branches of the brachial plexus
(*c*) coracobrachialis and biceps
(*d*) axillary lymph nodes and lymph vessels
(*e*) fat

THE AXILLARY ARTERY

This is the continuation of the subclavian artery beyond the outer border of the 1st rib and below the middle third of the clavicle. It arches downwards and laterally through the axilla and becomes the brachial artery at the lower border of teres major.

Relations
As the artery passes through the axilla it is surrounded by the cords and branches of the brachial plexus. The axillary vein lies medial to this neurovascular bundle; coracobrachialis muscle and the short head of biceps are lateral. The vessels and nerves are crossed superficially by pectoralis minor and lie on the muscles of the posterior wall of the axilla.

Branches
(i) **superior thoracic artery**—to the upper thoracic wall.
(ii) **acromiothoracic artery**—pierces the clavipectoral fascia and supplies branches to the anterior axillary wall and the front of the shoulder region.
(iii) **lateral thoracic artery**—supplies lateral thoracic wall. (In the female it contributes to the blood supply of the breast.)
(iv) **subscapular artery**—the largest branch of the axillary artery, follows the lower border of subscapularis. Its circumflex scapular branch passes backwards round the lateral border of the scapula into the infraspinous fossa.
(v) **anterior** and **posterior circumflex humeral arteries**—anastomose around the surgical neck of the humerus and supply the shoulder joint and deltoid.

Scapular anastomosis
This provides a collateral circulation to the upper limb from the beginning of the subclavian artery to the last part of the axillary artery by means of anastomotic connections between branches of the thyrocervical trunk (namely the suprascapular and the deep branch of the transverse cervical) proximally, and the subscapular artery and its circumflex branch distally.

THE AXILLARY VEIN

This is the continuation of the brachial vein above the lower border of teres major. It passes upwards and medially, medial to the axillary artery and reaches the outer border of the 1st rib where it becomes the subclavian vein. Most of its tributaries correspond to the branches of the axillary artery. The **cephalic vein**, which it receives, drains the area supplied by the acromiothoracic artery and the superficial structures on the lateral aspect of the limb.

THE BRACHIAL PLEXUS (Fig. 17.1)

This plexus of nerves supplies the upper limb. It is usually formed from the ventral rami of the lower four cervical and the greater part of the first thoracic nerves. These ventral rami are the *five roots* of the plexus and emerge between the anterior and middle scalene muscles in the neck. They unite to form *three trunks* in the posterior triangle of the neck. The upper two roots (C5 & 6) form the upper trunk, the middle root (C7) continues as the middle trunk and the lower two roots (C8 & T1) form the lower trunk. Behind the middle of the clavicle, where the neck and the axilla are in continuity, each trunk divides into *anterior* and *posterior divisions*. The three posterior divisions form the *posterior cord*. The anterior divisions of the upper and middle trunks form the *lateral cord*, and the anterior division of the lower trunk is continued as the *medial cord*. The posterior cord and its branches supply structures on the dorsal (extensor) surface

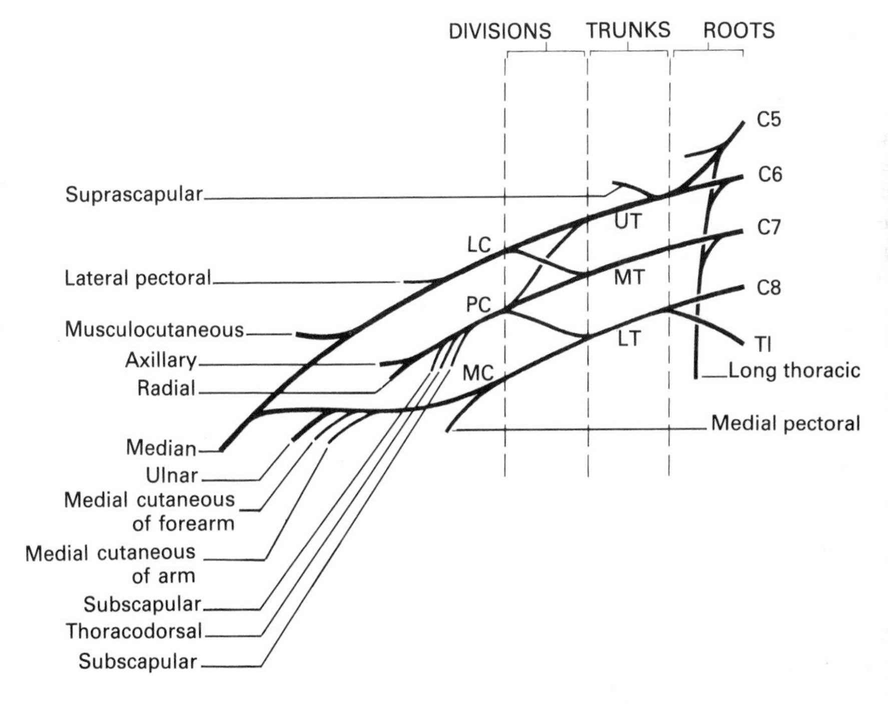

Fig. 17.1 Diagrammatic lay-out of the brachial plexus. UT, MT, LT upper, middle and lower trunks; LC, PC, MC lateral, posterior and medial cords.

of the limb. The medial and lateral cords and their branches supply structures on the ventral (flexor) surface. The three cords pass into the axilla; they are named according to their arrangement around the middle part of the axillary artery. The posterior cord ends on the posterior wall of the axilla by dividing into axillary and radial nerves. The lateral cord divides into the musculocutaneous nerve and the lateral head of the median nerve. The medial cord ends as the ulnar nerve and the medial head of the median nerve (which passes anterior to the lower part of the axillary artery and joins the lateral head).

Branches

1. ROOTS: A number of branches pass to the vertebral and back muscles. The C5 root contributes to the phrenic nerve (see p. 429). The **long thoracic nerve** (C5, 6, 7) passes downwards behind the trunks and vessels, reaches the medial wall of the axilla and supplies serratus anterior.

2. TRUNKS: The **suprascapular nerve** arises from the upper trunk, runs laterally across the root of the neck, passes through the suprascapular foramen and supplies supraspinatus and infraspinatus.

3. DIVISIONS: None.

4. CORDS: Lateral, medial and posterior

LATERAL

(i) **lateral pectoral nerve** supplies pectoralis major.

(ii) **musculocutaneous nerve**—see page 238.

(iii) lateral head of the **median nerve**—see pages 239 and 256

MEDIAL

(i) **medial pectoral nerve**—pierces pectoralis minor and reaches pectoralis major; it supplies both these muscles.

(ii) **ulnar nerve**—the largest branch of the medial cord (see pp. 240 and 257).

(iii) **medial head of the median nerve**—see pages 239 and 256.

(iv) and (v) **medial cutaneous nerves of the arm** and **of the forearm**—pass distally on the medial side of the limb, supplying the skin of this area. The former pierces the deep fascia halfway down the arm, the latter nearer the elbow.

POSTERIOR

(i) **subscapular nerves**—pass downwards on the posterior axillary wall and supply subscapularis and teres major.

(ii) **thoracodorsal nerve** (to latissimus dorsi)—enters the upper border of the muscle.

(iii) **axillary nerve**—passes downwards on subscapularis and then turns posteriorly round the surgical neck of the humerus, accompanied by the posterior circumflex humeral artery. It supplies deltoid and teres minor, the shoulder joint, and ends as the upper lateral cutaneous nerve of the arm.

(iv) **radial nerve**—see pages 239 and 255.

THE UPPER ARM

MUSCLES (Figs. 16.2, 16.3, 18.1)

ANTERIOR GROUP

1. Coracobrachialis
This muscle passes between the tip of the coracoid process of the scapula (medial to short head of biceps) and halfway down the medial side of the humerus. It adducts and flexes the humerus at the shoulder joint.

2. Biceps

Attachments

SCAPULAR
(i) *short head*—the tip of the coracoid process.
(ii) *long head*—the supraglenoid tubercle.

The tendon of the long head lies in the shoulder joint, in the intertubercular groove and in the arm. The synovial sheath which surrounds it inside the joint capsule is prolonged under the transverse humeral ligament into the intertubercular groove. This head unites with the short head in the distal third of the arm.

RADIAL
The posterior part of the radial tuberosity and also by the bicipital aponeurosis into the deep fascia over the medial side of the forearm and the posterior subcutaneous border of the ulna. The tendon passes down through the cubital fossa, lying in front of the elbow joint.

Action
It is a powerful supinator of the forearm and flexor at the elbow and shoulder joints.

3. Brachialis

Attachments
HUMERAL—distal half of the anterior surface of the humerus and the medial intermuscular septum.
ULNAR—anterior surface of the coronoid process.
Action: flexion at the elbow joint.
Nerve supply: the musculocutaneous nerve supplies the three muscles.

POSTERIOR GROUP

Triceps
This is a powerful muscle occupying the entire posterior compartment of the arm.

Attachments

PROXIMAL—by three heads.

(i) long head—the infraglenoid tubercle of the scapula.

(ii) lateral head—a linear attachment from the upper border of the radial groove of the humerus.

(iii) medial head—the whole posterior surface of the humerus below the radial groove and the adjacent medial and lateral intermuscular septa.

The medial head lies deep to the lateral and long heads. A bursa separates the tendon from the posterior aspect of the elbow joint.

DISTAL—by a strong tendon into the olecranon of the ulna.

Action: extension at the elbow joint.

Nerve supply: branches from the radial nerve.

THE BRACHIAL ARTERY

The brachial artery lies on the medial side of the upper arm and begins at the lower border of teres major as the continuation of the axillary artery. It descends in the anterior compartment of the upper arm and ends in the cubital fossa by dividing into radial and ulnar arteries.

Relations

Throughout its course it is covered only by skin and fascia and lies medial to biceps. It is accompanied by veins and the median nerve, the latter crossing over it from lateral to medial halfway down the upper arm. In the cubital fossa the artery lies medial to biceps tendon and is crossed by the medial cubital vein, separated from it only by the bicipital aponeurosis. As this vein is commonly used for venepuncture, this close relation should be specially noted.

Branches

(i) *profunda brachii artery*—accompanies the radial nerve in the radial groove and supplies the posterior muscles of the upper arm and the elbow joint.

(ii) *muscular and articular branches.*

(iii) *nutrient artery*—to the humerus.

THE MUSCULOCUTANEOUS NERVE

This is a terminal branch of the lateral cord of the brachial plexus and descends, deeply placed in the upper arm, to end anterior to the elbow joint as the lateral cutaneous nerve of the forearm.

Relations

In the axilla, it lies lateral to the axillary artery, pierces coracobrachialis and then

lies between biceps and brachialis. It emerges lateral to these muscles and pierc
the deep fascia in front of the elbow.

Branches

(i) muscular—to coracobrachialis, biceps and brachialis.
(ii) articular—to the elbow joint.
(iii) the nerve continues as the lateral cutaneous nerve of the forearm and supplies both surfaces of the radial side of the forearm as far as the thenar eminence.

THE RADIAL NERVE (in the axilla and upper arm. See also p. 255)

The radial nerve arises as a terminal branch of the posterior cord of the brachial plexus behind the axillary artery. It leaves the axilla and takes an oblique course through the posterior compartment of the arm. It pierces the lateral intermuscular septum, passes anterior to the elbow joint and enters the forearm.

Relations

IN THE AXILLA—at its origin it is behind the axillary artery, and lies on the posterior wall of the axilla. It then passes between the long head of triceps and the shaft of the humerus into—

THE POSTERIOR COMPARTMENT OF THE ARM. It runs with the profunda brachii artery in the radial groove of the humerus between medial and lateral heads of triceps. It pierces the lateral intermuscular septum and comes to lie anterior to the elbow joint between brachialis and brachioradialis.

Branches

IN THE AXILLA

(i) muscular—to triceps.
(ii) posterior cutaneous nerve of arm—to skin of posterior and medial aspects of upper arm.

IN THE POSTERIOR COMPARTMENT OF ARM

(i) muscular—to triceps, brachioradialis and extensor carpi radialis longus.
(ii) lower lateral cutaneous nerve of arm—to skin of the lower lateral aspect of arm.
(iii) posterior cutaneous nerve of forearm—to skin of the posterior aspect of forearm.

THE MEDIAN NERVE (in the axilla and upper arm. See also p. 256)

The median nerve arises in the lower axilla by the union of its medial and lateral heads, these being terminal branches of the corresponding cords of the brachial plexus. It descends through the anterior compartment of the upper arm and reaches the cubital fossa.

Relations
The nerve first lies lateral to the axillary and brachial arteries but halfway down the upper arm it passes in front of the latter and reaches its medial side. The nerve is anterior to triceps and brachialis. The median nerve has no branches above the elbow joint.

THE ULNAR NERVE (in the axilla and upper arm. See also p. 257)

This is a terminal branch of the medial cord of the brachial plexus and descends on the medial side of upper arm, initially in the anterior compartment but pierces the medial intermuscular septum and continues in the posterior compartment. It passes behind the medial epicondyle and enters the forearm.

Relations
The nerve lies medial to the axillary and brachial arteries. It leaves the latter by passing backwards and pierces the medial intermuscular septum halfway down the upper arm. It then descends between the medial head of triceps and the septum. It has no branches in the upper arm.

— 18 —

The elbow and forearm

ELBOW REGION

Radius (Fig. 18.1)
This is the lateral bone of the forearm; it has an upper end, body and lower end.

The **upper end** has a head, neck and tuberosity. The **head** is a cupped circular disc which articulates with the capitulum of the humerus and the radial notch of the ulna. On the medial side of the bone is the **radial tuberosity** for the attachment of biceps tendon.

The **body** has a slight lateral convexity and expands from above downwards. The upper part is cylindrical and is crossed by the anterior oblique line which runs from the radial tuberosity downwards and laterally to the point of maximum convexity. The areas of muscle attachments are indicated in Fig. 18.1.

The **lower end** is large and is extended distally into the **styloid process**. The posterior surface has a prominent **dorsal tubercle** around which hooks the tendon of extensor pollicis longus, and less prominent ridges, to which septa from the extensor retinaculum are attached. The medial surface has an ulnar notch for articulation with the head of the ulna. The inferior surface has facets for the scaphoid and lunate bones and medially it gives attachment to the base of an articular disc.

The radial head posteriorly, the lower end anteriorly and posteriorly, and the styloid process are all palpable. A fall on the outstretched hand may produce a fracture of the lower end of the radius in an older person, or a separation of the lower epiphysis in a young person.

Ulna (Fig. 18.1)
This is the medial bone of the forearm; it has an upper end, body and lower end.

The **upper end** has a deep **trochlear notch** anteriorly between two projections; above is the **olecranon** giving attachment to the triceps and below is the **coronoid process** to which brachialis is attached. On its medial side, the flexor digitorum superficialis and pronator teres are attached. The notch articulates with the trochlea of the humerus. On the lateral side of the coronoid process there is a shallow **radial notch** for articulation with the head of the radius; its

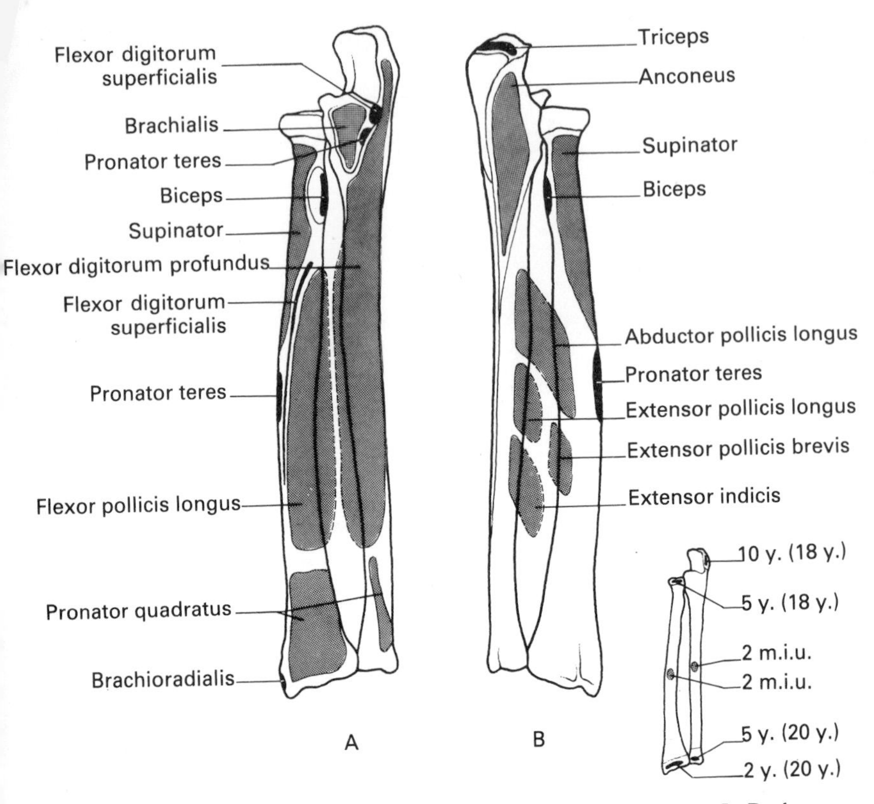

Fig. 18.1 The right radius and ulna: **A.** anterior and **B.** posterior aspects. In **B**, the dotted line of the ulna indicates the attachment of extensor carpi ulnaris. Inset: ossification (m.i.u., months in utero; y., years).

anterior and posterior borders give attachment to the annular ligament. Below the radial notch the bone is hollowed out and gives clearance to the radial tuberosity in pronation and supination. This area, limited posteriorly by the **supinator crest**, is for the attachment of supinator muscle. The medial aspect of the upper end gives attachment to flexor digitorum profundus.

The **body** narrows from above downwards. It has anterior, medial and posterior surfaces but towards the lower end it becomes cylindrical. Its numerous muscle attachments are shown in Figure 18.1. The interosseous border gives attachment to the interosseous membrane.

The **lower end** has a rounded **head** and a medial distal extension—the conical **styloid process**. The extensor carpi ulnaris tendon rests in the posterior groove between the two. The head articulates with the lower end of the radius and with the articular disc which takes origin from the base of the styloid process.

The posterior part of the olecranon, the posterior border of the shaft and lower medial surface are subcutaneous. The ulnar styloid is palpable on the posteromedial aspect of the wrist in the anatomical position and is 1 cm proximal to the radial styloid. In full pronation the anterior surface of the head of the ulna is prominent and palpable.

CUBITAL FOSSA

This triangular fossa is situated in front of the elbow joint.

Base—a line between humeral epicondyles; *lateral border*—brachioradialis; *medial border*—pronator teres; *floor*—brachialis and supinator overlie the capsule of the elbow joint; *roof*—deep fascia of forearm strengthened by the bicipital aponeurosis. Lying on the roof are the anterior branches of medial and lateral cutaneous nerves of the forearm, and the median cubital vein joining cephalic and basilic veins.

CONTENTS: from medial to lateral side, embedded in fat—

(i) median nerve.
(ii) brachial artery and its terminal branches the radial and ulnar arteries (the common interosseous branch of the ulnar artery is also present).
(iii) biceps tendon.
(iv) radial and posterior interosseous nerves (usually overlapped by the brachioradialis).

THE ELBOW JOINT

This is a synovial joint of the hinge variety, between the lower end of the humerus and upper ends of the radius and ulna. The humerus presents a rounded capitulum and a saddle-shaped trochlea for articulation with the cupped head of the radius and the trochlear notch of the ulna respectively. In the anatomical position, the forearm deviates laterally from the arm. This is called the carrying angle.

Ligaments

(a) CAPSULE

This is attached around the upper boundaries of the olecranon, coronoid and radial fossae on the humerus and to its articular margins medially and laterally. Inferiorly it is attached to the olecranon and coronoid process, the medial edge of the trochlear notch and laterally blends with the upper border of the annular ligament of the proximal radio-ulnar joint.

(b) CAPSULAR THICKENINGS

(i) **radial collateral ligament,** strong triangular-shaped with its apex attached to the lateral epicondyle and its base to the annular ligament.

(ii) **ulnar collateral ligament** is in three parts: the anterior and posterior bands radiate from the medial epicondyle to the bases of coronoid process and olecranon respectively, the oblique band joins the latter two attachments.

Intracapsular structures

The synovial membrane lines the inside of the capsule and is continuous with the synovial membrane of the proximal radio-ulnar joint. Fat pads are found between synovial membrane and capsular ligament in the olecranon and coronoid fossae.

Functional aspects

MOVEMENT: flexion and extension occur. (The humeroradial articulation is also involved in the movements of pronation and supination of the forearm.) Flexion—brachialis and biceps assisted by brachioradialis, pronator teres and the common flexor muscles. Extension—triceps muscle.

STABILITY: this is good and is contributed to by the shapes of the bones, the strong capsule and close proximity of brachialis and triceps.

Blood supply

Branches of brachial, radial and ulnar arteries which form a rich anastomosis.

Nerve supply

Median, radial, ulnar, musculocutaneous and posterior interosseous nerves.

Relations

Anterior—the cubital fossa and its contents; *posterior*—triceps; *medial*—the common flexor origin and ulnar nerve; and *lateral*—the common extensor origin and the radial nerve.

THE RADIO-ULNAR JOINTS

Proximal radio-ulnar joint

This is a synovial joint of the pivot variety. The sides of the disc-shaped head of the radius articulate within the osseoligamentous ring formed by the radial notch of the ulna and the annular ligament.

Ligaments

A strong **annular ligament**, which is lined by articular hyaline cartilage, encircles the head of the radius and is attached to the anterior and posterior borders of the radial notch on the ulna (the capsule of the elbow joint is attached to the upper border of the ligament).

Intracapsular structures

The synovial membrane lines the nonarticular surface of the joint and is continuous with that of the elbow joint.

Stability

The annular ligament is narrower below than above and holds firmly the reciprocally shaped head of the adult radius. Dislocation is more common in the child.

Distal radio-ulnar joint

This is a synovial joint of the pivot variety, the head of the ulna articulates with the ulnar notch on the radius. A triangular **fibrocartilaginous disc** lies distal to the ulna and contributes to the articular surface of the wrist joint. Its apex is attached to the base of the ulnar styloid and its base to the medial edge of the distal articular surface of the radius.

Ligaments

The capsular ligament is weak and is attached to the articular margin, except inferiorly where it gains attachment to the margins of the articular disc.

Intracapsular structures

The synovial membrane lines the nonarticular surface of the joint and is not usually continuous with the synovial membrane of the wrist joint.

Stability

This is dependent on the strength of the articular disc.

Interosseous membrane

This is a fibrous membrane connecting the adjacent interosseous borders of the two bones of the forearm, and giving attachment to the deep flexor and extensor muscles. Its fibres are directed downwards and medially. Thus a force passing upwards from the hand through the radius is transmitted to the elbow joint via the ulna as well as through the head of the radius, e.g. leaning on the hand, or punching.

MOVEMENTS AT RADIO-ULNAR JOINTS (Fig. 18.2)

The movements possible at these joints are supination and pronation. The movements occur about an axis joining the centre of the radial head above and the styloid process of the ulna below.

In the anatomical position, the forearm is in the fully **supinated** attitude. While the head of the radius rotates within the annular ligament, its lower end, bearing the hand and tethered by the articular disc, moves ventrally around the head of the ulna. This movement is called **pronation**. After full pronation from the anatomical position the palm faces dorsally. The range of these movements, usually above 180°, is greatly increased by simultaneous rotation of the humerus, provided the elbow joint is fully extended. Both movements are more effective when the elbow is flexed.

SUPINATION—is produced by biceps and supinator.

PRONATION—is produced by pronator teres and pronator quadratus.

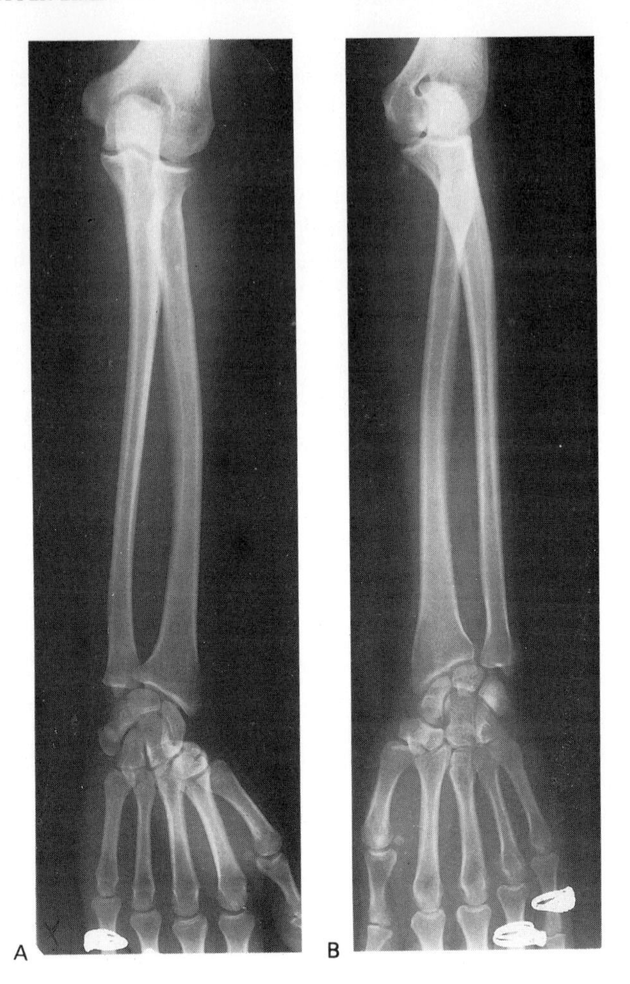

Fig. 18.2 The forearm in the supinated (**A**) and pronated (**B**) positions. Note how the distal end of the radius has rotated over the ulna.

FOREARM

ANTERIOR MUSCLES OF THE FOREARM (Figs. 18.1, 19.1, 19.2)

SUPERFICIAL GROUP

All these muscles are attached to the common flexor origin on the anterior aspect of the medial epicondyle of the humerus and some have extra attachments to the radius and ulna (Fig. 18.1). All are therefore weak flexors at the elbow joint as well as having other actions on the forearm and hand. Flexor carpi ulnaris is supplied by the ulnar nerve, but all the other superficial muscles are supplied by branches of the median nerve.

1. Pronator teres

Attachments

PROXIMAL—by two heads:
(i) common flexor origin.
(ii) medial side of the coronoid process of the ulna.
DISTAL—the middle of the lateral surface of the body of the radius.
Action: pronation of the forearm. The muscle forms the medial border of the cubital fossa, and the median nerve leaves the fossa between the two heads.

2. Flexor carpi radialis

Attachments

PROXIMAL—common flexor origin.
DISTAL—the palmar surface of the base of the 2nd and 3rd metacarpal bones.
Action: flexion and abduction at the wrist joint.

3. Palmaris longus

This extends from the common flexor origin on the humerus to the flexor retinaculum (see p. 266) and the apex of the palmar aponeurosis. It flexes the hand and tightens the palmar aponeurosis. It is a vestigial muscle and is often absent.

4. Flexor digitorum superficialis

This large muscle lies just deep to the other muscles of the group. It arises from the medial epicondyle and both bones of the forearm, and the tendinous arch formed between them covers the ulnar artery and median nerve. The median nerve is adherent to its deep surface.

Attachments

PROXIMAL

(i) the humero-ulnar head from the common flexor origin, the ulnar collateral ligament of the elbow joint and the medial aspect of the coronoid process.
(ii) the radial head from the anterior oblique line on the radius.

DISTAL

Its four tendons pass deep to the flexor retinaculum. Those to the middle and ring fingers lie anterior to the tendons to the index and little finger and so are more liable to be involved in lacerations of the front of the wrist. The tendons diverge in the palm, one passing to each finger. Over the proximal phalanx each tendon splits and encircles the corresponding tendon of flexor digitorum profundus. After partial decussation the tendon is attached to the sides of the middle phalanx.

Action: flexes the middle and then the proximal phalanx; it also flexes the hand at the wrist.

5. Flexor carpi ulnaris

This lies on the medial border of the forearm. The ulnar nerve enters the forearm between its two proximal heads of attachment.

Attachments

PROXIMAL

(i) common flexor origin.
(ii) upper two-thirds of the posterior subcutaneous border of the ulna.

DISTAL

To the pisiform bone and thence into the hamate and the medial aspect of the base of the 5th metacarpal bone.

Action

Flexion and adduction at the wrist joint.

DEEP MUSCLES OF THE FOREARM

1. Flexor pollicis longus

Attachments

PROXIMAL—anterior aspect of the radius between the anterior oblique line and the attachment of pronator quadratus, and the adjacent interosseous membrane.

DISTAL—the tendon passes deep to the flexor retinaculum and is attached to the palmar surface of the base of the distal phalanx of the thumb.

Action: flexes the thumb, especially the distal phalanx.

Nerve supply: branch of the median nerve.

2. Flexor digitorum profundus

Attachments

PROXIMAL—anterior and medial surfaces of the upper three-quarters of the ulna, and the adjacent interosseous membrane.

DISTAL—the four tendons pass under the flexor retinaculum, deep to flexor digitorum superficialis, the tendon for the index finger being separate from the others. Diverging in the palm, each tendon passes through the split tendon of superficialis and gains attachment to the palmar surface of the base of the terminal phalanx of the finger. (The tendons give attachment to the lumbrical muscles in the palm, see p. 269).

Action: flexion of the distal phalanges. It assists in flexion of the middle and proximal phalanges, and of the wrist.

Nerve supply: medial part—ulnar nerve; lateral part—median nerve.

3. Pronator quadratus

This small square muscle runs transversely between the lower quarters of the anterior surfaces of the radius and ulna. It is a pronator of the forearm and is supplied by a branch of the median nerve.

FLEXOR SYNOVIAL SHEATHS (Fig. 18.3)

At their points of maximum friction, beneath the flexor retinaculum and within the digital fibrous flexor sheaths, the long flexor tendons are invested with synovial sheaths. Deep to the flexor retinaculum, the tendons of profundus and superficialis have a common sheath, the **ulnar bursa**, and the tendon of flexor

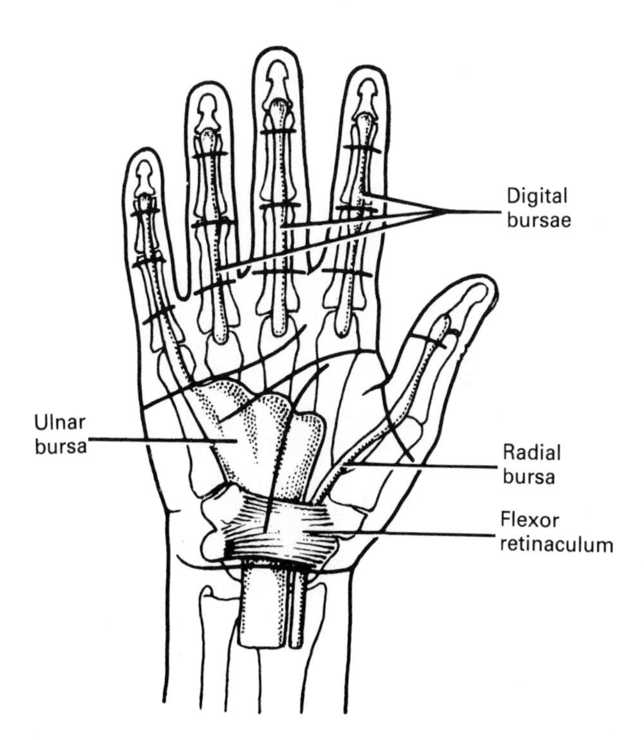

Fig. 18.3 Drawing from a radiograph of wrist and hand on which the skin creases have been accentuated with red lead paste. Note the relations of the creases to the underlying joints. The positions of the radial, ulnar and digital bursae and the flexor retinaculum are also indicated.

pollicis longus also has its own sheath, the **radial bursa**. They begin 2–3 cm above the wrist joint and occasionally communicate with each other. The ulnar bursa ends at the middle of the palm but a medial prolongation around the tendons of the little finger extends as far as the distal phalanx. The radial bursa continues within the fibrous flexor sheath of the thumb as far as the distal phalanx. The tendons of the index, middle and ring fingers have their own digital synovial sheaths extending from the heads of the metacarpals to the bases of the distal phalanges.

POSTERIOR AND LATERAL MUSCLES OF THE FOREARM
(Figs. 18.1, 19.1, 19.2)

SUPERFICIAL GROUP

Two superficial muscles are not attached to common extensor origin and are supplied by the radial nerve as it lies in the upper arm.

1. Brachioradialis

Attachments

PROXIMAL—upper two-thirds of the lateral supracondular ridge of the humerus.

DISTAL—lateral aspect of the lower end of the radius.

Action: flexion at the elbow joint (most strongly in the midprone position). It may be a supinator or a pronator of the forearm according to the position of the hand.

2. Extensor carpi radialis longus

Attachments

PROXIMAL—lower third of the lateral supracondylar ridge on the humerus.

DISTAL—dorsal surface on the base of the 2nd metacarpal bone.

Action: extension and abduction at the wrist joint.

Four muscles are attached to the common extensor origin which is situated on the anterior aspect of the lateral epicondyle of the humerus. These muscles are therefore weak flexors at the elbow joint as well as having actions at more distal joints. All are supplied by the posterior interosseous branch of the radial nerve.

1. Extensor carpi radialis brevis

Attachments

PROXIMAL—common extensor origin.

DISTAL—dorsal aspect of the bases of the 3rd metacarpal bone.

Action: extension and abduction at the wrist joint.

2. Extensor digitorum

Attachments

PROXIMAL—common extensor origin.

DISTAL—just above the wrist the muscle divides into four tendons which pass through a compartment of the extensor retinaculum (see p. 265) within a common synovial sheath. The tendons diverge on the back of the hand, one passing to each finger. The tendons to the middle, ring and little fingers are interconnected by a variable number of bands. As each tendon crosses a metacarpophalangeal joint it forms a triangular **dorsal expansion** which covers the proximal phalanx. The base of this expansion gives attachment to the corresponding interosseous and lumbrical muscles. At the apex of the triangle the long tendon reforms and then divides into three slips over the proximal interphalangeal joint. The middle slip is attached to the base of the middle phalanx, the outer two pass distally and reunite before being attached to the base of the terminal phalanx.

Action

(i) extension at the wrist joint.
(ii) extension of the proximal phalanges of the fingers.
(iii) extension of the middle and distal phalanges of the fingers when assisted by the interossei and lumbrical muscles.

3. Extensor digiti minimi

This passes from the common extensor origin on the humerus to the dorsal expansion of the little finger. It aids the extensor digitorum in its actions on the little finger.

4. Extensor carpi ulnaris

Attachments

PROXIMAL

(i) common extensor origin.
(ii) by an aponeurotic attachment to the posterior subcutaneous border of the ulna.

DISTAL

The tendon passes, in its own compartment of the extensor retinaculum and its own synovial sheath, to the ulnar side of the base of the 5th metacarpal bone.

Action

Extension and adduction at the wrist joint.

DEEP GROUP (Fig. 18.1B)

All these muscles are supplied by the posterior interosseous branch of the radial nerve.

1. Supinator

This muscle hooks round the upper end of the radius from the back and forms part of the floor of the cubital fossa. The posterior interosseous nerve passes between its humeral and ulnar heads of attachment.

Attachments

HUMERO-ULNAR—superficial fibres from the common extensor origin and deep fibres from the supinator area on the ulna.

RADIAL—the ulnar fibres first pass laterally, then anteriorly around the lateral aspect of the upper third of the radius, to be attached to its anterior surface as far medially as the anterior oblique line. The humeral fibres pass downwards to the same area.

Action: supination of the forearm, assisting biceps.

2. Abductor pollicis longus

Attachments

PROXIMAL—the posterior surface of the ulna and of the radius and the intervening interosseous membrane.

DISTAL—the long tendon passes downwards and laterally with the tendon of extensor pollicis brevis. They cross the two radial extensors of the carpus, the lower attachment of brachioradialis and the radial artery before passing, in separate synovial sheaths, under the most lateral compartment of the extensor retinaculum. Here they form the anterior boundary of the anatomical snuff box. The tendon is attached to the radial side of the base of the 1st metacarpal bone.

Action: abduction and extension at the carpometacarpal joint of the thumb (see p. 264).

3. Extensor pollicis brevis

Attachments

PROXIMAL—the posterior surface of the radius and adjacent interosseous membrane, distal to abductor pollicis longus.

DISTAL—the tendon accompanies abductor pollicis longus and is attached on the dorsal surface of the base of the proximal phalanx of the thumb.

Action: extends the proximal phalanx and the metacarpal bone of the thumb.

4. Extensor pollicis longus

Attachments

PROXIMAL—the posterior surface of the ulna and adjacent interosseous membrane distal to abductor pollicis longus.

DISTAL—the tendon passes downwards under the extensor retinaculum, grooving the medial side of the dorsal tubercle of the radius. It forms the posterior boundary of the anatomical snuff box as it crosses laterally over the wrist joint, the two radial extensors of the carpus and the radial artery. It has its own synovial sheath and is attached to the dorsal aspect of the base of the distal phalanx of the thumb.

Action: extends the distal phalanx, the proximal phalanx and the metacarpal bone of the thumb.

5. Extensor indicis

Attachments

PROXIMAL—the posterior surface of the ulna and adjacent interosseous membrane distal to extensor pollicis longus.

DISTAL—the tendon passes under the extensor retinaculum in a common synovial sheath with extensor digitorum; it lies medial to the index tendon of that muscle. It gains attachment to the dorsal expansion of the index finger.

Action: aids extensor digitorum in extending the index finger.

Anatomical snuff box

This is a depression formed on the lateral side of the wrist when the thumb is extended (see p. 264). Bounded anteriorly by the tendons of abductor pollicis longus and extensor pollicis brevis, and posteriorly by extensor pollicis longus, it overlies the radial styloid process, the wrist joint, the scaphoid and trapezium and the base of the 1st metacarpal bone. Crossing the floor are the two tendons of the radial extensors of the wrist and the radial artery. It is crossed superficially by the cephalic vein and branches of the radial nerve.

Extensor retinaculum and extensor synovial sheaths

The extensor retinaculum is a thickened band of deep fascia 2–3 cm in width. It passes obliquely over the dorsum of the wrist from the anterior border of the distal end of the radius to the medial side of the carpus. It is attached by fibrous septa to the bones of the forearm forming osseofascial tunnels for the passage of the tendons (see Fig. 19.5). The synovial sheaths round the tendons begin at the upper border of the retinaculum. Those of abductor pollicis longus and extensor carpi radialis longus and brevis reach the distal attachments of the muscles while the others end at about the middle of the dorsum of the hand.

THE RADIAL ARTERY

The radial artery is a terminal branch of the brachial artery and is formed near the apex of the cubital fossa. It descends through the anterior compartment of the forearm, over the lateral aspect of the wrist and on to the dorsum of the hand. It then passes between the heads of the first dorsal interosseous muscle and ends in the palm.

Relations
It leaves the cubital fossa and passes deep to the brachioradialis which covers its upper part. It descends near the regions of attachment of the forearm muscles to the radius (see Fig. 18.1A). In the lower forearm it is subcutaneously situated on the lower end of the radius. The radial nerve lies on its lateral side in the middle of the forearm.

The artery crosses the lateral aspect of the wrist joint, passes deep to the three tendons bordering the anatomical snuff box and on to the dorsum of the hand. It passes between the two heads of the first dorsal interosseous muscle and then the two heads of adductor pollicis and forms the deep palmar arterial arch.

Branches
(i) muscular—to muscles on the radial side of the forearm.
(ii) articular—to elbow and wrist joints.
(iii) superficial palmar—when present, completes the superficial palmar arch (p. 270).
(iv) to the thumb—arises deep to adductor pollicis and divides, giving palmar digital branches to each side of the thumb.
(v) to the index finger—arises deep to adductor pollicis and supplies the radial side of the index finger.
(vi) deep palmar arch (see p. 270).

THE ULNAR ARTERY

It is a terminal branch of the brachial artery. It arises near the apex of the cubital fossa, descends through the anterior compartment of the forearm, crosses anterior to the wrist and ends in the hand as superficial and deep palmar branches.

Relations (in the forearm)
After leaving the cubital fossa, it lies on flexor digitorum profundus and deep to the muscles attached to the common flexor origin. In the lower forearm, under flexor carpi ulnaris, the ulnar nerve lies on its medial side. At the wrist the artery lies between the tendons of flexor carpi ulnaris and flexor digitorum profundus, crosses anterior to the flexor retinaculum and ends on the lateral side of the pisiform bone.

Branches
(i) *muscular*—to muscles on the ulnar side of the forearm.
(ii) *articular*—to the elbow and wrist joints.
(iii) **common interosseous artery**—arises in the cubital fossa, passes to the upper border of the interosseous membrane and divides into an anterior and a smaller posterior interosseous artery. The **anterior** branch descends on the inter-

osseous membrane but pierces it above the wrist to end on the back of the carpus. It gives nutrient arteries to the radius and ulna, muscular branches and articular branches to the wrist. The **posterior interosseous artery** passes back above the interosseous membrane and then descends, supplying the deep extensor muscles of the forearm.

(iv) *superficial palmar branch*—passes laterally over the hook of the hamate, and deep to the palmar aponeurosis it forms the superficial palmar arch (p. 270).

(v) *deep palmar branch*—passes deeply between flexor and abductor digiti minimi on to the opponens digiti minimi, where it joins the deep palmar arch (p. 270).

THE RADIAL NERVE (in the forearm and hand)

This lies on the anterolateral aspect of the forearm overlapped by brachioradialis. Proximal to the wrist it turns posteriorly under the tendon of brachioradialis and ends on the dorsum of the hand as digital branches.

Relations

IN THE FOREARM

SUPERFICIAL—brachioradialis and its tendon.
DEEP—lies on the deep forearm muscles.
MEDIAL—radial artery throughout most of its length.

IN THE HAND

It reaches the dorsum of the hand by passing between the radius and brachioradialis tendon, crosses superficial to the tendons bounding the anatomical snuffbox and divides into terminal digital branches.

Branches

IN THE FOREARM

(i) articular—to elbow joint.
(ii) posterior interosseous nerve (see below).

IN THE HAND

Digital branches supply the lateral side of the dorsum of the hand and the posterior aspects of the lateral 2½ digits usually as far as the distal interphalangeal joints.

Posterior interosseous nerve

This arises under cover of brachioradialis at the level of the elbow joint, pierces supinator and in its substance curves laterally around the neck of the radius into the extensor compartment of the forearm. It reaches the interosseous membrane and ends on the back of the carpus.

Relations

SUPERFICIAL—muscles arising from common extensor origin and more distally, extensor pollicis longus and extensor indicis.

DEEP—the deepest muscles of the forearm, the interosseous membrane, wrist joint and posterior aspect of carpus.

Branches

(i) muscular—to all extensors except extensor carpi radialis longus.
(ii) articular—to elbow, wrist and intercarpal joints.

THE MEDIAN NERVE (in the forearm and hand)

The nerve lies in the anterior compartment of the forearm, crosses anterior to the wrist joint and ends in the palm as medial and lateral branches.

Relations

In the cubital fossa the nerve is medial to the brachial and ulnar arteries but as it passes between the heads of pronator teres it crosses in front of the ulnar artery, separated from it by the deep head of the muscle. The nerve then lies between flexor digitorum superficialis and flexor digitorum profundus, adherent to the former. It emerges lateral to superficialis and lies between the tendons of flexor carpi radialis and superficialis and under palmaris longus at the wrist. It enters the hand by passing deep to the flexor retinaculum, lying on the synovial sheath of the long flexor tendons and ends just distal to the retinaculum.

Branches

IN THE FOREARM

(i) *muscular*—to all superficial flexor muscles except flexor carpi ulnaris.

(ii) **anterior interosseous nerve**—arises after the median nerve has pierced pronator teres. It passes distally on the interosseous membrane between flexor pollicis longus and flexor digitorum profundus and then deep to pronator quadratus, ending on the front of the carpus. It supplies flexor pollicis longus, the radial half of flexor digitorum profundus, pronator quadratus and gives articular branches to the distal radio-ulnar joint and the joints at the wrist.

(iii) *palmar cutaneous nerve*—supplies the radial side of the palm.

IN THE HAND

(i) recurrent muscular branch to the abductor pollicis brevis, flexor pollicis brevis and opponens pollicis (lateral branch).

(ii) palmar digital nerves which pass to the palmar surface of the lateral $3\frac{1}{2}$ digits, and also supply the lateral two lumbrical muscles. The palmar digital nerves lie deep to the superficial palmar arch but cross the digital arteries to lie anterior to them in the fingers and thumb. These nerves innervate the finger tips and nail beds, and also give articular branches to the metacarpophalangeal and interphalangeal joints.

As it lies in the fibro-osseous tunnel formed by the flexor retinaculum and bones of the carpus the median nerve is occasionally subject to compression. This can produce numbness and pain in the thumb and lateral fingers and weakness of the hand muscles it supplies.

THE ULNAR NERVE (in the forearm and hand)

From behind the medial epicondyle the nerve descends deep to the flexor carpi ulnaris on the medial side of the forearm. It enters the hand lateral to pisiform bone and ends by dividing into superficial and deep branches.

Relations

It lies medial to the elbow joint, enters the forearm between the two heads of the flexor carpi ulnaris and continues deep to this muscle on flexor digitorum profundus. At the wrist it lies superficially, lateral to the tendon of flexor carpi ulnaris, and passes anterior to the flexor retinaculum. The ulnar artery lies on its lateral side in the lower forearm and hand.

Branches

IN THE FOREARM

(i) *articular*—to the elbow joint.

(ii) *muscular*—to flexor carpi ulnaris and the ulnar half of the flexor digitorum profundus.

(iii) *dorsal cutaneous branch*—arises 5 cm above the wrist and passes on to the back of the hand where it divides into digital branches which supply the ulnar half of the dorsum of the hand and the medial 2½ fingers usually as far as the distal interphalangeal joints.

(iv) *palmar cutaneous branch*—supplies the ulnar side of the palm.

IN THE HAND

(i) *superficial terminal branch*—divides into palmar digital nerves supplying the palmar surface of the medial 1½ fingers. The digital nerves also supply the finger tips, nail beds and the joints of the fingers.

(ii) *deep terminal branch*—accompanies the deep branch of the ulnar artery between abductor and flexor digiti minimi and then deep to the flexor tendons. It supplies the hypothenar muscles, all the interossei, 3rd and 4th lumbricals and adductor pollicis.

— 19 —

The wrist and hand

SKELETON

Carpus (Figs. 18.3, 19.1, 19.2)
The carpal bones are arranged as a proximal row of three bones, a distal row of four and an anteromedially situated pisiform. The proximal row is made up, from lateral to medial, of scaphoid, lunate and triquetral; collectively they form the distal articular surface of the wrist joint. The **scaphoid** has an anterior **tubercle** to which the flexor retinaculum is attached. The **lunate** has no muscular attachments. The **triquetral**, as well as articulating with the lunate and hamate, has an articular surface anteriorly for the pisiform.

The distal row has the ridged **trapezium** laterally, the small **trapezoid**, the large **capitate** and then the medially situated **hamate** completes the row. The trapezium gives attachment to the flexor retinaculum; in its groove lies the tendon of flexor carpi radialis. The large capitate has a rounded head proximally which articulates with the proximal carpals; its distal surface articulates with the 2nd, 3rd and 4th metacarpal bones. The hamate has an anterior hook which gives attachment to the flexor retinaculum. The **pisiform** is pea-shaped and articulates with the triquetral. It gives attachment to the flexor and extensor retinacula, and flexor carpi ulnaris. (It may be described as a sesamoid bone.)

The lateral surface of the scaphoid, the medial surface of the triquetral and the four bony points of attachment of the flexor retinaculum are all palpable. The scaphoid may be fractured in a fall on the outstretched hand, particularly in the younger age group. There is a high incidence of nonunion of this fracture because the blood supply to the bone is often impaired by such an injury.

Metacarpal bones (Figs. 19.1, 19.2)
These are five long bones. Their expanded bases articulate with the distal row of carpals and, in the case of the medial four, with each other. The slender bodies give attachment to the interossei, opponens and adductor pollicis muscles. The rounded heads articulate with the proximal phalanges. The first metacarpal is the shortest, strongest and most mobile. Its orientation differs from the others, being rotated almost through a right angle. Its base has a saddle-shaped surface for articulation with the trapezium.

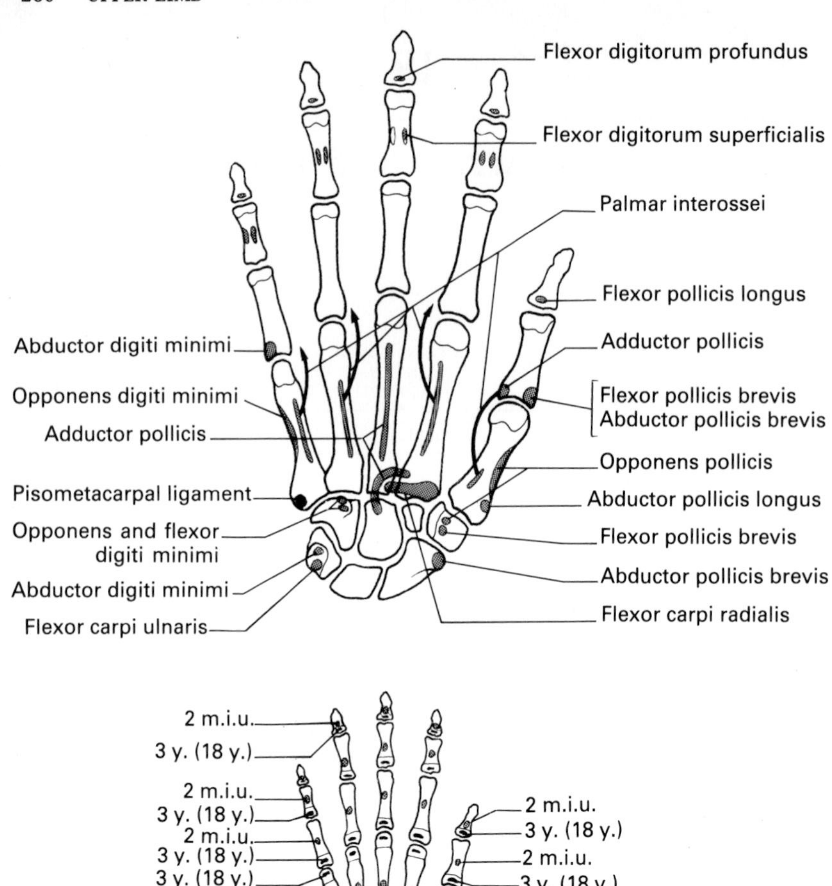

Fig. 19.1 The bones of the right wrist and hand, palmar surface. The attachments of the palmar interossei are indicated by the bold arrows. Inset: ossification (m.i.u., months in utero; y., years).

Phalanges (Figs. 19.1, 19.2)

These are long bones, two in the thumb, three in the medial four fingers. The bases of the proximal phalanges have cupped articular surfaces while those of the middle and distal have grooved trochlear surfaces. The bodies of the proximal and middle phalanges are flattened on their palmar surfaces and rounded dorsally; their heads have paired condyles. The distal phalanges taper distally, where they have anterior tuberosities.

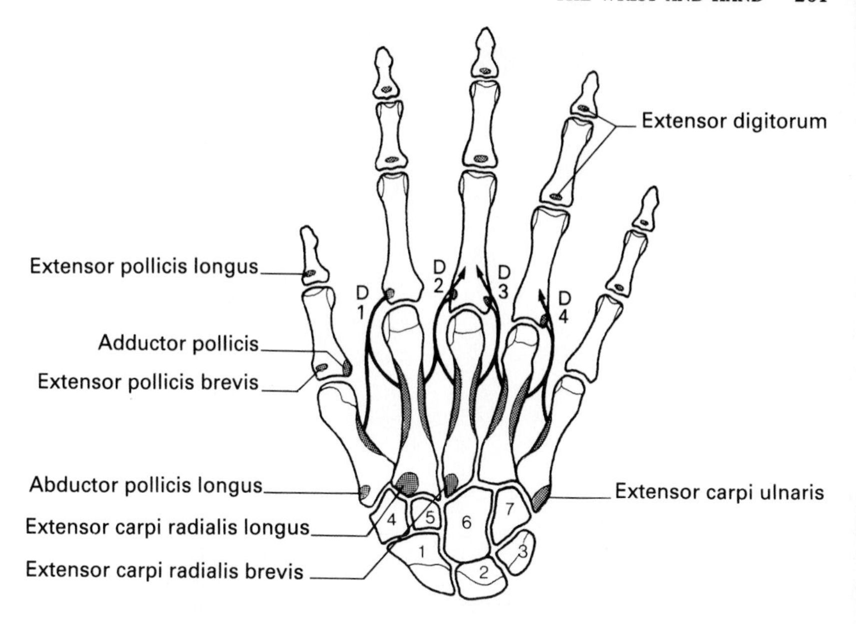

Fig. 19.2 The bones of the right wrist and hand, dorsal surface. The attachments of the dorsal interossei are indicated by the bold arrows (D1 to D4). The carpal bones are numbered: 1. Scaphoid 2. Lunate 3. Triquetrum 4. Trapezium 5. Trapezoid 6. Capitate 7. Hamate. The pisiform is on the palmar surface of the triquetrum (see Fig. 19.1). Distally are the metacarpal bones and the phalanges.

The phalanges give attachments to all the small muscles of the hand except the opponens muscles, and to the long flexor and extensor tendons of the thumb and fingers.

WRIST REGION (radiocarpal, intercarpal and carpometacarpal joints)

THE RADIOCARPAL (WRIST) JOINT

This is a synovial joint of the ellipsoid variety. The faceted lower end of the radius and fibrocartilaginous disc overlying the head of the ulna articulate with the rounded surfaces of the proximal row of carpal bones, the scaphoid, lunate and triquetral.

Ligaments

(*a*) CAPSULE: attached to the articular margins of the joint.

(*b*) CAPSULAR THICKENINGS: anterior and posterior ligaments and medial and lateral collateral ligaments.

Intracapsular structures

The synovial membrane lines the nonarticular surfaces of the joint. The triangular articular disc overlies the head of the ulna, its apex is attached to the base

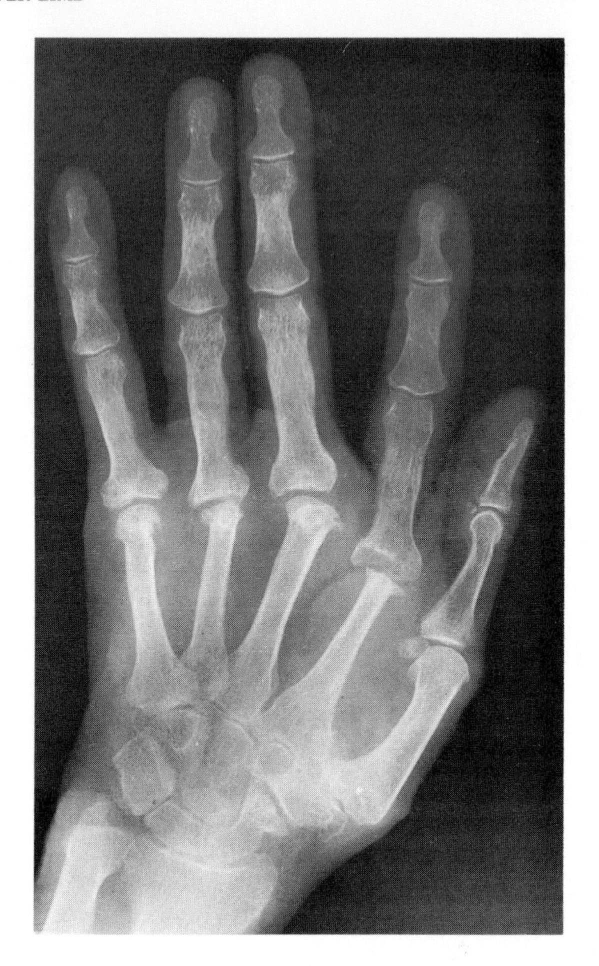

Fig. 19.3 Radiograph of the hand of a patient with severe rheumatoid arthritis. Note that the heads of the metacarpal bones have been eroded away. Compare this picture with Figure 1.1. Note also the different densities of the carpal bones in the two X-rays. Decalcification has also occurred in the phalanges, making the trabeculae more prominent.

of the ulnar styloid and its base to the lower end of the radius. Only on the rare occasions when this disc is perforated does the wrist joint communicate with the distal radio-ulnar joint.

The intercarpal joints

These are synovial joints. The extensive composite joint between the proximal and distal rows of carpal bones is called the midcarpal joint; it is of the condyloid variety. Other intercarpal joints are of the plane variety. The bones are bound

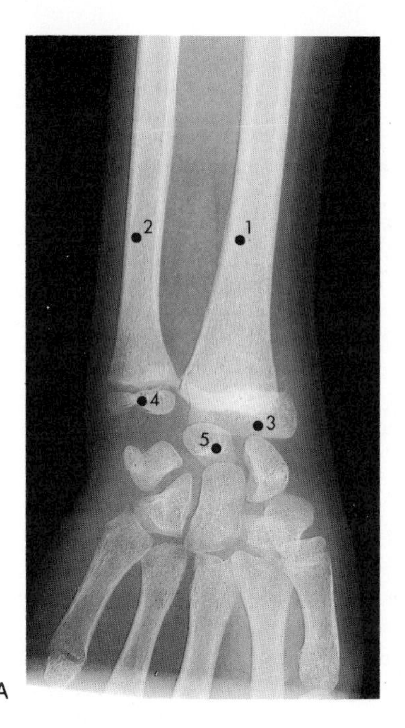

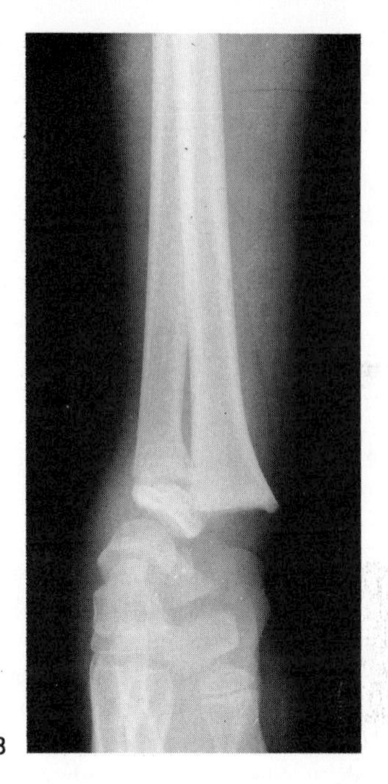

Fig. 19.4 Postero-anterior (**A**) and lateral (**B**) views of a child's wrist. Note that the epiphysis of the distal end of the radius has been dislocated posteriorly, the carpus and hand having moved backwards with the dislocated epiphysis. This does not show up clearly on the PA view.
1. Radius 2. Ulna 3. Radial epiphysis 4. Ulnar epiphysis 5. Lunate bone

together by palmar, dorsal and interosseous ligaments. The intercarpal joints greatly increase the mobility of the hand, the midcarpal having particular effect on the movements of flexion and abduction at the wrist.

The carpometacarpal joints of the fingers

These are synovial joints of the plane variety. The 2nd and 3rd are less mobile than the 4th and 5th and they are all less mobile than the joint of the thumb.

Functional aspects

MOVEMENT AT THE WRIST

The movements possible are flexion, extension, abduction, adduction and circumduction. Flexion is greater than extension and adduction greater than abduction. Since these movements take place at the radiocarpal, midcarpal and carpometacarpal joints simultaneously they are considered here together. The full analysis of these movements is dependent on radiological examination and detailed measurements.

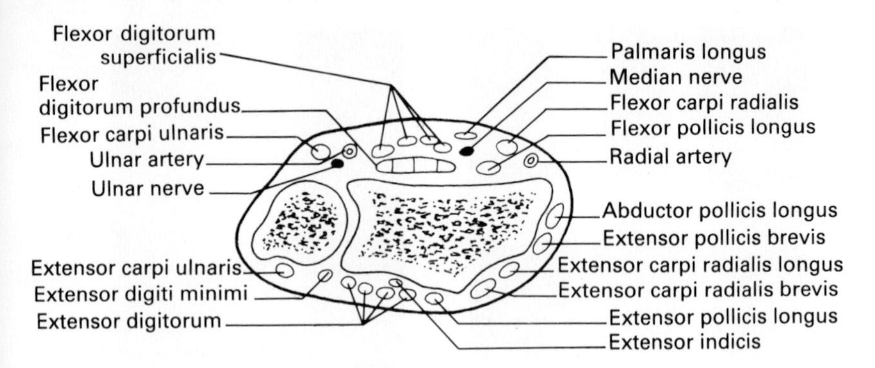

Fig. 19.5 Diagram showing principal structures seen in a transverse section of the forearm just proximal to the wrist joint.

FLEXION (mainly at the midcarpal joint)—flexor carpi radialis and flexor carpi ulnaris, assisted by the long digital flexors.

EXTENSION (mainly at radiocarpal joint)—extensor carpi radialis longus and brevis and extensor carpi ulnaris, assisted by the long digital extensors.

ABDUCTION (mainly at the midcarpal joint)—extensor carpi radialis longus and brevis and flexor carpi radialis.

ADDUCTION (mainly at the radiocarpal joint)—extensor carpi ulnaris and flexor carpi ulnaris.

STABILITY OF WRIST

This is mainly dependent on its ligaments, assisted by the many tendons crossing it and very few dislocations occur. (See Fig. 19.4)

The carpometacarpal joint of the thumb

This is a synovial joint possessing considerable mobility by virtue of its saddle-shaped articulation and lax capsule. The thumb is medially rotated at the carpometacarpal joint and therefore the terms used to describe its movements do not have their usual meaning. Flexion and extension occur in a plane at right angles to the thumb nail, and abduction and adduction in the same plane as the nail. Flexion, extension, abduction, adduction, circumduction and rotation are all possible. Because of the shapes of the bony surfaces flexion is always accompanied by medial rotation and extension by lateral rotation. The combined movement of flexion, medial rotation and adduction brings the thumb into contact with the other finger tips and is known as **opposition**.

FLEXION AND MEDIAL ROTATION—flexor pollicis longus and brevis and opponens pollicis.

EXTENSION AND LATERAL ROTATION—abductor pollicis longus and extensor pollicis longus and brevis.

ABDUCTION—abductor pollicis longus and brevis.

ADDUCTION—adductor pollicis.
OPPOSITION—opponens pollicis.

Blood supply: branches from the radial and ulnar arteries.
Nerve supply: anterior and posterior interosseous nerves.
Relations: a knowledge of the relations of the wrist joint assists greatly in understanding the anatomy of the forearm. The student should therefore study the diagram, Figure 19.5. In the superficial fascia in front and behind are prominent veins and the cutaneous nerves. A terminal branch of the radial nerve can be detected as it crosses the tendon of extensor pollicis longus.

The deep fascia dorsally is thickened above the wrist joint and forms the **extensor retinaculum** which prevents the long tendons 'bowstringing'. The retinaculum is firmly attached to the ulna and the radius. Fibrous septa pass between the extensor tendons with their surrounding synovial sheaths and are attached to the underlying bones. No large arteries are found dorsally. The tendons bounding the anatomical snuffbox (see p. 253) are clearly seen on the lateral side when the thumb is extended. The tendons of the extensors of the wrist and fingers are difficult to identify. The fascia of the ventral surface thickens and becomes continuous distally with the **flexor retinaculum** which crosses the front of the wrist (see p. 266). The radial artery is the most lateral structure as it lies on the radius and is easily palpated. Medial to the artery are the tendons of flexor carpi radialis and then palmaris longus. The median nerve lies deep to palmaris longus with flexor carpi radialis laterally and flexor digitorum superficialis medially. More medially is the tendon of flexor carpi ulnaris which passes to the pisiform bone and overlaps the ulnar artery and nerve (the nerve is medial to the artery). Lying in a deeper plane centrally are the tendons of flexor digitorum profundus and laterally the tendon of flexor pollicis longus.

The metacarpophalangeal joints

These are synovial joints of the ellipsoid variety. The rounded heads of the metacarpals articulate with the cupped bases of the proximal phalanges. The palmar thickenings of the joint capsules are strong and those of the medial four digits are joined together to form the deep transverse ligaments of the palm.

The movements possible are flexion, extension, abduction, adduction and circumduction. Active thumb movements are limited to flexion and extension.

FLEXION—long digital flexor muscles and flexor pollicis longus, assisted by the interossei, lumbricals, flexor pollicis brevis and flexor digiti minimi.

EXTENSION—in the fingers, extensor digitorum, extensor indicis and extensor digiti minimi; in the thumb, extensor pollicis longus and brevis.

ABDUCTION—abductor pollicis longus and brevis in the thumb and the dorsal interossei and abductor digiti minimi in the fingers.

ADDUCTION—palmar interossei and adductor pollicis.

The interphalangeal joints

These are synovial hinge joints. Their joint capsules are thickened by palmar and collateral ligaments and completed posteriorly by the extensor tendons.

Flexion at the joints is brought about by the long digital flexor muscles and extension mainly by the lumbrical and interosseous muscles.

The metacarpophalangeal and interphalangeal joints are supplied by adjacent nerves and vessels.

THE PALM

To facilitate grasping, the skin of the palmar surface of the hand is thick, ridged and has no hair or sebaceous glands. There are, however, many sweat glands. It has constant creases where it is bound to underlying fibrous tissue.

Palmar aponeurosis

The deep fascia of the palm consists of a central, strong, triangular part, the palmar aponeurosis, and weaker portions covering the hypothenar and thenar muscles. The palmar aponeurosis is firmly connected to the palmar skin and overlies the superficial palmar arch and the long flexor tendons. Its apex is continuous with the flexor retinaculum and receives the attachment of palmaris longus. Distally its base divides into four digital slips which bifurcate around the long flexor tendons and gain attachment to the fibrous flexor digital sheaths and the deep transverse ligaments of the palm.

Flexor retinaculum

The flexor retinaculum is a thickening of deep fascia, 2–3 cm square. Medially it is attached to the pisiform and the hook of the hamate, and laterally to the tubercle of the scaphoid and the ridge of the trapezium, crossing the concavity of the carpus to form an osseofascial **carpal tunnel** for the long flexor tendons. Superficially it gives origin to the muscles of the thenar and hypothenar eminences and is crossed by the ulnar nerve and artery, palmar branches of the ulnar and median nerves and the tendon of palmaris longus. It covers the median nerve, the long flexor tendons, and the radial and ulnar bursae.

Fibrous flexor sheaths

The deep fascia of the fingers is modified to form fibrous flexor sheaths which are in continuity with the digital slips of the palmar aponeurosis. They form osseofascial tunnels by arching over the tendons and gaining attachment to the sides of the phalanges. Each extends to the base of the distal phalanx and is weaker and more flexible in front of the joints. A similar sheath is present in the thumb. The continuity of the flexor retinaculum, the palmar aponeurosis and the fibrous flexor sheaths prevents the long flexor tendons 'bowstringing' across the palm during contraction. This would interfere with the power and efficiency of the grip.

THE PALMAR SPACES

Two septa pass deeply from the lateral and medial margins of the palmar aponeurosis to the shafts of the 1st and 5th metacarpal bones and divide the palm into three spaces:

(i) lateral, containing the thenar muscles.
(ii) medial, containing the hypothenar muscles.
(iii) central, containing the superficial palmar arch, median nerve, long flexor tendons and lumbricals, and the deep palmar arch.

Deep to the flexor tendons the central palmar space may be further divided by an intermediate palmar septum which extends from the deep surface of the flexor tendon sheaths to the palmar surface of the 3rd metacarpal. Through these palmar spaces the web-spaces of the fingers communicate with a fascial plane deep to the long flexor tendons in the forearm. Because of this communication infection in the fingers may spread to the forearm.

MUSCLES OF THE THENAR AND HYPOTHENAR EMINENCES

They are attached proximally to the radial and ulnar sides respectively of the flexor retinaculum and its supports (Figs. 19.1, 19.2).

THENAR MUSCLES

These are all supplied by a recurrent branch of the median nerve.

1. Abductor pollicis brevis

This is superficial to the other two muscles and therefore attached proximal to them.

Attachments

PROXIMAL—the tubercle of the scaphoid and the adjacent flexor retinaculum.

DISTAL—the radial side of the base of the proximal phalanx of the thumb alongside the flexor pollicis brevis.

Action: abduction of the thumb.

2. Opponens pollicis

Attachments

PROXIMAL—the ridge of the trapezium and adjacent flexor retinaculum.

DISTAL—the whole length of the radial margin of the 1st metacarpal bone.

Action: flexion, adduction and medial rotation of the 1st metacarpal bone.

This combined movement brings the thumb into opposition with the tips of the flexed fingers.

3. Flexor pollicis brevis

This is separated with difficulty from opponens pollicis.

Attachments

PROXIMAL—ridge of the trapezium and adjacent flexor retinaculum.

DISTAL—radial side of the base of the proximal phalanx of the thumb. The tendon contains a sesamoid bone and may be joined by a small tendon from adductor pollicis.

Action: flexion of the proximal phalanx and metacarpal bone of the thumb.

HYPOTHENAR MUSCLES

These are, in most respects, weaker mirror-images of the thenar muscles. They are all supplied by the deep branch of the ulnar nerve.

1. Abductor digiti minimi

Attachments

PROXIMAL—the pisiform bone.

DISTAL—the ulnar side of the base of the proximal phalanx of the little finger.

Action: abduction of the little finger.

2. Opponens digiti minimi

Attachments

PROXIMAL—the hook of the hamate and adjacent flexor retinaculum.

DISTAL—the whole length of the ulnar margin of the 5th metacarpal bone.

Action: flexion and a little rotation of the 5th metacarpal. It thus deepens the hollow of the palm.

3. Flexor digiti minimi

Attachments

PROXIMAL—the hook of the hamate and the adjacent flexor retinaculum.

DISTAL—the ulnar side of the base of the proximal phalanx of the little finger.

Action: flexion of the proximal phalanx and metacarpal bone of the little finger.

DEEP MUSCLES OF THE HAND (Figs. 19.1, 19.2)

1. Adductor pollicis

Attachments

LATERAL—into the ulnar side of the base of the proximal phalanx of the thumb and by a small tendon into the radial side of the base of the proximal phalanx. Both tendons contain a sesamoid bone (Fig. 1.1).

MEDIAL—by two heads:

(i) oblique head—the palmar surface of the bases of the 2nd and 3rd metacarpals and adjacent carpal bones.
(ii) transverse head—the palmar surface of the body of the 3rd metacarpal.

Action: adducts and assists in opposition of the thumb.

2. Interossei

The interossei lie deeply between the metacarpals; there are four dorsal and four palmar muscles. Each group is numbered one to four from the radial side.

Attachments

PROXIMAL—these muscles are attached to the shafts of the metacarpal bones, the palmar muscles by a single head to the palmar surfaces of the 1st, 2nd, 4th and 5th bones. The dorsal muscles are larger and more powerful and arise by two heads from adjacent sides of the metacarpal bones.

DISTAL—the muscles of both groups gain attachment to the base of the corresponding proximal phalanx and the extensor expansion.

Action: the attachments are such that the Palmar ADduct (PAD) and the Dorsal ABduct (DAB) the fingers about an axial line passing along the middle finger (see Figs. 19.1, 19.2). Both groups, acting with the lumbricals, flex the proximal phalanx, and through their attachments to the expansion of the extensor tendons help to extend the middle and distal phalanges.

3. Lumbricals

These four slender muscles arise in the palm from the radial side of each of the four tendons of flexor digitorum profundus. Distally each is attached to the radial side of the dorsal extensor expansion of the same finger.

Action

Acting with the interossei they flex the proximal phalanx and help to extend the middle and distal phalanges.

Nerve supply

The adductor pollicis and the interossei are supplied by the deep branch of the ulnar nerve. The lateral two lumbricals are supplied by the median nerve and the medial two by the ulnar nerve.

ARTERIES

Superficial palmar arch

This is a continuation of the ulnar artery in the hand. It is completed laterally by a branch of the radial artery. The summit of its convexity is at the level of the web of the thumb. The arch is deep to the palmar aponeurosis and lies on the long flexor tendons and the digital nerves.

Branches

Four palmar digital branches—three supply the clefts between, and then by bifurcating, the adjacent sides of the fingers. The medial digital branch supplies the ulnar side of the little finger. All the branches communicate with the deep palmar arch.

Deep palmar arch

This is a continuation of the radial artery in the hand. It is completed medially by the deep branch of the ulnar artery. It lies 1 cm proximal to the superficial arch on the metacarpal bones and interossei muscles, and deep to the long flexor tendons. The deep branch of the ulnar nerve lies within its convexity.

Branches

Three palmar metacarpal arteries (which anastomose with the palmar digital arteries), and perforating arteries to the dorsum of the hand.

The hand is an important sensory organ as well as a manipulator. Coarse grips (lifting a weight) and fine grips (writing) are enhanced by the very large areas of the brain which are concerned with the motor control of the hand in man.

— 20 —

The fascia, veins, lymph drainage and nerves

DEEP FASCIA

In the shoulder girdle region the deep fascia invests all the muscles and is drawn out as a tubular prolongation over the arm. The axillary floor is formed from this fascia. The fascia invests pectoralis minor and extends above to the clavicle (forming the clavipectoral fascia). In the arm the deep fascia invests the muscles and is condensed medially and laterally to form the medial and lateral intermuscular septa. Inferiorly these are attached to the respective supracondylar ridges of the humerus.

In the elbow region the deep fascia firmly ensheathes the muscles and provides an attachment for many of their fibres. It is reinforced by the bicipital aponeurosis. The deep fascia of the forearm is attached to the ulna posteriorly along its subcutaneous border. At the wrist it forms the flexor and extensor retinacula, and in the hand the palmar aponeurosis and fibrous flexor sheaths. The fascial spaces of the hand are described on page 267.

VENOUS DRAINAGE

The veins can be divided into superficial and deep groups with communicating vessels. Numerous valves are found in most veins.

SUPERFICIAL GROUP

The digital veins drain into a **dorsal venous arch** on the back of the hand. From this arise the cephalic vein laterally and the basilic vein medially. The **cephalic vein** ascends on the radial side of the forearm, crosses anterior to the cubital fossa and lies lateral to biceps in the upper arm. It turns medially in the deltopectoral groove, pierces the clavipectoral fascia and joins the axillary vein. The **basilic vein** ascends on the ulnar side of the forearm, crosses anterior to the cubital fossa and lies medial to biceps. In the middle of the upper arm it pierces the deep

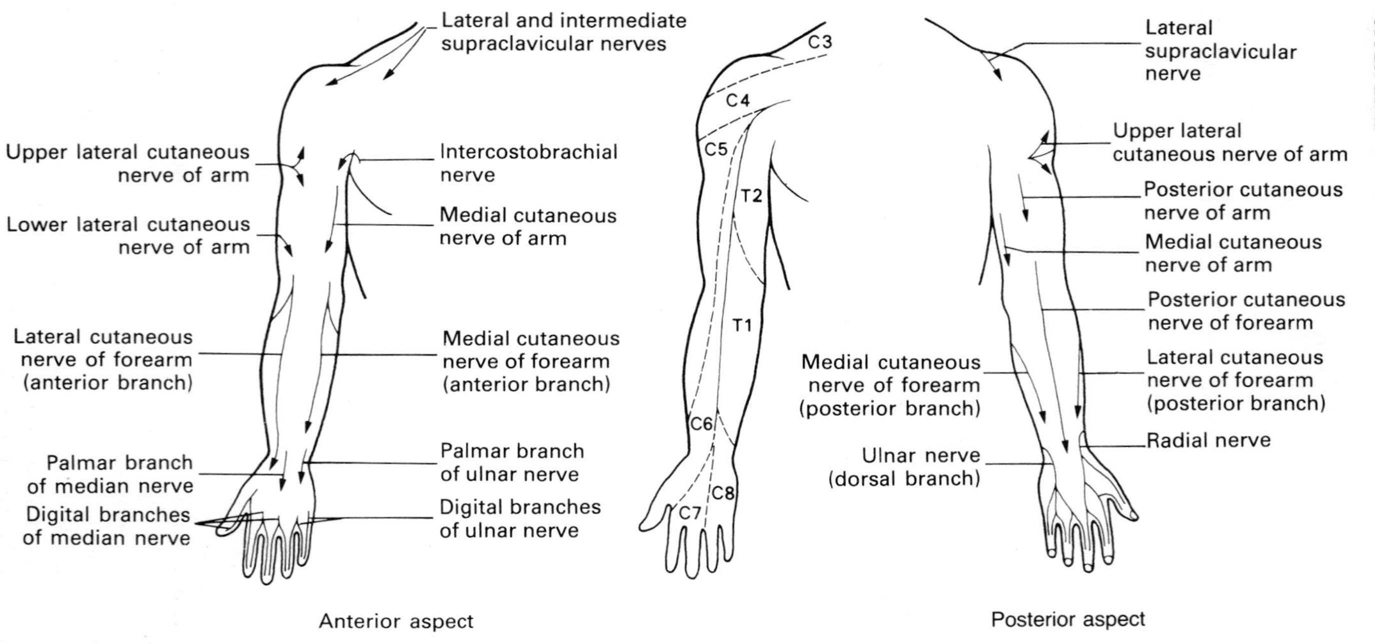

Fig. 20.1 Diagrammatic lay-out of the cutaneous innervation of the right upper limb.

fascia to join the brachial veins. The **median cubital vein**, usually chosen for venepuncture, joins the cephalic and basilic veins and runs across the cubital fossa, superficial to the bicipital aponeurosis. For intravenous infusions the veins of the forearm and dorsum of the hand are most frequently chosen. Stability of the needle and infusion tubing is more easily obtained when the needle puncture site is not too close to the wrist or elbow joint.

DEEP GROUP

Accompanying the arteries, the radial and ulnar veins join to form the brachial veins which continue as the axillary vein at the lower border of teres major (see p. 234).

LYMPH DRAINAGE

In the upper limb there are two groups of lymph vessels—superficial and deep.

The **superficial group** drains the skin. Medially, the vessels accompany the basilic vein and drain (some via the cubital node at the elbow), into the axillary nodes. The lateral vessels accompany the cephalic vein and drain into infraclavicular nodes and thence to the axillary nodes.

The **deep group**, draining bones and muscles, run with the deep veins and drain into the lateral group of axillary nodes.

THE AXILLARY NODES

These are arranged in five groups

(i) **pectoral**—deep to pectoralis major; drains the lateral and anterior chest wall, the mammary gland and the upper anterior abdominal wall.

(ii) **lateral**—on the lateral axillary wall; receives the efferents of the upper limb.

(iii) **subscapular**—on the posterior axillary wall; drains the back of the trunk.

(iv) **central**—a scattered group around the axillary vessels.

(v) **apical**—at the apex of the axilla; they are continuous with the inferior deep cervical nodes and drain the preceding groups. Blockage of the central or apical groups may lead to swelling of the limb—lymphoedema.

The **infraclavicular group**, around the termination of the cephalic vein, drains to the apical nodes. The efferent vessels from the apical nodes form the subclavian lymph trunk which, on the left side, joins the thoracic duct and on the right, opens into the internal jugular or subclavian yein.

NERVE SUPPLY

The **cutaneous nerve supply** is summarized in Figure 20.1. Note that the segmental innervation of the skin over the deltoid is C4 (c.f. the phrenic nerve and referred pain from the diaphragm), over the thumb is C6, over the little finger is C8, and over the axillary floor is T2 and 3. The anterior aspect of the limb is supplied by branches of the anterior divisions of the brachial plexus, and the posterior aspect by branches of posterior divisions. Autonomic nerves, mainly sympathetic, run with the branches of the brachial plexus and supply blood vessels, cutaneous glands and the arrector pili muscles.

NERVE INJURIES

Injury to the nerves of the upper limb at different sites produces characteristic disabilities.

BRACHIAL PLEXUS INJURIES

These produce signs of a segmental nature.

(*a*) *Upper trunk damage* may result from a birth injury. Deltoid and teres minor (axillary nerve), supraspinatus and infraspinatus (suprascapular nerve), biceps and brachialis (musculocutaneous nerve), are paralysed. The limb therefore hangs limply, medially rotated and fully pronated in the 'tip' position. There is sensory diminution on the radial side of the arm and the forearm.

(*b*) *Lower trunk injury* may be caused by pressure from a cervical rib and produces paralysis of the small muscles of the hand and the long digital flexors. This results in the characteristic 'clawed hand'. There is sensory diminution on the ulnar side of the arm and forearm.

PERIPHERAL NERVE INJURIES

(a) Radial nerve

(i) *in the forearm*—produces a small area of diminished sensitivity over the radial side of the dorsum of the hand. There is no paralysis.

(ii) *in the radial groove*—produces paralysis of the long digital and carpal extensors (posterior interosseous nerve) causing wrist-drop. Extension of the middle and distal phalanges by the interossei and lumbricals, the extension of the forearm by the triceps are retained. (Brachioradialis may escape injury). There is diminished sensitivity over the radial side of the dorsum of the hand.

(iii) *in the axilla*—paralysis of triceps and brachioradialis are added to the effects produced by a radial groove injury.

(b) Median nerve

(i) *at the wrist*—produces paralysis of the thenar and 1st and 2nd lumbrical muscles and loss of sensation over the front of the thumb and index fingers. No opposition of the thumb is possible and the index and middle fingers are flexed and partly 'clawed'.

(ii) *at the elbow*—paralysis of the forearm flexors is added to the effects produced by injury at the wrist. No pronation of the forearm is possible.

(c) Ulnar nerve

(i) *at the wrist*—produces paralysis of the interossei, adductor pollicis, the hypothenar muscles and the 3rd and 4th lumbricals. This results in a partly

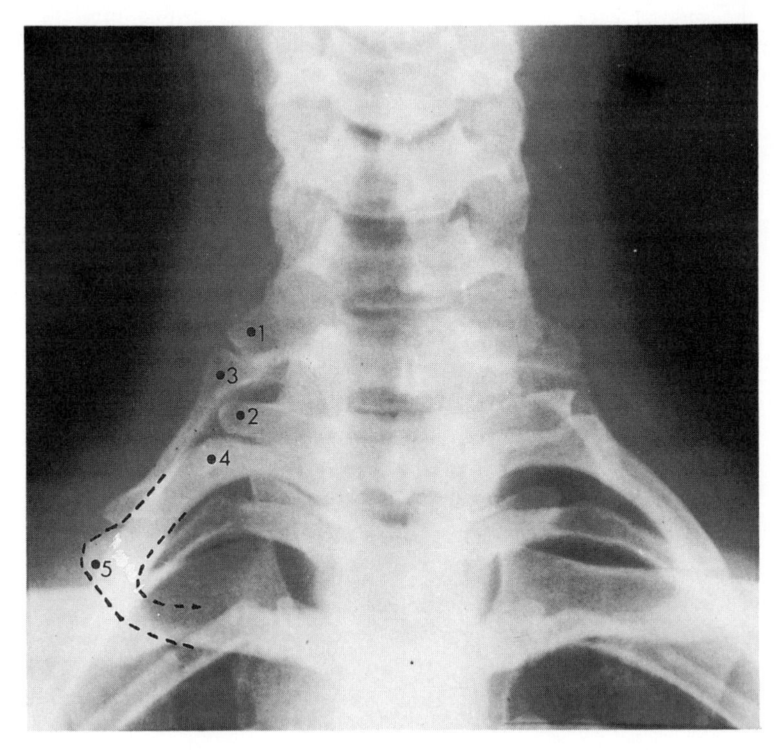

Fig. 20.2 Anteroposterior view of lower cervical spine showing bilateral abnormal cervical ribs. On the right side the transverse processes of the seventh cervical and first thoracic vertebrae, the abnormal rib and the first rib have been marked. The abnormal right cervical rib articulates with an unusually large prominence on the first rib. Note that the transverse process of the first thoracic vertebra slopes upwards and that of the seventh cervical vertebra downwards; this is a useful way of distinguishing the two vertebrae. An abnormal cervical rib can compress the subclavian artery and the first thoracic contribution to the brachial plexus.
1. Transverse process of the seventh cervical vertebra 2. Transverse process of the first thoracic vertebra 3. Cervical rib 4. First rib 5. Prominent tubercle on the first rib at the point of articulation with the cervical rib.

'clawed hand' in which the long digital flexors, unopposed by the lumbricals and the interossei, flex the middle and distal phalanges especially of the little and ring fingers. The consequent pull exerted on the long digital extensor tendons produces hyperextension of the metacarpophalangeal joints. Adduction of the thumb is lost. Sensation is diminished over the ulnar side of the palm.

(ii) *at the elbow*—paralysis of flexor carpi ulnaris and weakening of flexor digitorum profundus and loss of sensation on the ulnar side of the dorsum of the hand are added to the effects of the previous injury.

(d) Axillary nerve

This may be injured in downward dislocation at the shoulder joint and fractures of the surgical neck of the humerus, which produces deltoid paralysis and sensory loss over the humeral attachment of this muscle. Abduction of the humerus is greatly impaired.

PART FIVE

Lower limb

— 21 —

The hip and thigh

The lower limb is similar in its basic structure to the upper limb but is modified in accordance with its functions of supporting and propelling the body. During development the lower limb rotates medially on its long axis so that its ventral (flexor) surface comes to lie posteriorly and the sole of the foot to face backwards, and then later downwards.

The pelvic girdle, unlike the pectoral girdle, is firmly attached to the vertebral column and the body weight is transmitted through it to the lower limb whose bones, muscles and ligaments are correspondingly stronger than those of the upper limb. The centre of gravity of the body in the anatomical position is in the pelvis just in front of the second piece of the sacrum. When standing up straight the line of weight (passing through the centre of gravity) passes down behind the hip joints and in front of the knee and ankle joints. The weight is distributed between the heel and the balls of the toes. When standing upright, most of the body weight is carried by bones and ligaments, little muscle activity in the lower limbs being necessary to maintain stability. This is a very different situation from that in the upper limbs where weight has to be carried by muscles.

The pelvic girdle is described on page 175.

The blood supply of the limb comes in the adult from the external iliac artery, though in the early embryo the axial vessel is a branch of the inferior gluteal artery which arises from the internal iliac artery.

The nerves are derived from the lumbosacral plexus. Due to rotation of the limb the plan may not be so clear as in the upper limb. The plexus is formed by the anterior (ventral) rami of L1 to S5 spinal nerves. The three largest branches are the obturator nerve from the anterior divisions of L2, 3 and 4 supplying muscles on the inside of the thigh, the femoral nerve from the posterior divisions of L2, 3 and 4 supplying muscles in front of the thigh, and the sciatic nerve from L4, 5, S1, 2 and 3 supplying all other muscles. The anterior divisions of L4-S3 enter the tibial part of the sciatic nerve and the posterior divisions of L4-S2 enter the common peroneal part.

Injury to an intervertebral disc in the lumbar region, and protrusion of some of the semisolid nucleus may press on one of the above spinal nerves, producing pain, sensory loss and wasting of muscles.

Hip bone (Fig. 13.1 and p. 171)
The gluteal surface of the **ala of the ilium** is smooth and faces laterally. It is bounded above by the iliac crest, in front by the anterior border of the ilium, behind by the posterior border and below by the greater sciatic notch and the acetabulum. It is crossed by three curved **gluteal lines** which ascend from the greater sciatic notch, the inferior to the anterior border, and the middle and posterior to the iliac crest. They divide the alar surface into four areas; two large central areas for the gluteus medius (posteriorly) and minimus, a small posterior area for gluteus maximus and the sacrotuberous ligament, and a small antero-inferior area for the reflected head of rectus femoris and the iliofemoral ligament.

The greater sciatic notch is separated from the lesser notch by the **ischial spine**. Below the lesser sciatic notch is the **ischial tuberosity**. The sciatic notches are turned into foramina by the sacrotuberous and sacrospinous ligaments. The anterior border of the ilium has prominent **superior** and **inferior iliac spines** and is then continued into the **superior ramus of the pubis**. The acetabular fossa and the obturator foramen are readily identified. The **inferior ramus of the pubis** and the **ramus of the ischium** (together forming the **ischiopubic ramus**) join the body of the pubis (with the symphyseal surface) and the ischial tuberosity. The ilium, ischium and pubis contribute to the **acetabular fossa**. The **obturator foramen** is bounded by pubis and ischium. The obturator membrane is attached to the edges of the foramen except anteriorly where the obturator vessels and nerve pass in the obturator groove.

Femur (Fig. 21.1)
This is a long bone having an upper end, a body and a lower end.

The **upper end** consists of a head, a neck and greater and lesser trochanters. The **head** articulates with the acetabular fossa of the hip bone, it forms two-thirds of a sphere and near its centre there is a small pit (fovea) for the ligament of the head of the femur. The narrow elongated **neck** passes upwards, forwards and medially forming an angle of about 125° with the body. Its anterior surface and the larger part of its posterior surface lie within the capsule of the joint. The **greater trochanter** is a projection from the upper lateral part of the bone. To it are attached gluteus medius and minimus laterally, and the obturator internus and externus medially, the latter into a small pit. Piriformis is attached to the upper border. The gluteus minimus is attached along the anterior border of the trochanter and the medius to much of the lateral surface. The **lesser trochanter** lies medially at the junction of the body and the neck. It is joined to the greater trochanter anteriorly by a thin roughened ridge, the **intertrochanteric line**, and posteriorly by a thick rounded ridge, the **intertrochanteric crest**. A slight elevation on the middle of the crest is known as the quadrate tubercle and gives attachment to the quadratus femoris. The psoas is attached to the lesser trochanter and the iliacus and pectineus to the **body** below it.

The **body** inclines medially at an angle of about 20° to the vertical. Posteriorly there is a longitudinal ridge, the **linea aspera**, passing along the middle third of the bone. To the linea aspera are attached the adductor muscles, the short head

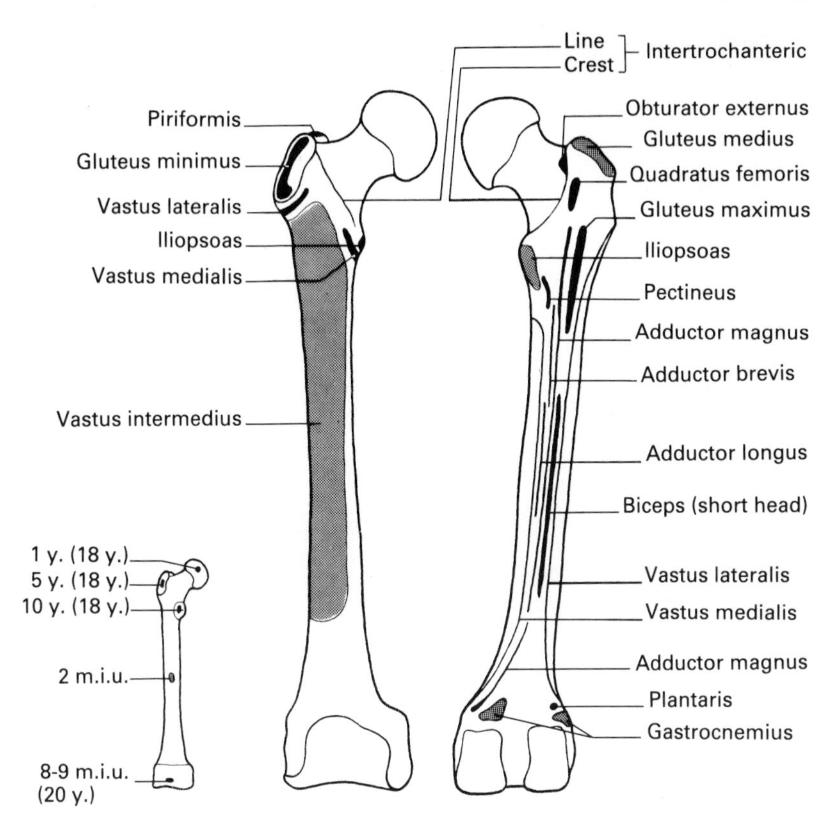

Fig. 21.1 The right femur. **A.** anterior and **B.** posterior surfaces. Inset: ossification (m.i.u., months in utero; y., years).

of biceps and the vastus medialis and lateralis. A roughened area for the attachment of gluteus maximus (the **gluteal tuberosity**) lies between the quadrate tubercle and the linea. In the lower third of the shaft, the edges of the ridge are continued as the medial and lateral supracondylar lines and limit the smooth popliteal surface of the bone. The medial supracondylar line ends at the **adductor tubercle** just above the medial condyle; the line and tubercle give attachment to the adductor magnus. The smooth medial, lateral and anterior surfaces of the body are covered by the vasti muscles.

The **lower end** is expanded and forms two large masses, the medial and lateral **condyles** for articulation with the tibial condyles and the patella. The femoral condyles are separated posteriorly by the intercondylar notch and united anterosuperiorly by the concave articular facet for the patella. The lateral side of this facet is wider and more prominent. The tibial articular surface of each condyle covers its inferior and posterior aspects and is markedly convex anteroposteriorly and slightly convex from side to side. The medial tibial surface is longer than the lateral. On the outer nonarticular surface of each condyle there is a small

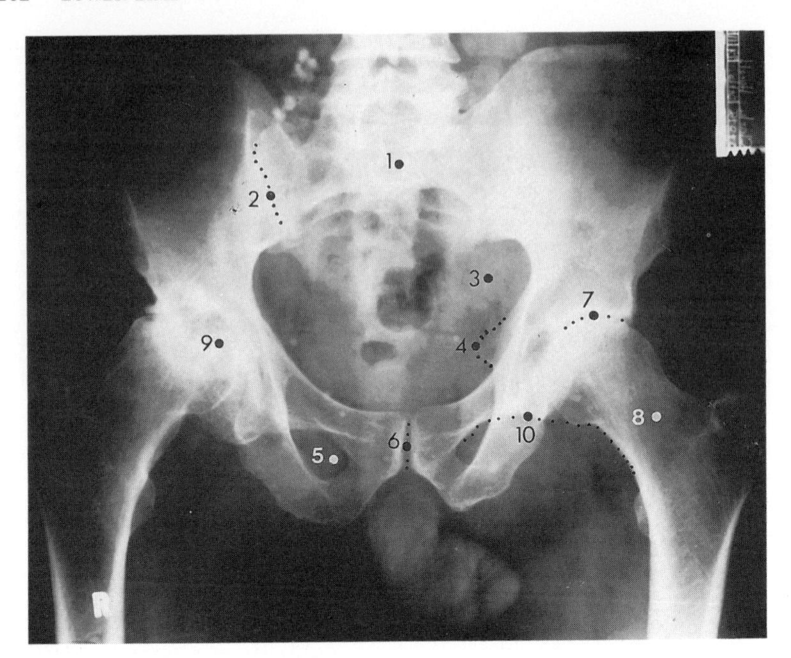

Fig. 21.2 Anteroposterior veiw of adult pelvis. The abnormal appearance of the right hip joint is due to severe osteo-arthritis. This condition affects mainly the larger joints and is due to degenerative changes in the articular cartilage secondary to old age and trauma.

1. Sacrum
2. Sacro-iliac joint
3. Greater sciatic notch
4. Spine of the ischium
5. Obturator foramen
6. Symphysis pubis
7. Joint space of left hip joint
8. Neck of femur
9. Osteo-arthritic changes
10. Smooth curvature of upper border of obturator foramen and lower border of femoral neck in normal joint (Shenton's line)

elevation, the **epicondyle**; these give attachments to the medial and lateral ligaments of the knee joint. A shallow groove below the lateral epicondyle gives attachment to the popliteus tendon.

Parts of the condyles, epicondyles and greater trochanter are subcutaneous and palpable. Fractures of the neck of the femur are common in the elderly and frequently complicated by nonunion due to a critical decrease in the blood supply of the femoral head. Fractures of the body of the femur are accompanied by a considerable amount of bleeding into the thigh due to associated injuries to the surrounding muscles and blood vessels.

THE HIP JOINT (Figs. 21.1, 21.3).

This is a synovial joint of the ball and socket variety between the acetabulum of the hip bone and the head of the femur. The articular surface of the acetabulum is ∩-shaped, being deficient inferiorly at the acetabular notch. The acetabulum is cupped and deepened by the **acetabular labrum**, a rim of fibrocartilage attached to its borders, and bridging the acetabular notch is the transverse acetabular ligament.

Ligaments

(a) CAPSULE

The capsule is strong and dense, attached proximally to the acetabular labrum and the edge of the notch. Distally it is attached to the femur along the intertrochanteric line and to the neck 1 cm above the intertrochanteric crest. Superiorly it crosses the neck medial to the tendinous insertions of the obturator muscles. Some of the capsular fibres turn back medially along the neck and are known as **retinacula**; they carry blood vessels towards the head.

(b) CAPSULAR THICKENINGS:

(i) **iliofemoral ligament**—is a strong λ-shaped band attached above to the anterior inferior iliac spine and bifurcating inferiorly, gains attachment to each end of the intertrochanteric line.

(ii) **pubofemoral ligament**—passes from the iliopubic eminence to the lower part of the capsule and undersurface of the neck.

(iii) **ischiofemoral ligament**—passes upwards from the acetabular margin to the upper end of the intertrochanteric line and the adjacent upper surface of the neck. Note that all three ligaments spiral in such a way as to limit extension at the joint.

(c) ACCESSORY LIGAMENTS:

(i) ligament of the head of the femur—passes from the fovea to the articular notch; it is lax and may contain some small vessels of special importance in the child.

(ii) acetabular labrum and transverse acetabular ligament.

Intracapsular structures

The **synovial membrane** lines the nonarticular surfaces of the joint and ensheaths the ligament of the head of the femur. The cavity of the joint communicates with the psoas bursa between the iliofemoral and pubofemoral ligaments. A pad of fat is found in the acetabular notch.

Functional aspects

MOVEMENT

The hip joint is capable of flexion, extension, abduction, adduction, circumduction, and medial and lateral rotation. In the anatomical position, the line of weight passes behind the axis of the joint and so gravity encourages extension at the joint.

FLEXION— iliopsoas, assisted by tensor fasciae latae, rectus femoris, sartorius

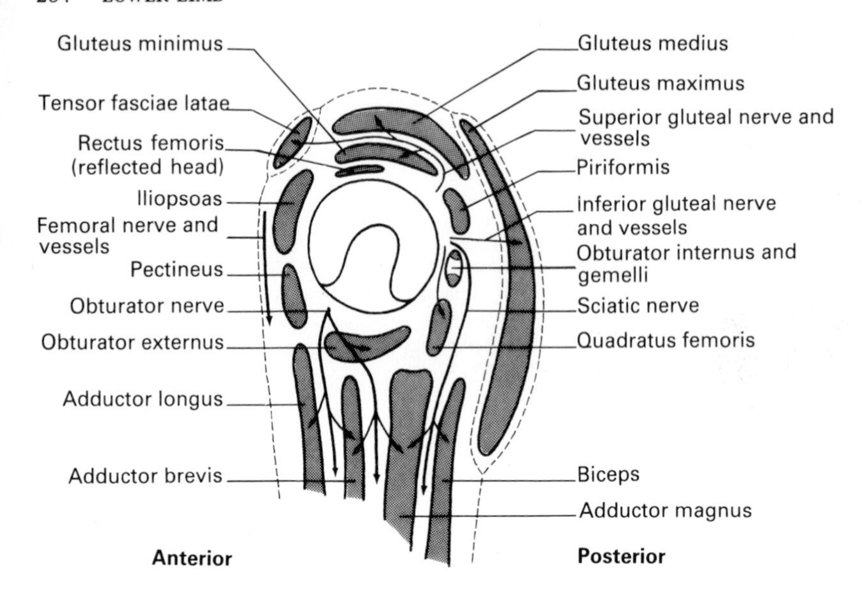

Fig. 21.3 Diagram showing the main relations of the left hip joint, viewed from the lateral aspect.

and pectineus. The movement is limited to about 90° when the knee is flexed, and is much less when the knee is extended, because of tension in the hamstrings. If the line of weight passes in front of the hip joint (as when leaning forwards slightly), then gravity is a major factor in flexion of the trunk on the lower limb.

EXTENSION—gluteus maximus, assisted by gravity, the hamstrings and the tensor fasciae latae. The movement is limited by the three capsular thickenings.

ABDUCTION—gluteus medius and minimus. This movement occurs in every step of walking when the pelvis is tilted on the femur of the grounded leg, thus allowing the foot of the swinging leg to clear the ground as it swings forwards.

ADDUCTION—adductor muscles of the thigh, gracilis and gravity.

Rotation occurs around an axis joining the centre of the head of the femur to the intercondylar notch of the femur.

MEDIAL ROTATION—Anterior fibres of gluteus medius and minimus assisted by the iliopsoas.

LATERAL ROTATION—Short muscles applied to the back of the joint—piriformis, obturator internus and quadratus femoris assisted by the gluteus maximus.

When the femur is fixed, as when the foot is on the ground, the same groups of muscles act on the pelvis assisting in flexion and extension of the trunk and in walking (p. 337).

STABILITY

In spite of its great mobility, it is a very stable joint. This is because of the deep cup of the acetabulum clasping the femoral head, the strong capsule and its thickenings, particularly the iliofemoral ligament, and the large number of closely applied short articular muscles. Dislocation does occur as a congenital abnormality; however, it is rare in the adult. When it occurs there is usually an associated fracture of the acetabular rim, and the closely related sciatic nerve may also be damaged. The joint may be affected in older people by degenerative joint disease (osteo-arthritis).

Blood supply
Rich anastomoses formed from the two gluteal and two circumflex femoral arteries. The ligament of the head of the femur carries a small branch of the obturator artery. The retinacula on the neck carry vessels to the head.

Nerve supply
Branches of the femoral, obturator and sciatic nerves.

Relations (Fig. 21.3)
ANTERIOR—iliopsoas and pectineus separate the joint from the femoral vessels and nerve.
POSTERIOR—piriformis, obturator internus and quadratus femoris separate the joint from the sciatic nerve and gluteus maximus.
SUPERIOR—gluteus minimus and the reflected head of the rectus femoris.
INFERIOR—obturator externus.

Bursae
The psoas bursa separates the iliopsoas tendon from the iliac fossa and superior pubic ramus. It may communicate with the cavity of the hip joint between the iliofemoral and pubofemoral ligaments.

THE GLUTEAL REGION

This fleshy region behind the pelvis is composed mainly of the three gluteal muscles which overlie the muscles, nerves and vessels leaving the inside of the pelvis and passing to the lower limb. The gluteal muscles are important extensors and abductors at the hip joint (see Fig. 13.1 and p. 172).

1. Gluteus maximus
This is the most superficial of the group. It is a large flat quadrilateral mass forming the prominence of the buttock.

Attachments
PELVIC—from an extensive area including the gluteal surface of the ilium behind the posterior gluteal line, and the posterior surfaces of the sacrum, coccyx and sacrotuberous ligament.

LOWER—the fibres pass downward and laterally into the iliotibial tract (see below and p. 331) and the gluteal tuberosity of the femur.

Actions
It is a powerful lateral rotator and extensor at the hip joint and, acting through the iliotibial tract, it extends and stabilises the knee joint.

Nerve supply
This is by the inferior gluteal nerve (Fig. 21.3).

Relations
It lies behind the hip joint and overlies numerous muscles and the structures passing through the greater and lesser sciatic foramina. The sciatic, gluteal and pudendal nerves are the largest.

2. Tensor fasciae latae
This is attached above to the anterior quarter of the outer lip of the iliac crest and below to the iliotibial tract. It is a flexor at the hip joint and through its action on the iliotibial tract it extends and stabilises the knee joint. It is supplied by the superior gluteal nerve.

The **iliotibial tract** is a broad thickening of the fascia lata passing from the outer lip of the iliac crest to the anterolateral aspect of the upper end of the tibia. The tract receives the attachments of the gluteus maximus and fasciae latae and these muscles, acting through it, extend and stabilise the knee joint.

3. Gluteus medius

Attachments
PELVIC—gluteal surface of the ilium between the middle and posterior gluteal lines.

FEMORAL—an oblique ridge on the lateral surface of the greater trochanter.

4. Gluteus minimus
This is the deepest and most anterior of the gluteal muscles. It is covered completely by gluteus medius.

Attachments
PELVIC—gluteal surface of the ilium between the middle and inferior gluteal lines.

FEMORAL—anterior border of the greater trochanter.

Actions and nerve supply: gluteus medius and minimus are powerful abductors and weak medial rotators at the hip joint. In most circumstances (e.g. walking) the pelvis is moved on the femur. They are supplied by the superior gluteal nerve.

5. Piriformis

This arises inside the pelvis from the anterior surface of the sacrum (Fig. 13.2 and p. 174) and passes laterally out of the greater sciatic foramen to be attached to the upper border of the greater trochanter of the femur. It is supplied by branches of the sacral plexus and stabilises the hip joint. It is a lateral rotator of the thigh at the hip joint.

6. Quadratus femoris

This small muscle is attached medially to the outer edge of the ischial tuberosity and laterally to a small tubercle on the intertrochanteric crest. It helps to stabilise the hip joint and is a weak lateral rotator of the femur. It is supplied by a branch of the sacral plexus.

7. Obturator externus

This short articular muscle lies close to the hip joint, above adductor brevis. It passes from the outer surface of the obturator membrane and the adjacent bone, and its tendon winds below and behind the hip joint to reach a pit on the medial surface of the greater trochanter.

Action: it supports and stabilises the hip joint and is a weak lateral rotator of the femur.

Nerve supply: obturator nerve.

VESSELS AND NERVES OF THE GLUTEAL REGION

These are branches of the internal iliac artery and the lumbosacral plexus respectively. They pass from the pelvis to the gluteal region above or below the piriformis in the greater sciatic foramen.

ABOVE PIRIFORMIS

The superior gluteal artery (p. 207) and nerve (p. 211) pass upwards and laterally between gluteus medius and minimus, supplying these muscles and tensor fasciae latae. The artery takes part in an anastomosis around the hip joint.

BELOW PIRIFORMIS

(i) *sciatic nerve*—emerges lateral to the ischial spine and lies about midway between the ischial tuberosity and the greater trochanter (p. 289).

(ii) *inferior gluteal nerve* (p. 211)—after a short course, enters the substance of gluteus maximus. The inferior gluteal artery (p. 207) ramifies in the lower part of the buttock and takes part in an anastomosis around the hip joint.

(iii) *internal pudendal vessels and the pudendal nerve*—descend over the ischial spine and sacrospinous ligament and pass through the lesser sciatic foramen into the perineum (p. 203).

(iv) *nerve to obturator internus* (p. 211)—descends across the ischial spine.

(v) *nerve to quadratus femoris*—emerges deep to the sciatic nerve and lies on the ischium and the hip joint.

(vi) *posterior femoral cutaneous nerve*—emerges superficial to the sciatic nerve and descends in the midline of the thigh, beneath the deep fascia. It supplies the skin of the buttock, the perineum, the posterior aspect of the thigh and the popliteal region.

Functional aspects

Intramuscular injections are often given into the buttock. The needle should be inserted in the upper lateral quadrant, as far away as possible from large nerves.

The roles of these muscles in standing and walking are discussed below (p. 337).

THE BACK OF THE THIGH

This is often known as the hamstring compartment of the thigh. It contains the three hamstring muscles (semimembranosus, semitendinosus and biceps femoris) which are attached to the ischial tuberosity and are supplied by the sciatic nerve. They are extensors at the hip joint and flexors at the knee joint (see Figs. 13.1B, 22.2).

1. Semimembranosus

This derives its name from its thin superior membranous tendon. It is the most medial muscle of the group.

Attachments

UPPER—a facet on the ischial tuberosity.

LOWER—a groove on the posteromedial aspect of the tibial condyle and by tendinous expansions upwards on to the lateral femoral condyle (oblique popliteal ligament) and downwards on to the soleal line of the tibia (forming the popliteal fascia over the popliteus muscle).

2. Semitendinosus

This derives its name from its very long inferior tendon. It lies lateral to semimembranosus.

Attachments

UPPER—a facet on the ischial tuberosity.

LOWER—upper fact of the medial surface of the tibia behind the attachment of the gracilis and sartorius tendons.

3. Biceps

The biceps has ischial and femoral heads.

Attachments

UPPER

(i) *long head* from the ischial tuberosity.

(ii) *short head*—the linea aspera near the lateral lip and the upper part of the lateral supracondylar line of the femur. The upper fibres are often continuous with those of gluteus maximus.

LOWER

The two heads join and form a single tendon attached to the head, including the apex, of the fibula.

Action

These three muscles act across both the hip and the knee joints. They are powerful flexors at the knee and extensors at the hip. When the knee is flexed, the muscles can produce a small amount of medial and lateral rotation at the knee joint. Tension in these muscles will restrict flexion at the hip joint.

Nerve supply

They are supplied by the sciatic nerve.

THE SCIATIC NERVE (Fig. 21.3)

This is the largest nerve in the body. It is derived from the lumbosacral plexus (L4, 5; S1, 2, 3) and is formed in the pelvis on the anterior surface of piriformis (p. 211). It descends through the greater sciatic foramen, enters the gluteal region and continues its descent through the posterior compartment of the thigh. It ends by dividing into tibial and common peroneal nerves, usually just above the popliteal fossa.

Relations

The nerve enters the gluteal region below piriformis. Anteriorly the nerve lies on the ischium, obturator internus, quadratus femoris and adductor magnus from above downwards, and is covered posteriorly by gluteus maximus and then the hamstring muscles.

Branches

(i) *articular*—to the hip and knee joints.

(ii) *muscular*—to the hamstrings and the ischial part of adductor magnus.

(iii) *tibial* and *common peroneal* nerves (p. 313 and 314).

There are no large vessels in the back of the thigh so that its structures are supplied by a deeply placed longitudinal anastomotic chain which receives blood from the inferior gluteal and the medial femoral circumflex arteries above, the four perforating branches of the profunda femoris artery, and the popliteal artery below.

THE FRONT AND MEDIAL SIDE OF THE THIGH

The oblique line of the sartorius passing downwards and medially separates the adductor muscles of the hip superomedially from the extensor muscles of the knee inferolaterally. The psoas and iliacus are described on pages 165 and 178 respectively. (See Figs. 13.1, 21.1).

THE ADDUCTOR GROUP OF MUSCLES

These are arranged in three vertical layers; anteriorly pectineus and adductor longus lie in the same plane in front of adductor brevis which separates them from the posteriorly placed adductor magnus (Fig. 21.3). Gracilis lies superficial to these on the medial side of the thigh.

1. Pectineus

Attachments

PELVIC—pectineal line and upper surface of the pubic bone.
FEMORAL—along a line joining the lesser trochanter to the linea aspera.

2. Adductor longus

Attachments

PELVIC—anterior surface of the body of the pubis.
FEMORAL—linea aspera.

3. Adductor brevis

Attachments

PELVIC—outer surface of the body and the inferior ramus of the pubis.
FEMORAL—upper half of the linea aspera.

4. Adductor magnus

This is a large triangular muscle of composite origin, part derived from the adductor muscles and part from the hamstrings.

Attachments

PELVIC—in a continuous line on the outer surface of the ischiopubic ramus and the ischial tuberosity.

FEMORAL—the length of linea aspera, the medial supracondylar line and by a tendon into the length of the adductor tubercle on the upper aspect of the medial femoral condyle. The attachment to the supracondylar line is interrupted about 10 cm above the knee joint and the opening is called the **adductor hiatus**. The femoral vessels pass through into the popliteal fossa.

5. Gracilis

This thin strap-like muscle passes from the inferior border of the ischiopubic ramus to the upper part of the medial surface of the tibia between the attachments of sartorius and semitendinosus.

Action: these muscles all adduct the femur. In addition the adductors laterally rotate the femur and adductor magnus extends it. Pectineus also helps adduct and flex the femur.

Nerve supply: pectineus is supplied by the femoral nerve, the hamstring (ischial) part of adductor magnus by the sciatic nerve. The rest of adductor magnus and the other muscles in this group are supplied by the obturator nerve.

THE OBTURATOR NERVE

This is a branch of the upper part of the lumbosacral plexus (anterior divisions of L2, 3, 4). Its course in the pelvis is described on page 167. It enters the thigh through the obturator groove where it divides into anterior and posterior branches.

Anterior division

This descends between pectineus and adductor longus anteriorly and adductor brevis posteriorly and ends in the subsartorial plexus on the medial side of the thigh.

Branches

(i) *muscular*—to gracilis, adductor brevis and adductor longus.
(ii) *articular*—to the hip joint.
(iii) *cutaneous*—to the medial side of the thigh.

Posterior division

This pierces obturator externus and descends between adductor brevis anteriorly and adductor magnus posteriorly. It later descends with the femoral and then popliteal arteries to the knee joint.

Branches

(i) *muscular*—to obturator externus and adductor magnus.
(ii) *articular*—to the knee joint.

The obturator artery

The obturator artery, a branch of the internal iliac, accompanies the nerve through the obturator groove and on emerging, divides into two branches which encircle the obturator membrane. It gives branches to the hip joint and to the adductor muscles.

THE ANTERIOR GROUP OF THIGH MUSCLES

1. Sartorius

This is a long strap-like muscle which, in its upper part, forms the lateral boundary of the femoral triangle (p. 293).

Attachments

PELVIC—anterior superior iliac spine.

TIBIAL—upper part of the medial surface of the tibia in front of the attachments of gracilis and semitendinosus.

Action: it is a weak flexor at both the hip and knee joints and a lateral rotator of the femur at the hip joint.

Nerve supply: femoral nerve.

2. Quadriceps

This extensive muscle mass forms the bulk of the anterior region of the thigh. It is divided into four separate muscles, all attached inferiorly to the upper border and the sides of the patella.

(a) **Rectus femoris**—this muscle lies superficially along the middle of the thigh.

Attachments:

SUPERIOR—by two heads.
(i) straight head—from the anterior inferior iliac spine.
(ii) reflected head—from the ilium just above the acetabulum.

INFERIOR—by a tendon into the upper border of the patella.

(b) **Vastus lateralis**—lies deep to rectus femoris and also covers the lateral side of the femoral body.

Attachments:

SUPERIOR—by an aponeurosis from the lateral side of the great trochanter and the gluteal tuberosity, and the lateral lip of the linea aspea.

INFERIOR—its tendon passes to the lateral side of the patella and blends with fibres of rectus femoris.

(c) **Vastus medialis**—lies deep to rectus femoris and also covers the medial side of the femoral body.

Attachments:

SUPERIOR—by an aponeurosis from the spiral line joining the lesser trochanter to the linea aspera, the medial lip of the linea aspera and the medial supracondylar line of the femur.

INFERIOR—the muscle passes to the medial side of the patella and blends with fibres of rectus femoris. The lower fibres are almost horizontal.

(d) **Vastus intermedius**—this muscle covers the front of the femur deep to rectus femoris and lies between the other two vasti.

Attachments:

SUPERIOR—the anterior and lateral surfaces of the upper part of the femoral body.

INFERIOR—to the upper border of the patella deep to the previous three muscles.

The four portions of the quadriceps are thus attached to the upper border and the sides of the patella forming a single musculotendinous expansion. From the apex of the patella a strong tendon, the **patellar ligament**, descends and is attached to the tibial tubercle. On each side of the patellar ligament the capsule of the joint is formed largely by downward fibrous expansions of the quadriceps (the **retinacula**) through which the muscles gain attachment to the tibial condyles.

Action

The whole muscle is a powerful extensor at the knee joint. Rectus femoris also flexes the hip joint. The lower fibres of vastus medialis prevent the patella moving too far laterally when the lower leg is being extended at the knee.

Nerve supply

Each part is supplied by branches of the femoral nerve.

THE FEMORAL TRIANGLE

This area on the front of the upper part of the thigh is bounded by the inguinal ligament superiorly, sartorius laterally and the medial border of adductor longus medially. Its floor is formed by the adductor longus, pectineus and iliopsoas from medial to lateral and its roof is the fascia lata of the thigh. There is an oval deficiency, the **saphenous opening**, in the fascia lata about 4 cm below and lateral to the pubic tubercle. This opening has a well-defined lateral margin and is filled by the cribriform fascia, a condensation of the superficial fascia. The opening is traversed by the great saphenous vein and some small cutaneous vessels. In the triangle are the femoral vessels contained within the femoral sheath, the femoral nerve and their branches.

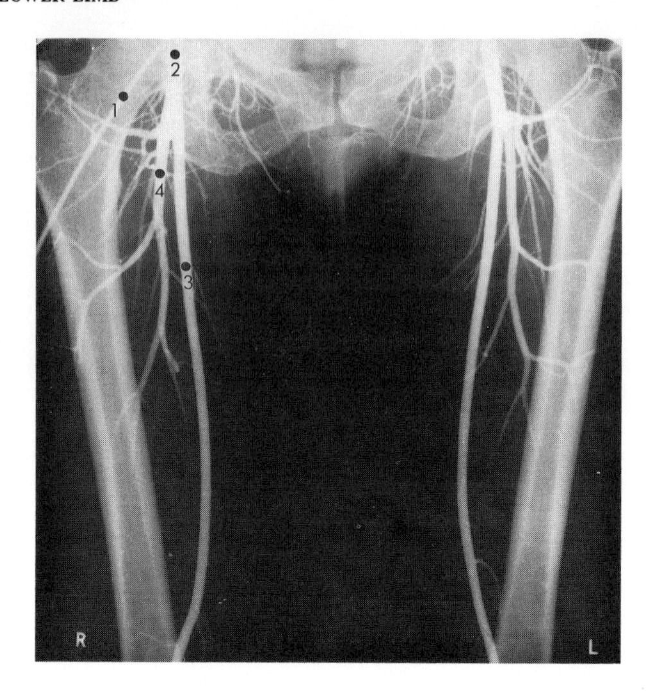

Fig. 21.4 Femoral arteries filled with contrast material which has been injected into the abdominal aorta. (See also Figures 15.1, 22.5, 22.6).
1. Catheter
2. and 3. Femoral artery
4. Profunda femoris artery

The **femoral sheath** is a tube-like downwards continuation of the extraperitoneal fascia into the thigh. It is formed anteriorly from the transversalis fascia and posteriorly from the iliac fascia. The femoral vessels evaginate the fascial envelope of the abdomen as they leave the cavity. Inferiorly the sheath fuses with the adventitia of the vessels about 3 cm below the inguinal ligament. Septa divide it into three compartments containing the femoral artery laterally, the femoral vein centrally, and fatty connective tissue and lymph nodes medially. The latter compartment is known as the **femoral canal** and is limited inferiorly by the blending of its medial wall with the adventitia over the laterally placed femoral vein. Superiorly the canal communicates with the extraperitoneal abdominal fascial space by an opening, the **femoral ring**. The ring is bounded anteriorly by the inguinal ligament, posteriorly by pectineus on the pubis, medially by the pectineal part of the inguinal ligament, and laterally by the femoral vein. These relations should be noted as herniation of abdominal contents into the femoral canal may occur and the femoral ring then provides a constriction around the hernial sac. During exercise, the femoral vein enlarges and is accommodated by the canal medially.

THE FEMORAL ARTERY (Fig. 21.4)

This is a continuation of the external iliac artery. It enters the thigh below the inguinal ligament, descends vertically and slightly backwards through the femoral triangle and the adductor canal. It ends by piercing adductor magnus about 10 cm above the knee joint. It then becomes the popliteal artery.

Relations

The artery begins deep to the inguinal ligament, halfway between the anterior superior iliac spine and the symphysis pubis (the midinguinal point). In the femoral triangle it is covered only by fascia and descends on the psoas tendon (which separates it from the hip joint), pectineus and adductor longus from above downwards. The artery leaves the triangle at its apex and enters the **adductor canal** in company with its vein, the saphenous nerve and the nerve to vastus medialis. The canal is a narrow intermuscular passage roofed by the middle third of sartorius. It is bounded posteriorly by adductor longus above and adductor magnus below, and anterolaterally by vastus medialis. The femoral artery leaves the lower end of the canal by passing through an opening in adductor magnus just above the adductor tubercle.

In the femoral triangle the femoral vein lies medial to the artery, in most of the adductor canal it is posterior, and in the popliteal fossa, the vein retains this relationship to the artery.

Branches

(i) *small cutaneous branches*—pierce the cribriform fascia and are distributed to the skin of the inguinal region and lower abdomen.

(ii) **profunda femoris artery**—arises in common with the medial and lateral femoral circumflex arteries in the femoral triangle and passes backwards and downwards between the adductor muscles. Just above the knee, it pierces the adductor magnus muscle and supplies the lower part of the back of the thigh. Its three main branches also pierce adductor magnus and supply the posterior part of the thigh. They are called perforating arteries.

(iii) **medial femoral circumflex artery**—passes posteriorly between pectineus and psoas and reaches the upper border of adductor magnus where it divides into muscular and articular branches to the hip joint.

(iv) **lateral femoral circumflex artery**—passes laterally between the branches of the femoral nerve, then deep to sartorius and rectus femoris and gives off muscular and articular branches to the hip joint.

(v) *articular branches*—to the knee joint.

The femoral artery is one of the commonest sites of peripheral arterial disease, particularly at the point where the artery passes under the tendon of the adductor magnus. The blood supply to the calf may be markedly reduced and exercise pain (intermittent claudication) and ischaemic changes of the lower leg may occur.

THE FEMORAL VEIN

This begins at the opening in the lower end of adductor magnus as the continuation of the popliteal vein. It accompanies the femoral artery and ends at the level of the inguinal ligament by becoming the external iliac vein. As it ascends it lies at first behind the artery but ends on its medial side.

Tributaries

(i) *profunda femoris vein.*

(ii) *great saphenous vein* (p. 332)—pierces the cribriform fascia and enters the femoral vein about 3 cm below the inguinal ligament.

(iii) and (iv) *medial* and *lateral femoral circumflex veins.*

(v) *small subcutaneous veins*—draining the lower abdominal wall and the upper thigh, enter the termination of the saphenous vein.

THE FEMORAL NERVE (Fig. 21.3)

This is a branch of the upper part of the lumbosacral plexus (posterior divisions of L2, 3, 4) (p. 167). It descends in the pelvis in the groove between psoas and iliacus, and enters the thigh deep to the inguinal ligament and outside the femoral sheath. Here it lies on iliacus lateral to the femoral artery. After a short course, it divides into a number of terminal branches.

Branches

(i) *anterior femoral cutaneous nerve.*

(ii) *medial femoral cutaneous nerve.*

(iii) *muscular branches*—to pectineus, sartorius, rectus femoris, vastus lateralis, vastus intermedius and vastus medialis. This last branch lies on the lateral side of the femoral artery through the femoral triangle and the adductor canal, supplying the muscle and the knee joint.

(iv) **saphenous nerve**—is the largest cutaneous branch of the femoral nerve. It accompanies the femoral artery through the femoral triangle and adductor canal and just above the knee emerges from under sartorius and descends subcutaneously over the medial side of the knee joint. In the leg it accompanies the great saphenous vein and passes in front of the medial malleolus into the foot. It supplies the skin over the medial border of the leg and foot as far as the ball of the big toe.

— 22 —

The knee, leg and dorsum of the foot

Patella (Fig. 22.1)
This is a sesamoid bone, triangular in shape with its apex inferiorly and having anterior and posterior surfaces. The roughened anterior surface is subcutaneous. The smooth posterior surface has a smaller medial and a larger lateral facet for articulation with the smooth articular surface on the front of the femoral condyles. The quadriceps is attached to the upper border, and the patellar ligament to the apex. Some muscle fibres from vastus medialis are attached to its medial border. Tendinous expansions from the vastus medialis and lateralis, known as retinacula, pass to the medial and lateral borders of the patella and to the tibial condyles.

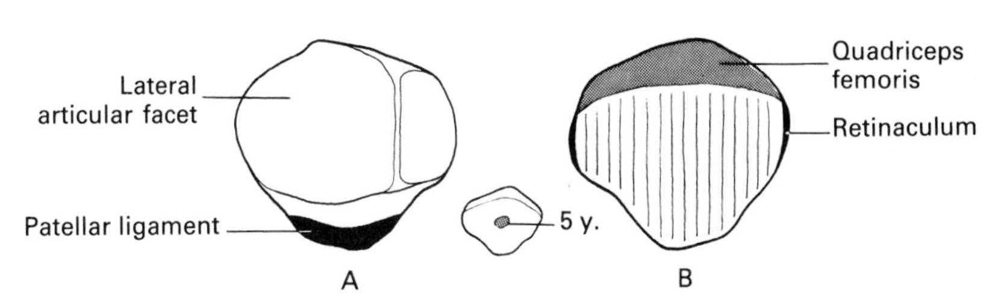

Fig. 22.1 The left patella: **A.** posterior and **B.** anterior surfaces. Inset: ossification (y., years).

Tibia (Fig. 22.2)
This is a long bone possessing an upper end, a body and a lower end.

The **upper end** is expanded to form two masses, the medial and lateral **condyles**. It is flattened superiorly and this surface has oval medial and lateral facets articulating with the femoral condyles. A central anteroposterior ridge across the upper surface, the intercondylar area, separates the articular surfaces and gives attachment to the ends of the cartilaginous menisci and to the anterior and posterior cruciate ligaments. A prominent **tibial tuberosity** is situated anteriorly below the superior surface and receives the attachment of the patellar ligament. Semimembranosus is attached to a wide groove along the side of medial condyle. The postero-inferior surface of the lateral condyle bears an oval facet for the head of the fibula.

The **body** is triangular in cross section having medial, lateral and posterior surfaces. The medial surface gives attachment superiorly to the tibial collateral ligament of the knee joint and to sartorius, gracilis and semitendinosus. Below these muscles the medial surface and the sharp anterior border lie subcutaneously. The lateral surface is hollowed for tibialis anterior. The posterior surface is bounded by the medial and interosseous (lateral) borders. It is crossed by an oblique ridge, the **soleal line**, passing downwards and medially, and below this is a prominent nutrient foramen. The ridge gives attachment to soleus and the surface above it to popliteus. The area below the ridge is divided by a vertical line; tibialis posterior is attached laterally and flexor digitorum longus medially. The interosseous membrane is attached to the lateral border.

The expanded **lower end** projects downwards on the medial side to form the **medial malleolus**. The inferior surface of the bone and the lateral surface of the malleolus articulate with the talus. The anterior surface is smooth and is crossed by the tendons of the muscles in the anterior compartment of the leg. The posterior surface is grooved medially by the tibialis posterior tendon. The lateral surface bears a roughened area for the interosseous ligament of the inferior tibiofibular joint.

The tibial condyles, the anterior border, the medial surface and the medial malleolus are palpable. Because of the subcutaneous position of the tibia, fractures are usually compound (p. 8) and prone to infection.

Fibula (Fig. 22.2)

This is a long bone possessing an upper end (head), a body and a lower end. It carries little or no weight but many muscles are attached to the bone which is an essential part of the ankle joint.

The expanded **head** bears an oblique anterosuperior oval articular facet for the tibia and, posterolateral to this, an upward projection, the **apex**. The fibular collateral ligament of the knee joint is attached to the top of the apex and the biceps tendon at its base. The head is attached to the body by a narrow **neck**.

The **body** is long and slender and has a slight spiral twist. The narrow anterior surface gives attachment to the extensor muscles; the posterior to the flexor muscles (its upper third above an oblique ridge gives attachment to soleus); and the lateral surface to the peroneal muscles.

The expanded **lower end** is flattened from side to side. It forms the **lateral malleolus** which extends to a lower level than the medial malleolus and articulates on its medial side by a triangular facet with the talus. There is a triangular roughening above the facet, for the interosseous ligament of the inferior tibiofibular joint. A marked **malleolar fossa**, situated on the posteromedial aspect behind the articular facet, gives attachment to the posterior talofibular ligament and forms a useful marker when orientating the bone.

The head and the lateral malleolus of the fibula are palpable. The common

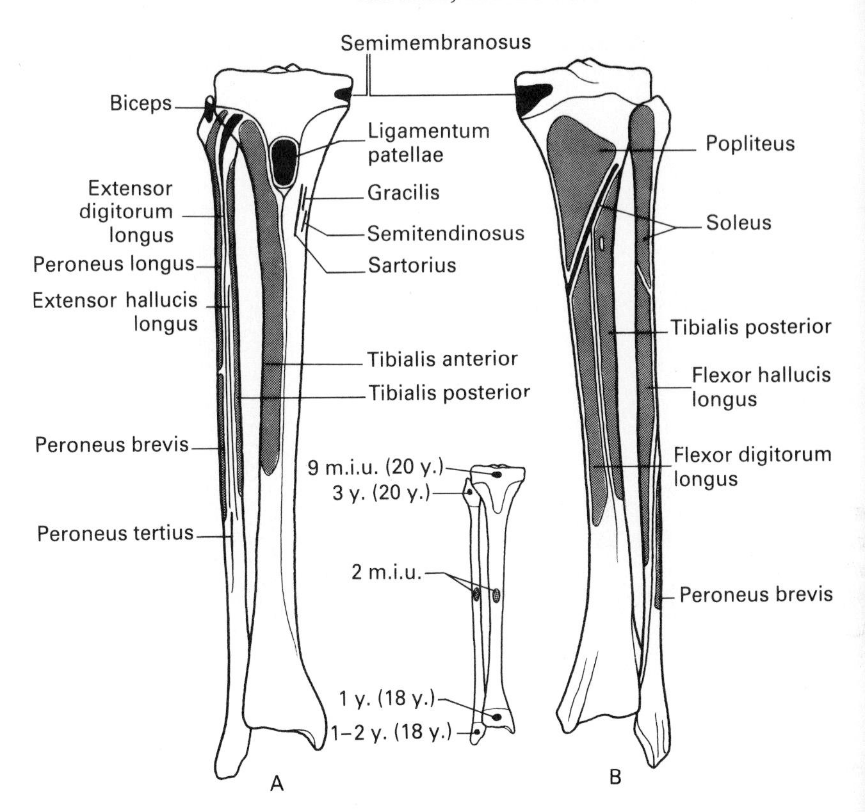

Fig. 22.2 The right tibia and fibula. **A.** anterior and **B.** posterior surfaces. Inset: ossification (m.i.u., months in utero; y., years).

peroneal nerve may be crushed as it passes round the neck. The lower end may be fractured above the interosseous ligament in injuries to the ankle joint.

THE KNEE JOINT (Figs. 22.3 and 22.4)

This is a synovial joint of a modified hinge (condyloid) variety between the lower end of the femur, the patella and the upper end of the tibia.

The femoral articular area is formed of three continuous surfaces, a middle anterosuperior concave surface for the patella, and medial and lateral condylar surfaces for the tibia. These condylar surfaces are markedly convex anteroposteriorly and mediolaterally. The posterior surface of the patella has a medial and a larger lateral facet for articulation with the sides of the femoral concavity. The facets are always in contact with the femur, their upper parts in flexion and their lower in extension. The upper surface of each tibial condyle is oval and slightly concave.

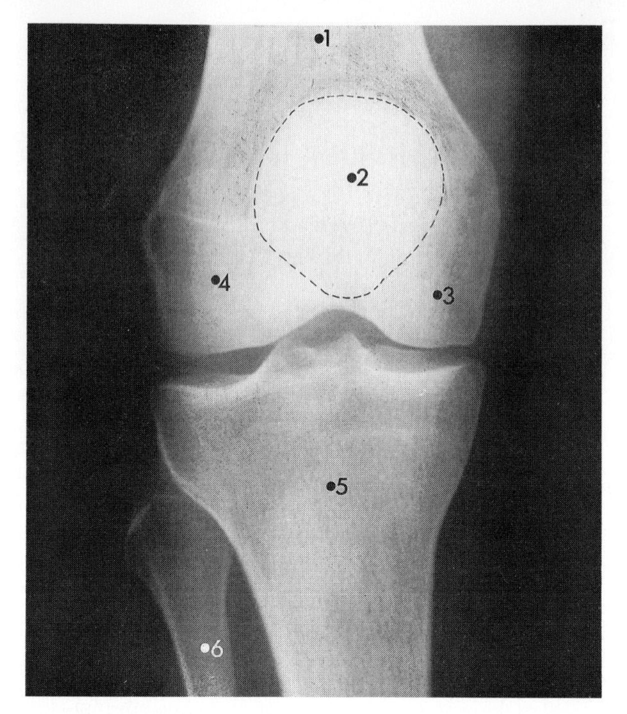

Fig. 22.3 Radiograph of knee. Anteroposterior view. Note that soft tissue shadows are seen and that overlapping of bones leads to a very white dense area indicating the position of the patella. Note the difference in density of the tibia and fibula.

1. Femur
2. Patella
3. Medial femoral condyle
4. Lateral medial condyle
5. Tibia
6. Fibula

Ligaments

(a) CAPSULE

On the femur this is attached near the articular margins medially and laterally. Posteriorly it is attached above the intercondylar notch. It is attached to the patellar margins and to the periphery of the upper end of the tibia except anteriorly where it descends to include the tibial tuberosity and posterolaterally where it is pierced by popliteus tendon. Anteriorly it is replaced by the quadriceps, patella and the patellar ligament. The capsule is variable in thickness.

(b) CAPSULAR THICKENINGS

Coronary ligaments—are the two parts of the capsule binding down the margins of the medial and lateral menisci to the outer margins of the tibial condyles.

(c) ACCESSORY LIGAMENTS

(i) **patellar ligament**—passes from the apex of the patella to the tibial tuberosity.

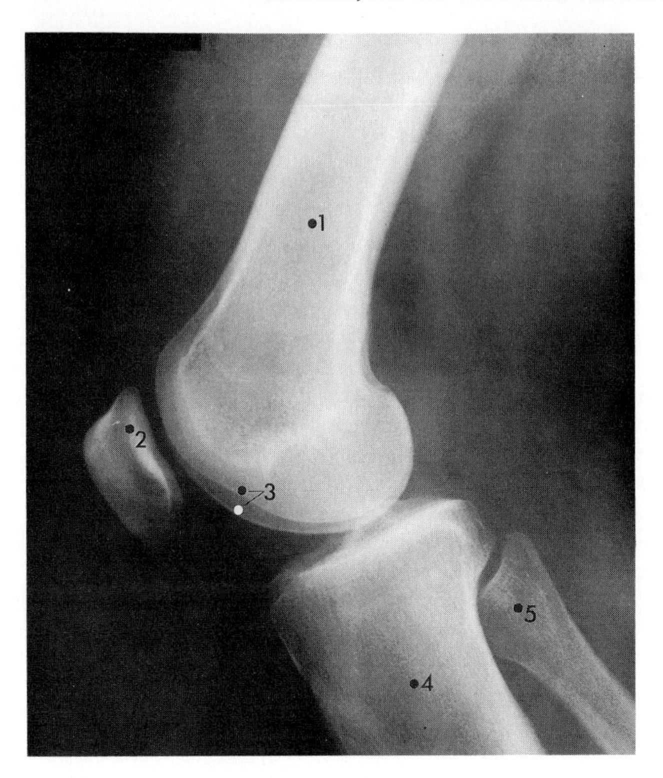

Fig. 22.4 Radiograph of knee. Lateral view. Note the soft tissue shadows.
1. Femur
2. Patella
3. Femoral condyles
4. Tibia
5. Fibula

(ii) **patellar retinacula**—are tendinous expansions of the vasti muscles passing to the sides of the patella and beyond to the tibial condyles.

(iii) **tibial collateral ligament**—is a broad flat band about 10 cm long passing from the medial epicondyle of the femur to the upper medial surface of the tibial shaft. It is attached to the capsule.

(iv) **fibular collateral ligament**—is a rounded cord passing from the lateral epicondyle of the femur to the head of the fibula in front of the apex and is separate from the capsule.

(v) **oblique popliteal ligament**—this extension of the semimembranosus tendon passes upwards and laterally from the medial tibial condyle across the posterior surface of the capsule.

(vi) **anterior** and **posterior cruciate ligaments**—are strong ligamentous bands found inside the capsule. The anterior ligament passes upwards, backwards and laterally from the front of the intercondylar area to the medial surface of the lateral femoral condyle. The posterior ligament passes forwards, upwards and medially from the back of the intercondylar area to the lateral surface of the medial femoral condyle.

INTRACAPSULAR STRUCTURES

(*a*) The **synovial membrane** develops from the linings of the three separate joint cavities that originally form the knee joint: the anterior between the patella and the femur, and the two posterior each between a femoral and a tibial condyle. The cavity of the patellofemoral joint later communicates on each side with the tibiofemoral joints but the inferior portion of synovial membrane, connecting the inferior border of the patella to the femoral intercondylar notch, persists as the **infrapatellar fold**. Remnants of the partitions originally separating the anterior and two posterior cavities persist as two fringes, the **alar folds**, which pass horizontally away from each side of the infrapatellar fold. The anterior and posterior cruciate ligaments develop in the fibrous septum between the tibiofemoral joints and a communication between these two joint cavities is formed across the septum in front of the ligaments.

The adult shape of the synovial membrane (Fig. 22.5) may be likened to a vertically running sleeve of synovia passing between the femur and tibia, horseshoe-shaped in transverse section with its concavity facing posteriorly. Superiorly the sleeve is attached to the periphery of the articular area of the femur. Inferiorly the attachment includes the two tibial articular facets and the front of the intercondylar area.

This arrangement is modified when: (i) an anterior central portion of the sleeve is replaced by the articular surface of the patella. (ii) The membrane below the patella is invaginated and gains attachment to the front of the intercondylar notch of the femur. The fold so formed is named the infrapatellar fold. (iii) The membrane above the patella is extended upwards between the quadriceps and the femur as the suprapatellar bursa. (iv) The two menisci are attached by their outer margins to the capsule, thus breaking the continuity of the synovial membrane on each side. (v) The joint cavity communicates with the suprapatellar, popliteal and gastrocnemius bursae.

(*b*) The **menisci** are two crescentic pieces of fibrocartilage with thickened outer margins. Each lies on a tibial condyle with its ends, the anterior and posterior horns, attached to the intercondylar area and its outer margin attached to the capsule. The medial cartilage is larger and semicircular; its central attachments embrace those of the lateral. The lateral cartilage is smaller and forms three-fifths of a circle. It is separated from the capsule posteriorly by the tendon of the popliteus which is partly attached to the cartilage, thus increasing its mobility.

(*c*) The anterior and posterior cruciate ligaments, the tendon to the popliteus, and a fat pad within the infrapatellar fold between the synovial membrane and the capsule are also present.

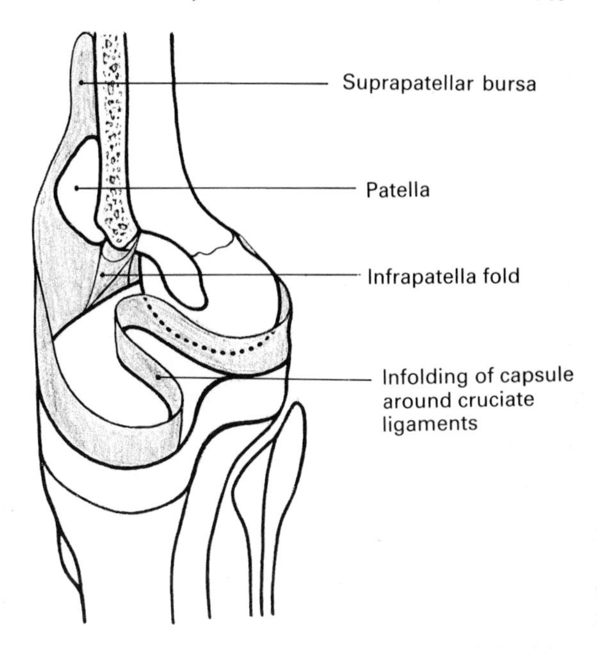

Fig. 22.5 Diagram showing some of the relations of the synovial membrane of the right knee joint.

The tendon of popliteus pierces the back of the capsule laterally. It becomes covered with a sheath of synovial membrane and is attached in a pit below the lateral epicondyle. (The pit is intracapsular in position.)

Functional aspects

MOVEMENTS

The knee joint is capable of flexion, extension and a little rotation.

FLEXION—hamstrings, aided by gastrocnemius. The movement is limited by the apposition of the surfaces of the calf and thigh.

EXTENSION—quadriceps and iliotibial tract muscles. The femoral condyles roll on the tibial condyles and also glide backwards. The capsular and cruciate ligaments become taut when the leg is nearly straight and so prevent further backward gliding of the lateral femoral condyle. Further extension (hyperextension) is therefore accompanied by backward movement only of the medial femoral condyle (around the axis of the taut anterior cruciate ligament). This movement is medial rotation of the femur on the tibia. The direction of the fibres of the collateral and oblique popliteal ligaments is such that they limit the rotation and extension at the knee. When the limit is reached the knee is said to be in the close-packed ('locked') stable position. At the start of flexion of the knee the femur rotates laterally on the tibia. The popliteus muscle, which is attached to the lateral meniscus, pulls the meniscus backwards thus preventing it from being crushed between the lateral femoral and tibial condyles.

A small amount of lateral and medial rotation may be produced by the hamstrings when the knee is flexed at a right angle.

STABILITY

The bony surfaces contribute little to the stability of the joint, this being dependent on strong ligaments and powerful muscles. As it is predominantly a hinge joint, the inextensible ligaments are placed medially and laterally. Anteriorly the joint is stabilised by the quadriceps and posteriorly by hamstrings and gastrocnemius muscles. The anterior and posterior cruciate ligaments limit anteroposterior gliding and distraction of the bones.

Quadriceps tends to pull the patella laterally as well as superiorly because of the obliquity of the femur. This lateral displacement is minimised by the anterior prominence of the lateral femoral condyle and by those fibres of the vastus medialis which are attached to the medial side of the patella. When standing on the extended knee the line of weight passes in front of the axis of rolling of the joint. The posterior ligaments take the strain and when these are stretched, reflex contraction of the hamstrings limits hyperextension.

Tears of the menisci occur when the slightly flexed weight-bearing knee is subjected to severe lateral strain as in sporting injuries. Removal of the damaged meniscus may be required to restore painless function to the knee joint.

Blood supply

This is from the anastomosis around the knee joint which receives a number of branches from the popliteal, femoral and anterior tibial arteries.

Nerve supply

This is from branches from the tibial, common peroneal, obturator and femoral nerves.

Relations

The joint is mainly subcutaneous, being separated from the skin anteriorly by quadriceps, the patella and the patellar ligament, laterally by the biceps tendon, and medially by semimembranosus and semitendinosus. Posteriorly it is related to popliteus, the popliteal vessels, the tibial and common peroneal nerves, and the muscular boundaries of the popliteal fossa. The popliteal artery lies next to the capsule and is separated from the the tibial nerve by its vein.

Bursae

There are many bursae around the knee joint, the more important being:

ANTERIOR
(i) suprapatellar, deep to quadriceps.
(ii) prepatellar.
(iii) superficial and deep infrapatellar, related to the patellar ligament.

POSTERIOR
(i) popliteus, deep to the muscle.
(ii) gastrocnemius, deep to the medial head of this muscle.
(iii) semimembranosus, between this muscle and the medial head of gastrocnemius.

A bursa may be present between each collateral ligament and the adjacent tendons. The suprapatellar, popliteus and gastrocnemius bursae communicate with the joint cavity and the semimembranosus may communicate with the gastrocnemius bursa and thus with the joint cavity.

THE TIBIOFIBULAR JOINTS

(i) **superior tibiofibular joint**—this is a synovial joint of the plane variety between the posteroinferior surface of the lateral tibial condyle and the oval facet on the head of the fibula. The capsule is thickened anteriorly and posteriorly, and the synovial membrane lining the nonarticular surface of the joint may be continuous with that of the knee joint.

(ii) **inferior tibiofibular joint**—is a fibrous joint between the adjacent inferior surfaces of the two bones. The surfaces are united by a short strong band, the interosseous tibiofibular ligament, and reinforced by anterior and posterior inferior tibiofibular ligaments.

(iii) **interosseous membrane**—is a fibrous membrane uniting the interosseous borders of the two bones. It separates and gives attachment to the deep flexor and extensor muscles of the leg.

Little movement occurs between the tibia and the fibula.

THE POPLITEAL FOSSA

This is a diamond-shaped fossa behind the knee joint. It is bounded superomedially by semitendinosus and semimembranosus, superolaterally by biceps, and inferiorly by the medial and lateral heads of gastrocnemius. It is roofed in by thickened fascia lata. Its floor is formed from above downwards by the popliteal surface of the femur, the posterior surface of the capsule of the knee joint and the fascia covering the popliteus muscle.

The fossa contains the popliteal vessels and branches (p. 311), the tibial and common peroneal nerves and branches (p. 313), much fat and a few lymph nodes.

THE LEG AND DORSUM OF THE FOOT

The muscles of the leg are divided into anterior, lateral and posterior groups by the tibia, interosseous membrane, and anterior and posterior intermuscular septa which pass inwards from the thick investing deep fascia to the fibula. The anterior intermuscular septum passes to the anterior border of the fibula and separates the anterior (dorsiflexor) compartment from the lateral (evertor) compartment. The posterior septum passes to the posterior border of the fibula and separates the lateral from the posterior (plantar flexor) compartment.

THE ANTERIOR (DORSIFLEXOR) GROUP OF MUSCLES
(Figs. 22.2A, 23.1)

Only tibialis anterior is attached to the tibia, the others to the fibula. During walking, these muscles pull the leg forward over the grounded foot. When the foot is not bearing weight, they dorsiflex the foot and the toes.

1. Tibialis anterior

Attachments

UPPER—upper two-thirds of the lateral surface of the tibia and adjoining interosseous membrane.

LOWER—the fibres converge on to a central tendon which descends in front of the ankle joint deep to the superior and inferior extensor retinacula and ends on the medial surface of the medial cuneiform and of the base of the 1st metatarsal bone.

Action: dorsiflexion and inversion of the foot.

2. Extensor hallucis longus

Attachments

UPPER—middle two-quarters of anterior surface of the fibula and adjoining interosseous membrane.

LOWER—its long tendon passes under the extensor retinacula, in front of the ankle joint and, crossing the dorsum of the foot, reaches the base of the distal phalanx of the great toe.

Action: dorsiflexion of the great toe and it assists in dorsiflexion of the foot.

3. Extensor digitorum longus

Attachments

UPPER—upper two-thirds of the anterior surface of the fibula lateral to the attachment of extensor hallucis longus, and from the adjacent intermuscular septum.

LOWER—above the ankle the muscle divides into four tendons which pass under the extensor retinacula. Diverging on the dorsum of the foot a tendon passes to each of the lateral four toes and is attached in a similar manner to extensor digitorum in the hand, i.e. there is a dorsal expansion which is attached to the dorsum of the middle and terminal phalanges.

Action: dorsiflexion of the toes and it assists in dorsiflexion of the foot.

4. Peroneus tertius

This is a thin sheet of muscle attached to the lower third of the anterior surface of the fibula. Its tendon descends under the extensor retinacula and is attached to the body of the 5th metatarsal. It is a weak dorsiflexor and evertor of the foot.

5. Extensor digitorum brevis

This is the only short muscle on the dorsum of the foot.

Attachments

PROXIMAL—anterior part of the upper surface of the calcaneum.

DISTAL—the muscle divides into four small tendons which pass obliquely forwards to the medial four toes. The 1st tendon is attached to the base of the proximal phalanx of the great toe, the remainder join the lateral side of the extensor tendons passing to the 2nd, 3rd, and 4th toes.

Action: dorsiflexion of the medial four toes.

Nerve supply: the deep peroneal branch of the common peroneal nerve supplies all the muscles of this group.

The extensor retinacula

The extensor retinacula lie across the extensor tendons in the lower part of the leg and in front of the ankle joint:

(i) **superior extensor retinaculum**—is a thickening of deep fascia stretching between the anterior borders of the tibia and fibula about 3 cm above the ankle joint. It covers the tendons, the anterior tibial vessels and the deep peroneal nerve.

(ii) **inferior extensor retinaculum**—is a thicker bifurcate structure extending medially from the anterior part of the upper surface of the calcaneus over the front of the ankle joint and dividing into two limbs. The upper limb is attached to the medial malleolus and the lower curves around the medial border of the foot to blend with the plantar aponeurosis.

The **synovial sheaths** of the long extensor tendons lie deep to the inferior retinaculum; only that of tibialis anterior extends proximally under the superior retinaculum. Extensor hallucis longus has its own sheath and there is a common sheath for extensor digitorum longus and peroneus tertius.

THE LATERAL (EVERTOR) GROUP OF MUSCLES

(Figs. 22.2, 23.1, 23.2)

When the foot is bearing all the body weight, these muscles assist in preventing the leg and the body tilting medially. At other times they evert the foot.

1. Peroneus longus

Attachments

UPPER—upper two-thirds of the lateral surface of the fibula.

LOWER—its tendon descends behind the lateral malleolus, separated from it by the tendon of peroneus brevis, crosses the lateral surface of the calcaneus below the peroneal tubercle and reaches the cuboid. It passes medially in the groove on the under surface of the cuboid, runs obliquely across the sole of the foot deep to the long plantar ligament (p. 325) and gains attachment to the lateral side of the base of 1st metatarsal and the adjacent area of the medial cuneiform bone.

2. Peroneus brevis

Attachments

UPPER—lower two-thirds of the lateral surface of the fibula.

LOWER—in its descent, the tendon lies anterior to that of peroneus longus. It passes behind the lateral malleolus, across the lateral surface of the calcaneus above the peroneal tubercle, to the tuberosity on the base of the 5th metatarsal.

Action: both these muscles evert possibly plantar flex the foot. Peroneus longus also helps to maintain the lateral longitudinal arch of the foot.

Nerve supply: the superficial peroneal nerve supplies both muscles.

The tendons of both these muscles are bound down to the lateral malleolus and the side of the calcaneus by two condensations of deep fascia, the **superior and inferior peroneal retinacula**. The tendons are enclosed in a common synovial sheath which is prolonged over each tendon to its distal attachment.

THE POSTERIOR (PLANTAR FLEXOR) GROUP OF MUSCLES

(Figs. 22.2, 23.1, 23.2)

Towards the end of the supporting phase of walking, the grounded foot is plantar flexed by these muscles, thus helping to propel the body forward.

A. SUPERFICIAL MUSCLES

These comprise gastrocnemius, soleus and plantaris.

1. Gastrocnemius
This is the most superficial muscle in the calf.

Attachments

UPPER—by medial and lateral heads to the back of the corresponding femoral condyles.

LOWER—the two fleshy bellies form the inferior boundaries of the popliteal fossa and unite to form a tendon which joins that of soleus in the lower calf and forms the conjoined **tendo calcaneus** which is attached to the middle of the posterior surface of the calcaneus. A bursa and a pad of fat separate the tendon from the upper part of the calcaneal surface.

2. Soleus
This lies deep to gastrocnemius.

Attachments

UPPER—in a continuous line from the upper part of the posterior surface and head of the fibula, a tendinous arch over the posterior tibial vessels and tibial nerve, and the soleal line on the tibia.

LOWER—the tendon unites with that of gastrocnemius and forms the tendo calcaneus.

3. Plantaris
This vestigial muscle is attached above to the lower end of the lateral supracondylar line and by a long tendon below to the medial side of the tendo calcaneus.

Action
Both gastrocnemius and soleus are powerful plantar flexors of the foot and as such are important in both posture and locomotion. In gastrocnemius and soleus, and between them, is a plexus of veins. Contraction of the muscles pumps the blood upwards (Fig. 24.2).

Nerve supply
Branches from the tibial nerve supply all the calf muscles.

B. DEEP MUSCLES

These comprise popliteus above the soleal attachment and the flexor digitorum longus, tibialis posterior and flexor hallucis longus from medial to lateral below soleus.

1. Popliteus

This triangular muscle lies deep in the lower part of the popliteal fossa.

Attachments

TIBIAL—posterior surface of the tibia above the soleal line.

FEMORAL—the muscle converges on to a tendon which pierces the capsule of the knee joint and gains attachment to the pit below the lateral epicondyle. It also sends a slip to the lateral meniscus. The popliteal bursa lies between the muscle and the upper part of the tibia.

Action

(i) lateral rotation of the femur on the tibia.
(ii) retraction of the lateral meniscus.

These two actions occur simultaneously in flexing the extended knee joint (p. 303). The backward movement of the lateral femoral condyle is preceded by the retraction of the cartilage which thus avoids injury.

Nerve supply: a branch from the tibial nerve.

2. Flexor digitorum longus

Attachments

UPPER—medial side of the posterior surface of the tibia below the soleal line.

LOWER—the tendon crosses over tibialis posterior and may groove the lower end of the tibia lateral to the tendon of tibialis posterior. It then passes deep to the flexor retinaculum and enters the sole of the foot where it divides into four slips. These slips pass to the terminal phalanges of the lateral four toes (their attachment being similar to that of the tendons of flexor digitorum profundus in the hand). In the foot, the main tendon gives attachment to flexor accessorius and the four slips to the four lumbrical muscles.

Action: plantar flexion of the interphalangeal and metatarsophalangeal joints. It is also a weak plantar flexor of the foot.

3. Tibialis posterior

Attachments

UPPER—posterior surface of the interosseous membrane and the adjoining surfaces of the tibia and fibula.

LOWER—the tendon inclines medially deep the flexor digitorum longus, grooves the back of the medial malleolus, passes under the flexor retinaculum and then lies on the deltoid ligament. In the sole it is attached mainly to the tuberosity of the navicular but also to all other tarsal bones except the talus.

Action: inversion of the foot and helps to support the medial longitudinal arch.

4. Flexor hallucis longus

Attachments

UPPER—lower two-thirds of the posterior surface of the fibula.

LOWER—the muscle becomes tendinous beehind the lower end of the tibia where it lies lateral to the tendons of flexor digitorum longus and tibialis posterior. It grooves the back of the tibia and the talus and the under surface of the sustentaculum tali, and enters the sole. Here it is crossed over by the tendon of flexor digitorum longus and runs forwards between the two heads of the flexor hallucis brevis. It passes in the fibrous flexor sheath of the great toe to be attached to the plantar surface of the base of the distal phalanx.

Action: powerful plantar flexor of the great toe. It also helps in plantar flexion of the foot and is of importance in walking (p. 337) and in the maintenance of the medial longitudinal arch.

Nerve supply: these three deep muscles are supplied by the tibial nerve.

The flexor retinaculum—is a thickened band of deep fascia passing from the medial malleolus to the medial tubercle of the calcaneus. It overlies from medial to lateral, the tendons of tibialis posterior and flexor digitorum longus, posterior tibial vessels, tibial nerve and the flexor hallucis longus tendon. Each tendon is enclosed in a separate synovial sheath, those of flexor hallucis longus and flexor digitorum longus passing into the sole of the foot.

VESSELS AND NERVES OF THE LEG

THE POPLITEAL ARTERY (Fig. 22.6)

This is a continuation of the femoral artery. It descends from the opening in the adductor magnus to the lower border of popliteus where it divides into anterior and posterior tibial arteries.

Relations

It descends on the floor of the popliteal fossa (p. 305), deep to its vein which separates it throughout its course from the tibial nerve.

Branches

(i) *muscular branches*.

(ii) *articular branches*—to the knee joint.

(iii) **anterior tibial artery**—passes forwards above the interosseous membrane and accompanies the deep peroneal nerve in its descent through the extensor compartment of the leg. Below the ankle joint it continues as the **dorsalis pedis artery** which passes forwards on the dorsum of the foot and reaches the space between the 1st and 2nd metatarsal bones. It enters the sole of the foot between these bones and anastomoses with the lateral plantar artery (p. 328).

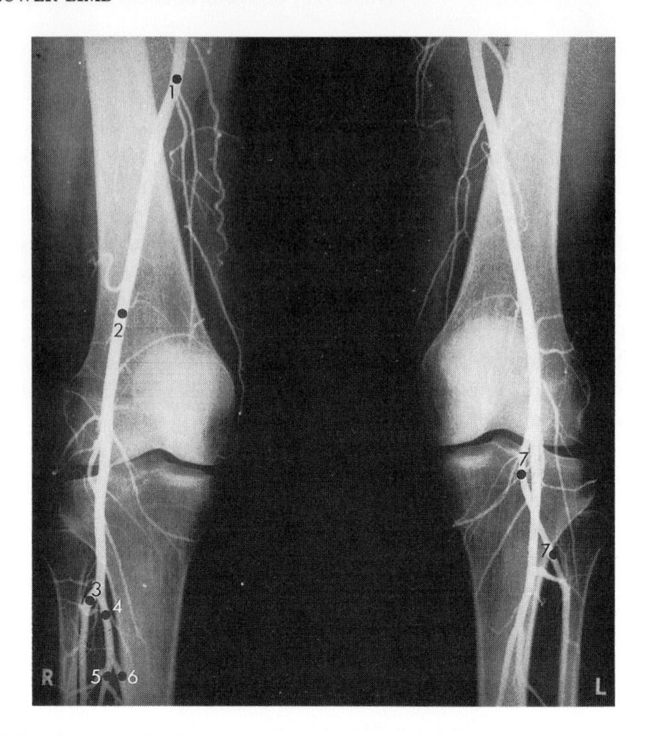

Fig. 22.6 Popliteal arteries filled with contrast material which has been injected into the abdominal aorta. (See also Figures 15.1, 21.4, 22.7).

1. Femoral artery
2. Popliteal artery
3. Anterior tibial artery
4. Tibio-peroneal trunk
5. Peroneal artery
6. Posterior tibial artery
7. Abnormally high origin of the left anterior tibial artery

(iv) **posterior tibial artery**—is also a terminal branch of the popliteal artery. It begins at the lower border of popliteus and descends through the flexor compartment of the leg with the tibial nerve. It ends behind the medial malleolus by dividing into medial and lateral plantar arteries. Its largest branch, the **peroneal artery**, descends between the fibula and flexor hallucis longus. It gives off muscular branches and an articular branch to the ankle joint.

THE POPLITEAL VEIN (Fig. 24.2)

This is formed by the union of the anterior and posterior tibial veins. Throughout its course it lies between the popliteal artery and tibial nerve. Its tributaries correspond mainly to the branches of the artery but the **small saphenous vein** enters it after piercing the fascial roof of the popliteal fossa. It receives blood which has been pumped up by contraction of the calf muscles.

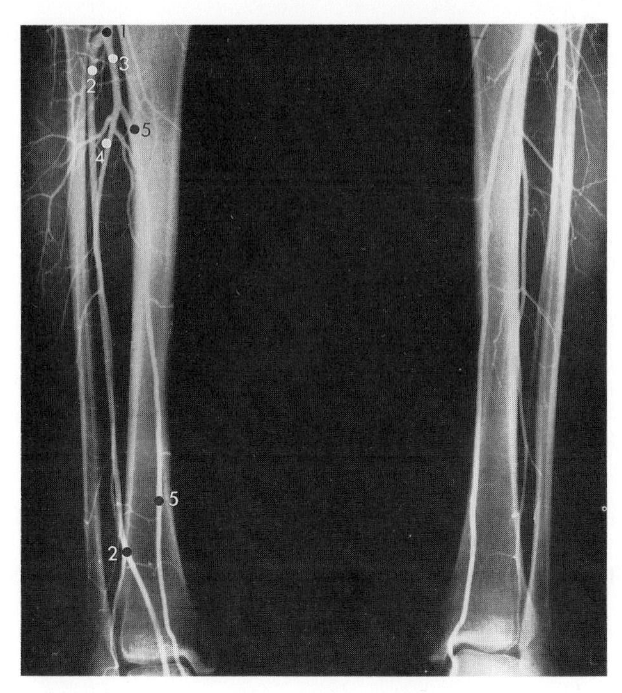

Fig. 22.7 Arteriogram of lower leg arteries. (See also Figures 15.1, 21.4, 22.6.)
1. Distal popliteal artery
2. Anterior tibial artery
3. Tibio-peroneal trunk
4. Peroneal artery
5. Posterior tibial artery

THE TIBIAL NERVE

This is a terminal branch of the sciatic nerve. It is formed just above the popliteal fossa, descends almost vertically through the fossa and passes deep to soleus. It descends through the flexor compartment of the leg and reaches the back of the medial malleolus where it ends by dividing into medial and lateral plantar nerves.

Relations

In the popliteal fossa: in most of its course it lies subcutaneously in the popliteal fat superficial to the popliteal vessels, the vein separating it from the artery.

In the calf: it descends deep to soleus lying between flexor digitorum longus medially and flexor hallucis longus laterally. It lies on tibialis posterior superiorly and the capsule of the ankle joint inferiorly and ends deep to the flexor retinaculum.

Branches

(i) **sural nerve**—this cutaneous branch descends on the posterior surface of gastrocnemius and unites with a branch of the common peroneal nerve halfway down the leg. It continues in company with the small saphenous vein behind the lateral malleolus and along the lateral side of the foot. It supplies the skin over the back of the leg, the ankle joint and the lateral surface of the heel, foot and 5th toe.

(ii) *articular branches*—to the knee and superior tibiofibular joints.

(iii) *muscular branches*—to the medial and lateral heads of gastrocnemius, plantaris, popliteus, soleus and the more deeply placed tibialis posterior, flexor hallucis longus and flexor digitorum longus.

(iv) *medial and lateral plantar nerves* (p. 329).

THE COMMON PERONEAL NERVE

This also is a terminal branch of the sciatic nerve formed just above the popliteal fossa. It descends along the lateral margin of the fossa, and passes into peroneus longus where it divides into superficial and deep peroneal nerves.

Relations

In its descent it lies on the lateral head of gastrocnemius and then the neck of the fibula. It is overlapped by biceps tendon in its upper course but is subcutaneous on the neck of the fibula.

Branches

(i) a branch which pierces the roof of the popliteal fossa and joins the sural nerve halfway down the leg.

(ii) lateral cutaneous nerve of the leg—descends over the lateral head of gastrocnemius and supplies the lateral aspect of the leg.

(iii) *articular branches*—to the knee and superior tibiofibular joints.

(iv) **superficial peroneal nerve**—is a terminal branch of the main trunk. It begins in the substance of peroneus longus and descends deep to this muscle, between it and peroneus brevis. In the lower part of the leg it emerges between peroneus brevis and peroneus tertius and divides into cutaneous branches.

Branches:

(*a*) muscular branches—to peroneus longus and brevis.

(*b*) cutaneous branches—to the skin over the lower lateral part of the leg.

(*c*) dorsal cutaneous branches—these descend over the extensor retinacula on to the dorsum of the foot and there divide into dorsal digital branches. The medial branch supplies the medial side of the great toe and the 2nd interdigital cleft; the lateral branch supplies the 3rd and 4th interdigital clefts.

(v) **deep peroneal nerve**—is also a terminal branch of the main trunk. It continues around the neck of the fibula, pierces the anterior intermuscular septum and then descends in the extensor compartment in company with the anterior tibial artery initially between extensor digitorum longus laterally and tibialis anterior medially and then between extensor hallucis longus laterally and tibialis anterior medially. At the ankle joint it passes under the extensor retinaculum lateral to the artery and divides into medial and lateral terminal branches.

Branches:

(*a*) muscular—to the long dorsiflexor muscles.

(*b*) lateral terminal branch—passes deep to extensor digitorum brevis, supplying it and the ankle joint.

(*c*) medial terminal branch—continues on the dorsum of the foot lateral to the dorsalis pedis artery and supplies the skin of the 1st interdigital cleft.

— 23 —

The foot

The foot is an arched platform consisting of a number of separate bones bound together by ligaments and muscles. The arched nature of the foot supports the body's weight, acts as a lever which is sufficiently rigid to propel the body forwards, and yet is resilient enough to absorb sudden shocks. Less important functions of the foot concern the obtaining of proprioceptive information, and under special circumstances its use in grasping. As long as an artificial foot will carry the body's weight and permit reasonable walking, it is functionally adequate and can be aesthetically acceptable. In the hand, however, sensory and manipulative abilities are of prime importance and most artificial substitutes are functionally very inadequate.

SKELETON

TARSUS (Figs. 23.1, 23.2)

This comprises the talus, calcaneus, navicular, cuboid and the three cuneiform bones.

Talus

The talus forms the connecting link between the bones of the leg and those of the foot. It has a body, a neck and a head, and has no muscular attachments.

The **body** lies posteriorly. Its articular superior surface is markedly convex anteroposteriorly, slightly concave from side to side, and articulates with the inferior surface of the tibia. This facet is widest in front and is continuous with a facet on each side of the bone for articulation with the medial and lateral malleoli; the medial facet is comma-shaped and the lateral is triangular. These three articular surfaces are collectively known as the **trochlear surface** of the talus. The inferior surface of the body is concave and articulates with the posterior facet on the calcaneus. The posterior margin bears posterior and medial tubercles; the posterior gives attachment to the posterior talofibular ligament and the groove between the tubercles lodges the tendon of flexor hallucis longus. The

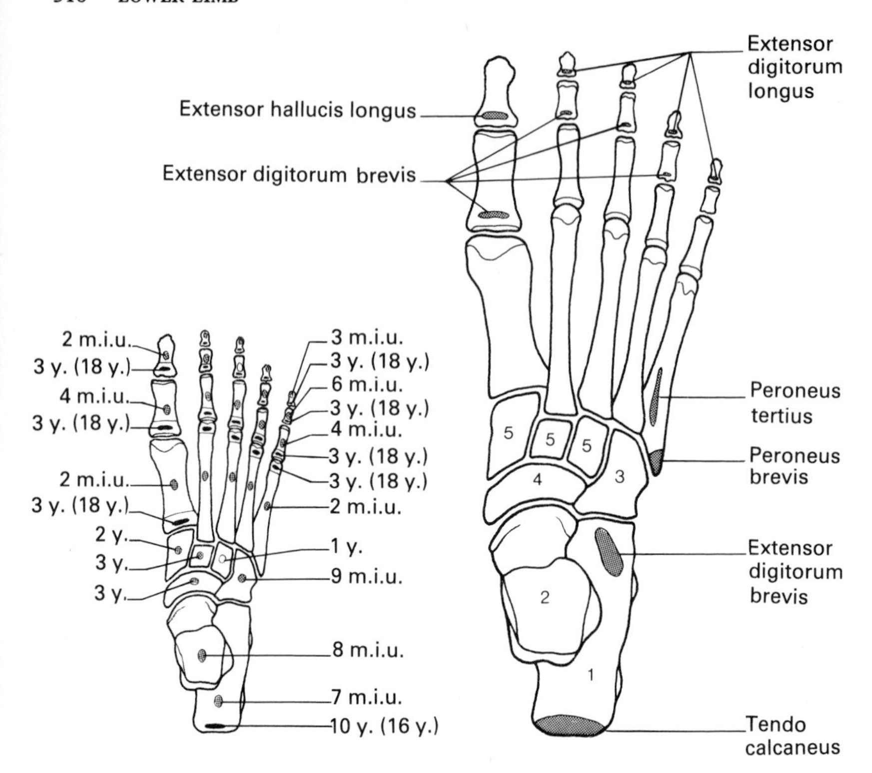

Fig. 23.1 The bones of the right foot, dorsal aspect. The tarsal bones are numbered: 1. Calcaneus; 2. Talus; 3. Cuboid; 4. Navicular; 5. Medial, middle and lateral Cuneiforms. In front are the metatarsal bones and the phalanges. Inset: ossification (m.i.u., months in utero; y., years).

neck is constricted and inferiorly bears a deep groove, the sulcus tali. The **head** is hemispherical and is inclined medially. It has articular facets for (*a*) the navicular bone anteriorly, (*b*) the anterior and middle calcaneal facets inferiorly, and (*c*) the plantar calcaneonavicular (spring) ligament inferomedially. The facets are all in continuity.

Calcaneus

This is an irregular bone roughly rectangular in shape; its posterior part forms the prominence of the heel. The upper surface has posterior, middle and anterior articular facets for the talus. The posterior facet is convex and covers the middle third of the bone; it receives the inferior concavity of the body of the talus. The middle and anterior facets articulate with the head of the same bone. The middle facet lies on a medial extension of the bone, the **sustentaculum tali** which gives attachment to the plantar calcaneonavicular ligament. The middle facet is separated from the posterior by a deep groove, the **sulcus calcanei**. (The sulcus

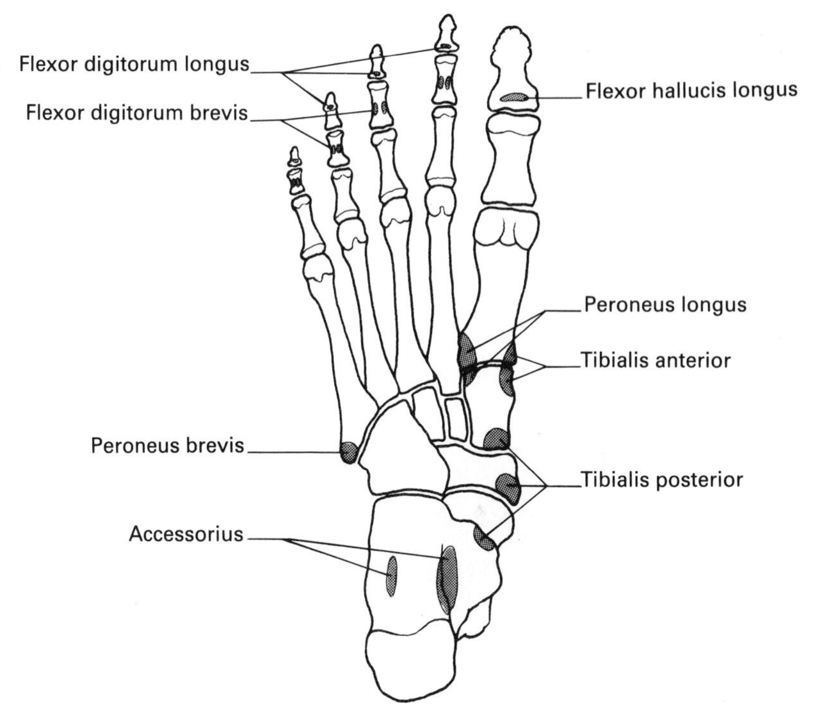

Fig. 23.2 The bones of the right foot, plantar aspect. On the cuboid is a groove for the peroneus longus tendon.

calcanei and the sulcus tali together form a narrow tunnel, the **sinus tarsi**, which houses the strong interosseous talocalcaneal ligament.)

The inferior surface has an **anterior tubercle**, to which is attached the short plantar ligament, and medial and lateral **processes** on the **tuberosity**. The region gives attachment to the long plantar ligament, some short muscles of the sole and the plantar aponeurosis. The inferior surface of the sustentaculum tali is grooved by the flexor hallucis longus tendon.

The lateral surface has a short ridge, the **peroneal tubercle**, which separates the peroneus longus and brevis tendons. The posterior surface gives attachment to the tendo calcaneus (Achilles' tendon) in its middle third, and the smaller anterior surface articulates with the cuboid bone.

Navicular

This lies on the medial side of the foot between the talus and the cuneiform bones. Posteriorly it has a concave facet for the head of the talus and anteriorly three facets for the cuneiform bones. Its medial surface is extended downwards and medially to form the prominent navicular **tuberosity** to which the tendon of tibialis posterior and the plantar calcaneonavicular ligament are attached.

Cuboid

The cuboid lies on the lateral side of the foot. It is roughly wedge-shaped and narrower inferiorly. Posteriorly it articulates with the calcaneus, anteriorly with the bases of the 4th and 5th metatarsal bones, and medially with the lateral cuneiform and sometimes the navicular. The anterior part of the inferior surface has a prominent **groove** (for the peroneus longus tendon) situated between anterior and posterior ridges which give attachment to the long plantar ligament.

Cuneiforms

These are three wedge-shaped bones. The medial is narrower superiorly and the intermediate and lateral bones narrower inferiorly. Posteriorly they articulate with the navicular and anteriorly with the bases of the medial three metatarsal

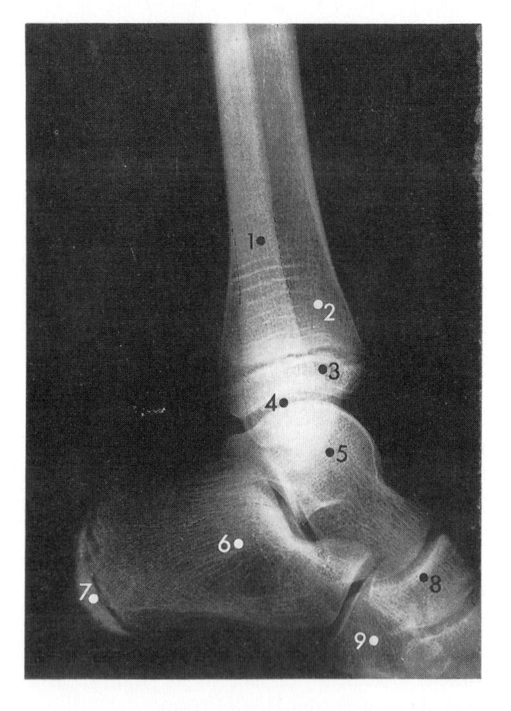

Fig. 23.3 Lateral view of the ankle region of a girl aged 12 years. Note the presence of the epiphyseal cartilage at the distal end of the tibia. Above the epiphyseal line are eight transversely running white lines. These are 'scars' (Harris's lines) and indicate periods of arrested longitudinal growth during a series of illnesses over a period of four years.

1. Outline of fibula overlapping tibia on this view
2. Tibia
3. Tibial epiphysis
4. Line of ankle joint
5. Talus
6. Calcaneus
7. Calcaneal epiphysis
8. Navicular
9. Cuboid

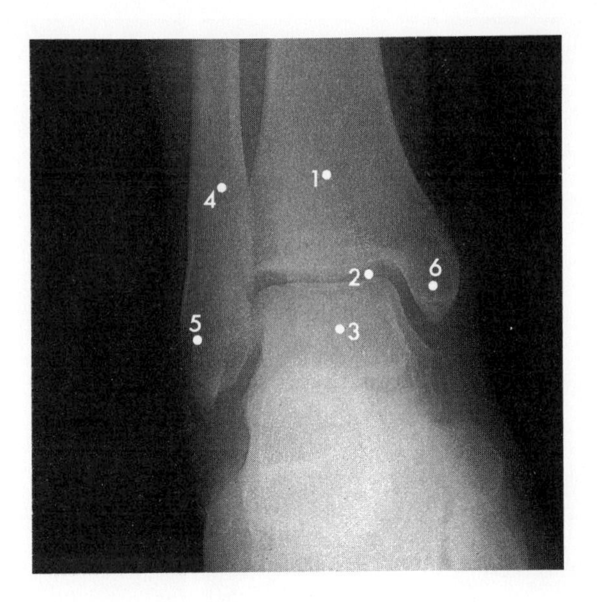

Fig. 23.4 Anteroposterior view of normal adult ankle.

1. Tibia
2. Line of ankle joint
3. Talus
4. Fibula
5. Lateral malleolus
6. Medial malleolus

bones. The medial is the largest and the intermediate the shortest. The base of the 2nd metatarsal is wedged between the medial and lateral bones. The lateral cuneiform bone has a lateral articular facet for the cuboid. Parts of the tendons of tibialis anterior, tibialis posterior and peroneus longus are attached to the medial cuneiform.

Metatarsal bones and phalanges (Figs. 23.1, 23.2)
These are long bones resembling the metacarpals and phalanges of the hand. However, (*a*) the metatarsals are longer and more slender than the metacarpals, (*b*) the base of the 1st metatarsal is not saddle-shaped, (*c*) the base of the 5th metatarsal bears a posterior projection from its lateral side, (*d*) the phalanges are shorter in the foot and the 5th toe frequently has only two.

THE ANKLE JOINT (Figs. 23.3–23.6)

This is a synovial joint of the hinge variety between the mortise formed by the inferior surface of the tibia and the facing surfaces of the medial and lateral malleoli and, inferiorly, the upper, medial and lateral surfaces of the talus.

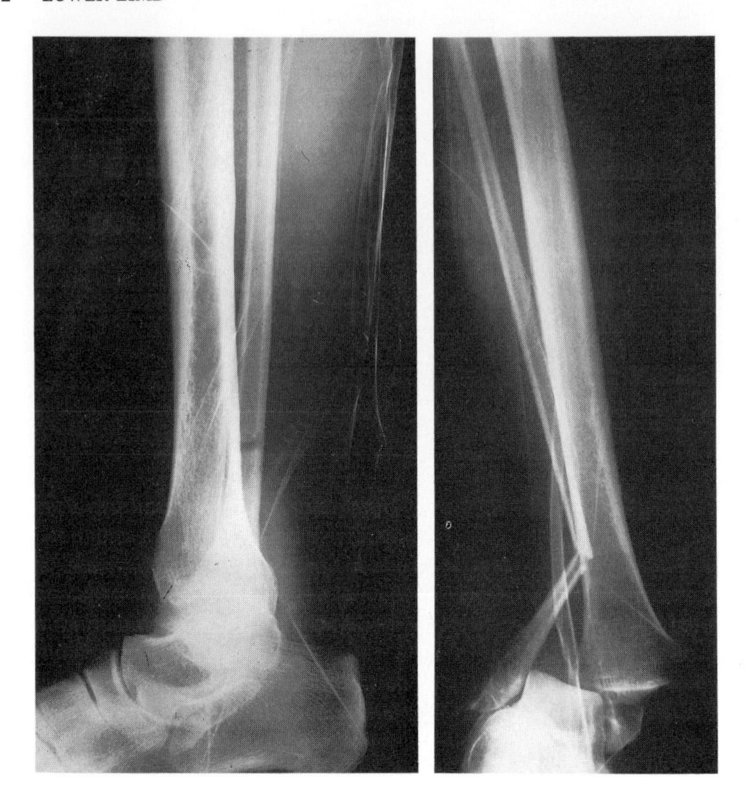

Fig. 23.5 Lateral (**A**) and anteroposterior (**B**) views of the distal end of the right tibia and fibula and the ankle joint. Note the fractures of the tibial malleolus and the distal body of the fibula. The pair of radiographs demonstrates the importance of obtaining views in at least two planes. **A**. shows the fibular fracture but does not indicate the gross displacement and disruption of the ankle joint. Compare with Figures 23.3 and 23.4.

Ligaments

(a) CAPSULE

This is attached to the articular margins except anteriorly where it extends on to the neck of the talus.

(b) CAPSULAR THICKENINGS

(i) **medial (deltoid) ligament**—is triangular in shape with its apex attached to the medial malleolus and its base to the navicular, the neck of the talus and the plantar calcaneonavicular ligament, the sustentaculum tali and the body of the talus from before backwards.

(ii) **lateral ligament**—is composed of three separate parts; anterior talofibular ligament passing from the lateral malleolus horizontally forwards to the neck of the talus, the calcaneofibular ligament passing backwards from the tip of the malleolus to the lateral side of the calcaneus, and the posterior talofibular ligament passing horizontally from the malleolar fossa on the lateral malleolus to the posterior tubercle of the talus.

(iii) **anterior and posterior ligaments**—are weak flat bands.

(c) ACCESSORY LIGAMENTS

Inferior transverse tibiofibular ligament is a thick band passing between the upper part of the malleolar fossa on the fibula and the posterior edge of the articular surface on the tibial malleolus. It forms part of the articular surface for the talus posteriorly.

Intracapsular structures

The synovial membrane lines the nonarticular surfaces of the joint.

Functional aspects

MOVEMENT

The joint may be plantar flexed and dorsiflexed. Plantar flexion is downward movement of the foot. It is produced by gastrocnemius and soleus assisted by other muscles of the flexor and possibly the peroneal compartments of the leg. Dorsiflexion is upward movement of the foot. It is produced by tibialis anterior assisted by the other muscles of the extensor compartment.

When the foot is plantar flexed slight passive side-to-side movement of the talus is possible as its articular surface is narrower posteriorly.

STABILITY

This is a stable joint maintained by the mortise arrangement of its bones and the powerful ligaments. The joint is least stable in the plantar flexed position. The line of weight passes anterior to the joint and as the leg dorsiflexes on the grounded foot, the tibiofibular mortise firmly grips the wider anterior talar surface. Severe inversion or eversion injuries to the ankle frequently cause a fracture dislocation of the joint, the treatment of which is difficult and complex. Less severe strains may be associated with disabling tears of the medial or the lateral ligaments.

Blood and nerve supply

Anterior and posterior tibial vessels and tibial and deep peroneal nerves.

Relations (Fig. 23.6)

Medially and laterally the joint lies subcutaneously. Anteriorly it is crossed from medial to lateral by tendons of tibialis anterior and extensor hallucis longus, the anterior tibial vessels and deep peroneal nerve and the tendons of extensor digitorum longus and peroneus tertius. Posteromedially from medial to lateral it is crossed by the tendons of tibialis posterior and flexor digitorum longus, the

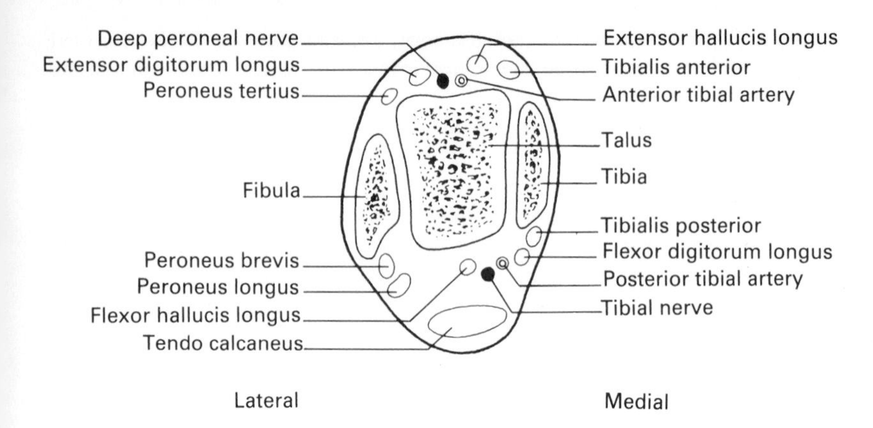

Fig. 23.6 Diagrammatic lay-out showing some of the main structures surrounding the left ankle joint.

posterior tibial vessels and tibial nerve, and the tendon of flexor hallucis longus. The peroneus longus and brevis tendons cross the joint posterolaterally. All the tendons are surrounded by synovial sheaths. The joint is separated by a posterior pad of fat from the tendo calcaneus.

THE INTERTARSAL AND OTHER JOINTS

These comprise the subtalar (posterior talocalcaneal), the talocalcaneonavicular (including the anterior talocalcaneal joint), the calcaneocuboid, the smaller joints around the cuneiform and metatarsal bones and the phalanges.

Subtalar (posterior talocalcaneal) joint—is a synovial joint between the inferior concave facet on the body of the talus and the posterior convex facet of the calcaneus. The capsule is partly thickened by talocalcaneal bands but the main bond of union between the two bones is the **interosseous talocalcaneal ligament** lying in the sinus tarsi.

Talocalcaneonavicular joint—is a synovial joint of the ball and socket variety between the hemispherical head of the talus and the concavity formed by (*a*) the anterior and middle facets on the upper surface of the calcaneus posteriorly, (*b*) the concavity of the navicular anteriorly, and (*c*) the inner surface of the plantar calcaneonavicular (spring) ligament inferomedially. The capsule is reinforced by the spring ligament inferomedially, the deltoid ligament superomedially and the medial band of the bifurcate ligament (see below) laterally.

The plantar calcaneonavicular (spring) ligament is a strong thick band passing from the sustentaculum tali to the navicular tuberosity; it forms an articular surface for the head of the talus and is here covered by articular hyaline cartilage. It is an important factor in maintaining the medial longitudinal arch of the foot.

Calcaneocuboid joint—is a synovial joint of the saddle variety between the reciprocal concavoconvex facets of the two bones; the capsule is reinforced by the bifurcate ligament, and the long and short plantar ligaments are accessory to the joint.

The bifurcate ligament is attached proximally to the anterior part of the upper surface of the calcaneus and divides distally to gain attachment to the adjacent upper surfaces of the cuboid and navicular bones. The **long plantar ligament** passes from the posterior calcaneal processes to the posterior and anterior ridges on the cuboid and then on to the bases of the lateral metatarsals. The **short plantar ligament** is more deeply placed and passes from the anterior calcaneal tubercle to the adjacent surface of the cuboid.

Functional aspects

The movements of inversion and eversion occur at these joints. In **inversion** the sole of the foot is turned to face inwards, thus raising its medial border; in **eversion** the sole is turned to face outwards, thus raising its lateral border. Both movements occur mainly at the subtalar and the talocalcaneonavicular joints. The calcaneus and the navicular, carrying the forepart of the foot with them, move medially on the talus by a combination of rotary and gliding movements. The combined movements produced at these joints are similar to those obtained at a saddle-shaped joint.

However, (*a*) the range of inversion and eversion is increased by movement at the midtarsal (talonavicular and calcaneocuboid) joint.

(*b*) Inversion is increased during plantar flexion on account of a small amount of rocking of the narrower posterior end of the talar articular facet at the ankle joint. (*c*) Eversion is increased in dorsiflexion.

Inversion is produced by the tibialis anterior and posterior, and is limited by tension in the peroneal muscles and in the interosseous talocalcaneal ligament. Eversion is produced by the peroneal muscles, and limited by tension in the tibialis anterior and posterior, and in the deltoid ligament.

The movements have been described when the foot is off the ground. The same movements occurring when the foot is on the ground adjust the foot and lower limb to uneven and sloping surfaces.

The **tarsometatarsal joints** and the remaining **intertarsal joints** are small plane synovial joints tightly bound together by interosseous ligaments of which those on the plantar aspect of the foot are the stronger. These joints are capable only of slight gliding movements.

Metatarsophalangeal and interphalangeal joints—resemble those of the fingers but are less mobile, the heads of the five metatarsals being bound together by deep transverse ligaments which unite the plantar thickenings of all the joint capsules. (In the hand, the thumb is freely opposable.) Plantar flexion and dorsiflexion occur at all these joints, and also a very limited amount of abduction and adduction at the metatarsophalangeal joints may be possible. Plantar flexion is produced by flexor hallucis longus and brevis and flexor digitorum longus and brevis. Flexor accessorius maintains the toes in a plantar flexed position when the ankle joint is being dorsiflexed. Dorsiflexion is produced by extensor hallucis longus and extensor digitorum longus and brevis muscles. Abductor hallucis and abductor digiti minimi may spread the toes, the interossei bring them together.

The first metatarsophalangeal joint in man is much larger than the others. Inferiorly a sesamoid bone is found in each tendon of flexor hallucis brevis, the tendons and bones being incorporated in the joint capsule. In the groove between the sesamoid bones, and so protected from pressure, is the tendon of the flexor hallucis longus. The natural angle between the long axis of the first metatarsal and that of the first phalanx may be accentuated by the wearing of poorly fitting shoes. A bursa is often present on the medial side of the joint and when it becomes inflamed, the condition is called a bunion.

THE ARCHES OF THE FOOT

The arched nature of the foot is a predominantly human characteristic and is present from birth, although masked in the baby by a large amount of subcutaneous fat. The arches allow the body's weight to be distributed over a larger area, and prevent crushing of the vessels and nerves crossing the sole.

The jointed pattern of the arch gives the foot resilience well suited to absorb the impact of the body's weight when the foot comes into contact with the ground, while still allowing it to act as a semirigid lever for propelling the body forwards. The arched form comprises separate medial, lateral and transverse arches which, although to be considered individually, are *collectively involved* in the functions already enumerated.

The maintenance of each arch is dependent on bony, ligamentous and muscular factors which act in a complementary manner, it being unlikely that a single factor could maintain the arch for any length of time. The short ligaments and muscles tie the adjacent bones of the arch together and the long ligaments, the plantar aponeurosis and muscle tendons tie the two ends of the arch together. Some long tendons act in a sling-like manner, supporting the centre of the arch. The ligaments and muscles found on the plantar surface of the foot are stronger and more numerous than those on the dorsum.

Medial longitudinal arch

The medial longitudinal arch extends from the medial process of the calcaneus to the heads of the medial three metatarsals. The bones forming it are the calcaneus, talus, navicular, three cuneiforms and the medial three metatarsal bones. The arch is maintained by muscles and ligaments.

The *ligamentous supports* are:

(i) the plantar calcaneonavicular (spring) ligament supporting the head of the talus.
(ii) the interosseous ligaments.
(iii) the plantar aponeurosis tying together the ends of the arch.

The *muscular supports* are:

(i) the short muscles of the foot, especially abductor hallucis, flexor hallucis brevis and flexor digitorum brevis.
(ii) tibialis anterior attached near to the centre of the arch.
(iii) flexor hallucis longus and the medial part of flexor digitorum longus passing between the ends of the arch.
(iv) tibialis posterior tying together the posterior bones of the arch.

The bony contribution is limited to the support given by the sustentaculum tali to the talus and to the wedge-shape of some of the bones.

Lateral longitudinal arch

The lateral longitudinal arch is lower than the medial and extends from the lateral process of the calcaneus to the heads of the 4th and 5th metatarsal bones. It is formed from the calcaneus, cuboid and 4th and 5th metatarsal bones. The arch is maintained mainly by ligaments.

The *ligamentous supports* are:

(i) the plantar aponeurosis.
(ii) the long and short plantar ligaments uniting the calcaneus to the cuboid and the bases of the 4th and 5th metatarsal bones.
(iii) the interosseous ligaments.

The *muscular supports* are:

(i) the short muscles of the foot, especially abductor digiti minimi and flexor digitorum brevis.
(ii) peroneus longus, passing below the cuboid and supporting the centre of the arch.

The transverse arch

The transverse arch lies across the distal row of tarsal bones and the adjacent metatarsal bones and is formed by the cuneiform and cuboid bones and the metatarsal bases. The arch is maintained by the shape of the bones, the interosseous ligaments, and peroneus longus, tibialis posterior and adductor hallucis muscles.

THE SOLE OF THE FOOT

The skin over the weight-bearing areas of the heel and the 'ball' of the foot is thickened in its superficial layers and firmly attached to the deep fascia by fibrous septa passing through the superficial fascia. The skin is thin in the flexure creases. The fatty superficial fascia insulates and cushions the deeper structures.

The deep fascia consists of a thickened central portion, the **plantar aponeurosis,** and thin medial and lateral portions. The aponeurosis is attached posteriorly to the back of the under-surface of the calcaneus and divides anteriorly into five digital expansions. These are continuous with the fibrous flexor tendon sheaths along the plantar aspect of the toes, and are also attached to the ligaments binding together the heads of the metatarsals. When the toes are dorsiflexed, the plantar aponeurosis is tightened and the longitudinal arch is accentuated.

MUSCLES OF THE SOLE

The small muscles of the sole have a similar pattern of attachment to those of the hand. Short flexors, abductors, and adductor of the big toe, lumbrical and interosseous muscles are present but there is no opponens muscle. The small muscles help to maintain the arches and move the toes but have little abductor or adductor action. A large sesamoid bone is present on each side of the head of the 1st metatarsal. The sesamoids are embedded in the tendons of the short muscles passing to the big toe and in the capsule of the 1st metatarsophalangeal joint. The sesamoids carry much of the weight and form a protective groove for the tendon of flexor hallucis longus.

The **flexor accessorius** has no counterpart in the hand. It is attached to the calcaneus posteriorly and to the lateral side of the tendons of flexor digitorum longus and flexor hallucis longus anteriorly. Tension in the digital tendons and flexion of the toes can be maintained by accessorius when the long flexors have to relax. This occurs when the leg is being pulled forwards over the ankle joint by tibialis anterior and the extensor muscles during the supporting phase of walking.

All the short muscles of the sole are supplied by the lateral plantar nerve except for abductor hallucis, flexor hallucis brevis, flexor digitorum brevis and the 1st lumbrical which are supplied by the medial plantar nerve. (Compare this distribution with that of the ulnar and median nerves in the palm.)

The long flexor tendons pass along the sole of the foot into the flexor tendon sheaths and are attached to the base of the terminal phalanges. The tendons of three large muscles are inserted close together on the medial side of the foot. The tendon of tibialis posterior is attached to the tuberosity of the navicular and then spreads out to all the other tarsal bones except the talus and to some metatarsal bases. The tendon of peroneus longus, having crossed the foot, is attached to the lateral side of the medial cuneiform and of the 1st metatarsal base. The tendon of tibialis anterior is attached to the medial side of the same two bones. These three muscles are strong supporters of the arches of the foot.

PLANTAR VESSELS

1. Lateral plantar artery

This is the larger terminal branch of the posterior tibial artery and arises deep to the flexor retinaculum. It runs in a superficial plane obliquely across the sole, to the base of the 5th metatarsal bone where it turns medially and passes deeply back across the foot and anastomoses with the dorsalis pedis artery in the 1st metatarsal space, thus completing the plantar arch.

Branches

(i) digital branch to lateral side of 5th toe.

(ii) from the plantar arch—four plantar metatarsal arteries. They pass distally and divide into plantar digital arteries supplying the sides of adjacent toes, and anastomose with the dorsal metatarsal branches of the dorsalis pedis artery by means of perforating branches which pass through the metatarsal spaces.

2. Medial plantar artery

This is the smaller terminal branch of the posterior tibial artery. It passes anteriorly, supplying the medial side of the great toe and anastomoses with the medial three plantar metatarsal arteries.

The veins accompany the arteries and anastomose with the dorsal venous arch.

PLANTAR NERVES

1. Medial plantar nerve

This begins under the flexor retinaculum and passes forwards accompanied by its vessels. It supplies muscular branches to flexor hallucis brevis, abductor hallucis, flexor digitorum brevis and the 1st lumbrical, and cutaneous branches to the medial part of the sole and the medial $3\frac{1}{2}$ toes.

2. Lateral plantar nerve

This is the smaller terminal branch of the tibial nerve. It begins under cover of the flexor retinaculum and, accompanied by its artery, passes obliquely across the sole, to the base of the 5th metatarsal bone where it divides into deep and superficial branches. It supplies all the other muscles of the sole of the foot.

Branches

(i) *cutaneous branches*—to the skin over the lateral side of the foot.

(ii) *superficial terminal branch*—gives muscular branches and then cutaneous branches to the lateral $1\frac{1}{2}$ toes.

(iii) *deep terminal branch*—this turns medially in company with the lateral plantar artery and supplies the small deep muscles.

The plantar digital nerves supply the plantar surface of the toes and also the dorsal surface of the terminal segment of the toes, including the nail bed.

— 24 —

The fascia, veins, lymph drainage, nerves and posture

THE DEEP FASCIA

The deep fascia of the gluteal region and thigh forms a firm investing layer, the **fascia lata**, which surrounds the limb and provides medial and lateral intermuscular septa in the thigh. Superiorly, the fascia lata is attached, in a continuous line, to the pubic tubercle, inguinal ligament, iliac crest, back of the sacrum, sacrotuberous ligament, ischiopubic ramus, and the front of the body of the pubis as far as the pectineal line and pubic tubercle.

About 3 cm below and lateral to the pubic tubercle there is an oval deficiency, the **saphenous opening** (p. 293) which transmits the great saphenous vein and many vessels. The fascia lata is thickened on the lateral side of the thigh and forms the **iliotibial tract** (p. 286). Gluteus maximus and tensor fasciae latae are attached to the iliotibial tract and so help extend and stabilise the knee joint. Over the popliteal fossa the deep fascia is reinforced by transverse bundles of fibres. It is pierced by the small saphenous vein.

The deep fascia of the lower leg ensheaths the muscles and forms anterior and posterior intermuscular septa which pass to the fibula and separate anterior, lateral and posterior groups of muscles. Superiorly it is continuous with the fascia lata. It is attached to bone around the borders of the patella, the condyles of the tibia, the medial tibial surface (the shin), and inferiorly to both malleoli. It is thickened round the ankle and forms the retinacula tethering the tendons entering the foot. On the dorsum of the foot the deep fascia is thin, but on the sole it is thickened and forms the plantar aponeurosis.

The deep fascia is pierced by perforating veins joining the superficial and deep veins, and by cutaneous nerves, arteries and lymph vessels.

THE VENOUS DRAINAGE

As in the upper limb the veins may be divided into superficial, deep and intercommunicating groups. Valves are present in the larger vessels including most communicating veins (Fig. 24.1).

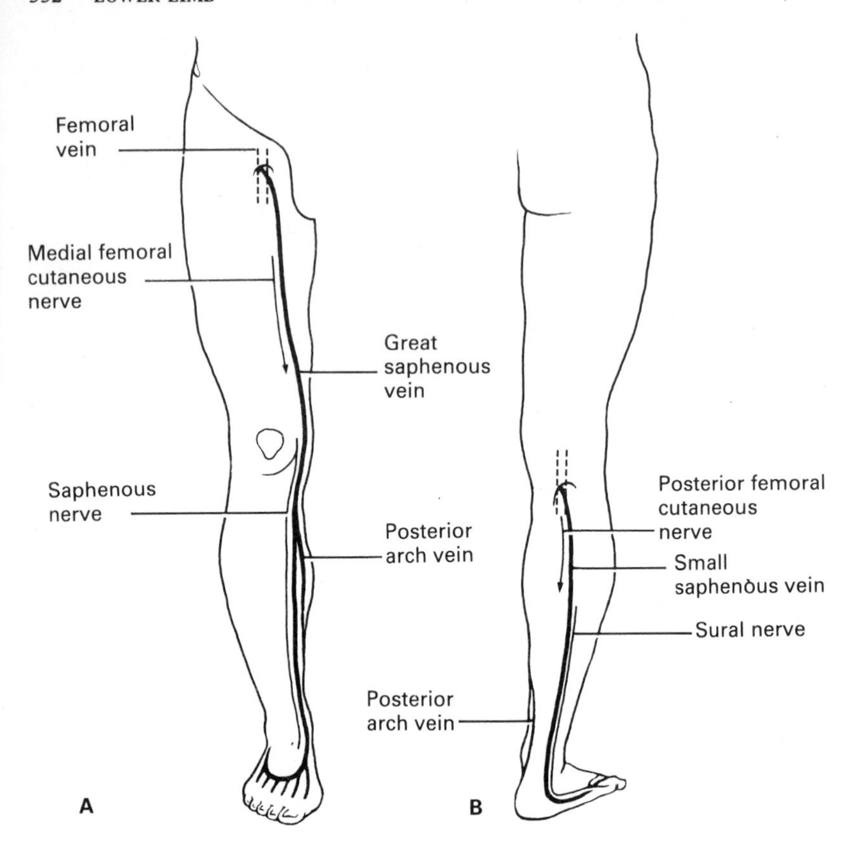

Fig. 24.1 The main superficial veins are shown and the accompanying cutaneous nerves. The posterior arch vein gives off most of the communicating veins with the soleal plexus. The small saphenous vein is seen entering the popliteal vein in B.
A. Anterior aspect **B**. Posterior aspect

Superficial veins

The superficial veins drain the subcutaneous tissues through two main longitudinal channels, the great and small saphenous veins:

(*a*) in the foot, communicating **dorsal and plantar venous arches** receive the digital and metatarsal veins of the sole and dorsum of the foot. The great and small saphenous veins originate from the medial and lateral ends respectively of the dorsal arch.

(*b*) The **great saphenous vein** originates from the medial end of the dorsal venous arch. It ascends just in front of the medial malleolus and subcutaneously along the medial side of the calf and thigh, accompanied in the calf by the saphenous nerve. It passes through the saphenous opening and enters the femoral vein.

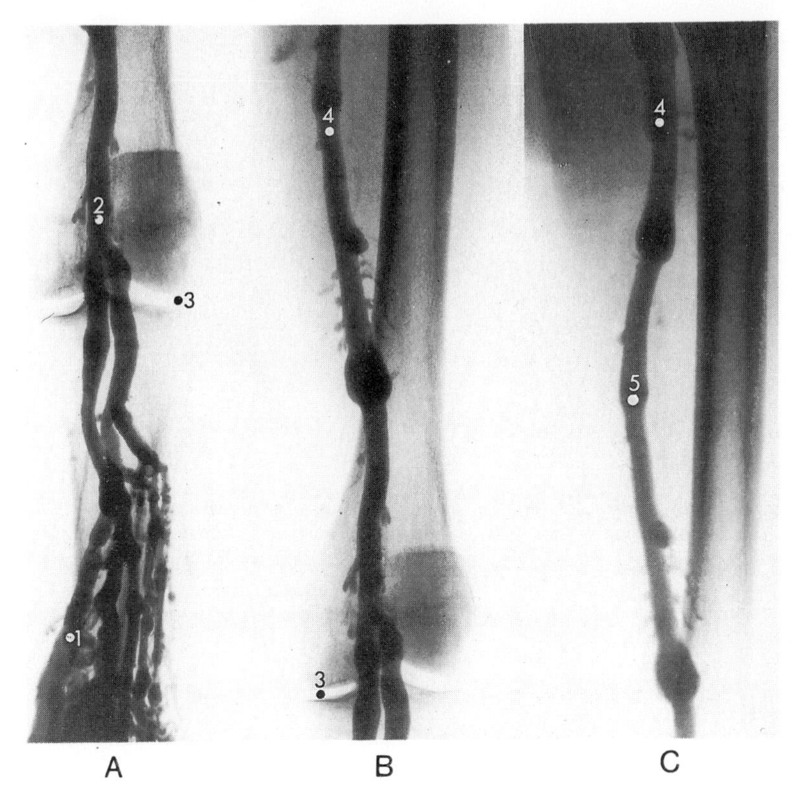

Fig. 24.2 Venogram of the leg. The three views show progression of the contrast material up the leg. Note the soleal plexus of veins in A.

1. Soleal plexus of veins
2. Popliteal vein
3. Knee joint
4. Femoral vein
5. Valve

Tributaries

(i) communicating branches with the small saphenous vein and the deep veins of the calf and thigh.

(ii) A fairly constant tributary (posterior arch vein) ascends behind the saphenous vein and joins it just below the knee. Most of the veins communicating with the deep plexus are tributaries of this vein.

(iii) subcutaneous branches draining the skin of the inguinal and pubic regions join it just prior to its passage through the saphenous opening.

(*c*) The **small saphenous vein** originates from the lateral end of the dorsal venous arch. It ascends behind the lateral malleolus, along the back of the calf, and ends by piercing the fascial roof of the popliteal fossa and entering the popliteal vein.

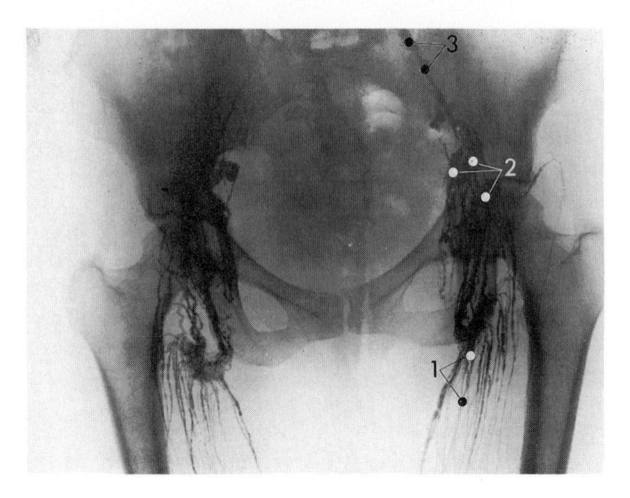

Fig. 24.3 Lymphangiogram of lower limbs. The contrast material has been introduced into a lymph vessel over the dorsum of each foot. (See also Figs. 15.2 and 15.3)

Its tributaries communicate with the great saphenous vein and the deep veins of the limb.

Deep veins

The deep veins comprise the plantar, the soleal plexus of veins, the popliteal vein (p. 312) and the femoral vein (p. 296). They communicate with the superficial veins by tributaries, the valves of which are so arranged as to allow blood to pass only towards the deep veins. Pressure of contracting muscles on the deep veins is an important factor in the return of blood from the lower limb, e.g. **the soleal pump**. Incompetence of these valves or blockage of the deep veins leads to varicosity of the superficial veins. Defective blood supply to the skin above the ankle may give rise to venous ulcers. A blood clot from the deep veins may become dislodged and travel to the lungs—a pulmonary embolus. The long saphenous vein is commonly used as a vascular autograft.

THE LYMPH DRAINAGE (Fig. 24.3)

There are two main groups of lymph nodes in the lower limb—superficial and deep.

1. The superficial groups

These drain mainly superficial tissues and lie in the inguinal region:

(i) **upper superficial inguinal group**—lies just below the inguinal ligament and drains lymph from the lower part of the anterior abdominal wall, the perineum, external genitalia, anal canal and the gluteal region. Its efferent vessels pass to the lower superficial inguinal and deep inguinal groups of nodes.

(ii) **lower superficial inguinal group**—lies around the saphenous opening and drains lymph from the previous group of nodes, and the skin over the thigh and medial part of the leg and foot. Its efferent vessels pass to the deep inguinal group through the saphenous opening.

2. The deep groups

These are found under the deep fascia in the popliteal and inguinal regions.

(i) **popliteal group**—this small group lies in the popliteal fossa around the termination of the small saphenous vein. It drains skin over the lateral side of the calf and foot, and also deeper tissues of the leg. Its efferent vessels pass to the deeper inguinal group.

(ii) **deep inguinal group**—this group comprises one to three nodes lying in the femoral canal. It receives lymph from all superficial nodes and deep vessels from the entire limb. Its efferent vessels pass deep to the inguinal ligament to the external iliac group of nodes.

Maldevelopment of the lymph vessels or blockage of the lymph nodes by disease may produce swelling of the limb due to interference with lymph flow (lymphoedema).

NERVE SUPPLY

The cutaneous nerve supply is summarised in Figure 24.4. Note that we kneel on skin supplied by L3 and 4 segments, the dorsum of the foot is supplied by L5: we stand on skin supplied by S1 and sit on skin supplied by S3. Autonomic nerves, mainly sympathetic, run with branches of the lumbosacral plexus, and supply blood vessels, cutaneous glands and arrector pili muscles.

Peripheral nerve injuries in the leg

(i) **femoral nerve**—this produces loss of knee extension (quadriceps), some loss in hip flexion (iliacus and pectineus) and loss of sensation over the front and medial side of the thigh, leg and foot (anterior and medial femoral cutaneous nerves and the saphenous nerve).

(ii) **obturator nerve**—some loss in the power of adduction (adductor magnus is partly supplied by the sciatic nerve) but often there is not loss in sensation.

(iii) **sciatic nerve**—a proximal injury, e.g. following a posterior dislocation of the hip joint, produces an almost flail limb. A distal injury may spare the nerves supplying the hamstrings and allow the knee to be flexed.

(iv) **tibial nerve**—loss of plantar flexion and sensory loss over the sole of the foot. An injury near the ankle denervates the small muscles of the sole and the unopposed action of the long flexor and extensor muscles produces a highly

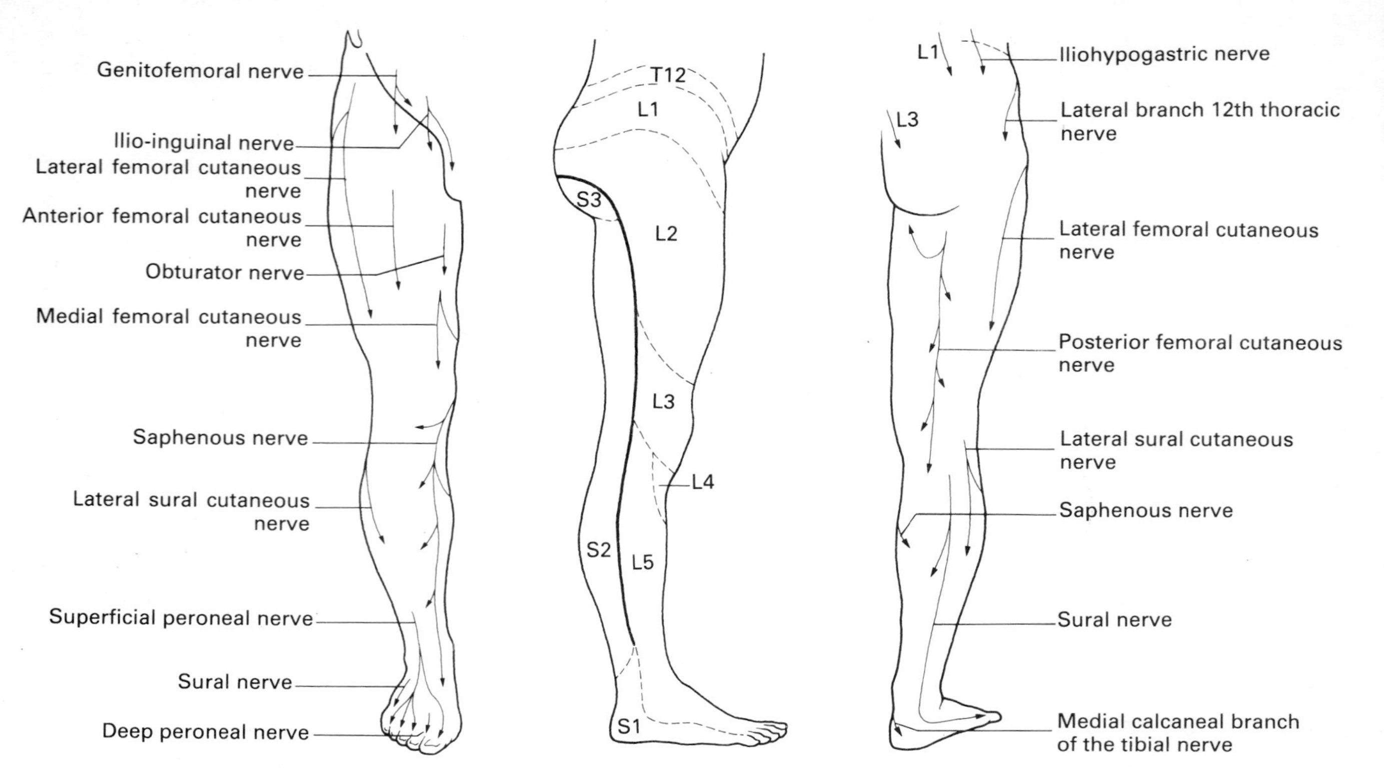

Fig. 24.4 Diagrammatic layout of the cutaneous innervation of the right lower limb.

arched foot. Injury proximal to the origin of the sural nerve also produces loss of sensation over the lateral side of the leg and foot.

(v) **common peroneal nerve**—this is the most frequent nerve to be injured in the leg, often associated with fractures of the neck of the fibula or a badly fitting leg plaster. The power of dorsiflexion (extensor muscles) and eversion (peroneal muscles) is lost—the foot drops and becomes inverted. There is sensory loss over the medial side of the dorsum of the foot.

(vi) **superficial peroneal nerve**—the power of eversion (peronei) is lost and the foot becomes inverted. There is loss of sensation over the medial part of the dorsum of the foot.

(vii) **deep peroneal nerve**—the power to dorsiflex the foot and toes is lost and there may be loss of sensation between the 1st and 2nd toes. The foot becomes inverted by the unopposed action of the tibialis posterior.

POSTURE IN STANDING AND WALKING

The body's weight is transmitted from the vertebral column through the pelvic girdle to both femoral heads. In each limb, the weight is transmitted through the bones, whose internal architecture is well adapted to withstand compressive stresses and strains, to the foot whose arched nature distributes the weight equally between the metatarsal heads and the tubercles of the calcaneus. The head of the 1st metatarsal normally takes more than a third of the total metatarsal load.

In the **normal stance** the line of weight passes through the centre of gravity which, in the anatomical position, is just in front of the 2nd piece of the sacrum. The line passes slightly behind the hip joints and just in front of the knee and ankle joints. The erect posture is maintained by a combination of muscular and ligamentous factors. Normally the effect of the body's weight is to cause hyper-extension at the hip and knee joints. This is resisted at the hip by the anteriorly placed iliofemoral ligament and by contraction of iliopsoas. At the knee it is resisted by most of the ligaments and contraction of the hamstring muscles and gastrocnemius. At the ankle joint, the shape of the mortise between the tibia and fibula above and the talus below contributes more than the ligaments to the stability of this joint. Slight changes in posture may bring the line of weight behind the knee and flexion is then resisted by contraction of quadriceps femoris. The body's weight tends to dorsiflex the body over the foot but this is resisted by the calf plantar flexor muscles, especially soleus. The long digital flexors help by holding the toes firmly on the ground.

During **walking**, the cycles of movement of each limb can be divided into a stance (supporting) phase and a swing phase. Each cycle begins as the heel touches the ground and finishes as the same heel again touches the ground.

The stance (supporting) phase
In this phase, the weight of the body is taken by the grounded leg, and muscle activity maintains or increases the forward momentum of the body. The phase starts as the heel touches the ground (heel strike). At the hip joint, extension gives a forward thrust, and medial rotation maintains the correct direction of progress. Meanwhile, abduction at the hip joint by gluteus medius and minimus tilts the pelvis on the supporting leg and allows the swinging leg to move forwards. The knee joint is in the extended position, locked so that weight is not carried by the muscles but by the bones and ligaments. (At the end of this phase, the knee joint is unlocked by lateral rotation of the femur on the tibia.) At the ankle joint plantar flexion (partly passive as the foot goes on the ground) gives way to an active dorsiflexion which pulls the body weight over the ankle. The foot becomes slightly everted. The weight of the body is transferred forwards from the heel along the outer border of the foot to the metatarsal heads, especially the first.

The swing phase
In this, the leg swings through from a trailing to a leading position. At the hip joint, there is flexion, lateral rotation and adduction. At the knee joint, flexion occurs until the swinging foot has passed the supporting foot and then extension occurs, at the end of which some lateral rotation of the tibia may lock the knee joint in preparation for weight bearing. At the ankle joint, the foot is first dorsiflexed but becomes plantar flexed as the knee extends. The foot itself becomes inverted and the heel touches the ground (heel strike) in readiness for the next supporting phase.

Starting to walk involves a forward tilting of the body thus advancing the centre of gravity and the line of weight. The supporting limb is extended, the pelvis is tilted on the supporting leg, and the heel is raised by the calf muscles. The opposite limb flexes and then extends until the heel touches the ground. Thus it becomes the supporting limb after a very short swing phase. Turning to left or right involves most of the trunk muscles as well as the rotators at the hip joint. They largely determine the route followed by the swinging leg and the position of the touch-down of the swinging heel. In running, the centre of gravity is usually further forward due to flexion of the vertebral column, thus altering the total posture. There is considerable time between the end of the stance phase and the heel strike of the foot of the swinging leg, so that the body is clear of the ground (unsupported) for this time. In running, active plantar flexion at the ankle joint at the end of the stance phase increases the forward momentum of the body.

Although described separately these movements merge into one another indistinguishably. Limitation of the movement at any joint results in a marked alteration in gait. The individuality of the gait depends on the length of the stride, the extent of pelvic rotation (swinging the hips) and the impetus at the end of the stance phase.

The antigravity muscles of the whole limb are all strong, and are very active when rising from sitting or squatting positions and when climbing up stairs or an incline.

PART SIX

Head and neck

— 25 —

The skull, scalp and face

The head and neck are involved in many important functions. As well as protecting the brain and upper spinal cord, the region contains the end organs of the special senses, and the upper (rostral) parts of the alimentary and respiratory systems. Communication by speech and facial expression is another important social activity of this region.

THE SKULL

The skull will be considered as viewed from above, in front, from the side, from below, and from behind, after which individual bones will be examined. The osteology of the intracranial region, and of the orbit, nose and ear is considered under the relevant sections.

Superior aspect. The vault is crossed by three sutures. The **coronal suture** separates the frontal bone anteriorly from the two parietal bones posteriorly. The midline **sagittal suture** separates the parietal bones. Its junction with the coronal suture is known as the **bregma** and in the fetal skeleton, when the frontal bone is incompletely ossified, a diamond-shaped bony deficiency, the **anterior fontanelle**, is present. The anterior fontanelle closes at about 18 months. The **lambdoid suture** lies between the two parietal bones and the occipital bone posteriorly; it meets the sagittal suture at the **lambda**. At the lambda there is, in the fetal skull, a triangular bony deficiency, the **posterior fontanelle**. It is usually closed by the 2nd or 3rd month after birth.

Anterior aspect (Figs. 25.1, 26.1). This is formed superiorly by the smooth convexity of the frontal bone. The point of maximum convexity on each side is known as the frontal tuberosity. The inferior part is irregular and contains the openings of the orbital, nasal and oral cavities. Above the supra-orbital margins are rounded superciliary arches. The supra-orbital margin, which is formed by the frontal bone, possesses a supra-orbital notch or foramen (at the junction of its medial third and lateral two-thirds) which transmits the supra-orbital vessels and nerve. The lateral orbital margin is formed by the frontal and zygomatic bones (the frontozygomatic suture is palpable in the living). The medial orbital margin is formed by the frontal bone and the frontal process of the maxilla; the inferior margin by the maxillary bone medially and the zygomatic bone laterally.

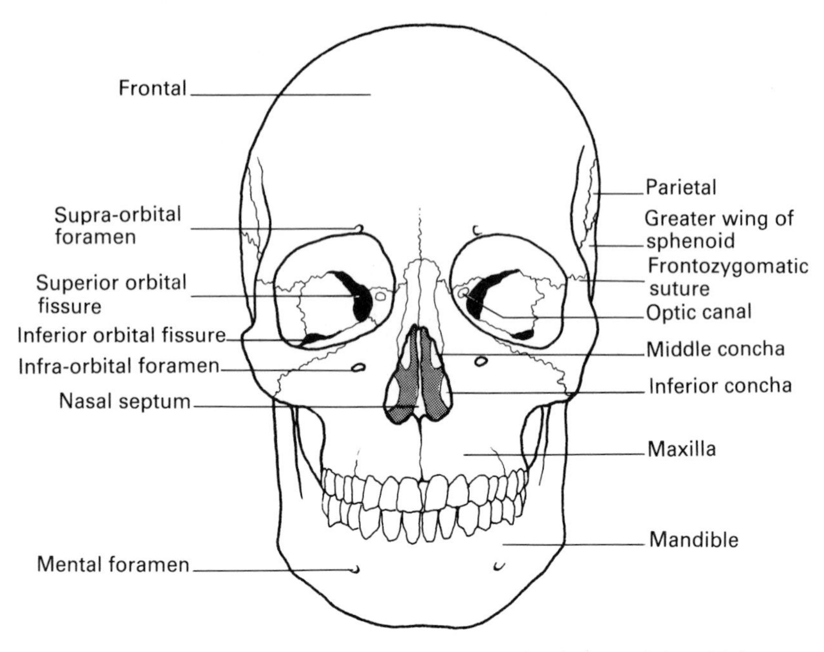

Fig. 25.1 The skull, frontal aspect. (See also Fig. 26.1 for skeleton of the orbit.)

The prominence of the cheek, below and lateral to the orbital margin is produced by the zygomatic bone. On the maxilla 1 cm below the orbit, in line with the supra-orbital notch, is the infra-orbital foramen from which emerge the infra-orbital vessels and nerve. The pear-shaped nasal aperture is bounded superiorly by the nasal bones, and below and laterally by the maxillae. The alveolar margins of the maxillae and mandible surround the opening of the oral cavity and bear the sockets for the teeth. The mental foramen on the mandible is in line with the supra-orbital and infra-orbital foramina, and from it emerge the mental vessels and nerve.

Lateral aspect (Fig. 25.2). The temporal line curves upwards and backwards from the zygomatic process of the frontal bone across the parietal bone and downwards and forwards over the squamous temporal bone to end as the supra-mastoid crest above the external acoustic meatus. Below the temporal line and deep to the zygomatic arch is the temporal fossa. It is roofed in by the temporal fascia, itself attached to the temporal line and upper border of the zygomatic arch. The **zygomatic arch** is formed by the zygomatic process of the squamous temporal bone and the temporal process of the zygomatic bone. The medial wall of the fossa is the smooth convexity formed by the frontal, parietal, temporal and the greater wing of the sphenoid bones. Their H-shaped point of union, the **pterion**, lies about 3.5 cm behind and 1.5 cm above the frontozygomatic suture. This point indicates the position where the middle meningeal artery grooves, or runs through a canal in, the inner surface of the skull. The fossa is limited

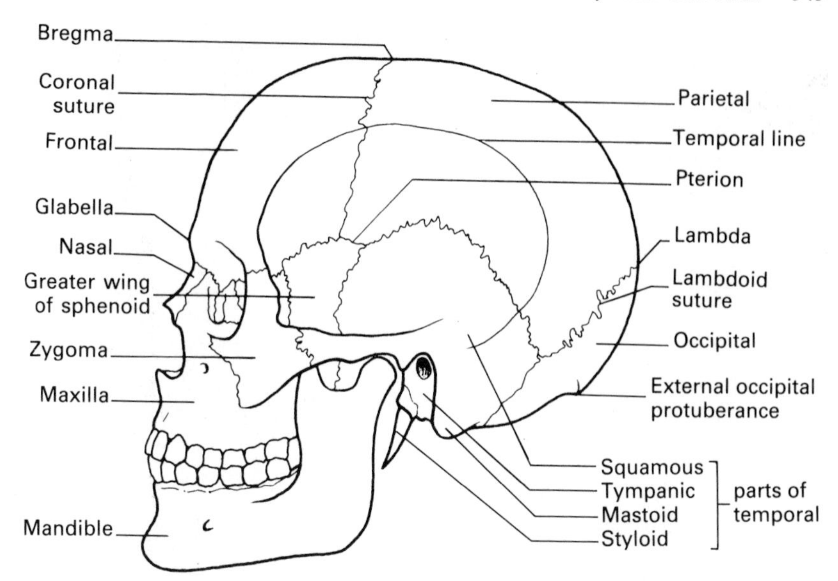

Fig. 25.2 The skull, lateral aspect. The external acoustic meatus (opening) is surrounded by the squamous and the tympanic parts of the temporal bone.

inferiorly by the infratemporal crest on the greater wing of the sphenoid and the squamous temporal bones. Below and medial to the crest is the **infratemporal fossa** which is limited medially by the lateral pterygoid plate. This fossa communicates medially with the pterygopalatine fossa through the pterygomaxillary fissure (between the pterygoid process and the back of the maxilla) and superiorly with the orbit through the inferior orbital fissure (between the greater wing of the sphenoid and the maxilla). The external acoustic meatus opens below the posterior end of the zygomatic arch. The meatus is formed mainly by the tympanic plate of the temporal bone; the roof and upper part of the posterior wall are formed by the squamous temporal bone. The palpable mastoid process passes downwards behind the meatus.

Inferior aspect (Fig. 25.5). Anteriorly is the hard palate. It is formed from the palatine processes of the maxillae in front of the horizontal plates of the palatine bones. It is bounded anteriorly and laterally by the alveolar processes of the maxillae containing the sockets for the teeth; the rounded posterior part of the process on each side is called the maxillary tuberosity. Anteriorly a midline incisive foramen communicates with the nose and transmits the greater palatine arteries and nasopalatine nerves. Opening on to the posterolateral part of the palate are the greater and lesser palatine foramina; these convey the vessels and nerves of the same names from the pterygopalatine fossa. The posterior nasal apertures open above the palate. They are bounded superiorly by the body of the sphenoid bone, inferiorly by the horizontal plates of the palatine bones, and laterally by the medial pterygoid plates. The openings are separated in the

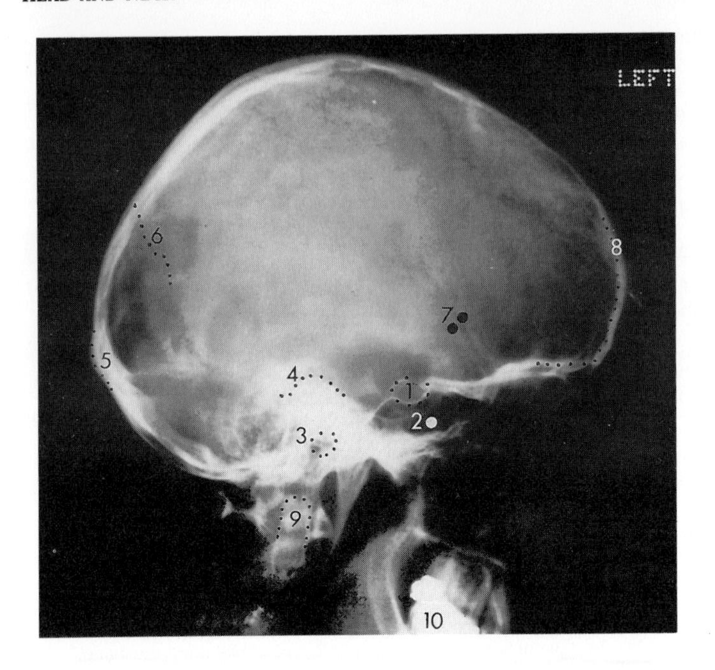

Fig. 25.3 Lateral view of adult skull. Identify the pituitary fossa just above the sphenoidal air sinus and compare the region as shown in Figure 25.4. Note how the metallic dental fillings show up as opaque structures on radiographs.

1. Pituitary fossa
2. Sphenoidal air sinus
3. External acoustic meatus
4. Petrous temporal bone
5. External occipital protuberance
6. Parieto-occipital suture
7. Marking for meningeal vessels
8. Frontal bone
9. Dens of axis bone
10. Dental fillings.

midline by the vomer. The medial pterygoid plate projects downwards as the pterygoid hamulus, which gives attachment to the pterygomandibular raphé; the tensor veli palatini tendon hooks around the hamulus. The lateral pterygoid plates enclose the pterygoid fossa. Behind the root of the lateral pterygoid process is the foramen ovale which transmits the mandibular nerve. Posterolateral to the foramen is the spine of the sphenoid and between them is the foramen spinosum which transmits the middle meningeal artery. The spine gives attachment to the sphenomandibular ligament. On the squamous temporal bone, lateral to the spine of the sphenoid, is the mandibular fossa for the condyle of the mandible. It is bounded anteriorly by the articular tubercle and posteriorly by the squamotympanic fissure and the tympanic plate of the temporal bone. The fissure is subdivided by a thin projection of the petrous temporal bone into a petrosquamous and a petrotympanic portion; the chorda tympani nerve emerges through the latter. The irregular inferior surface of the petrous temporal bone projects anteromedially between the sphenoid and occipital bones; its apex is separated from them by the foramen lacerum (see p. 460). The bone contains

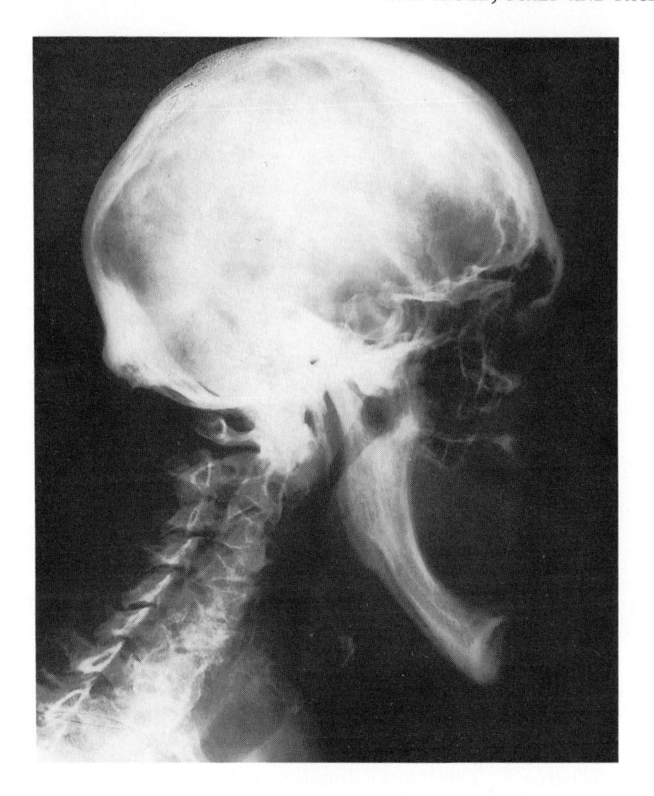

Fig. 25.4 Lateral view of the skull of a patient with excessive secretion of growth hormone due to a pituitary tumour (acromegaly). Note the enlarged pituitary fossa (compare with Fig. 25.3) and the large elongated (prognathic) lower jaw.

the carotid canal whose inferior opening lies behind the spine of the sphenoid. The jugular foramen, transmitting the internal jugular vein, the inferior petrosal sinus, and glossopharyngeal, vagus and accessory nerves, lies immediately posterolateral to the carotid opening. The styloid process projects downwards lateral to the jugular foramen. The stylomastoid foramen transmits the facial nerve and lies between the styloid and mastoid processes. Medial to the mastoid process are two grooves, the medial for the occipital artery and the lateral for the attachment of the posterior belly of digastric muscle. The occipital bone contains the large foramen magnum bounded on each side by the occipital condyles, each with two associated foramina. The anterior (hypoglossal foramen) transmits the hypoglossal nerve, and the posterior (condylar foramen) transmits an emissary vein and opens into a fossa behind the condyle.

Posterior aspect. Behind the foramen magnum are the superior and inferior nuchal lines. A number of postvertebral muscles gain attachment to the roughened bone in this region. The superior line arches upwards from the mastoid process on each side, over the occipital bone to the palpable midline external

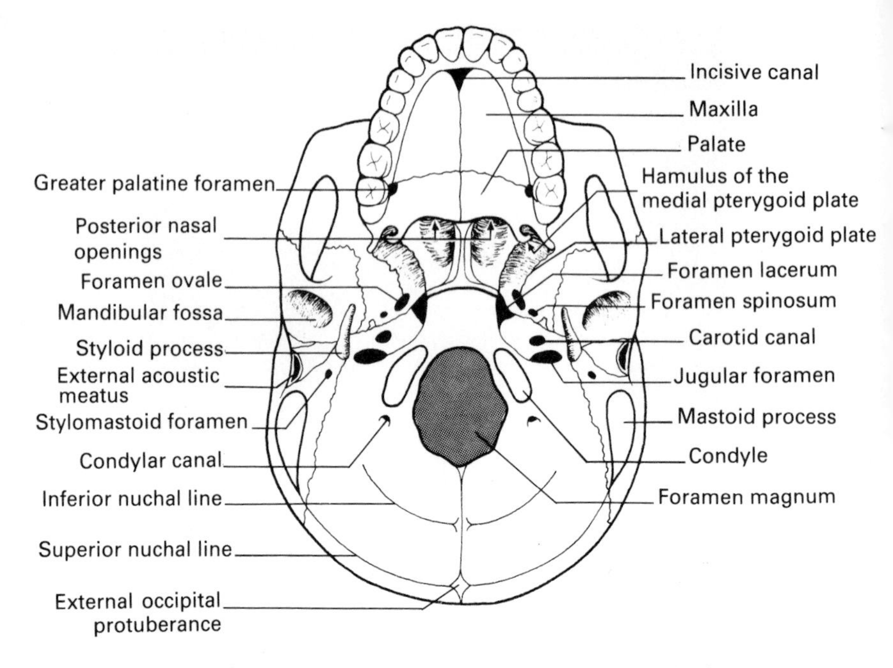

Fig. 25.5 The skull, inferior aspect. Between the styloid process and the foramen spinosum is the spine of the sphenoid.

occipital protuberance. The inferior line lies midway between the superior and the foramen magnum. Above the superior line, the bone is smooth.

Development and growth of the skull (See also p. 29)
The bones of the vault of the skull and many face bones are formed in mesenchyme while those of the base are developed mainly in cartilage (the chondrocranium).

At birth the face is relatively small. It increases in size with the appearance of the deciduous teeth and still more after the sixth year with the eruption of the first permanent teeth and the enlargement of the nasal cavities and paranasal air sinuses. Growth in length of the base of the skull occurs mainly at the cartilaginous unions between the body of the sphenoid and the basilar part of the occipital bone. Fusion of the two bones occurs after the appearance of the last molar teeth (17–25 years). Growth in width occurs mainly round the edges of the foramen lacerum which is filled with fibrocartilage. This remains throughout life after growth has ceased.

Although the bony protective characteristics of the skull have obvious advantages, its inextensibility means that an expanding space-occupying lesion, such as a tumour, or a blood clot from a damaged intracranial vessel, will produce serious pressure effects on the brain. In particular, the brain stem may be compressed and vital centres become irreversibly damaged.

THE INDIVIDUAL BONES OF THE SKULL

Mandible (Fig. 25.6)
This bone is formed by the midline union of two halves. Each half has a horizontal body and a vertical ramus. The junction of the posterior border of the ramus and the lower border of the body forms the prominent **angle** of the mandible.

The **ramus** is flat and rectangular in shape; superiorly it forms two processes which are separated by the **mandibular notch**. The anterior (**coronoid process**) gives attachment to temporalis, and the posterior (**condylar process**) has an articular head and a narrower neck. The lateral pterygoid muscle is attached to the fossa on the front of the neck. The lateral surface of the ramus is roughened by the attachment of masseter. In the centre of the medial surface is the **mandibular foramen**. The sharp upward projecting inferior lip of the foramen is the **lingula**; it gives attachment to the sphenomandibular ligament. The medial pterygoid muscle is attached to the roughened area on the ramus below the foramen.

The **body** has a smooth rounded inferior border and an upper alveolar border containing the sockets for the teeth. The **mental foramen** is situated on the lateral surface at the level of the interval between the premolar teeth, and from it emerge the mental vessels and nerve. On the inner surface the **mylohyoid line** passes downwards and forwards from below the last molar tooth to the midline. It gives attachment to mylohyoid, and separates the sublingual fossa (for the sublingual gland) above from the submandibular fossa (for the submandibular gland) below. Near the midline the digastric fossa for the anterior belly of digastric lies below the line and the **mental spine** is above it.

Each half of the mandible ossifies in mesenchyme from a centre appearing in the sixth week of intra-uterine life. The centre lies lateral to Meckel's cartilage (the cartilage of the first pharyngeal arch) which becomes invaded and surrounded by the mesenchyme bone. Secondary cartilages appear later for the condyle, the coronoid process and the region of the midline. They are ossified from the adjacent mesenchyme bone. Bony union of the two halves takes place in the second year and the symphysis menti disappears. Subperiosteal deposition of bone increases the width, whilst absorption helps to mould the shape.

The length of the body is increased by absorption of the bone from the anterior border of the ramus and deposition along its posterior border. Vertical growth of the body is mainly by deposition of bone along the alveolar margin (associated with tooth eruption) and on the inferior border. The condylar secondary cartilage contributes to the height of the ramus. Loss of the teeth results in absorption of the alveolar bone. Thus the mental foramen, which lies midway between the upper and lower borders in adult life, is nearer the lower border at birth and the alveolar border in later life. Also associated with changes in the alveolar margin are changes in the angle of the jaw, it being about 140° at birth and in old age, and 110° in the adult.

Growth in size ceases at about the 25th year though moulding of the shape

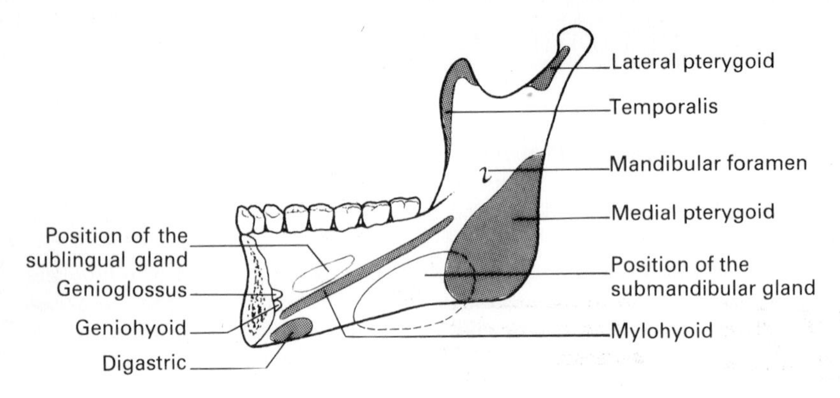

Fig. 25.6 The mandible, medial aspect.

may continue. In fractures of the mandible, many of the related structures may be damaged.

Maxilla

Each maxilla consists of a body and four processes. The **body** contains the **maxillary air sinus** (p. 8) and has four surfaces. On the medial (nasal) surface there is a large opening which, in the articulated skeleton, is closed largely by other bones of the lateral wall of the nose (p. 365). The anterior surface forms part of the facial skeleton and contains the **infra-orbital foramen**. The posterior surface is separated from the pterygoid process of the sphenoid by the pterygo-maxillary fissure. The thin superior (orbital) surface forms the larger part of the orbital floor.

The **zygomatic process** extends laterally to the zygomatic bone. The **alveolar process** projects downwards and carries the upper teeth; a rounded posterior extension beyond the molars forms the **maxillary tuberosity**. The **horizontal (palatine) process** passes medially and contributes to the hard palate. The **frontal process** ascends and articulates with the frontal bone and forms part of the medial margin of the orbit.

The maxilla develops in mesenchyme. The major part of the bone develops from a centre above the canine tooth. A second (premaxillary) centre forms that part of the bone carrying the incisor teeth. The infra-orbital nerve is usually damaged in fractures of the maxilla and gives rise to numbness over the cheek. An associated fracture of the zygomatic bone is common and then there may be difficulty in moving the temporomandibular joint. Injury to the orbital floor may give rise to double vision because of changes in the position of the eyeball.

Sphenoid bone

This consists of a central body from which two pairs of processes (the greater and lesser wings) extend laterally and two processes (the pterygoid processes) pass inferiorly.

The cube-shaped **body** encloses the paired sphenoidal air sinuses. It articulates anteriorly with the ethmoid bone, and posteriorly with the occipital bone. The inferior surface roofs in the nasopharynx. The superior surface is indented to form the **pituitary fossa** (sella turcica), anterior to which is the shallow **optic groove**, and posterior to which is the **dorsum sellae** projecting upwards and forwards over the fossa. At each end of the upper border of the dorsum sellae is a **posterior clinoid process**.

The **greater wing** has four surfaces. The upper surface is concave and supports the temporal lobe of the brain. The lower surface overlies the infratemporal fossa and carries the spine of the sphenoid. The lateral surface forms part of the medial wall of the temporal fossa and is separated from the lower surface by the infratemporal crest. The front of the wing is thickened to form an anterior surface which contributes to the lateral wall of the orbit. It is separated from the maxilla by the inferior orbital fissure. The root of the greater wing contains from before backwards the foramen rotundum, the foramen ovale the foramen spinosum.

The **lesser wing** extends laterally from the anterior part of the body and forms the posterior limit of the floor of the anterior cranial fossa. The **anterior clinoid process** is the projecting medial end of the posterior free border. Between the lesser wing and the body is the optic canal and between the lesser and greater wings is the superior orbital fissure.

The **pterygoid process** extends downwards behind the maxilla. The process bears medial and lateral **pterygoid plates** which are united anteriorly.

The pterygoid processes and the lateral part of the greater wings ossify in mesenchyme, the remainder of the bone in cartilage.

Temporal bone

Each consists of four parts: the squamous, petromastoid, tympanic and styloid.

The **squamous part** is a thin plate of bone on the lateral aspect of the skull articulating anteriorly with the greater wing of the sphenoid, superiorly with the parietal bone and posteriorly with the occipital bone. Its zygomatic process extends forwards from the lateral surface and contributes to the zygomatic arch. Below the process is the **mandibular fossa** for the condyle of the mandible. The fossa is continued anteriorly on to the **articular tubercle**, and posteriorly the squamotympanic fissure separates the articular surface (squamous temporal) from the nonarticular surface (tympanic plate). The squamous bone forms the roof and upper posterior part of the bony external acoustic meatus and lies lateral to the petrous part, forming the lateral wall of the middle ear.

The **petromastoid part** contains the middle and inner ears, and extends medially between the sphenoid and occipital bones. Its upper (intracranial) surface forms part of the floor of the middle and posterior cranial fossae. On the apex, in the middle fossa, is a depression for the trigeminal ganglion, and lateral to this are openings for the greater petrosal nerve (posteriorly) and the lesser petrosal nerve (more anteriorly). The part forming the front of the posterior fossa is pierced by the **internal acoustic meatus** transmitting the facial and vestibulocochlear nerves. The irregular inferior surface contains the **carotid canal** for the internal carotid artery, and the **jugular fossa** opens between it and the occipital bone. The bony **auditory tube** opens anteriorly on to the inferior surface of the petrous temporal bone. The mastoid process extends down posteriorly and contains the tympanic antrum and mastoid air cells.

The **tympanic part** is a curved bony plate forming the anterior wall and floor of the bony external acoustic meatus. It also forms the nonarticular posterior wall of the mandibular fossa.

The **styloid process** is a slender downward projection from the petrous part. It is about 2 cm long. The **stylomastoid foramen**, between it and the mastoid process, transmits the facial nerve.

The squamous and tympanic parts of the temporal bone develop in mesenchyme, the petromastoid and styloid parts in cartilage (the styloid from the cartilage of the second pharyngeal arch). The four parts do not fuse until after birth.

Occipital bone

This has a basilar, a squamous and two lateral portions situated around the **foramen magnum**.

The **basilar portion** lies anterior to the foramen magnum and articulates with the body of the sphenoid bone. In the midline near the foramen is the **pharyngeal tubercle**.

The **squamous part** is a flat plate behind the foramen magnum extending at first backwards and then upwards to articulate with the parietal and squamous temporal bones. The lower part of the outer surface carries the superior and inferior **nuchal lines** and the midline **external occipital protuberance**; it gives attachment to the postvertebral muscles. The intracranial surface is grooved by the superior sagittal and transverse venous sinuses.

The **lateral parts** bear, on their inferior surface, the **occipital condyles** for articulation with the atlas. Lateral to each condyle there is a projecting jugular process which forms the posterior wall of the jugular foramen. Anteriorly, the **hypoglossal canal** carries the hypoglossal nerve and posteriorly, an emissary vein passes through the **condylar canal**.

The smooth upper portion of the squamous part of the bone develops in membrane, the remainder of the bone in cartilage.

Parietal bone

These are two convex quadrilateral plates of bone forming the posterior part of the vault of the skull. They meet at the sagittal suture and articulate anteriorly with the frontal bone at the coronal suture, laterally with the temporal bone, and posteriorly with the occipital bone at the lambdoid suture. The area of greatest convexity of each bone is called the parietal tuberosity.

The bone develops in membrane.

Frontal bone

This has a domed superior portion forming the anterior part of the vault of the skull; the two areas of maximum convexity are called the frontal tuberosities. The inferiorly placed horizontal **orbital plates** contribute to the roof of each orbit. The supra-orbital margin, with its notch or foramen is formed where the vault meets the orbital plate. The bone articulates superiorly with the parietal bones at the coronal suture, and inferiorly with the greater wing of the sphenoid, the zygomatic, nasal and ethmoid bones and the frontal process of the maxilla. The orbital plates meet the lesser wings of the sphenoid posteriorly. The paired **frontal air sinuses** are situated in the bone above the nose and the medial part of the orbit.

The bone develops in mesenchyme in two halves which fuse about the 5th year. The suture between the halves is known as the metopic suture when it persists in the adult.

Zygomatic bone

These are irregular bones forming the prominence of each cheek.

The orbital surface forms the anterior part of the lateral wall of the orbit; the lateral surface forms the prominence of the cheek; the temporal surface forms the anterior limit of the temporal fossa.

The temporal process forms the anterior part of the zygomatic arch, the frontal process articulates with the frontal bone; the zygoma articulates with the maxilla and with the greater wing of the sphenoid in the lateral wall of orbit.

The bone ossifies in mesenchyme.

Nasal bone

The two nasal bones meet in the midline and form the skeleton of the upper part of the external nose. They articulate superiorly with the frontal bone, laterally with the frontal processes of the maxillae and inferiorly with the nasal cartilages. They are formed in mesenchyme.

Ethmoid bone

This consists of a midline **perpendicular plate** and a perforated horizontal **cribriform plate** uniting the upper extremities of two lateral portions, the ethmoidal labyrinths. The perpendicular plate forms part of the nasal septum below the cribriform plate and an upward projection above, the **crista galli**. The cribriform plate transmits the olfactory nerves.

Each **ethmoidal labyrinth** is thin-walled and contains the ethmoidal air cells. It separates the medial wall of the orbit from the upper part of the lateral wall of the nose, lies in front of the body of the sphenoid and is covered superiorly by the orbital plate of the frontal bone. The superior and middle nasal **conchae** project from its medial wall into the nasal cavity and below the middle concha it forms the **bulla ethmoidalis**, a bulge containing the middle ethmoidal air cells.

The bone develops in cartilage at the front of the chondrocranium.

Lacrimal bone

These are small thin bones. On the medial wall of the orbit, each forms, with the frontal process of the maxilla, a hollow for the lacrimal sac and the upper part of the nasolacrimal duct.

The bone develops in mesenchyme.

Palatine bone

Each consists of a horizontal and a vertical plate. The two plates unite at right angles and the posterior part of the union expands to form a **tubercle** which separates the maxilla from the pterygoid process of the sphenoid bone.

The **horizontal plates** form the posterior third of the hard palate. Each **vertical plate** overlies the medial wall of the maxilla enclosing a canal for the greater palatine vessels and the anterior palatine nerve. The upper end of the vertical plate is notched and forms orbital and sphenoid processes which articulate with the body of the sphenoid and form the **sphenopalatine foramen**.

The bone develops in mesenchyme.

Inferior concha bone

Each is a thin plate of bone which curls downwards and medially from the lower part of the lateral wall of the nose. It forms the lower medial wall of the maxillary air sinus and of the nasolacrimal duct which opens below it.

The bone develops in mesenchyme.

Vomer

This is a thin midline plate of bone forming the posterior part of the nasal septum. It is quadrilateral in shape articulating with the body of the sphenoid above, the hard palate below, and the perpendicular plate of the ethmoid and the septal cartilage anteriorly. It has a free posterior border.

The bone develops in mesenchyme.

Hyoid bone

This is a U-shaped bone situated just below and behind the mandible and slung from the styloid process by the stylohyoid ligament which passes to its lesser horn.

The bone consists of an anterior horizontal midline bar, the body, from each end of which extend narrow greater and lesser horns. The greater horn projects backwards, and the short lesser horn, above, upwards and slightly backwards. The hyoid gives attachment superiorly to muscles passing to the tongue, mandible and base of the skull. Inferiorly, a membrane passes to the thyroid cartilage, and muscles pass to this cartilage, and to the sternum, clavicle and scapula.

The bone develops from the 2nd and 3rd pharyngeal arch cartilages.

THE SCALP

The scalp is the term given to the soft tissue over the vault of the skull. It consists of five layers.

(*a*) The skin is thick, containing many hair follicles and sweat glands.

(*b*) The superficial fascia is fibrous and adherent to the skin and underlying aponeurosis. It is richly supplied with vessels and nerves.

(*c*) The epicranial aponeurosis is a musculofibrous layer containing the **occipitalis** muscle behind and the **frontalis** muscle in front. The occipitalis is attached to the superior nuchal line. Laterally the aponeurosis is attached to the temporal fascia. Anteriorly the frontalis has no bony attachment but blends with the fibres of the orbicularis oculi muscles. The occipito-frontalis muscle draws the scalp backwards and forwards and raises the eyebrows; it is innervated by the facial nerve.

(*d*) A layer of loose areolar tissue allows free movement of the aponeurosis and adherent skin over the periosteum.

(*e*) The periosteum (pericranium) of the bones of the vault is continuous through the sutures with the endocranium.

The scalp is richly supplied by anastomosing branches of the external carotid artery (occipital, posterior auricular and superficial temporal) and of the ophthalmic artery (supra-orbital and supratrochlear). Scalp wounds bleed profusely. The veins pass with the arteries and reach the jugular vessels but may communicate with the diploic veins of the skull (p. 8) or through emissary veins (p. 18) with the intracranial venous sinuses.

The sensory innervation of the scalp is from the trigeminal, lesser and greater occipital nerves, and other branches of the dorsal rami of the upper three cervical nerves (see Fig. 30.2).

The lymph vessels pass to adjacent nodes in the circular chain around the base of the skull (see p. 433).

THE FACE

The face contains the anterior openings of the mouth, nose and orbit. The skin of the face is thin, sensitive, and hairy (especially in men). Voluntary muscles (of facial expression) are attached to it. The subcutaneous tissues are very vascular (heat loss and blushing) and have a varying amount of fat; there is usually no fat in the eyelids. There is no deep fascia over most of the face. Sweat and sebaceous glands are abundant. The mucocutaneous junction is on the facial aspect of the lips, their red margin. Mucous and small salivary glands are present on the inner aspects of the lips and cheeks.

FACIAL MUSCLES (Fig. 25.7)

These are the muscles of facial expression; they are arranged as sphincters and dilators around the openings of the mouth, nose and orbit. The muscles are derived from the mesoderm of the 2nd pharyngeal arch and are supplied by the facial nerve.

1. Orbicularis oris

This is the sphincter surrounding the mouth and forms the greater part of the substance of the lips. When it contracts it produces a small orifice, as in whistling. Other muscles blend with and reinforce the orbicularis oris, particularly the buccinator and the levator and depressor muscles. These last muscles are attached to the angles of the mouth (anguli) and also to the middle of the lips (labii).

2. Buccinator

Buccinator forms the greater part of the substance of the cheek. It has a continuous lateral attachment to the pterygomandibular raphé and the outer surfaces of the maxillae and mandible adjacent to the last molar teeth. (The buccinator and superior constrictor have a common attachment to the pterygomandibular raphé.) The fibres pass forwards and medially, decussate behind the angle of the mouth and enter the lips, blending with orbicularis oris. The muscle is used to keep food between the teeth in chewing, and also to increase the pressure in forced blowing.

3. The compressor and dilator nares

These muscles pass from the maxilla to the nasal cartilages and are vestigial.

4. Orbicularis oculi

This surrounds the opening of the orbit and has palpebral and orbital portions. The central palpebral fibres lie within the eyelids and are attached by the medial palpebral ligament to the frontal process of the maxilla. The peripheral orbital fibres surround the orbital margin and are attached medially to the medial palpebral ligament and directly to the frontal process of the maxilla. The fibres blend with frontalis above the orbit.

The palpebral part closes the eyelids in sleep, winking and blinking; the orbital part produces more positive closure of the eyelids and is used in frowning (see p. 363).

5. Platysma

This is a broad flat sheet of muscle situated in the superficial fascia on each side of the neck. Superiorly it passes over the mandible to blend with the muscles around the mouth; inferiorly it blends with the superficial fascia over the upper part of the chest. The muscle is a weak depressor of the jaw.

The attachments of the above muscles and also of the corrugator, zygomaticus major and minor, levator labii superioris, mentalis, depressor labii inferioris and depressor anguli oris are shown in Figure 25.7.

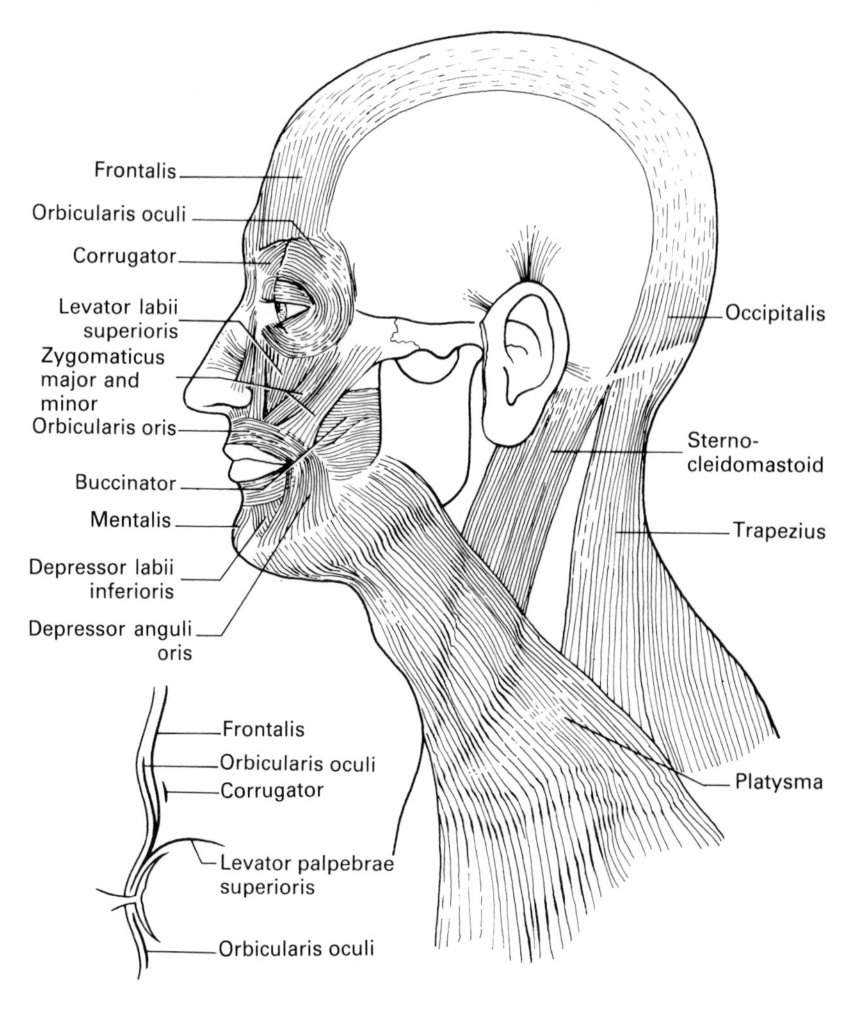

Fig. 25.7 The superficial muscles of the head and neck. Inset; the muscle relations in the eyelids.

VESSELS AND NERVES

The face is supplied mainly by the facial and transverse facial arteries and is drained by the facial and retromandibular veins. The lymph drainage of the chin is to the submental nodes, that of the medial part of the cheek and face to the submandibular nodes, and of the forehead and lateral part of the cheek to the parotid nodes. Sensory nerves pass to the three divisions of the trigeminal nerve and, over the angle of the jaw, to the greater auricular nerve (see Fig. 30.2).

Motor innervation is by the facial nerve. Paralysis leads to loss of protection for the cornea of the eye and difficulties in eating and speaking. The unopposed raising of the upper eyelid by the levator palpebrae superioris makes blinking impossible, so airborne dust can deposit on the cornea. The drooping of the lower eyelid allows tears to run down the face instead of moistening the cornea and conjunctiva.

DEVELOPMENT OF THE FACE

The face develops from a median frontonasal process and paired maxillary and mandibular processes. The frontonasal process is formed between the roof of the mouth and the forebrain. Its upper part forms the forehead. Two thickenings, the nasal placodes develop on its anterior aspect and sink to form nasal pits. The two pits divide the frontonasal process into a median and two lateral nasal processes. The nasal pits deepen to form the nasal sacs which extend backwards through the head mesoderm separated by a median nasal septum. The median nasal processes become the premaxilla which is largely submerged and forms the part of the upper jaw carrying the incisor teeth.

The mandibular processes comprise the first pharyngeal arch structures (p. 388); they form the lower jaw. The maxillary processes of the mandibular arch pass medially below the nasal pits and fuse with each other near the midline, forming the upper lip. The nasolacrimal duct develops along the line of fusion of the frontonasal process and the maxillary process of each side. Palatine processes from the two maxillary processes fuse in the midline and form the hard and soft palates. Incomplete growth or failure to fuse gives rise to defects such as cleft chin, hare lip or cleft palate which interfere with suckling, swallowing and later with speech.

— 26 —

The orbit, nose and mouth

THE ORBITAL CAVITY AND EYEBALL

The two **orbital cavities** (Figs. 26.1, 26.2) lie within the facial skeleton and contain the eyeballs, extra-ocular muscles, vessels, nerves, glands and fat. Each cavity is pyramidal in shape with the apex posterior. It has a roof, floor, and medial and lateral walls. Anteriorly the cavity opens on to the face.

The roof is formed mainly by the orbital plate of the frontal bone and, posteriorly, by the lesser wing of the sphenoid. It separates the orbit from the anterior cranial fossa. The floor, formed largely by the maxilla and zygoma, separates the orbit from the maxillary air sinus and has an inferior orbital fissure posteriorly which passes forwards into the infraorbital canal. The lateral wall is formed by the zygomatic bone and the greater wing of the sphenoid, and separates the orbit from the temporal fossa. The medial wall consists of the

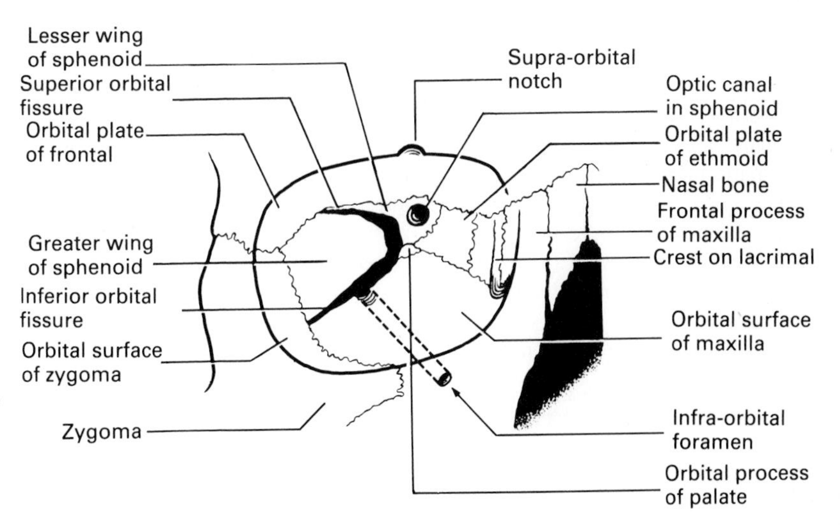

Fig. 26.1 Skeleton of the right orbit.

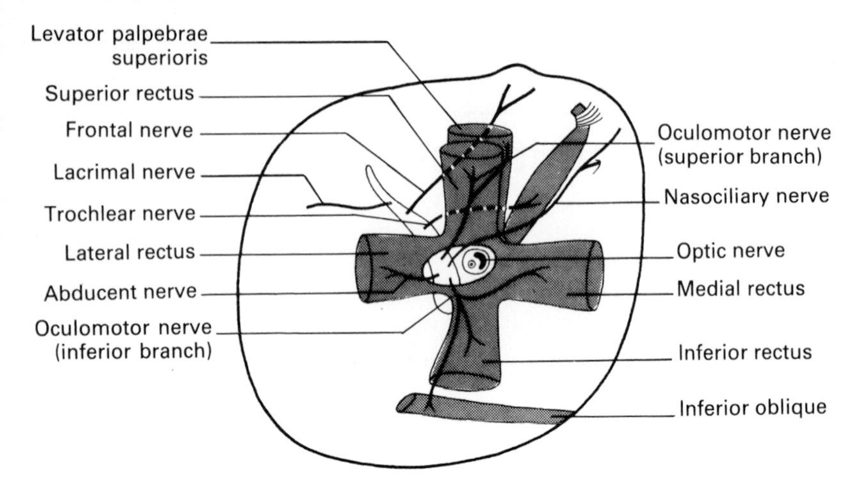

Fig. 26.2 Diagram indicating the relations of the extra-ocular muscles and the nerves in the right orbital cavity, viewed from in front.

frontal process of the maxilla, the lacrimal bone, the orbital plate of the ethmoid, and the body of the sphenoid bone, from before backwards. It separates the orbit from the ethmoidal air sinuses and the nasal cavity.

There are three posterior openings in the cavity, the superior and inferior orbital fissures laterally and the optic canal medially. The **superior orbital fissure** lies between the greater and lesser wings of the sphenoid and opens into the middle cranial fossa. Ophthalmic veins pass through and enter the cavernous sinus. It is divided by the common tendinous attachment of the extra-ocular muscles into a narrow lateral and a wider medial part. The lateral part transmits the lacrimal, frontal and trochlear nerves, and the medial part the abducent, the superior division of the oculomotor, the nasociliary, and the inferior division of the oculomotor nerves, all from lateral to medial. The **inferior orbital fissure** lies between the floor and lateral wall and opens into the pterygopalatine fossa medially and the infratemporal fossa laterally. The maxillary nerve and its zygomatic branch, together with communicating veins to the pterygoid venous plexus, pass through the fissure. The **optic canal** opens into the apex of the cavity from the middle cranial fossa; it transmits the optic nerve with the meningeal coverings and the ophthalmic artery.

Anteriorly a wide canal passes downwards from the inferomedial angle of the orbital cavity to the nose; it contains the nasolacrimal duct. Superolaterally is the fossa for the lacrimal gland.

THE EYEBALL (Fig. 26.3)

The eyeball is the visual organ. It is situated in the anterior part of the orbital cavity and is roughly spherical in shape, being distorted by an anterior projecting clear area. The optic nerve leaves it posteromedially.

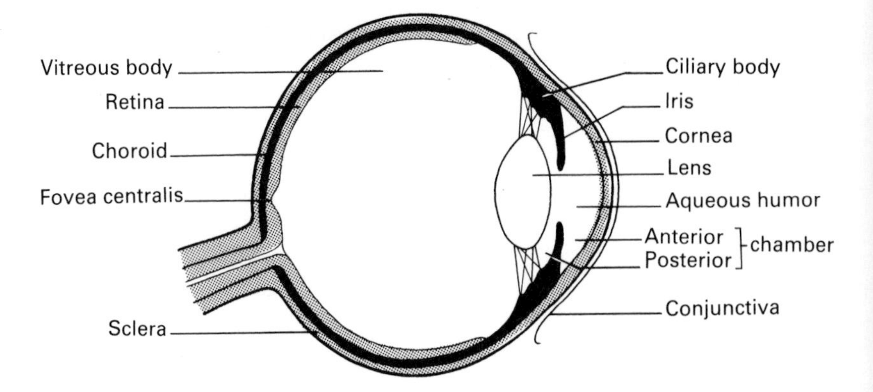

Fig. 26.3 Horizontal section through the left eyeball.

The walls of the sphere have three coats: an outer fibrous (supporting) layer, a middle choroid (vascular) and an inner retinal (nervous). The eyeball is modified anteriorly to allow light to enter.

The fibrous coat forms the dense white **sclera** posteriorly, and anteriorly the transparent **cornea**. The **choroid** is vascular and pigmented. Anteriorly it has a circular thickening, the **ciliary body**, which contains the smooth ciliary muscle. In front of the ciliary body the choroid thins and forms the **iris** with its central aperture, the **pupil**. The iris contains the circularly placed sphincter pupillae and radially arranged dilator pupillae. They alter the size of the pupil.

The **retina** contains the light receptors from which information is passed eventually along the optic nerve. The retina consists of 10 layers as shown in Figure 26.4. To the medial side of the posterior wall there is a paler area, the **optic disc**, where the optic nerve leaves the eyeball. About 3 mm lateral to the disc, where the visual axis projects on to the retina, there is a small depression, the **fovea centralis**. The area around it is called the **macula** and is used mainly in daylight vision.

The anteriorly placed biconvex **lens** divides the cavity of the eyeball into two parts. The posterior part is filled with a jelly-like **vitreous body**, the anterior with the more fluid **aqueous humor**. The iris further divides the anterior part into the anterior chamber between the iris and the cornea, and the posterior chamber between the iris and the lens. The two chambers are in continuity through the pupil.

The lens is held in position by its suspensory ligament (zonular fibres) which passes from the periphery of the lens to the ciliary body. Accommodation is the process by which near objects are focussed on to the retina (see p. 480). The ciliary muscle contracts, draws forwards the ciliary body and relaxes the suspensory ligament, thus allowing the lens, because of its inherent elasticity, to become more convex.

Blood supply

The central artery of the retina is a branch of the ophthalmic artery. Ciliary branches of the ophthalmic artery pass forwards mainly in the choroid. The veins collect to form four venae vorticosae which leave the eyeball posteriorly and pass to the ophthalmic veins.

Nerve supply

The optic nerve carries sensory fibres from the retina. The long ciliary nerves (from the nasociliary) and short ciliary nerves (from the ciliary ganglion) pierce the sclera posteriorly. They carry autonomic fibres (to the sphincter and dilator pupillae and the ciliary muscles see p. 481), and sensory fibres from the conjunctiva over the cornea and sclera.

Embryology

The optic nerve, retina, ciliary body and iris develop as an outgrowth of the primitive forebrain; the cornea, sclera and choroid are condensations of mesoderm; and the lens is derived from ectoderm. (See also p. 40.)

Orbital (bulbar) fascia

The eyeball is closely surrounded, behind the cornea, by the orbital fascia which separates it from the orbital fat. The fascia is pierced by, and forms a sheath around, the muscles attached to the sclera. Thickenings of the fascia are attached to the lacrimal and zygomatic bones and form the medial and lateral **check ligaments** respectively. A thickening below the eyeball forms the **suspensory ligament**.

EXTRA-OCULAR (BULBAR) MUSCLES (Fig. 26.2)

1. Rectus muscles

These include superior, inferior, medial and lateral.

Attachments

ANTERIOR—the muscles pass forwards, in the positions implied by their names, to the sclera just in front of the equator.

POSTERIOR—a common tendinous ring which surrounds the optic canal and the medial end of the superior orbital fissure.

2. Superior oblique

This is attached posteriorly above the common tendinous ring. It passes forwards and its tendon hooks around a fibrocartilaginous pulley, the trochlea, on the superomedial border of the front of the orbit. It then passes backward and laterally between the superior rectus and the eyeball and gains its posterolateral surface behind the equator.

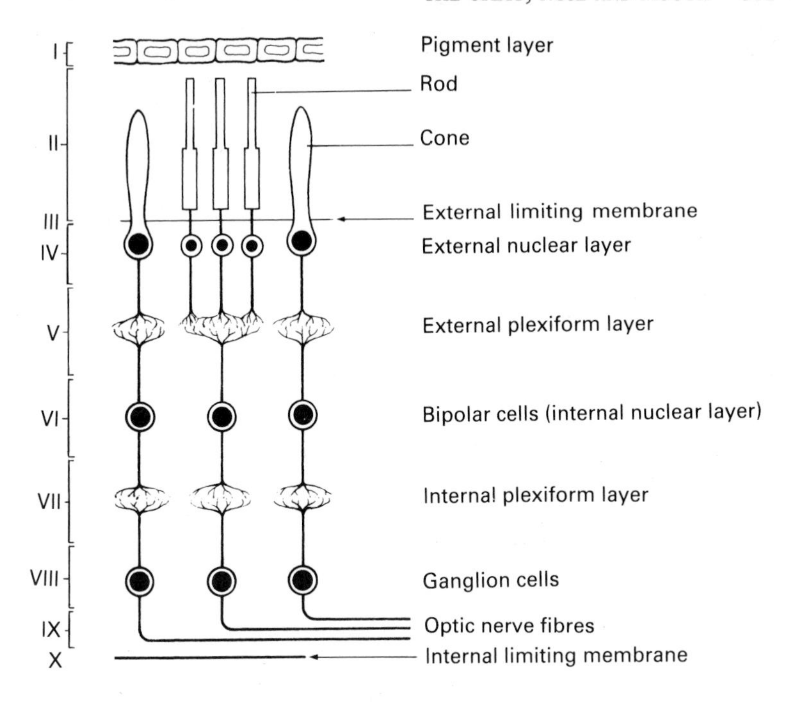

Fig. 26.4 Diagram of the layers of the retina.

3. Inferior oblique

This is attached to the anteromedial aspect of the floor of the orbit. It passes backwards and laterally below the inferior rectus to gain the posterolateral surface of the eyeball behind the equator.

Action

There is co-ordinated contraction and relaxation of the different muscles of each eye as they work together. There is also, under most circumstances, a co-ordinated movement of the two eyes to fix on an object. Because of the angle between the visual axis and the orbital axis movements other than horizontal to left or right involve some rotation and angular deviation of the eyeball. Looking towards the left involves contraction of the left lateral rectus and the right medial rectus. Looking down and to the left involves the left superior oblique and the right inferior rectus. Looking up and to the left involves the left inferior oblique and the right superior rectus.

Inco-ordination of the muscles of the two eyes leads to a squint and double vision (diplopia).

Nerve supply

The lateral rectus is supplied by the abducent nerve, the superior oblique by the trochlear nerve, the remaining muscles by the oculomotor nerve (LR_6 SO_4).

4. Levator palpebrae superioris
This is attached posteriorly above the common tendinous origin and anteriorly to the conjunctiva and tarsal plate.

Action
It raises the upper eyelid and is supplied by the oculomotor nerve.

THE OPHTHALMIC ARTERY

This is a branch of the internal carotid artery after it has pierced the dural roof of the cavernous sinus. It arises in the middle cranial fossa, enters the orbit through the optic canal and ends behind the medial side of the upper eyelid by dividing into supratrochlear and dorsal nasal branches.

Relations
In the middle cranial fossa, the artery lies above the cavernous sinus, medial to the anterior clinoid process, inferolateral to the optic nerve. It passes through the optic canal inferior to the optic nerve. In the orbit, it passes medially over the optic nerve (in company with the nasociliary nerve) and passes forwards along the medial wall of the cavity.

Branches
(i) *central artery of the retina*—arises near the apex of the orbit and enters the optic nerve.
(ii) *ciliary branches*—to the eyeball.
(iii) *muscular branches.*
(iv) *further branches*—accompany the supra-orbital, supratrochlear, anterior and posterior ethmoidal and lacrimal nerves.

The ophthalmic veins
Superior and inferior ophthalmic veins drain the orbital contents and pass through the superior orbital fissure to the cavernous sinus. They communicate posteriorly, through the inferior orbital fissure with the pterygoid venous plexus. Anteriorly, they communicate with the facial vein near the medial angle of the eye and thus infection may pass from the face through the orbit to the cavernous sinus.

Nerves of the orbit
The nerves are described more fully in Chapter 28. Their relations are best understood by appreciating their positions within the superior orbital fissure

(Fig. 26.2). Nerves passing outside the common tendinous ring remain outside the cone of extra-ocular muscles and those within the ring remain inside the cone.

THE EYELIDS

The orbital cavity is limited anteriorly by two movable skin folds, the larger upper and smaller lower eyelids. They are united medially and laterally and limit the palpebral fissure at its medial and lateral angles. Within the medial angle there is a pinkish elevation, the **lacrimal caruncle**. On the medial end of the margin of each lid there is a small elevation, the **lacrimal papilla**, on the apex of which is the **lacrimal punctum** (the opening of the lacrimal canaliculus).

Each eyelid comprises five layers, from without inwards:

(*a*) Skin.

(*b*) Superficial fascia, usually devoid of fat.

(*c*) Palpebral fibres of obicularis oculi and levator palpebrae superioris (Fig. 25.7).

(*d*) **Tarsal plate**—this is an elliptical plate of dense fibrous tissue which is attached to the margin of the orbit by the **medial** and **lateral palpebral ligaments**. On the deep surface of the plate lie the tarsal (Meibomian) glands which are modified sebaceous glands. Their ducts open on to the lid margin.

(*e*) **Conjunctiva**—this is a thin layer of stratified columnar epithelium which lines the inner surface of both lids and is reflected over the front of the eyeball, so that when the lids are approximated, it encloses a narrow sac containing a small amount of fluid. Upper and lower pouches between the lids and the eyeball are known as the superior and inferior conjunctival fornices. A crescentic fold of conjunctiva, the semilunar fold, lies lateral to the lacrimal caruncle on the surface of the eyeball. Over the cornea, there is stratified squamous epithelium which is firmly attached.

The skin and conjunctiva are in continuity around the free border of the lids. The eyelashes emerge through the skin on the margin of the lids and numerous small sebaceous glands are associated with the hair follicles. The eyelids are closed by the action of orbicularis oculi and the upper lid is raised by levator palpebrae superioris.

THE LACRIMAL APPARATUS

The front of the eyeball, with its conjunctival covering, is continually being washed by lacrimal secretions which are drained by blinking, when the increased intraconjunctival pressure, produced by the closed lids, forces the tears into the puncta. The structures concerned with the production of the lacrimal fluid and its subsequent passage to the nasal cavity are collectively called the lacrimal apparatus. This comprises: the lacrimal gland and its ducts, the conjunctival sac, the lacrimal canaliculi, the lacrimal sac and the nasolacrimal duct.

Lacrimal gland

This is a serous gland situated in the superolateral angle of the orbit behind the upper lid. It is almond-shaped and has a palpebral process which descends between the conjunctiva and the tarsal plate. It has 6–12 ducts which open into the superior conjunctival fornix.

It is supplied by the lacrimal branch of the ophthalmic artery and innervated by the facial nerve through the greater petrosal nerve and the pterygopalatine ganglion; postganglionic fibres pass in the zygomatic and lacrimal nerves. Its lymph drainage is to the parotid lymph nodes.

Lacrimal canaliculi

Each canal is about 10 mm long and passes from the lacrimal punctum in each eyelid medially, deep to the medial palpebral ligament, to the lacrimal sac. They are lined by stratified squamous epithelium.

Lacrimal sac

This is a thin fibrous sac situated on the medial side of the orbit in the lacrimal fossa. It receives the lacrimal canaliculi and continues downwards as the nasolacrimal duct. The lacrimal sac is lined by stratified columnar epithelium.

Nasolacrimal duct

This descends, having the maxillary bone laterally and the lacrimal bone and inferior concha medially, and opens into the inferior meatus of the nose. The opening is guarded by a fold of mucous membrane which prevents air passing upwards when blowing the nose. The duct is lined by columnar epithelium which is ciliated in places.

THE NOSE

The nose comprises the external nose and the two nasal cavities. The **external nose**, a prominent feature of the face, bears the anterior nasal openings, (the nostrils, nares). It is pyramidal in shape but varies in width and in the direction of the anterior nares in different individuals. Its skeleton is formed superiorly by the nasal bones which articulate with the frontal bone, and the frontal processes of the maxillae. Inferiorly it is formed by several cartilages which articulate with the nasal and maxillary bones and surround the anterior nares. The skin covering the external nose is freely movable over the nasal bones but more firmly attached over the cartilages. The part of the nasal cavities just within the external nose is the **vestibule**, the expanded lateral wall of which is called the **ala** of the nose.

The **nasal cavities** form the first part of the respiratory passage and extend from the anterior nares to the nasopharynx. They are lined mainly by respiratory

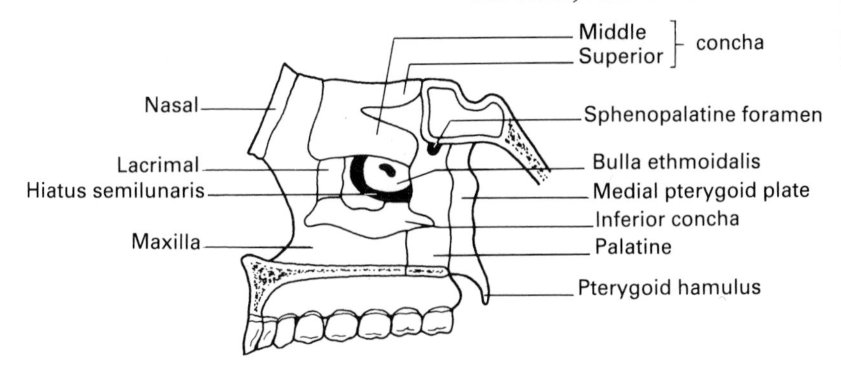

Fig. 26.5 Diagram of the bones forming the lateral wall of the nasal cavity.

(ciliated columnar) epithelium, but superiorly with olfactory epithelium. The cavities are separated by a midline septum and each possesses a roof, a floor, lateral and medial walls, and anterior and posterior apertures.

The narrow **roof** lies below the anterior cranial fossa, it is arched anteroposteriorly. It is formed from the nasal cartilages, the nasal and frontal bones, the cribriform plate of the ethmoid and the body of the sphenoid from before backwards.

The horizontal **floor** forms part of the roof of the oral cavity, and is formed by the palatine process of the maxilla and the horizontal plate of the palatine bone. Anteriorly, in the midline, the incisive canal transmits the greater palatine artery and nasopalatine nerve.

The **medial wall** of the cavity is the nasal septum. It is formed by the septal cartilage, the perpendicular plate of the ethmoid and the vomer from before backwards, with small peripheral contributions from the palatine, maxillary and sphenoid bones.

The **lateral wall** (Fig. 26.5) lies medial to the orbit, the ethmoid and maxillary air sinuses and the pterygopalatine fossa. Its surface area is greatly increased by three horizontal bony projections, the superior, middle and inferior nasal **conchae**, and by a series of diverticulae, the **paranasal air sinuses**. The inferior concha is the largest and lies about 1 cm above the floor of the nose. The superior and middle conchae are united anteriorly. The space below each concha is called a **meatus** and the space above the superior concha the spheno-ethmoidal recess. The sphenoid air sinus opens into the spheno-ethmoidal recess; the posterior ethmoidal air cells open into the superior meatus; the nasolacrimal duct opens into the inferior meatus. The middle meatus possesses a domed bulge, the **bulla ethmoidalis**, containing the middle ethmoidal air cells which open on to it. A semicircular groove below the bulla, the **hiatus semilunaris**, has the openings of the frontal, anterior ethmoidal and maxillary air sinuses, from before backwards.

The skeleton of the lateral wall is formed mainly by the maxillary and ethmoid bones. The labyrinth of the ethmoid, containing the ethmoidal air sinuses, lies superiorly; the superior and middle conchae are both part of this bone. The maxilla, containing the maxillary air sinus, is deficient medially but is overlapped by the bulla ethmoidalis, the lacrimal bone, the vertical plate of the palatine bone and the inferior concha. The sphenopalatine foramen, between the vertical plate of the palatine bone and the body of the sphenoid in the roof of the cavity, leads into the pterygopalatine fossa. The greater palatine canal descends from the fossa, between the palatine and maxillary bones to the greater palatine foramen (Fig. 25.5). The nasolacrimal duct descends between the maxilla laterally and the lacrimal bone and inferior concha medially. The lateral wall is completed anteriorly by the nasal bone and cartilages, and posteriorly by the medial pterygoid plate articulating with the vertical plate of the palatine bone.

Nerve supply

This is from branches of the maxillary nerve and the anterior ethmoidal branch of the ophthalmic nerve. The lateral wall may be divided into quadrants; the upper anterior quadrant is supplied by the anterior ethmoidal nerve, the upper posterior by the nasal nerves, the lower posterior quadrant by the greater palatine nerve and the lower anterior by the anterior superior alveolar nerve. The roof and upper anterior half of the nasal septum are innervated by the anterior ethmoidal nerve, the floor and lower posterior half of the septum by the nasopalatine nerve. The olfactory mucosa, in the upper part of the cavity, is also supplied by the olfactory nerves.

Blood supply

This is from the ophthalmic and maxillary arteries. The names of the arteries largely correspond to the nerves, the anterior inferior part of the cavity receiving additional branches from the facial artery. The veins drain to the pterygoid venous plexus and the facial vein.

Lymph drainage

The anterior part of the cavity drains to the submandibular nodes and the posterior part to the retropharyngeal nodes.

Histology

The nose is lined by vascular mucoperiosteum. The superior concha, the roof and adjoining nasal septum are covered by yellow olfactory epithelium consisting of supporting cells around the endings of the bipolar olfactory cells. The rest of the cavity is lined by respiratory epithelium. The nose warms and moistens the inhaled air, some of which passes over the olfactory endings. Owing to its connection with the oropharynx, it modifies sounds produced by the larynx.

THE PARANASAL AIR SINUSES

Paired frontal, maxillary, ethmoidal and sphenoidal air sinuses are found within the corresponding bones. The sinuses develop as a series of diverticula from each nasal cavity and are lined by extensions of its mucoperiosteum. They are small at birth, enlarge mainly during the eruption of the second dentition and reach their adult size soon after puberty. They lighten the front of the skull and may increase the resonance of the voice. Their presence may be due to an overgrowth of the diploë. The sinuses are often the site of infection, drainage is poor and their lymph vessels pass mainly to the deeply placed retropharyngeal nodes.

1. Frontal

This sinus is situated above the medial end of the superciliary arch separated by a bony septum from its fellow. It lies anterior to the anterior cranial fossa and above the orbit. Each opens into the middle meatus of the nasal cavity through the long, narrow frontonasal duct which traverses the anterior ethmoidal air cells. The sinus is supplied by the supra-orbital vessels and nerves.

2. Maxillary

This is the largest of the paranasal air sinuses and is pyramidal in shape. Its base forms part of the lateral wall of the nose. Its apex projects laterally into the zygomatic process of the maxilla. The roof separates the sinus from the orbit and transmits the infra-orbital vessels and nerve in a canal. The floor is the alveolar margin containing the molar teeth, the roots of which project into the cavity covered only by a thin layer of bone. The posterior wall contains the posterior superior alveolar nerve and lies in front of the infratemporal and pterygopalatine fossae. The anterior wall is the facial surface of the maxilla which is covered by the mucous membrane of the vestibule of the mouth. The medial wall (base) forms part of the lateral wall of the nose and is completed by the inferior concha and parts of the ethmoid, lacrimal and palatine bones (Fig. 26.5). The sinus opens into the hiatus semilunaris posteriorly. The opening is high on the base of the sinus and drainage is difficult. The mucoperiosteum is supplied by the anterior and posterior superior alveolar vessels and nerves. Its lymph vessels pass to the submandibular and retropharyngeal nodes.

In a healthy state the sinuses contain air but in the event of fluid formation, because of infection or a fracture, the fluid level may be demonstrated radiologically.

3. Ethmoidal

Each ethmoidal labyrinth, lying between the orbit and the upper part of the nose, contains numerous air cells which are divided into anterior, middle and posterior groups. The anterior group drains into the anterior part of the hiatus semilunaris: the middle group opens on to the bulla ethmoidalis, within which the air cells are situated; the posterior group drains into the superior meatus. The sinuses are supplied by the anterior and posterior ethmoidal vessels and nerves.

4. Sphenoidal

These are contained within the body of the sphenoid bone and usually communicate with each other through an incomplete bony septum. They lie below the sella turcica and the pituitary gland, medial to the cavernous sinus and its contents, and above the nasal cavity. Each sinus opens into the spheno-ethmoidal recess and is supplied by the posterior ethmoidal vessels and nerve.

THE ORAL CAVITY

This is the first part of the alimentary tract and extends from the lips to the isthmus of the fauces. It contains the tongue, alveolar arches with the gums and teeth, and receives the openings of the salivary glands. Its mucous membrane is lined throughout by stratified squamous epithelium. The alveolar arches, gums and teeth divide the cavity into an outer vestibule and an inner mouth cavity proper.

The **vestibule** is a slit-like cavity limited externally by the lips and cheeks, and opens on to the face between the lips at the oral fissure. Internally it is limited by the gingivae and teeth. It communicates with the mouth cavity proper either between the teeth or, when the teeth are occluded, through the retromolar spaces behind the last molars. The cavity is limited superiorly and inferiorly by the mucous membrane passing from the lips and cheeks on to the maxilla and mandible. The parotid duct opens into the vestibule just above the crown of the second upper molar tooth.

The **mouth cavity proper** is limited anteriorly and laterally by the maxilla and mandible and teeth, and possesses a roof, a floor and a posterior opening. The roof is formed by the palate which separates the mouth from the nasal cavities. Much of the floor is occupied by the tongue from which the mucous membrane extends to the body of the mandible. A midline fold of mucous membrane, the **frenulum**, connects the under surface of the tongue to the floor. On each side of the base of the frenulum is a **sublingual papilla** on to which opens the submandibular duct. Passing backwards from the papillae, the **sublingual folds** of mucous membrane cover the sublingual glands. The posterior opening of the oral cavity is the **fauceal isthmus** which is bounded laterally by the palatoglossal fold (the anterior arch of the fauces), superiorly by the soft palate and inferiorly by the back of the tongue.

Gums (gingivae)

The gums surround the necks of the teeth and are composed of fibrous tissue covered with a vascular mucous membrane. They are firmly attached to the alveolar margins and the adjacent bone but less so to the necks of the teeth.

THE TEETH

Each tooth consists of a projecting crown, a neck surrounded by the gum and one or more roots embedded in the alveolar bone. It has a central cavity which opens on to the apex of the root and is filled with **pulp** (loose connective tissue containing vessels and nerves). The cavity is surrounded by **dentine**, a hard calcified material which forms the bulk of the tooth. The dentine of the crown is covered by **enamel**, a very hard inorganic layer, and that of the root by **cementum**, a bone-like material. The modified periosteum of the alveolar socket (**periodontal membrane**) holds the root in position.

The teeth are names, from in front laterally: incisors (I); canines (C); premolars (P) and molars (M). They vary in shape, size and function.

Crowns—incisors have sharp cutting edges used for biting food. Canines are large and conical. Premolars (bicuspid teeth) and molars (with three to five tubercles) are used in chewing.

Roots—the incisors, canines and premolars, with the exception of the 1st upper premolar, possess a single root; the 1st upper premolar and lower molars have two roots and the upper molars three roots.

Man has two sets of teeth which erupt at different times. The deciduous (milk) teeth appear between the 6th and 24th months; they are replaced by the permanent teeth which appear between the 6th and 25th years. Each half of each jaw contains five deciduous (I2, C1, M2) and eight permanent teeth (I2, C1, P2, M3).

The average eruption times of each half of the upper and lower jaws are:

		I	C	M	
Deciduous:	upper	7 8	18	12 24	Months
	lower	6 9	18	12 24	

		I	C	P	M	
Permanent:	upper	7 8	12	9 10	6 12 18+	Years
	lower	7 8	12	9 10	6 12 18+	

N.B. The first deciduous tooth is a lower central incisor, the first permanent tooth is a 1st molar. The lower permanent teeth appear slightly earlier than the upper. The wisdom teeth (3rd molars) appear between the 17th and 25th years and are usually the first to be shed.

Nerve and blood supply

The teeth of the upper jaw are innervated by the anterior and posterior superior alveolar nerves, the lower by the inferior alveolar nerve. The central incisors have a bilateral innervation. The blood supply is by the corresponding arteries.

THE UPPER GUMS—the labial surface is supplied by the infra-orbital and posterior superior alveolar nerves, the lingual surface by the nasopalatine and greater palatine nerves.

THE LOWER GUMS—the labial surface is supplied by the mental and buccal nerves, the lingual surface by the lingual nerve.

Lymph drainage

Drainage from the maxillary teeth and gums is to the upper deep cervical nodes, and from the mandibular teeth and gums to the submandibular, submental and deep cervical nodes.

Embryology

An ectodermal fold grows from the mouth cavity into the mesoderm of the maxillary and mandibular alveolar arches. This is the primary dental lamina, and cup-shaped buds from it produce the enamel of each tooth. The mesoderm within the cup forms the dentine, cementum and pulp. Both deciduous and permanent teeth are formed in this manner and, after the formation of the latter, the primary dental lamina disappears. X-rays have shown many of the details of postnatal tooth development.

THE PALATE

This forms the roof of the mouth and separates the mouth from the nasal cavities. It consists of the hard palate anteriorly and the soft palate posteriorly.

HARD PALATE

This is formed by the palatine processes of the maxillae and the horizontal plates of the palatine bones. The anterior midline **incisive foramen** transmits the nasopalatine nerve and the terminal branch of the greater palatine artery. The greater and lesser palatine foramina are situated posterolaterally and transmit the greater and lesser palatine vessels and nerves. The hard palate is covered by a mucoperiosteum which is rich in mucous glands and is continuous anteriorly and laterally with the gums, and posteriorly with the mucous membrane over the soft palate. It is supplied by the greater and lesser palatine vessels and nerves except for the area behind the incisor teeth which is supplied by the nasopalatine nerve. Its lymph vessels pass to the deep cervical lymph nodes. The palate develops from the palatine process of the maxillary process on each side and the midline median nasal process of the frontonasal process.

SOFT PALATE

This is a movable partition between the naso- and oropharynx. It consists of a number of paired muscles covered by a fold of mucous membrane. Anteriorly it is attached to the back of the hard palate. Its posterior free border has a midline conical projection, the **uvula**. On each side two vertical folds, the anterior and posterior arches of the fauces, pass from its inferior surface to the tongue and pharynx. They contain the palatoglossus and palatopharyngeus muscles respectively.

Muscles of the soft palate

1. **Tensor veli palatini**—is attached superiorly to the auditory tube and the base of the skull between the pterygoid fossa and the spine of the sphenoid. Its fibres converge on to a narrow tendon which hooks medially around the hamulus. In the palate the tendon fans out, joins that of the opposite side and forms the palatine aponeurosis. The aponeurosis, which is attached to the back of the hard palate, is the 'skeleton' of the palate and the other palatine muscles are attached to it.

2. **Levator veli palatini**—is attached superiorly to the auditory tube and the apex of the petrous temporal bone. It descends in the lateral wall of the nasopharynx and blends with the palate.

3. **Palatoglossus**—passes downwards and forwards from the palate to the side of the tongue.

4. **Palatopharyngeus**—descends from the palate, blends with the inner surface of the constrictor muscles and gains attachment to the posterior border of the thyroid cartilage. Some of the fibres pass horizontally around the inner surface of the superior constrictor and form the palatopharyngeal sphincter and the pharyngeal isthmus.

Movement of these muscles plays an important role in swallowing, and defects of the palate interfere with swallowing and speech (p. 389).

Nerve supply

The tensor palatini is innervated by the medial pterygoid branch of the mandibular nerve, and the other muscles by the vagus nerve via the pharyngeal plexus (p. 386). The cell bodies lie in the nucleus ambiguus. The mucous membrane is innervated by the greater and lesser palatine branches of the maxillary nerve and branches of the glossopharyngeal nerve. Taste fibres may also be present in these nerves.

Blood supply

Palatine branches of the maxillary, facial, and lingual arteries; the venous drainage is to the pharyngeal venous plexus.

Lymph drainage

This is to the deep cervical lymph chain.

Histology

The upper surface is covered by respiratory epithelium and the lower by stratified squamous. Numerous mucous glands are present on both surfaces. The development is outlined on page 356. Incomplete merging of the sides may lead to degrees of cleft palate and is often associated with hare lip. Suckling, swallowing and speech are then impaired.

THE TONGUE (Fig. 26.6)

This is a muscular organ situated in the floor of the mouth and oropharynx. It is the shape of an inverted shoe, attached mainly to the hyoid bone and the mandible, and resting on the geniohyoid and mylohyoid muscles. The tip is free and projects anteriorly. The upper surface (dorsum) is covered by mucous membrane and is divided by a V-shaped groove, the **sulcus terminalis**, into an anterior two-thirds and a posterior third. There is a small pit, the **foramen caecum**, at the apex of the sulcus. The two parts differ in their nerve supply, histology and development. The mucous membrane which is continued from the dorsum to the anterior surface of the epiglottis, is raised in a midline glosso-epiglottic fold. The fold separates two shallow fossae, the **valleculae**, which are limited laterally by the pharyngeal wall. The lower surface of the tongue is also covered by thin mucous membrane and is attached by a midline fold of mucous membrane, the **frenulum**, to the floor of the mouth.

MUSCLES OF THE TONGUE

Intrinsic and extrinsic muscles lie on each side of a midline fibrous septum. The fine intrinsic muscles are arranged in vertical, horizontal and transverse bundles. The coarse extrinsic muscles connect the tongue to the mandible, hyoid bone, styloid process and palate. The extrinsic muscles mainly alter the position of the tongue and the intrinsic muscles its shape.

1. Genioglossus is attached to the mental spine on the mandible. The fibres pass backwards and, fanning out vertically near the midline, pass to the tip and the whole length of the dorsum of the tongue.

2. Hyoglossus: this thin quadrilateral muscle is attached to the body and

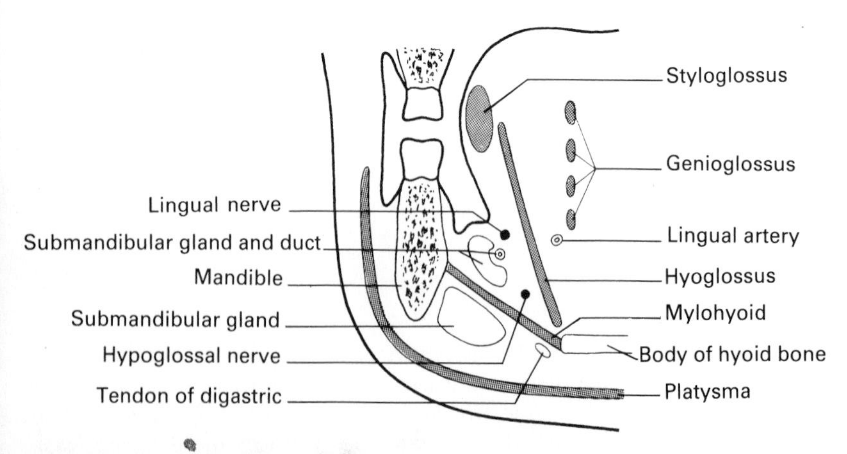

Fig. 26.6 Diagrammatic layout of the main structures seen in a coronal section at the level of the 2nd molar teeth.

greater horns of the hyoid bone, and passes upwards and forwards into the side of the tongue.

Relations: medially from before backwards are the intrinsic muscles of the tongue, the lingual artery, the middle constrictor, the stylohyoid ligament, and the glossopharyngeal nerve. Laterally, from above downwards, are styloglossus, the lingual nerve and the submandibular ganglion, the submandibular gland and its duct, the sublingual gland and the hypoglossal nerve. Lateral to these is the mylohyoid muscle.

3. Styloglossus (see p. 388) and **palatoglossus** (see p. 371).

Blood supply

Lingual artery, whose dorsal lingual branch passes to the posterior third and whose deep branch passes to the tip and the rest of the tongue. The deep lingual vein passes along the inferior surface of the tongue, lateral to hyoglossus and then enters the lingual vein accompanying the lingual artery. The dorsal lingual vein is also a tributory of the lingual vein which then enters the internal jugular vein.

Nerve supply

The mucous membrane of the anterior two-thirds is supplied by the lingual nerve. Taste fibres leave the nerve in the chorda tympani and are then conveyed by the facial nerve to the nucleus of the tractus solitarius. The glossopharyngeal nerve carries both taste and common sensation from the posterior third. All the extrinsic and intrinsic muscles of the tongue, except palatoglossus, are supplied by the hypoglossal nerve. Palatoglossus is supplied by the vagus nerve, through the pharyngeal plexus (p. 386).

Lymph drainage

Vessels from the tip of the tongue pass bilaterally to the submental lymph nodes. The sides drain to the ipsilateral submandibular lymph nodes, and the dorsum drains to these nodes and the jugulodigastric nodes of both sides. The posterior third drains to the retropharyngeal and jugulodigastric nodes of both sides. From all these nodes, efferent vessels pass to the jugulo-omohyoid and other nodes of the deep cervical lymph chain.

Histology

The mucous membrane of the anterior part of the dorsum is thin and possesses numerous papillae. The papillae are of three types.

(*a*) Fungiform papillae—numerous small rounded projections mainly on the tip and margins of the tongue. They have a few taste buds.

(*b*) Filiform papillae—the most numerous, conical and arranged in rows.

(*c*) Seven to twelve circumvallate papillae lie just in front of the sulcus terminalis; they are surrounded by a moat, the walls of which have many taste buds and are innervated by the glossopharyngeal nerve.

Taste buds are found all over the dorsum and sides of the tongue (and also on the epiglottis and the soft palate). The mucous membrane of the inferior surface resembles that over the rest of the oral cavity and contains scattered small salivary glands. The mucous membrane over the posterior third of the tongue is thick and loose with numerous mucous glands and aggregations of lymphoid tissue, the **lingual tonsil**.

Embryology

The mucous membrane of the anterior two-thirds is formed of endoderm from the two lateral lingual swellings on the floor of the pharynx over the 1st arch. A midline swelling between the 1st and 2nd arches, the tuberculum impar, forms a small part of the tongue in front of the foramen caecum. The posterior third is formed from a midline swelling over the 3rd and 4th arches, the copula. The muscles of the tongue develop from suboccipital myotomes which migrate forwards around the pharynx carrying the hypoglossal nerve with them.

Function

The tongue is concerned in chewing, swallowing, speaking and (in the infant) suckling, as well as being a sensory organ of taste and common sensation (e.g. touch and pain). When one side is paralysed, the protruded tongue deviates to the paralysed side.

MUSCLES OF THE FLOOR OF THE MOUTH

1. Mylohyoid—the two mylohyoid muscles form a diaphragm across the floor of the mouth. Each is attached anterolaterally to the mylohyoid line on the mandible. Its fibres pass downwards and medially, the posterior to the body of the hyoid bone, the anterior to a midline raphé between the hyoid bone and the midline of the mandible. It is supplied by the mylohyoid branch of the infeior alveolar nerve.

2. Geniohyoid—lies above the mylohyoid and passes from the mental spine on the mandible to the body of the hyoid bone. It is supplied by C1 fibres carried in the hypoglossal nerve.

3. Digastric—consists of two fleshy bellies united by an intermediate tendon which is slung by a fibrous sheath to the greater horn of the hyoid bone. The posterior belly is attached to a groove on the medial side of the mastoid process and the anterior belly to the digastric fossa on the mandible. The posterior belly is supplied by the facial nerve, the anterior belly by the mylohyoid branch of the inferior alveolar nerve.

Relations (of the posterior belly): it crosses, from behind forwards, the transverse process of the atlas, the accessory nerve, the internal jugular vein, the hypoglossal nerve, the internal and external carotid arteries, the facial and lingual arteries, and hyoglossus. The posterior auricular artery runs along its upper border, the occipital artery along its lower. Stylohyoid splits around its upper border which grooves the lower part of the parotid gland (Fig. 26.7).

Action
The three muscles on each side raise the hyoid bone in swallowing, or when the hyoid bone is fixed, depress the mandible thus opening the mouth.

SALIVARY GLANDS

There are three paired glands—the parotid, submandibular and sublingual—and numerous smaller ones around the mouth cavity. Their secretions clean and moisten the mouth, assisting in chewing, digestion, swallowing and phonation.

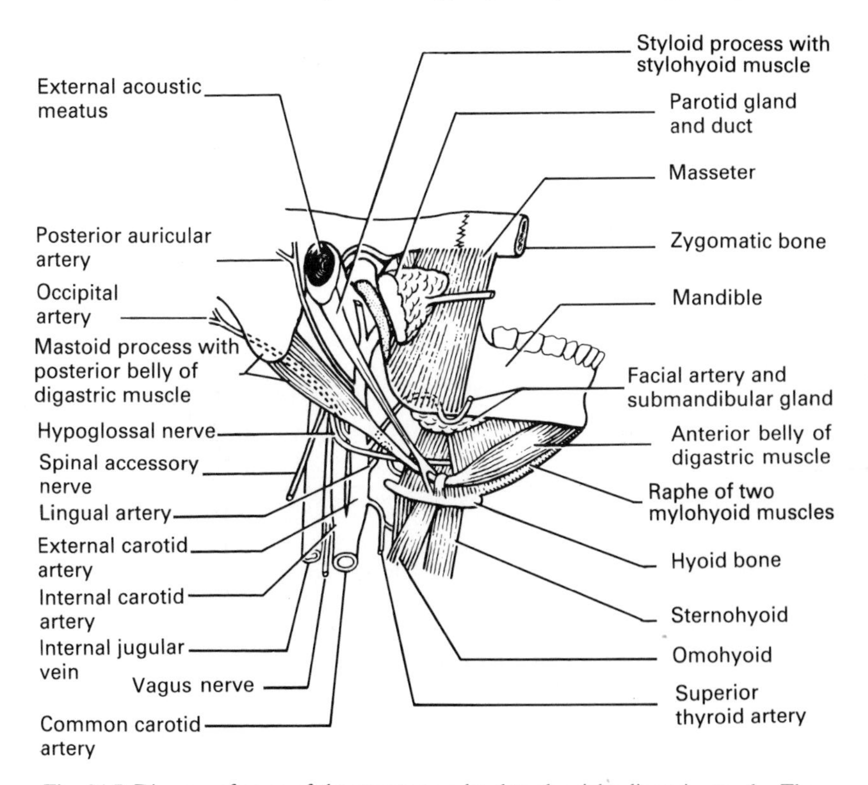

Fig. 26.7 Diagram of some of the structures related to the right digastric muscle. The carotid arteries and the internal jugular vein lie on the constrictor muscles of the pharynx. Most of the space between the mandible, the mastoid process and the styloid process is occupied by the parotid gland through which pass branches of the facial nerve.

The duct systems of the parotid and submandibular glands can be demonstrated in the living by injecting a contrast medium into the opening of the main ducts.

THE PAROTID GLAND (Figs. 26.7, 26.8, 26.9)

This is the largest of the salivary glands. It is wedged between the ramus of the mandible in front, the mastoid process behind and the styloid process medially. Each of these bones has muscles attached to it. The gland has an irregular shape and is described as having anteromedial, posteromedial and superficial surfaces. The anterior border overlies the masseter and from it the duct passes forwards across the muscle to turn medially around its anterior border and pierce buccinator. The duct opens on to the mucous membrane of the cheek just above the crown of the second upper molar tooth; it is about 5 cm long. A detached part of the gland, the accessory lobe, is often found lying on the masseter above the duct. The deep (investing) layer of cervical fascia covers the superficial surface of the gland. The fascia is less well defined deep to the gland but forms the stylomandibular ligament which separates the parotid and submandibular glands.

Relations

The superficial surface is covered by skin and fascia. The superficial parotid lymph nodes are partly embedded in the surface. The anteromedial surface is

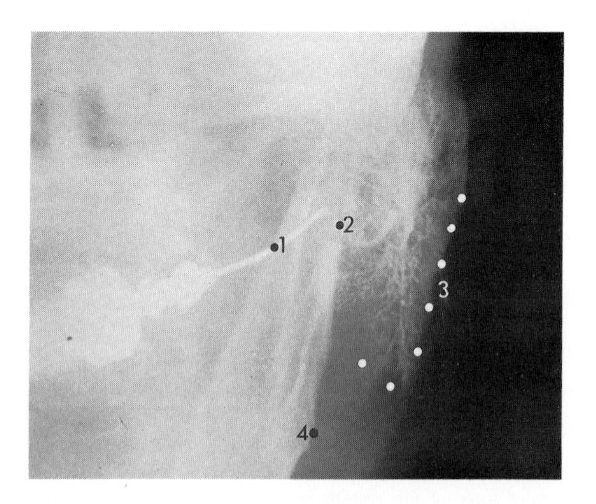

Fig. 26.8 Left parotid sialogram. The duct system of the gland is outlined by the contrast medium injected through the cannula (1).

1. Cannula
2. Main parotid duct
3. Outline of gland as shown by duct system
4. Angle of mandible

wrapped around the ramus of the mandible and extends on to masseter laterally and medial pterygoid medially. The posteromedial surface lies on the mastoid process, sternocleidomastoid and the posterior belly of digastric. The deep medial part of the gland is separated by the styloid process and its attached muscles from the carotid sheath and the pharynx. The upper part of the gland lies between the temporomandibular joint and the external acoustic meatus. The inferior part lies on the posterior belly of digastric and the stylomandibular ligament. The gland is traversed by the external carotid artery deeply, the retromandibular vein, and the facial nerve, superficially.

Blood supply
This is by branches of the external carotid artery. Venous blood passes to the retromandibular vein.

Nerve supply
Sympathetic fibres from the superior cervical ganglion pass along the external carotid artery; parasympathetic fibres come from the glossopharyngeal nerve via the otic ganglion and the auriculotemporal nerve. As in most glands, the parasympathetic nerves are secretomotor (producing saliva) and the sympathetic nerves are vasoconstrictor (giving rise to a dry mouth).

Lymph drainage
Drainage from the superficial part is to the parotid nodes and from the deep part to the retropharyngeal nodes.

Histology and embryology
The gland is lobulated and consists of serous alveoli with cubical-lined collecting ducts. It develops as a tubular ectodermal outgrowth from the inside of the cheek.

THE SUBMANDIBULAR GLAND (Figs. 26.6, 26.7)

This is situated in the floor of the mouth and has a fibrous capsule. The gland is divided into superficial and deep parts which communicate round the posterior border of mylohyoid. The submandibular duct passes forwards from the deep part of the gland, between mylohyoid and hyoglossus, medial to the sublingual gland and opens on to the sublingual papilla in the floor of the mouth at the base of the frenulum. It is crossed laterally, from above downwards, by the lingual nerve which then turns up medial to it.

Relations
SUPERFICIAL PART—superomedially are mylohyoid and the anterior belly of digastric; superolaterally are the body of the mandible and medial pterygoid;

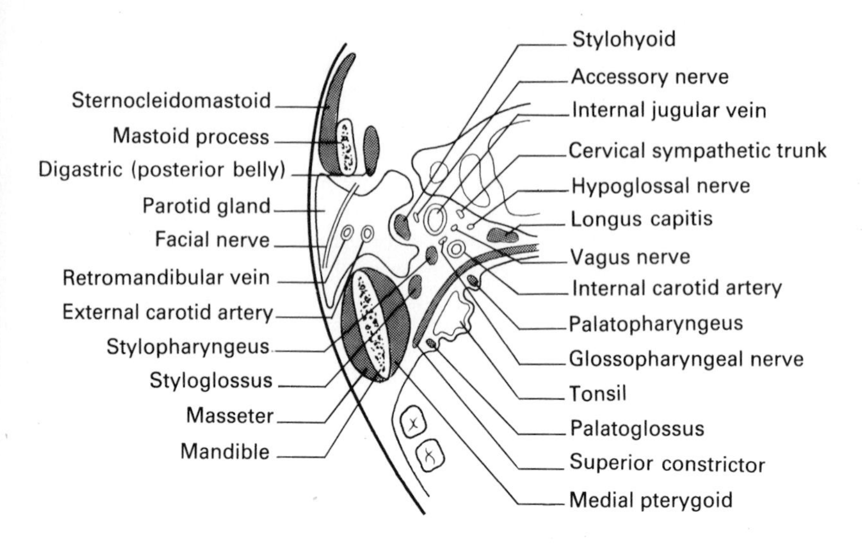

Fig. 26.9 Diagrammatic lay-out of the main structures seen in an oblique section through the parotid gland and the palatine tonsil.

inferiorly are the deep investing layer of fascia of the neck, the submandibular lymph nodes, superficial fascia with platysma and skin. The facial artery, having grooved the posterior surface of the gland, passes laterally over it and reaches the inferior border of the body of the mandible.

DEEP PART—is wedged between mylohyoid and hyoglossus. On the latter it is related to the lingual nerve and the submandibular ganglion above and the hypoglossal nerve below. Posteriorly it lies on the middle constrictor, where the hypoglossal nerve crosses the loop of the lingual artery.

Blood supply

This is from facial and lingual arteries. Venous blood passes to the facial and lingual veins.

Nerve supply

Sympathetic fibres from the superior cervical ganglion pass along the arteries of the gland; parasympathetic fibres come from the facial nerve via the chorda tympani and the submandibular ganglion.

Lymph drainage

This is to the overlying submandibular lymph nodes.

Histology and embryology

The submandibular is a mixed salivary gland possessing both serous and mucous cells. The serous cells form darkly-staining crescents opposite the openings of the alveolar ducts. The gland develops as a tubular endodermal outgrowth from the floor of the mouth.

THE SUBLINGUAL GLAND

This is situated below the mucous membrane of the floor of the mouth. Its secretions (mainly mucous) pass either into the submandibular duct or directly into the mouth cavity by 15–20 small ducts. Its resembles the submandibular gland in its blood and nerve supply, lymph drainage, histology and development.

Relations

Superiorly—the sublingual fold of mucous membrane; *inferiorly*—mylohyoid; *laterally*—the body of the mandible; *medially*—the submandibular duct, the lingual nerve, and hyoglossus and genioglossus muscles.

— 27 —

The temporomandibular joint, pharynx and larynx

THE TEMPOROMANDIBULAR JOINT

This is a synovial joint of a condyloid (modified hinge) variety between the condyle of the mandible and the mandibular fossa and articular tubercle of the squamous temporal bone. The hemicylindrical condyle is directed medially and slightly backwards, the temporal articular surface is concavoconvex from behind forwards, extending from the squamotympanic fissure to the front of the articular tubercle. The articular surfaces are covered with fibrocartilage, unlike most other synovial joints.

Ligaments

(a) CAPSULAR
The capsular ligaments are attached around the neck of the mandible and to the articular margins on the temporal bone.

(b) CAPSULAR THICKENINGS
The **lateral ligament**—a strong band passing downwards and backwards from the inferior surface of the zygomatic arch to the back of the mandible so limiting backward movement.

(c) ACCESSORY LIGAMENTS
(i) **sphenomandibular ligament**—passes from the spine of the sphenoid to the lingula of the mandible.
(ii) **stylomandibular ligament**—a thickening of the deep layer of the cervical fascia passing from the styloid process to the posterior border of the ramus of the mandible.

Intracapsular structures
A fibrocartilaginous disc, whose margins are attached to the capsule, completely divides the joint cavity into superior and inferior compartments each with its own synovial membrane. The disc has a concave undersurface and a concavoconvex

upper surface. The lateral pterygoid tendon is attached to the anterior aspect of the capsule and to the disc.

Functional aspects

MOVEMENT

The mandible may be depressed, elevated, protruded and retracted as in opening and closing the mouth, and moved from side to side as in grinding movements. However, movement at the joints has two components, each occurring in a separate compartment of the joint.

(*a*) Gliding occurs in the upper compartment. The condyle of the mandible and the articular disc move together over the articular surface.

(*b*) Hinge movement occurs in the lower compartment. The condyle of the mandible articulates with the undersurface of the articular disc.

In mandibular movements these components are never completely dissociated and movement at one joint must be accompanied by movement at the other. In depression and elevation both components are equally involved but in protrusion, retraction and grinding movements the gliding component is predominant. Some of these details may be demonstrated by radiological means.

Depression takes place about an axis passing through the lingulae. The condyles are pulled forwards by lateral pterygoid and the body is pulled downwards by geniohyoid, digastric and mylohyoid acting from the fixed hyoid bone. The movement is assisted by gravity.

Elevation is around the same axis and is produced by masseter, temporalis and medial pterygoid.

Protrusion is produced by lateral and medial pterygoid muscles.

Retraction is produced by the posterior fibres of temporalis.

Grinding movements are composed of alternating protrusion and retraction on the two sides and are brought about by the appropriate muscles. In chewing, the mandible is moved from side to side and at the same time is elevated by masseter and temporalis, the muscles of the tongue and cheek keeping the food between the teeth.

STABILITY

This varies with the position of the mandible. The joint is most stable when the mandible is fully elevated because (*a*) the condyle lies within the articular fossa, (*b*) occlusion of the teeth prevents further upward movement, (*c*) the lateral ligament is taut in this position, preventing posterior dislocation. Contraction of the muscles is important in maintaining this position, and in sleep or under anaesthesia the jaw drops open. In the resting position the teeth are separated slightly.

When the mandible is depressed the above factors are absent. The condyle lies on the articular tubercle and the integrity of the joint is dependent on the tone of the muscles and the strength of the capsule. Forward dislocation is the commonest form of displacement.

Blood supply

This is from superficial temporal and maxillary arteries.

Nerve supply
This is from auriculotemporal and masseteric nerves.

Relations (Fig. 26.7)
LATERALLY—skin and superficial fascia.
MEDIALLY—sphenomandibular ligament, auriculotemporal nerve, and the middle meningeal artery.
ANTERIORLY—lateral pterygoid.
POSTERIORLY—the parotid gland and related structures, the auriculotemporal nerve and the tympanic plate of the temporal bone.
The chorda tympani emerges from the petrotympanic fissure just behind the capsule of the joint.

MUSCLES OF MASTICATION

These muscles are all attached to the mandible and are supplied by the mandibular division of the trigeminal nerve.

1. Masseter

Attachments
SUPERIOR—lower border of the anterior two-thirds of the zygomatic arch.
INFERIOR—the fibres descend to the lateral surface of the ramus and angle of the mandible.
Action: elevation of the mandible.

2. Temporalis

Attachments
SUPERIOR—lateral surface of the skull below the temporal line.
INFERIOR—from their wide upper attachment, the fibres converge on to a narrow tendon which is attached to the coronoid process and the anterior border of the ramus of the mandible. The anterior fibres are vertical, the posterior are horizontal.
Action: the anterior fibres elevate and the posterior fibres retract the mandible.
Temporal fascia—a dense sheet covering temporalis muscle. It has a continuous attachment to the temporal line above and the zygomatic arch below, and is continuous with the epicranial aponeurosis.

3. Medial pterygoid

Attachments
SUPERIOR—by two heads which embrace the lower fibres of lateral pterygoid—

the deep from the medial surface of the lateral pterygoid plate, and the superficial from the maxillary tuberosity.

INFERIOR—the medial surface of the ramus and angle of the mandible.

Action: elevation and protrusion of the mandible.

4. Lateral pterygoid

Attachments

ANTERIOR—by two heads—the upper from the infratemporal surface of the greater wing of the sphenoid, and the lower from the lateral surface of the lateral pterygoid plate.

POSTERIOR—the neck of the mandible, and the capsule and intra-articular disc of the temporomandibular joint.

Action: protrusion and depression of the mandible.

INFRATEMPORAL FOSSA

This fossa lies below the base of the skull behind the maxilla and pterygoid plates of the sphenoid bone and in front of the carotid sheath. It is limited medially by the pharynx and laterally by temporalis and the ramus of the mandible. Inferiorly it is continuous with the fascial spaces of the neck.

The fossa contains the lateral and medial pterygoid muscles, the mandibular division of the trigeminal nerve and its branches, the maxillary artery and branches, the pterygoid venous plexus, and the sphenomandibular ligament.

THE PHARYNX

This is a muscular tube extending from the base of the skull to the level of the 6th cervical vertebra where it is continuous with the oesophagus. It is about 14 cm long. It lies behind, and communicates with, the nose, mouth and larynx, from above downwards, and is correspondingly divided into the naso-, oro- and laryngo-pharynx. Posteriorly it lies on the prevertebral fascia and muscles. Laterally it is related from above downwards to the auditory tube, the styloid process and its muscles, the carotid sheath and its contents, and the thyroid gland.

The walls of the pharynx have mucous, submucous and muscular coats.

(*a*) The mucous coat is continuous with that of the nose, auditory tube, oral cavity, larynx and oesophagus; in the nasopharynx it is lined with respiratory epithelium, and in the remainder with stratified squamous epithelium.

(*b*) The submucosa is thickened superiorly, where the muscular coat is deficient, to form the **pharyngobasilar fascia** which is attached to the base of the skull.

(*c*) The muscular coat comprises the superior, middle and inferior constrictor muscles, and the salpingo-, stylo- and palato-pharyngeus muscles.

The **superior constrictor** is attached anteriorly to—(i) the hamulus of the medial pterygoid plate, (ii) the retromolar fossa on the mandible, and (iii) between these attachments, to the pterygomandibular raphé.

The **middle constrictor** is attached anteriorly to—(i) the lower part of the stylohyoid ligament, (ii) the lesser horn of the hyoid bone and (iii) the greater horn of the hyoid bone.

The **inferior constrictor** is attached anteriorly to—(i) the oblique line on the thyroid cartilage, (ii) the fascia over cricothyroid and (iii) the lateral aspect of the cricoid cartilage.

Each constrictor fans out from its anterior attachment, passing posteriorly around the pharynx to join its fellow of the opposite side in a fibrous midline raphé. The raphé extends from the pharyngeal tubercle on the basilar part of the occipital bone down to the oesophagus. Owing to the posterior spreading of the muscles, the pairs overlap. The inferior lies outside the middle, which is itself outside the superior.

The salpingopharyngeus, stylopharyngeus and palatopharyngeus muscles are attached superiorly to the auditory tube, the styloid process and the soft palate respectively. They descend to blend with the inner surfaces of the constrictors and gain attachment to the posterior aspect of the lamina of the thyroid cartilage.

The muscles of the pharynx are concerned in swallowing (see p. 389). They are all supplied by the vagus nerve via the pharyngeal plexus, with the exception of stylopharyngeus (supplied by the glossopharyngeal nerve). The cell bodies are all in the nucleus ambiguus.

Between the superior constrictor and the base of the skull, the pharyngobasilar fascia is pierced by the auditory tube; between the superior and middle constrictors, stylopharyngeus and styloglossus and the glossopharyngeal nerve enter the pharyngeal wall; between the middle and inferior muscles pass the internal laryngeal nerve and the superior laryngeal artery to supply the larynx. The recurrent laryngeal nerve enters the region by ascending deep to the inferior constrictor.

THE INTERIOR OF THE PHARYNX

The nasopharynx

The nasopharynx lies behind the two posterior nasal openings (**choanae**). On each lateral wall are the opening of the auditory tube and a ridge produced by salpingopharyngeus. A small diverticulum behind the ridge is the pharyngeal recess and longus capitis bulges into the posterior wall. A submucous aggregation of lymphoid tissue, the pharyngeal tonsil (adenoids), is present. Superiorly the cavity is limited by the base of the occipital bone and the body of the sphenoid. Below, it communicates with the oropharynx through the pharyngeal isthmus, a constriction which may be closed by the backward-projecting, mobile soft palate.

The oropharynx

This extends down to the level of the upper border of the epiglottis where it is continuous with the laryngopharynx. Anteriorly it communicates through the **faucial isthmus** with the oral cavity. Below the faucial isthmus is the tongue which lies partly in the pharynx and partly in the oral cavity. In the lateral wall of the faucial isthmus are two folds of mucous membrane, the anterior and posterior arches of the fauces. They are produced by palatoglossus (extending from the undersurface of the palate to the side of the tongue) and palatopharyngeus (extending from the undersurface of the palate to the lateral wall of the pharynx) respectively. The **palatine tonsil** is situated between the anterior and posterior arches.

The laryngopharynx

This extends from the oropharynx to the oesophagus. Anteriorly are the opening of the larynx and the posterior surfaces of the arytenoid and cricoid cartilages. The cavity extends forwards on each side of the larynx and the recesses so formed are the **piriform fossae**. Each fossa is bounded laterally by the mucous membrane over the lamina of the thyroid cartilage and medially by the mucous membrane over the ary-epiglottic folds and the arytenoid and cricoid cartilages. The lumen of the pharynx is narrowest near its union with the oesophagus, the cricopharyngeal sphincter.

Blood supply: branches from the ascending pharyngeal, superior thyroid, lingual, facial and maxillary arteries. Venous blood passes through the pharyngeal venous plexus to the internal jugular vein.

Nerve supply: mainly through the **pharyngeal plexus** on the outer surface of the middle constrictor. The plexus is formed by:

(i) sensory fibres in pharyngeal branches of the glossopharyngeal and vagus nerves.
(ii) the pharyngeal branch of the vagus carrying motor fibres from the nucleus ambiguus to all pharyngeal muscles except stylopharyngeus. (The fibres to the stylopharyngeus pass from the nucleus ambiguus in the glossopharyngeal nerve.)
(iii) branches from the cervical sympathetic chain.

The nasopharynx receives an added sensory supply from the pharyngeal branch of the maxillary nerve (through the pterygopalatine ganglion) and the laryngopharynx receives sensory branches from the internal and recurrent laryngeal nerves.

Lymph drainage: the nasopharynx drains to the retropharyngeal lymph nodes and the remainder of the pharynx to nodes in the deep cervical lymph chain.

THE TONSIL (Fig. 26.9)

The tonsil (palatine tonsil) is a mass of lymph tissue situated in a fossa on the lateral wall of the faucial isthmus. It is variable in size, being usually larger in children. It is oval in shape, possessing upper and lower poles. The pitted medial surface is covered by stratified squamous epithelium and there is usually a well marked supratonsillar cleft between the upper pole and the superior wall of the fossa. The deep surface of the tonsil has a fibrous capsule and by this it is firmly attached antero-inferiorly (where vessels enter it) to its fossa. The tonsil is related anteriorly to the palatoglossal arch, posteriorly to the palatopharyngeal arch, superiorly to the soft palate and inferiorly to the tongue. Its medial surface projects into the oropharynx and the lateral surface lies on the superior constrictor which separates it from the facial artery and the carotid sheath.

Blood supply: tonsillar branches from the facial, lingual and ascending pharyngeal arteries. Venous blood passes to the pharyngeal venous plexus.

Lymph drainage: to the deep cervical lymph chain especially the jugulodigastric node.

The palatine tonsil develops from the second pharyngeal pouch. Aggregations of lymph tissue surround the upper part of the pharynx: the palatine tonsil below the palate, the lingual tonsil in the posterior third of the tongue, the tubal tonsil around the opening of the auditory tube, and the pharyngeal tonsil (adenoids) beneath the mucous membrane of the roof and posterior wall of the nasopharynx. The palatine tonsils and adenoids are frequently enlarged in children as the result of recurrent throat infections. In some cases enlargement may interfere with breathing and phonation. Surgical removal is then indicated.

THE AUDITORY TUBE

This passes from the lateral wall of the pharynx to the anterior wall of the middle ear. It is about 4 cm long and is directed upwards, backwards and laterally. The lateral third lies within the petrous temporal bone, the medial two-thirds is mainly cartilaginous but is completed inferiorly by fibrous tissue. The tube is lined by ciliated columnar epithelium with many mucous glands. It is supplied by the ascending pharyngeal artery and its veins pass to the pterygoid venous plexus. Its nerve supply is from the pharyngeal branch of the maxillary nerve and its lymph vessels pass to the retropharyngeal lymph nodes.

The tube serves to equalize the pressure on the two sides of the tympanic membrane. Salpingopharyngeus and tensor and levator veli palatini are attached to and pull on the medial part, and thus swallowing opens the tube. In the child the tube is wider and more horizontal, and infection is more likely to spread from the nasopharynx to the middle ear.

STYLOID MUSCLES

The styloid process is a bony projection of the temporal bone; it gives attachment to three muscles and a ligament. The muscles are active in swallowing; they are described in their order of attachment from above downwards.

Stylopharyngeus descends between the internal and external carotid arteries and enters the pharynx between the superior and middle constrictors. It blends with the middle constrictor and is attached to the posterior border of the thyroid cartilage. It is supplied by the glossopharyngeal nerve. **Stylohyoid** descends to the hyoid bone, splitting around the intermediate tendon of digastric. It is supplied by the facial nerve. **Styloglossus** passes anteriorly between the internal and external carotid arteries and between the superior and middle constrictors to the side of the tongue. It is supplied by the hypoglossal nerve. The **stylohyoid ligament** passes from the tip of the styloid process to the lesser horn of the hyoid bone and gives attachment to the middle constrictor. The styloid process, the stylohyoid ligament and the lesser horn are remnants of the 2nd pharyngeal arch cartilage.

Development of the pharynx and pharyngeal (branchial) arches

In the walls of the primitive foregut, from which the pharynx develops, six pharyngeal arches are formed on each side. They are numbered from before backwards. The arches consist of a mesenchymal mass lined internally with pharyngeal endoderm and covered externally with ectoderm. The furrows between the arches externally form the pharyngeal clefts and internally the pharyngeal pouches. Each arch is supplied by a cranial nerve and a central artery joins the ventral and dorsal aortae. Its mesenchyme forms a cartilaginous bar and striated muscle. During early development the 5th arch is obliterated, while the others undergo considerable modification, contributing to the formation of the face, mouth, pharynx and larynx. The nerve of the 1st arch is the trigeminal; 2nd, the facial; 3rd, the glossopharyngeal; 4th, the superior laryngeal; and the 6th, the recurrent laryngeal nerves.

Arch derivatives

1. ECTODERMAL

The 1st pharyngeal cleft forms the external acoustic meatus. The other clefts are obliterated by an ectodermal fold from the 2nd arch (the operculum) growing caudally over them. If it remains unfused, a cervical sinus develops. Remnants may also persist as a congenital defect known as a branchial cyst.

2. MESENCHYMAL

(*a*) *Cartilage*: 1st arch—the malleus and incus, the sphenomandibular ligament and Meckel's cartilage; 2nd arch—the stapes, styloid process, stylohyoid ligament and the lesser horn and upper part of the body of the hyoid bone; 3rd arch—the greater horn and the lower part of the body of the hyoid bone; 4th and 6th arches—cartilages of the larynx.

(*b*) *Muscle*: 1st arch—the muscles of mastication and the tensor veli palatini and tensor tympani; 2nd arch—the muscles of facial expression and other muscles supplied by the facial nerve; 3rd arch—stylopharyngeus; 4th and 6th arches—muscles of the pharynx, soft palate and larynx.

(c) *Arteries*: The arteries form the paired aortic arches uniting the ventral and dorsal aortae. The 1st and 2nd arches disappear; the 3rd forms part of the common and internal carotid arteries; the 4th forms the subclavian artery on the right and the aortic arch on the left. The 6th arch forms the pulmonary arteries; on the right the artery loses its connection with the dorsal aorta but this persists on the left as the ductus arteriosus of the fetus.

3. ENDODERMAL

The 1st pharyngeal pouch forms the auditory tube, the middle ear, and mastoid antrum; the 2nd pouch contributes towards the tonsil; the 3rd, the thymus and inferior parathyroid gland; and the 4th, the superior parathyroid gland and possibly the C-cells of the thyroid gland which produce calcitonin.

Modifications in the ventral parts of the upper arches in the floor of the mouth give rise to the tongue (see p. 374), the thyroid gland (see p. 418) and the respiratory tube (see p. 37).

Deglutition

Swallowing (deglutition) is the act by which fluid or a bolus of food is passed from the mouth to the stomach. X-ray examination, using radio-opaque material, has greatly helped in understanding the processes involved in swallowing.

The first stage (oral phase) is voluntary and involves passage of food from the mouth to the oropharynx. The mandible is fixed, with the teeth occluded, by the muscles of mastication. The tongue is tensed, by its intrinsic muscles, and raised by the palatoglossus and muscles acting on the hyoid bone (geniohyoid, mylohyoid, stylohyoid and digastric). The tongue, from the tip backwards, is pressed against the hard palate so forcing the food into the oropharynx.

In the second stage (pharyngeal phase), which is largely involuntary, food is passed to the lower end of the pharynx. It is initiated by food touching the arches of the fauces, the soft palate, the posterior wall of the oropharynx or the epiglottis. It involves closing off the nasopharynx, closing off the larynx, cessation of respiration and shortening of the pharynx.

The pharyngeal isthmus is closed by tensing and raising the soft palate with tensor and levator veli palatini, and the simultaneous approximation of the walls of the pharynx to the posterior free border of the palate. The latter movement is produced by contraction of palatopharyngeus and the upper fibres of superior constrictor.

The larynx is closed by the approximation of the epiglottis and arytenoid cartilages. Contraction of the aryepiglottic muscles is assisted by passive bending back of the epiglottis by the bolus of food. Liquids pass round the sides of the epiglottis. The larynx is raised by contracting the palatopharyngeus, salpingopharyngeus, stylopharyngeus and thyrohyoid muscles. The mucous membrane also rises, and when the larynx drops back some of the mucous membrane lags behind, so helping to close the opening. The pharynx is shortened by the pharyngeal muscles raising the larynx. The vocal folds are adducted and respiration is inhibited.

Finally food is passed by the inferior constrictor into the oesophagus. This stage (oesophageal phase) is involuntary and is initiated by food stimulating the walls of the lower pharynx and oesophagus. Peristaltic waves pass food on to the stomach. The soft palate and larynx return to their initial positions and respiration is recommenced. When food is in the mouth, it can be spat out; in the oropharynx it can be coughed out; beyond this, it has to be vomited out.

THE LARYNX (Fig. 27.1)

This is the part of the respiratory passage between the pharynx and the trachea. It forms a sphincter protecting the inlet of the trachea and includes the vocal folds. It is situated in the neck anterior to the laryngopharynx and opposite the 3rd-6th cervical vertebrae. A lobe of the thyroid gland and the carotid sheath lie on each side, and it is covered anteriorly by the infrahyoid muscles and the deep (investing) layer of fascia of the neck.

The skeleton of the larynx is formed of a number of cartilages which surround its lumen. These are the unpaired thyroid, cricoid and epiglottic and the paired arytenoid cartilages. Articulations occur at the cricothyroid and crico-arytenoid joints which are acted upon by the muscles of the larynx. The epiglottis and part of the arytenoid cartilages are yellow elastic cartilage, the rest hyaline cartilage and tend to calcify in later adult life.

CARTILAGES

Thyroid cartilage—this comprises two thin quadrilateral laminae united anteriorly at an angle of 120° in the female and 90° in the adult male. The angle forms the median subcutaneous **laryngeal prominence** and above this the laminae are separated by the palpable V-shaped **thyroid notch**. The free posterior border of each lamina extends superiorly and inferiorly as the superior and inferior horns respectively. On the outer surface of the lamina, an oblique line

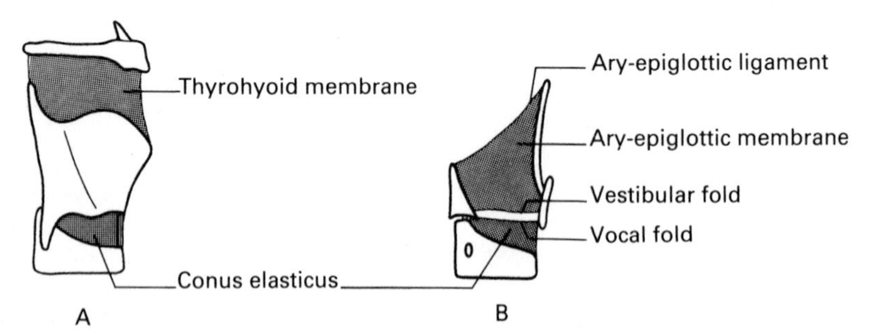

Fig. 27.1 The larynx: **A.** lateral aspect and **B.** lateral aspect after removal of right half of the thyroid cartilage and the hyoid bone. In **B**, the pyramidal-shaped arytenoid cartilage is shown articulating with the cricoid cartilage.

passes downwards and forwards from the superior horn to the inferior border. The inferior horn articulates with the cricoid cartilage.

The upper border of the cartilage gives attachment to the thyrohyoid membrane. The thyrohyoid, sternothyroid and inferior constrictor muscles are attached to the oblique line, and the palatopharyngeus, salpingopharyngeus and stylopharyngeus muscles to the posterior border. Cricothyroid is attached to the inferior horn and adjacent lower border of the lamina. To the posterior aspect of the laryngeal prominence are attached the epiglottic cartilage above, and below it the vocal ligament and thyro-arytenoid muscle, and the cricothyroid ligament.

Cricoid cartilage—this is situated at the lower border of the larynx and completely encircles it. It is the shape of a signet ring, possessing a broad posterior lamina and a narrow anterolateral arch. Each lateral surface has an articular facet for the inferior horn of the thyroid cartilage; the upper border of the lamina has two facets for the arytenoid cartilages.

To the upper border of the arch are attached the conus elasticus and the cricothyroid ligament; cricothyroid and lateral crico-arytenoid muscles are attached to the other lateral surfaces. The posterior crico-arytenoid muscle passes to the back of the lamina.

Epiglottic cartilage (epiglottis)—this is a leaf-shaped cartilage. Its narrow inferior end is attached to the back of the laryngeal prominence of the thyroid cartilage and its free broad upper end projects upwards behind the tongue. The sides give attachment to the ary-epiglottic membrane and the thyro-epiglottic and ary-epiglottic muscles. It is covered on its upper anterior and all its posterior surfaces by mucous membrane.

Arytenoid cartilages—each arytenoid cartilage is the shape of a three-sided pyramid, possessing medial, posterior and anterolateral surfaces; the base is inferior and articulates with the lamina of the cricoid. It projects anteriorly as the vocal process and laterally as the muscular process. The anterior border gives attachment to the aryepiglottic membrane; the muscular process to the lateral and posterior crico-arytenoid muscles; the vocal process, to the vocal ligament; the anterolateral surface, to the thyro-arytenoid muscle. The aryepiglottic muscle passes to the apex and continues downwards and medially to the back of the cartilage of the opposite side as the oblique arytenoid muscle. The transverse arytenoid muscle passes between the posterior surfaces of the two cartilages.

JOINTS, MEMBRANES AND LIGAMENTS

Cricothyroid joints—are plane synovial joints between the inferior horns of the thyroid cartilage and the lateral aspect of the cricoid. They possess a lax capsule. Rotary movement of the thyroid cartilage about the transverse axis through the two joints, and some anteroposterior gliding occur.

Crico-arytenoid joints—are plane synovial joints between the base of the arytenoid cartilages and the facets on the upper border of the cricoid lamina. Rotation and gliding of the arytenoid cartilages occur; lateral gliding is accompanied by downward displacement because of the obliquity of the joint.

Conus elasticus—this ligament is formed of yellow elastic tissue. Each half is triangular in shape, having an upper free border deep to the thyroid lamina and a lower attachment to the cricoid cartilage. In front it is continuous with the ligament of the opposite side. The union is reinforced anteriorly and forms the cricothyroid ligament. The upper free border of the conus extends from the back of the laryngeal prominence of the thyroid cartilage to the vocal process of the arytenoid cartilage and is thickened to form the vocal ligament. This is covered with mucous membrane and forms the vocal fold (true vocal cord).

Aryepiglottic (quadrangular) membrane—extends from the side of the epiglottis to the anterior border of the arytenoid cartilage. The free superior border forms the aryepiglottic ligament and the free inferior border the vestibular ligament. The lower thickening is in the **vestibular fold** (false vocal cord).

Thyrohyoid membrane—is a fibrous membrane uniting the upper border of the thyroid cartilage to the posterior surface of the body and greater horns of the hyoid bone. Between the hyoid bone and the membrane is a small midline bursa. The lateral extremities of the membrane are thickened to form the lateral thyrohyoid ligaments. The membrane is pierced by the superior laryngeal artery and the internal laryngeal nerve.

MUSCLES

With the exception of the transverse arytenoid these are paired muscles.

Cricothyroid—extends from the inferior horn and adjacent lower border of the thyroid cartilage to the outer anterolateral surface of the cricoid.

Posterior crico-arytenoid—passes from the posterior surface of the lamina of the cricoid to the muscular process of the arytenoid.

Lateral crico-arytenoid—passes from the outer lateral surface of the cricoid arch to the muscular process of the arytenoid.

Transverse arytenoid—is attached to the posterior surface of both arytenoid cartilages.

Aryepiglottic—passes from the lateral border of the epiglottis along the upper border of the ary-epiglottic membrane to the apex of the arytenoid cartilage. The muscle continues to the back of the arytenoid of the opposite side and is here called the oblique arytenoid muscle.

Thyro-arytenoid—passes from the back of the laryngeal prominence along the outer surface of the conus elasticus to the anterolateral surface of the arytenoid cartilage. Some fibres are attached to the free border of the ligament and form the **vocalis muscle**, others pass upwards to the side of the epiglottis forming the **thyro-epiglottic muscle**.

INTERIOR OF THE LARYNX

The superior opening of the larynx is bounded anteriorly by the epiglottis, laterally by the aryepiglottic folds, and posteriorly by the apices of the arytenoid cartilages and the transverse arytenoid muscle. Within the larynx two pairs of parallel horizontal folds are present in the lateral walls. The upper is the vestibular fold (false vocal cord), the lower is the vocal fold (true vocal cord). The gap between the vocal folds is known as the **rima glottidis**. A lateral recess of mucous membrane between the vestibular and vocal folds is known as the **sinus** of the larynx.

Blood supply: laryngeal branches of the superior and inferior thyroid arteries and their accompanying veins.

Nerve supply: the mucous membrane is supplied by the internal laryngeal nerve above the vocal folds and the recurrent laryngeal nerve below. The former enters the larynx by piercing the thyrohyoid membrane, the latter by ascending posterior to the cricothyroid joint and deep to the inferior constrictor muscle. Cricothyroid is supplied by the external laryngeal nerve, the other muscles by the recurrent laryngeal nerve.

Lymph drainage: efferent vessels pass to both upper deep cervical and pretracheal nodes.

Histology: the vocal folds are lined by stratified squamous epithelium; the larynx above and below is lined by respiratory epithelium.

Embryology: the pulmonary diverticulum arises ventrally in the pharyngeal floor between the 4th and the 6th arches. It passes caudally in front of the oesophagus. The larynx develops from the cartilages and muscles of the 4th and 6th pharyngeal arches. The cricoid may be a modified tracheal ring.

MOVEMENTS OF THE LARYNX

During vocalisation movements of the thyroid and arytenoid cartilages alter the length, tension and position of the vocal folds. In swallowing, the whole larynx is raised and then lowered, and the epiglottis is approximated to the arytenoid cartilages.

(*a*) *Movements of the thyroid cartilage*—forward rotation at the cricothyroid joints lengthens the vocal folds and is produced by cricothyroid; the movement is reversed by thyro-arytenoid.

(*b*) *Movements of the arytenoid cartilages*—occur at the crico-arytenoid joints, and are gliding and rotatory:

(i) GLIDING—lateral gliding separates the cartilages, thus abducting the vocal folds and producing a ∧-shaped rima glottidis. The combined action of the lateral and posterior crico-arytenoid muscles are involved, and counteracted by the transverse arytenoid muscle.

(ii) ROTATION—lateral rotation separates the vocal process and thus abducts the vocal folds. In this case a <>-shaped rima glottidis is produced. The movement is produced by posterior crico-arytenoid and counteracted by lateral crico-arytenoid. The posterior crico-arytenoid is the only muscle which, acting alone, abducts the vocal folds.

(*c*) *Vocalis increases the tension* in the vocal fold.

(*d*) *Raising and lowering of the larynx*: the larynx is raised by palatopharyngeus, salpingopharyngeus, stylopharyngeus and thyrohyoid. It is lowered by the infrahyoid muscles aided by gravity.

(*e*) *Approximation of the epiglottis to the arytenoid cartilage* may be produced by the aryepiglottic muscles but probably occurs passively in swallowing.

The larynx functions as a sphincter and in vocalisation. Direct visualisation of the region is possible with the help of a laryngoscope. The sphincter action prevents food entering the trachea and also closes the larynx when increased intrathoracic pressure (e.g. before coughing) or intra-abdominal pressure (e.g. in micturition) is required. Sound is produced by columns of air passing through the larynx usually in expiration. The frequency of the sounds emitted varies with the movement, length and tension of the vocal folds, and the intensity with the volume of air expired. The resonance of the sound is increased by the large spaces in the lungs, pharynx, mouth and nose. Hoarseness or loss of voice may result from improper function of the laryngeal muscles.

— 28 —

The ear, intracranial region and cranial nerves

EAR

The ear is the organ of hearing and balance. It is divided into an external ear directing sound waves to the tympanic membrane, a middle ear relaying the vibrations of the membrane to the internal ear, and the cochlea of the internal ear which translates these vibrations into nerve impulses. The organ of balance, the semicircular canals, is also in the inner ear.

THE EXTERNAL EAR

This consists of the auricle and the external acoustic meatus.

The **auricle** is formed of an irregularly shaped piece of fibrocartilage covered by firmly adherent skin. It has a dependent lobule and an anterior tragus which overlaps the opening of the meatus. To the auricle are attached three vestigial muscles supplied by the facial nerve. They are of little functional value in man. The **external acoustic meatus** is cartilage laterally and bone medially. It passes nearly horizontally from the tragus to the tympanic membrane and is about 4 cm long. The cartilaginous lateral third is continuous with the cartilage of the auricle. The bony medial two-thirds is formed mainly by the tympanic part of the temporal bone completed posterosuperiorly by the squamous temporal bone. The direction of the canal is slightly posterior in the lateral portion; slightly anterior in the medial portion. The canal is lined with skin richly endowed with wax secreting (ceruminous) glands. It receives its blood supply from the maxillary and superficial temporal arteries and is innervated by the auriculotemporal nerve anteriorly and the vagus nerve posteriorly.

The translucent **tympanic membrane** (ear drum) separates the external ear from the middle ear. It is oval in shape and lies obliquely with its lateral surface facing downwards and forwards. It consists of an incomplete fibrous layer covered laterally by modified skin and medially by modified mucous membrane of the middle ear. The handle of the malleus, attached to the medial surface, can be seen through the drum extending upwards from the centre to the superior flaccid part of the membrane where the fibrous layer is deficient. The lower end of the handle produces a small elevation on the drum and on inspection with an auriscope, a cone of light is seen passing downwards and forwards from this point.

THE MIDDLE EAR (TYMPANIC CAVITY)

The middle ear lies within the temporal bone and may be likened to a narrow room. The height and length are about 15 mm.The lateral and medial walls curve inwards and are about 2 mm apart at the middle and about 6 mm apart at the roof. The cavity is lined in places by ciliated columnar epithelium and contains three small bones (auditory ossicles).

The **lateral wall** is formed mainly by the tympanic membrane. The chorda tympani passes forwards between the membrane and the handle of the malleus laterally and the incus medially. Above and behind the membrane there is an upward extension of the cavity, the **epitympanic recess** which accommodates the head of the malleus and body of the incus.

The **medial wall** has a central bulge, the **promontory**, formed by the first turn of the cochlea. Above and behind the bulge is the **fenestra vestibuli** (oval window) and below and behind it the **fenestra cochleae** (round window), both foramina leading into the inner ear. The fenestra vestibuli is closed by the foot-plate of the stapes and the fenestra cochleae by a fibrous disc. The facial nerve, in a bony canal, crosses the medial wall superiorly from before backwards and then descends on the posterior wall.

The **anterior wall** has two openings inferiorly, separated by a bony shelf. The upper transmits tensor tympani whose tendon turns laterally around a projection of the bony shelf and is attached to the handle of the malleus. The lower is the opening of the auditory tube through which the cavities of the middle ear and nasopharynx are in continuity.

High on the **posterior wall** is the **aditus**, the opening into the mastoid (tympanic) antrum. Below it there is a projection of bone, the pyramid, out of which the tendon of the stapedius muscle passes to the stapes. Lateral to this there is a small foramen out of which passes the chorda tympani. The canal for the facial nerve passes down the medial side of the wall.

The **roof** is covered by a thin plate of the petrous temporal bone (the tegmen tympani) which separates the cavity from the middle cranial fossa. The bony **floor** separates the cavity from the carotid canal in front and the jugular foramen behind.

The middle ear contains three small ossicles, the malleus, the incus and the stapes; they extend in line from the tympanic membrane to the fenestra vestibuli and transmit the vibrations of the membrane to the inner ear. The **malleus** has a handle which is attached to the tympanic membrane, and a rounded superior head which lies in the epitympanic recess and articulates with the body of the incus. The **incus** has a body and two processes; one process rests on the posterior wall of the middle ear, the other articulates with the **stapes** (stirrup-shaped), the footplate of which occupies the fenestra vestibuli. The malleus and incus develop from the cartilage of the 1st pharyngeal arch, the stapes from the 2nd. Decreased mobility of the ossicles in middle and old age produces degrees of conductive deafness.

Two small muscles, the **tensor tympani** (supplied by the mandibular nerve) and the **stapedius** (supplied by the facial nerve) are attached to the malleus and the stapes respectively; they modify the transmission of the sound waves.

The middle ear receives its blood supply from the internal carotid and maxillary arteries and is innervated by the glossopharyngeal nerve through the tympanic plexus. Lymph passes to the retropharyngeal lymph nodes.

Mastoid (tympanic) antrum

This is a cavity within the mastoid part of the petrous temporal bone. It communicates with the cavity of the middle ear through the aditus. At birth the antrum is small and lies very superficially. As the mastoid process enlarges, the antrum becomes more deeply placed and mastoid air cells develop as diverticula around it. Posteromedially the antrum is separated by a thin plate of bone from the posterior cranial fossa with the cerebellum and the sigmoid venous sinus. Superiorly it is related to the temporal lobe of the brain. Laterally are the muscles attached to the mastoid process.

The communication of the nasopharynx with the middle ear along the auditory tube provides a pathway for infection to pass from the throat to the mastoid antrum and mastoid air cells. Here an abscess may develop and this is of serious consequence for it is related to the temporal lobe of the brain and the cerebellum, being separated from them only by thin plates of bone. Meningitis and brain abscess are thus occasional complications of a mastoid infection.

THE INTERNAL EAR

The internal ear lies within the petrous temporal bone. It consists of a complicated membranous sac (the membranous labyrinth) filled with a clear fluid (endolymph) and contained within a larger bony cavity (the bony labyrinth). The space between the membranous and bony labyrinths is filled with fluid (perilymph).

The bony labyrinth

The bony labyrinth comprises the cochlea anteriorly, the vestibule in the middle and the semicircular canals posteriorly. The cavities are in continuity. The **cochlea** is a spirally coiled bony tube of $2\frac{3}{4}$ turns, around a central bony pillar, the **modiolus**. The medial wall of the coil has a narrow plate of bone, the **spiral lamina**, which projects into the cavity and partly divides it. The basal turn of the cochlea opens into the **vestibule** which lies anterolateral to the internal acoustic meatus. The lateral wall of the vestibule opens on to the middle ear at the vestibular and cochlear fenestrae. The anterior, posterior and lateral **semi-**

circular canals each form two-thirds of a circle and open at both ends into the vestibule; as the anterior and posterior canals have a conjoined medial end, there are only five openings into the vestibule. The lateral end of each canal possesses a dilatation, the ampulla. The canals lie in three planes at right angles to each other; the anterior and posterior canals are set vertically, the lateral horizontally.

The membranous labyrinth

The membranous labyrinth (Fig. 28.1) comprises the duct of the cochlea, the saccule, the utricle and the three semicircular ducts from before backwards; all are in continuity. The **duct of the cochlea** lies within the bony cochlea placed between the projecting edge of the spiral lamina and the opposite wall. It divides the cavity into three: an upper **scala vestibuli** containing perilymph, a middle **scala media** (the cavity of the duct) containing endolymph, and a lower **scala tympani** containing perilymph. The scala vestibuli extends from the fenestra vestibuli to the apex of the coil where it communicates by a small opening, the **helicotrema,** with the scala tympani. The latter extends to the fenestra cochleae. Lying on the basilar membrane which separates the scala media and scala tympani, are hair cells supporting the **membrana tectoria**. The hair cells are stimulated by the disturbance of perilymph in the scala vestibuli and scala tympani set up by movement of the stapes at the fenestra vestibuli. The **spiral organ** (end organ of hearing) consists of the hair cells amongst which are the endings of the cochlear nerves, and the tectorial membrane (Fig. 28.2).

The **saccule** and **utricle** are two sacs within the vestibule; they are united by the narrow Y-shaped endolymphatic duct. The saccule is joined to the duct of the cochlea by the narrow ductus reuniens, the utricle receives the five openings of the semicircular ducts. Both sacs possess a small thickened area of columnar cells with projecting hairs. These are the **maculae,** the end organs of static balance.

The narrow **semicircular ducts** lie within their respective bony canals. They each have a dilated ampulla containing a sensory area, the **crista**. The cristae are

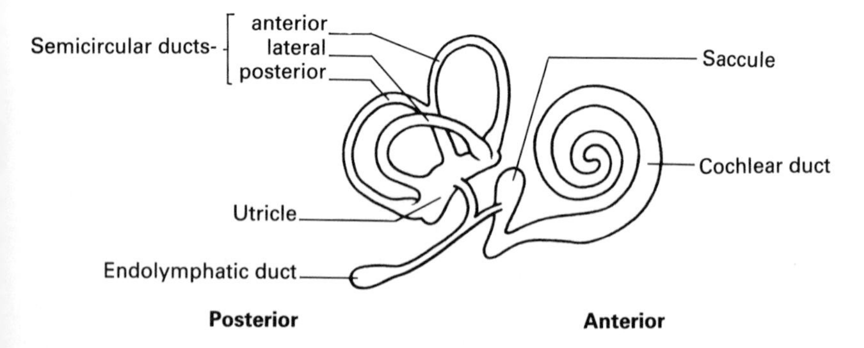

Fig. 28.1 Diagrammatic lay-out of the right membranous labyrinth viewed from the lateral side.

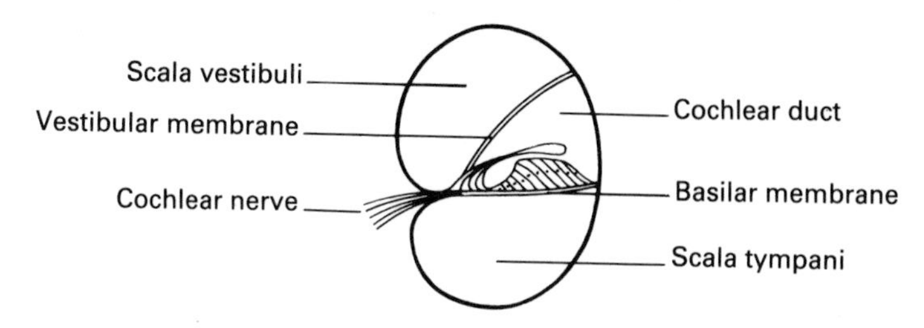

Fig. 28.2 Diagrammatic lay-out of a section through one of the coils of the cochlea. In the cochlear duct, the hair cells lie on the basilar membrane, and the tectorial membrane lies against their free surface.

similar to the maculae but are the end organs of kinetic balance. The maculae and cristae contain nerve fibres which are stimulated when the supporting hair cells are deformed either by movement of the endolymph or of the otoliths, small calcium carbonate crystals in the fluid.

Blood is supplied by the labyrinthine branch of the basilar artery. Venous blood passes to the superior and inferior petrosal sinuses. The maculae and cristae are innervated by the vestibular division of the eighth cranial nerve, the spiral organ of the cochlea by the cochlear division of the nerve. The cochlear ganglion is in the modiolus and the vestibular ganglion is in the internal acoustic meatus.

INTRACRANIAL REGION

The cranial cavity contains the brain and its membranes. The vault of the skull covers the hemispheres superiorly, and inferiorly the base of the skull forms the anterior, middle and posterior cranial fossae which contain the frontal lobes, temporal lobes and hindbrain respectively.

The vault

The vault is formed from the frontal, parietal, squamous temporal and occipital bones (Fig. 25.2). There are two foramina for the parietal emissary veins near the vertex. The inner surface shows shallow impressions for the cerebral gyri and is grooved in the sagittal plane by the superior sagittal sinus as far back as the internal occipital protuberance. On each side of the groove are pits for the **lateral recesses** containing the **arachnoid granulations**. The lateral side of the vault shows markings for the anterior and posterior branches of the middle meningeal vessels.

The anterior cranial fossa

This is formed by the orbital plates of the frontal bone, the cribriform plate of the ethmoid bone with its upward midline projection, the **crista galli**, and the lesser wings of the sphenoid which project posteromedially to form the anterior clinoid processes. The fossa houses the frontal lobes of the brain and overlies the nasal and orbital cavities. The cribriform plates underlie the olfactory bulbs and the foramina transmit bundles of the olfactory nerves with their meningeal coverings, and the anterior ethmoidal vessels and nerves. The orbital plates of the frontal bone show markings for the orbital gyri of the frontal lobes and contain the posterior parts of the frontal air sinuses.

The middle cranial fossa

This has a median portion and two larger lateral compartments. The median part is formed by the body of the sphenoid which is hollowed out to form the **pituitary fossa** (the sella turcica). Anteriorly a transverse ridge, the tuberculum sellae, separates the fossa from the shallow optic groove which leads to the optic canal on each side. Posteriorly a plate of bone, the dorsum sellae, projects upwards and forwards; its two corners are the posterior clinoid processes. On each side of the body a shallow carotid groove leads posteriorly to the foramen lacerum. The body of the sphenoid contains the sphenoidal air sinuses, lies under the pituitary gland and cavernous venous sinuses, and above the nasopharynx. Each **optic canal** transmits the optic nerve with its meningeal coverings and the ophthalmic artery. The **foramen lacerum** lies between the occipital bone behind, the greater wing of the sphenoid bone in front, and the apex of the petrous temporal laterally. It is irregular in shape, and in life it is filled with fibrocartilage. The cartilage in the foramen forms the floor of the medial end of the carotid canal. The internal carotid artery, and the greater and deep petrosal nerves pass medially over the cartilage.

The lateral compartments are formed, from before backwards, by the lesser and greater wings of the sphenoid, and the squamous and petrous parts of the temporal bone. They support the temporal lobes of the brain, separating them from the temporal and infratemporal regions. One fissure and three foramina are found towards the medial side of each compartment.

1. **The superior orbital fissure** lies between the greater and lesser wings of the sphenoid. It opens into the orbit and transmits the oculomotor, trochlear, branches of the ophthalmic, and abducent nerves and the ophthalmic veins (p. 363).

2. **The foramen rotundum** is in the greater wing of the sphenoid behind the medial end of the superior orbital fissure. It opens into the pterygopalatine fossa and transmits the maxillary nerve.

3. **The foramen ovale** is also in the greater wing and is about 1.5 cm behind the foramen rotundum. It opens into the infratemporal fossa and transmits the mandibular nerve, the lesser petrosal nerve and an accessory meningeal artery.

4. **The foramen spinosum** lies just posterolateral to the foramen ovale. It transmits the middle meningeal vessels and meningeal branches of the mandibular nerve. The vessels groove the bone as they pass laterally and divide into anterior and posterior branches. The anterior branch lies deep in the pterion (p. 342).

The anterior surface of the petrous temporal bone has a depression near its apex for the trigeminal ganglion. Further laterally the surface has an elevation, the **arcuate eminence**, produced by the anterior semicircular canal. Anterior and lateral to the eminence a thin plate of bone, the **tegmen tympani**, overlies the mastoid antrum, the middle ear and the auditory tube. Grooves for the greater and lesser petrosal nerves are found medial to the eminence.

The posterior cranial fossa

This is the largest and deepest of the three fossae. It is formed from the sphenoid, temporal, occipital and parietal bones. Anterolaterally the fossa is limited by the dorsum sellae and the upper border of the petrous temporal bones. Posterolaterally the grooves for the transverse venous sinus on the occipital and parietal bones form the upper limit of the fossa. The fossa contains the cerebellum, pons and medulla. It is roofed in by the tentorium cerebelli on which lie the occipital lobes of the cerebral hemispheres. The following foramina are present:

1. **The foramen magnum** is large and lies in the midline in the floor of the fossa, bounded on each side by the occipital condyles. It transmits the spinal cord with its coverings, the vertebral and spinal arteries, and the accessory nerves.
2. **The jugular foramen** lies between the occipital and petrous temporal bones. It is irregular in outline and contains the upper bulb of the internal jugular vein, and transmits the inferior petrosal sinus and the glossopharyngeal, vagus and accessory nerves.
3. **The internal acoustic meatus** lies on the posterior aspect of the petrous temporal bone. Each is directed laterally and transmits the facial and vestibulocochlear nerves and the labyrinthine artery, all surrounded by a meningeal sheath. The ganglion of the vestibular nerve lies in the meatus.
4. **The hypoglossal canal** lies above and anterior to the occipital condyle; it transmits the hypoglossal nerve.

5 & 6. **The condylar canal** (behind the condyle) and the **mastoid foramen** transmit emisssary veins.

The posterior surface of the petrous temporal bone is grooved above by the superior petrosal sinus, medially by the inferior petrosal sinus and laterally by the sigmoid part of the transverse sinus. The internal occipital protuberance lies at the point of union of the two transverse sinuses in the midline posteriorly.

CRANIAL NERVES

The 12 pairs of cranial nerves originate in the brain and leave the cranial cavity through its basal foramina. The nerves have motor and sensory components of the somatic, visceral and special visceral (branchial) varieties in varying proportions. The olfactory and optic nerves are not usually included in this classification. The attachments of the nerves to the brain stem are shown in Fig. 31.1 and the central connections of nerves 3–12 are outlined in Chapter 33.

The somatic motor component supplies muscles derived from somatic mesoderm and is present in the oculomotor, trochlear, abducent and hypoglossal nerves. The general somatic sensory component supplies the skin of the face by the trigeminal nerve. The special somatic sensory component supplies the organ of balance and hearing by the vestibulocochlear nerve. The general visceral (parasympathetic) motor components are in the oculomotor, facial, glossopharyngeal and vagus nerves. The special visceral motor component supplies muscles derived from pharyngeal arch mesoderm and is present in the trigeminal, facial, glossopharyngeal, and vagus nerves. The special visceral sensory component provides taste fibres in the facial and glossopharyngeal nerves. The general visceral sensory (reflex) components are in the glossopharyngeal and vagus nerves.

The course of the sensory nerves (the olfactory, optic and vestibulocochlear) will be described as approaching the brain, and motor and mixed nerves as leaving it.

1. OLFACTORY NERVES

They supply the olfactory mucous membrane in the upper part of the nasal cavity. The nerve fibres originate in the bipolar olfactory cells of the mucosa and join to form 15–20 olfactory bundles which pass through the cribriform plate of the ethmoid bone to reach the olfactory bulb. (For central connections see p. 449).

2. OPTIC NERVE (Figs. 26.2, 26.3, 26.4)

It is the sensory nerve of the retina. Its fibres originate in the ganglion layer and converge on the posterior part of the eyeball. The nerve passes backwards through the orbit and optic canal into the middle cranial fossa where it unites with the nerve of the opposite side to form the **optic chiasma**.

Relations

Within the orbit the nerve is enclosed in a meningeal sheath containing cerebrospinal fluid, and surrounded by the cone of extra-ocular muscles embedded in fat. The ciliary ganglion lies on its lateral side and the ophthalmic artery and nasociliary nerve pass forwards and medially above it. The central artery of the retina (a branch of the ophthalmic artery) enters the substance of the nerve in this part of its course. *In the optic canal* the nerve with its meningeal sheath is accompanied by the ophthalmic artery. *The short intracranial portion* lies on the sphenoid bone and is medial to the internal carotid artery. The optic chiasma lies anterosuperior to the pituitary gland. (For central connections see p. 480).

3. OCULOMOTOR NERVE (Figs. 26.2, 36.9)

This nerve has somatic motor and general visceral (parasympathetic) motor fibres. The somatic fibres supply the bulbar muscles, except superior oblique and lateral rectus. The parasympathetic fibres synapse in the ciliary ganglion and supply the sphincter pupillae and ciliary muscle.

The nuclei of the nerve are situated in the upper midbrain in the peri-aqueductal grey matter. The nerve fibres pass forwards through the midbrain and leave it between the cerebral peduncles. The nerve passes through the posterior and middle cranial fossae and divides into superior and inferior divisions near the superior orbital fissure.

Relations

In the posterior cranial fossa—the nerve passes between the posterior cerebral and superior cerebellar arteries, then medial to the trochlear nerve and lies near the edge of the tentorium cerebelli. In the middle cranial fossa—it pierces the cerebral layer of dura to pass forwards on the lateral wall of the cavernous sinus where it is at first above but later descends medial to the trochlear nerve.

Branches

(i) the superior division passes through the superior orbital fissure within the tendinous ring of the extra-ocular muscles and supplies superior rectus and levator palpebrae superioris.

(ii) the inferior division also passes through the tendinous ring and supplies the medial and inferior recti and inferior oblique. The nerve to inferior oblique gives the parasympathetic branch to the ciliary ganglion (see p. 431).

Because of its close relationship to the free edge of the tentorium cerebelli the oculomotor nerve may be damaged by a lateral shift of the brain. One of the earliest localising signs of rapidly increasing intracranial pressure may be dilatation of the pupil due to damage of the parasympathetic fibres passing to the ciliary ganglion.

4. TROCHLEAR NERVE (Figs. 26.2, 36.9)

This is the somatic motor nerve supply to the superior oblique. Its nucleus lies in the lower midbrain in the peri-aqueductal grey matter. The fibres pass posteriorly and undergo a dorsal decussation with the nerve of the opposite side caudal to the inferior colliculi. The nerve then passes forwards through the posterior and middle cranial fossae, enters the orbit through the superior orbital fissure and supplies superior oblique.

Relations

In the posterior cranial fossa the nerve passes forwards around the midbrain, then between the posterior cerebral and superior cerebellar arteries, following the edge of the tentorium, lateral to the oculomotor nerve. *In the middle cranial fossa* it pierces the cerebral layer of dura and lies between the oculomotor and ophthalmic nerves in the lateral wall of the cavernous sinus. The oculomotor later descends medial to it. It passes through the superior orbital fissure outside the tendinous ring, passes medially between levator palpebrae superioris and the roof of the orbit and reaches the superior oblique.

5. TRIGEMINAL NERVE (Figs. 26.2, 26.6, 36.9)

This nerve has general somatic sensory and special visceral motor fibres. The sensory fibres supply the anterior part of the scalp and dura, the face, the nasopharynx, the nasal and oral cavities, and the paranasal air sinuses. The motor fibres supply the muscles of mastication. The nerve has four nuclei: the spinal (in the medulla), the mesencephalic (in the midbrain), and the superior sensory and motor (in the pons). The first three are sensory in function.

The nerve arises in the pons and emerges at the junction of the pons and the middle cerebellar peduncle as a large sensory and smaller motor root. It passes forwards to the trigeminal ganglion on the apex of the petrous temporal bone in the middle cranial fossa. The ganglion is semilunar in shape and lies on the anterior aspect of the petrous temporal bone lateral to the cavernous sinus. It is partly surrounded by a dural sheath carried forwards by the nerve. The ophthalmic, maxillary and mandibular divisions emerge from the anterior border of the ganglion. The motor root of the nerve passes medial to the ganglion and joins the mandibular division.

Ophthalmic division

This is the smallest division; it passes forwards on the lateral wall of the cavernous sinus below the oculomotor and trochlear nerves and above the maxillary division. Near the superior orbital fissure it divides into the lacrimal, frontal and nasociliary nerves which pass through the fissure into the orbit.

(a) **Lacrimal nerve**—passes through the lateral part of the superior orbital fissure, outside the tendinous ring and above lateral rectus to reach the lacrimal gland. It then enters the upper eyelid and supplies the skin and conjunctiva on the lateral side. A communicating branch from the zygomaticotemporal nerve conveys parasympathetic fibres from the pterygopalatine ganglion to the lacrimal gland.

(b) **Frontal nerve**—passes through the superior orbital fissure just outside the tendinous ring and then forwards between levator palpebrae superioris and the roof of the orbit. In the orbit it divides into supra-orbital and supratrochlear nerves. The former leaves the orbit through the supra-orbital notch and its branches supply the upper eyelid, the frontal sinuses and the scalp as far back as the vertex. The latter passes forwards above the trochlea for the superior oblique tendon, and supplies the skin of the upper eyelid and the medial part of the forehead.

(c) **Nasociliary nerve**—passes through the superior orbital fissure within the tendinous ring, forwards and then medially above the optic nerve and ends on the medial wall of the orbit as the anterior ethmoidal and infratrochlear nerves.
BRANCHES:
(i) anterior ethmoidal nerve—passes through a foramen on the medial wall of the orbit and enters the anterior cranial fossa. It then descends through the cribriform plate, passes through the nasal cavity, and emerges between the nasal bone and the nasal cartilages as the external nasal nerve. It supplies the dura of the anterior cranial fossa, the anterior ethmoidal air cells, the upper anterior part of the nasal cavity and the skin over the sides and tip of the nose.
(ii) posterior ethmoidal nerve—supplies the posterior ethmoidal and sphenoidal air sinuses.
(iii) infratrochlear nerve—passes forwards below the trochlea and supplies the medial part of the upper eyelid, conjunctiva and adjacent part of the nose.
(iv) long ciliary nerves—two or three branches pass to the sclera and cornea. They also convey sympathetic fibres to the dilator pupillae from the internal carotid plexus.
(v) branches via the ciliary ganglion to the eyeball.

Maxillary division

This passes forwards through the middle cranial fossa, the foramen rotundum, the pterygopalatine fossa and the inferior orbital fissure where it becomes the infra-orbital nerve.

Relations

In the middle cranial fossa the nerve lies for a short distance in the lateral wall of the cavernous sinus below the ophthalmic division; it leaves the fossa through the foramen rotundum. In the pterygopalatine fossa the pterygopalatine ganglion is suspended from it and it is related to the terminal branches of the maxillary artery.

Branches

(a) Fibres to the **pterygopalatine ganglion**, through which they pass without synapsing to the nose, palate and nasopharynx. The nasal and nasopalatine nerves pass through the sphenopalatine foramen into the nose; the greater and lesser palatine nerves descend between the maxillary and palatine bones to the palate; and the pharyngeal nerves pass backwards to the nasopharynx.

(b) **Zygomatic nerve**—passes forwards on the lateral wall of the orbit and divides into temporal and facial branches which, passing through small foramina in the zygomatic bone, supply the skin over the temple and cheek. In the orbit a communication with the lacrimal nerve carries parasympathetic fibres from the pterygopalatine ganglion to the lacrimal gland.

(c) **Posterior superior alveolar nerve**—three or four branches descend through foramina in the posterior aspect of the maxilla to the upper molar and premolar teeth.

(d) **Infra-orbital nerve**—is the continuation of the maxillary nerve out of the orbit. It lies in the infra-orbital groove and canal and emerges at the infra-orbital foramen on to the face where its terminal branches supply the lower eyelid and conjunctiva, the side of the nose and the upper lip. Its **anterior superior alveolar branch**, given off in the canal, supplies the canine and incisor teeth, the lower part of the lateral wall of the nose and the maxillary air sinus.

Mandibular division

This carries the motor root of the trigeminal nerve. It leaves the middle cranial fossa through the foramen ovale and soon divides into its branches. The nerve lies between tensor veli palatini and lateral pterygoid and has the otic ganglion on its medial side.

1. **Motor branches** to the muscles of mastication, tensor tympani and tensor veli palatini. The fibres to the tensor muscles leave the nerve to medial pterygoid and pass through the otic ganglion without synapsing. Fibres to the mylohyoid and the anterior belly of digastric are described below.

2. **Sensory branches** have a wide distribution.

(a) **Meningeal branch**—a small twig which passes through the foramen spinosum and supplies the dura of the middle cranial fossa.

(b) **Buccal nerve**—passes forwards between the two heads of lateral pterygoid and medial to the mandible. It gives off cutaneous branches then pierces buccinator and supplies the mucous membrane of the cheek.

(c) **Auriculotemporal nerve**—passes laterally between the sphenomandibular ligament and the neck of the mandible, and above the parotid gland. It then ascends behind the joint, with the superior temporal artery, to the temporal region of the scalp. The nerve carries sensory fibres to the anterior part of the tympanic membrane, external acoustic meatus and the auricle, to the joint and to the temporal region of the scalp. It conveys parasympathetic fibres to the parotid gland from the otic ganglion and sympathetic fibres to the gland from the plexus on the middle meningeal artery, around which the nerve often divides.

(d) **Inferior alveolar nerve**—descends to the mandibular foramen where it enters the mandibular canal. It lies between the medial pterygoid and the ramus of the mandible, and the sphenomandibular ligament lies medial to it as it enters the foramen.

BRANCHES

(i) mylohyoid nerve—leaves the nerve before it enters the mandibular foramen, passes forwards close to the mandible and supplies mylohyoid and the anterior belly of digastric.

(ii) dental branches—arise in the mandibular canal and supply adjacent gums and teeth.

(iii) incisor nerve—supplies the canine and incisor teeth and the central incisor of the opposite side.

(iv) mental nerve—emerges through the mental foramen and supplies the skin and mucous membrane of the lower lip and chin, and the gum above the foramen.

(e) **Lingual nerve**—arises anterior to the previous nerve and passes forwards along the side of the tongue. The nerve supplies sensory branches to the anterior two-thirds of the tongue, the floor of the mouth and the lingual gum. It receives the chorda tympani branch of the facial nerve about 3 cm below the base of the skull. The chorda tympani carries parasympathetic and taste fibres: the former synapse in the submandibular ganglion and supply the submandibular and sublingual salivary glands (p. 432); the latter are from the anterior two-thirds of the tongue and have their cell bodies in the genicular ganglion of the facial nerve.

Relations: the nerve lies between four pairs of structures from above downwards—(*a*) tensor veli palatini and lateral pterygoid, (*b*) medial pterygoid and ramus of the mandible, (*c*) mucous membrane of the mouth and the roots of the 3rd lower molar tooth, (*d*) hyoglossus and mylohyoid. Between the last pair of muscles the nerve is overlapped by the submandibular gland and then passes from lateral to medial below the submandibular duct.

6. ABDUCENT NERVE (Figs. 26.2, 36.9)

It is a somatic motor nerve supplying lateral rectus. Its nucleus is situated in the lower pons. The nerve leaves the inferior border of the pons near the midline, passes forwards through the posterior and middle cranial fossae, the cavernous sinus and the orbit, and supplies lateral rectus.

Relations

In the posterior cranial fossa the nerve lies between the pons and the basilar part of the occipital bone. It pierces the inner layer of dura covering the dorsum sellae, passes over the apex of the petrous temporal bone and through the cavernous sinus lateral to the internal carotid artery. It enters the orbit through the superior orbital fissure within the tendinous ring and supplies lateral rectus. This long thin nerve is prone to damage in patients with increased intracranial pressure.

7. FACIAL NERVE (Figs. 26.9, 33.5)

This nerve carries special visceral motor, special visceral sensory, and general visceral (parasympathetic) motor fibres. The special visceral motor fibres arise in the facial nucleus in the lower pons and upper medulla, and are distributed to the muscles of the face region. The special visceral sensory fibres carry taste sensation from the anterior two-thirds of the tongue to the nucleus of the tractus solitarius in the medulla. The cell bodies of the fibres are in the genicular ganglion of the facial nerve. The parasympathetic fibres are secretomotor to the submandibular and sublingual glands (synapses are in the submandibular ganglion) and to the lacrimal gland and the mucous membrane of the nose, pharynx and mouth (synapses are in the pterygopalatine ganglion). The preganglionic parasympathetic fibres arise in the superior part of the salivary nucleus in the medulla. (Intracranially the sensory and parasympathetic fibres are usually gathered together in a separate root, the nervus intermedius.)

The nerve leaves the lower lateral surface of the pons and passes laterally, with the vestibulocochlear nerve, through the internal acoustic meatus to the facial (genicular) ganglion. It here turns sharply posteriorly through a bony canal on the medial wall of the middle ear and then downwards behind the same cavity to emerge from the temporal bone at the stylomastoid foramen. The nerve then runs forward into the parotid gland where it divides into a number of branches. The genicular ganglion contains the cell bodies of the taste and other sensory fibres.

Branches

(a) **greater petrosal nerve**—carries secretomotor fibres to the pterygopalatine ganglion. It leaves the facial nerve at the genicular ganglion, and passes through the petrous temporal bone into the middle cranial fossa. It traverses the fossa, reaches the foramen lacerum and, joining the deep petrosal nerve (a sympathetic branch from the internal carotid plexus), forms the nerve of the pterygoid canal. The nerve passes through its canal in the sphenoid bone to the pterygopalatine ganglion in the pterygopalatine fossa. Some sensory fibres may be present in the nerve and have their cell bodies in the genicular ganglion.

(b) nerve to stapedius muscle.

(c) **chorda tympani**—arises just above the stylomastoid foramen, passes forwards through the middle ear between the tympanic membrane and the handle of the malleus laterally and the incus medially. It emerges from the petrotympanic fissure of the temporal bone and joins the lingual nerve about 3 cm below the base of the skull. It contains taste fibres from the anterior two-thirds of the tongue (the cell bodies are in the genicular ganglion) and parasympathetic fibres which synapse in the submandibular ganglion.

(d) nerves to the posterior belly of digastric and stylohyoid, and the posterior auricular nerve to occipitalis and the auricular muscles arise just below the stylomastoid foramen.

(e) within the parotid gland the temporal, zygomatic, buccal, mandibular and cervical motor branches are formed; they supply the muscles of facial expression and also buccinator and platysma muscles.

Lower motor neurone facial paralysis may result from virus infections, from trauma, e.g. during surgical removal of a parotid tumour, or from neoplastic infiltration.

8. VESTIBULOCOCHLEAR NERVE (Figs. 28.2, 33.1, 33.5)

This is a special somatic sensory nerve consisting of two parts, the vestibular nerve concerned with balance and the cochlear nerve concerned with hearing.

The nerve is formed in the internal acoustic meatus by the union of the two parts, and passes medially to enter the cerebellomedullary angle of the brain stem. Fibres from the cochlear duct pass to the bipolar cells of the spiral ganglion within the modiolus and the central connections pass to the dorsal and ventral cochlear nuclei in the upper medulla. Vestibular fibres pass from the semicircular ducts, saccule and utricle to the bipolar cells of the vestibular ganglion in the internal acoustic meatus, and the central fibres pass to the vestibular nuclei in the floor of the 4th ventricle. (For central connections see p. 464 and 481.)

9. GLOSSOPHARYNGEAL NERVE (Fig. 26.9)

This has general and special visceral sensory, special visceral motor and general visceral motor (parasympathetic) fibres. The general visceral sensory fibres come from the middle ear, pharynx, posterior third of the tongue and the carotid sinus and carotid body. The special visceral sensory fibres (taste) come from the posterior one-third of the tongue. Their cell bodies are in the ganglion of the glossopharyngeal nerve and their central fibres enter the medulla (nucleus of the tractus solitarius). The parasympathetic (secretomotor) fibres arise in the salivary nucleus (inferior part), synapse in the otic ganglion and supply the parotid gland. The special visceral motor fibres arise in the nucleus ambiguus and supply stylopharyngeus.

The nerve rootlets leave the medulla lateral to the olive. The nerve passes through the jugular foramen medial to the vagus and accessory nerves and the internal jugular vein, and lateral to the inferior petrosal sinus. It passes forwards between the internal and external carotid arteries, pierces the pharyngeal wall between the superior and middle constrictor muscles near the lower pole of the palatine tonsil and ends in the posterior one-third of the tongue. The nerve has two small sensory ganglia situated in the jugular foramen.

Branches

(a) tympanic nerve—passes to the tympanic plexus on the promontory of the middle ear and supplies the mucous membrane of this cavity. The lesser petrosal nerve arises from the plexus and conveys the parasympathetic to the parotid gland via the otic ganglion (p. 432).

(b) nerve to stylopharyngeus.

(c) sensory branches, including taste, from the mucous membrane of the oropharynx, tonsil, soft palate and posterior one-third of the tongue (including the vallate papillae).

(d) sensory branches from the carotid body and carotid sinus.

10. VAGUS NERVE (Figs. 6.1, 6.3, 6.6, 6.8, 26.9, 29.1)

This nerve has general visceral motor (parasympathetic), special visceral motor, general and special visceral sensory, and somatic sensory fibres. The parasympathetic fibres supply the heart, lungs and alimentary canal nearly to the splenic flexure. The special visceral motor fibres innervate the striated muscles of the larynx, pharynx and palate. The general visceral sensory fibres come from the mucous membrane of the palate, pharynx and larynx, and from the heart, lungs and alimentary canal. The special visceral sensory (taste) fibres come from the valleculae and epiglottis. The somatic sensory fibres supply the posterior part of the external acoustic meatus and the tympanic membrane. The nuclei, which lie in the medulla, are the dorsal motor and sensory (parasympathetic), the nucleus ambiguus (special visceral motor) and nucleus of the tractus solitarius (visceral sensory). The somatic sensory fibres probably pass to the sensory nucleus of the trigeminal nerve.

The nerve emerges from the medulla lateral to the olive as rootlets and leaves the skull through the jugular foramen with the inferior petrosal sinus and the glossopharyngeal nerve medially and the accessory nerve and internal jugular vein laterally. It descends through the neck and thorax to the oesophageal plexus where it joins with its fellow of the opposite side to form the anterior and posterior vagal trunks. Near the base of the skull the vagus has superior and inferior sensory ganglia.

Relations (in the neck)

The nerve lies deeply in the carotid sheath, between the internal jugular vein laterally and the internal and common carotid arteries medially. It lies on the prevertebral fascia and muscles. In the root of the neck the right nerve passes forwards in front of the subclavian artery and enters the thorax. The left nerve runs between the common carotid and subclavian arteries as it enters the thorax.

Branches (in the neck)

(a) *auricular nerve*—supplies the posterior part of the external acoustic meatus and the tympanic membrane.

(b) *pharyngeal nerves*—pass between the internal and external carotid arteries to reach the pharyngeal plexus and supply the muscles and mucous membrane of the pharynx and soft palate.

(c) **superior laryngeal nerve**—descends between the pharynx medially and the internal and external carotid arteries, and divides below the hyoid bone into internal and external branches. The internal nerve pierces the thyrohyoid membrane and supplies the mucous membrane of the larynx above the vocal folds. The external nerve descends on the larynx and supplies cricothyroid muscle.

(d) *cervical cardiac branches*— on each side they descend to the cardiac plexuses.

(e) **right recurrent laryngeal nerve**—loops around the right subclavian artery and then ascends in the groove between the oesophagus and trachea. It enters the larynx and supplies the intrinsic muscles except cricothyroid, and the mucous membrane below the vocal folds.

Vagus nerve, in the thorax (see p. 85; central connections, see also p. 466).

11. ACCESSORY NERVE (Fig. 26.9)

This has somatic motor fibres which supply sternocleidomastoid and trapezius and have their nuclei in the upper five cervical segments of the spinal cord. The somatic motor fibres constitute the spinal root of the accessory nerve. The nerve arises in the vertebral canal; it ascends behind the ligamenta denticulata and through the foramen magnum into the posterior cranial fossa. It then leaves the fossa through the jugular foramen and below the skull it passes laterally in front of the internal jugular vein and the transverse process of the atlas into the substance of sternocleidomastoid. It crosses the posterior triangle of the neck lying on levator scapulae and reaches the deep surface of trapezius.

Inside the skull, fibres from the nucleus ambiguus join the accessory nerve but quickly return to the vagus. (These special visceral fibres are called the cranial root of the accessory nerve.)

12. HYPOGLOSSAL NERVE (Figs. 26.6, 26.7, 26.9)

This nerve is the somatic motor supply to all the intrinsic and extrinsic muscles of the tongue, except palatoglossus. Its nucleus is situated in the medulla in the floor of the 4th ventricle.

The nerve emerges from the medulla between the pyramid and the olive, and leaves the posterior cranial fossa through the hypoglossal canal. It descends behind the carotid sheath before passing forwards around the pharynx to the tongue.

Relations

Below the base of the skull the nerve lies behind the internal carotid artery and the vagus nerve on the prevertebral fascia. It passes forwards between the internal carotid artery and the internal jugular vein and crosses in turn the internal and external carotid arteries, the loop of the lingual artery on the middle constrictor muscle and then hyoglossus. As it runs forwards it is covered laterally by the posterior belly of digastric, the submandibular gland and the mylohyoid. Anterior to hyoglossus, the nerve turns medially to ramify in the substance of the tongue.

The hypoglossal nerve receives fibres from the ventral ramus of the 1st cervical nerve near the base of the skull. It conveys some of them to geniohyoid and thyrohyoid, and then forms the superior root of the ansa cervicalis which (with the inferior root from the 2nd and 3rd cervical nerves) supplies the infra-hyoid group of muscles.

— 29 —

The neck

FASCIA OF THE NECK

The **superficial fascia** is a fatty areolar layer between the skin and the more obvious deep fascia. It contains the platysma muscles and the external jugular veins and their tributaries. In an older person the fat is lost and the anterior edges of platysma show clearly through the skin. The **deep fascia** consists of: (1) the cervical fascia, (2) the prevertebral fascia, (3) the pretracheal fascia, (4) the carotid sheath (Fig. 29.1).

The **cervical fascia** forms an investing sleeve around all the deep structures of the neck. It splits to enclose sternocleidomastoid and trapezius, and posteriorly it is attached to the ligamentum nuchae. Superiorly it gains attachment to the superior nuchal lines, the mastoid processes and the inferior border of the body of the mandible. Between the mastoid and mandible it splits into two layers which form a sheath for the parotid gland; the superficial layer is attached above to the zygomatic arch; a thickening in the deep part, the stylomandibular ligament, passes to the styloid process. Inferiorly the fascia is attached to the manubrium sterni, the clavicles, the acromial processes and spines of the scapulae and between the scapulae it descends over the postvertebral muscles. Above the manubrium, a split in the fascia contains the jugular venous arch. The infrahyoid group of muscles is embedded in the deep surface of this fascia.

The **prevertebral fascia** lies in front of the vertebral column, the prevertebral muscles and the cervical and brachial plexuses. Laterally it turns backwards covering the postvertebral muscles on the floor of the posterior triangle and blends with the cervical fascia. Superiorly it is attached to the base of the skull and inferiorly it blends with the anterior longitudinal ligament in front of the body of the 4th thoracic vertebra. Medial to scalenus anterior, the fascia is deficient and the subclavian artery here passes deep to it. Laterally, the subclavian artery carries the fascia into the arm and forms the axillary sheath. The cervical sympathetic trunk is embedded in the prevertebral fascia.

The **pretracheal fascia** is attached superiorly to the hyoid bone and the oblique lines on the thyroid cartilage. It splits around the thyroid gland to provide a sheath, and below the gland the fascia passes downwards to blend with the carotid sheaths.

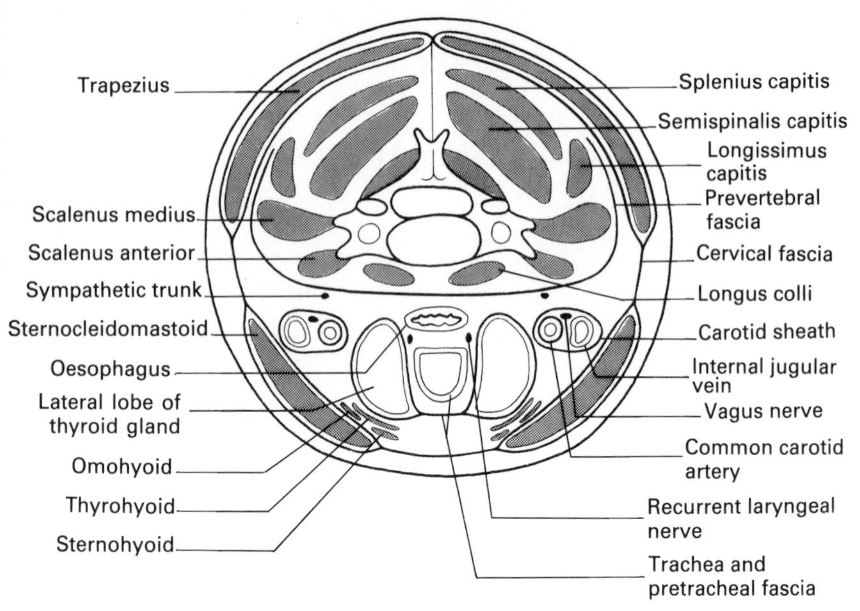

Fig. 29.1 Diagram showing some of the principal structures seen in a horizontal section of the neck at the level of the 6th cervical vertebra.

The **carotid sheath** is a condensation of fascias of the neck, and encloses the internal jugular vein laterally, the common and internal carotid arteries medially and the vagus nerve posteriorly. Above, it is attached to the base of the skull and below it fuses with the fibrous pericardium.

MUSCLES OF THE NECK AND BACK

THE TRIANGLES OF THE NECK

Classically each side of the neck is divided for descriptive purposes into anterior and posterior triangles by sternocleidomastoid. It must be remembered that many important structures lie deep to this muscle.

The posterior triangle is bounded by sternocleidomastoid anteriorly, trapezius posteriorly and the clavicle inferiorly. The roof is formed by the cervical fascia, the floor by the prevertebral fascia over splenius capitis, levator scapulae, and scalenus medius from above downwards. The triangle is crossed by the posterior belly of omohyoid and contains the accessory nerve, branches of the cervical plexus, the upper and middle trunks of the brachial plexus, and the transverse cervical, suprascapular and subclavian arteries.

The anterior triangle is bounded laterally by sternocleidomastoid, superiorly by the body of the mandible and anteriorly by the midline. It contains structures related to the floor of the mouth, the larynx, the trachea and the pharynx.

Sternocleidomastoid
This muscle lies on the anterolateral aspect of the neck enclosed in the cervical fascia.

Attachments

INFERIORLY—by two heads:
(i) the upper anterior surface of the manubrium sterni.
(ii) the upper surface of the medial one-third of the clavicle.

SUPERIORLY—the lateral surface of the mastoid process and the lateral half of the superior nuchal line.

Action
(i) each muscle turns the chin to the opposite side and laterally flexes the neck to its own side, as in looking under a table.
(ii) together, the two muscles flex the head against resistance but extend the neck as in jutting out the chin.
(iii) when the head is fixed the muscles may be used as accessory muscles of respiration.

Nerve supply
This is from the spinal accessory nerve and a branch from the 2nd and 3rd cervical nerves.

Trapezius
See page 229.

THE MUSCLES OF THE BACK

These comprise muscles attaching the shoulder girdle to the trunk (see p. 228) and a group which extends the head and vertebral column.

The latter group forms a large composite mass, lying deep to trapezius and the other girdle muscles and extending from the back of the sacrum to the base of the skull. They may be subdivided from without inwards into: (a) the superiorly placed splenius capitis and cervicis, (b) the erector spinae muscles including longissimus, iliocostalis, spinalis and semispinalis and (c) the deep short muscles—levator costae, multifidus, rotatores, interspinous and intertransversus. These short muscles are replaced between the axis and the base of the skull by the suboccipital muscles.

Splenius capitis—passes upwards and laterally from the ligamentum nuchae and the spines of the upper thoracic vertebrae to the mastoid process and the lateral part of the superior nuchal line deep to the sternocleidomastoid.

Semispinalis—passes upwards and medially from the transverse processes to the spines above. The upper part of the muscle, semispinalis capitis, is attached to the occipital bone between the superior and inferior nuchal lines.

Action

The muscles of the back for the most part maintain the upright position of the body. When standing at rest, the centre of gravity lies just in front of the 2nd piece of the sacrum. Movement of the body frequently carries the centre of gravity much further forwards and this large mass of muscles is required to restore the upright position. The muscles attached to the skull produce extension, lateral flexion and rotation of the head.

Nerve supply

All these muscles are supplied by the dorsal rami of the spinal nerves.

Suboccipital muscles

Rectus capitis posterior major—passes from the spine of the axis to the occipital bone below the inferior nuchal line.

Rectus capitis posterior minor—lies anteromedial to the major and passes from the posterior tubercle of the atlas to the occipital bone behind the foramen magnum.

Obliquus capitis superior—passes from the tip of the transverse process of the atlas to the occipital bone between the superior and inferior nuchal lines, lateral to semispinalis capitis.

Obliquus capitis inferior—passes from the tip of the transverse process of the atlas to the spine of the axis.

The suboccipital muscles extend the skull at the atlanto-occipital joints and rotate it at the atlanto-axial joints. They are all supplied by the dorsal ramus of the 1st cervical nerve. Accurate positioning of the head is necessary for stereoscopic vision.

The **suboccipital triangle** is bounded laterally by the superior oblique, medially by the rectus capitis posterior major and inferiorly by the inferior oblique. Its floor is formed by the posterior atlanto-occipital membrane and the posterior arch of the atlas. The vertebral artery, with the dorsal ramus of the 1st cervical nerve below it, passes deep to the membrane. The roof is crossed by the greater occipital nerve and covered by semispinalis capitis and longissimus capitis. The triangle is filled with fat and the suboccipital plexus of veins which drains to the vertebral veins.

Ligamentum nuchae

In animals possessing a protruding head and neck this is a powerful midline elastic ligament passing from the occipital bone to all the cervical spines. In man it is smaller, mainly fibrous tissue and has little supporting action.

Prevertebral and scalene muscles

These muscles lie in front of all the cervical and the upper thoracic vertebrae. They are covered anteriorly by the prevertebral fascia.

Prevertebral muscles (rectus capitis anterior and lateralis, longus capitis and longus colli)—pass between the skull and vertebrae, and between vertebrae.

Scalenus anterior—is attached superiorly to the anterior tubercles of the 3rd-6th cervical vertebrae and inferiorly to the scalene tubercle on the 1st rib.

Relations: the anterior surface is crossed obliquely by the phrenic nerve and separated by the prevertebral fascia with the sympathetic trunk from the carotid sheath and its contents, the transverse cervical and suprascapular arteries and the subclavian vein. The posterior surface lies on the scalenus medius (separated from it by the ventral rami of the lower cervical nerves), the dome of the pleura and the subclavian artery.

Scalenus medius—is attached superiorly to the posterior tubercles of the 2nd-7th cervical vertebrae and inferiorly to the upper surface of the 1st rib behind the groove for the subclavian artery.

Scalenus posterior—is the part of the previous muscle which is attached to the 2nd rib.

Action

These muscles are weak flexors of the head and neck; the scalene muscles also raise and fix the upper two ribs.

Nerve supply

All are supplied by the ventral rami of the cervical nerves.

Infrahyoid muscles

There are four pairs of 'strap' mucles lying anteriorly in the neck and enclosed in a fascial sheath which is adherent to the deep surface of the cervical fascia:

Sternohyoid—is attached superiorly to the body of the hyoid bone and inferiorly to the back of the manubrium sterni and the medial end of the clavicle.

Omohyoid—has superior and inferior bellies united by an intermediate tendon which is held to the medial end of the clavicle by a fascial sling. Superiorly the muscle is attached on the body of the hyoid bone and inferiorly to the lateral end of the superior border of the scapula.

Thyrohyoid—is attached superiorly to the back of the body and greater horn of the bone and inferiorly to the oblique line on the thyroid cartilage.

Sternothyroid—is attached superiorly to the oblique line on the thyroid cartilage and inferiorly to the back of the manubrium sterni and 1st costal cartilage. The muscle lies deep to sternohyoid.

These muscles fix or depress the hyoid bone in swallowing and speech. They are supplied by the ventral rami of the upper three cervical nerves (via the superior or inferior roots of the ansa cervicalis, the ansa, or the hypoglossal nerve).

THE THYROID GLAND (Fig. 29.1)

This is a bilobed endocrine gland situated on each side of the trachea and oesophagus, with a communicating isthmus anterior to the trachea. It is enclosed in pretracheal fascia. Each lobe is about 5 cm long and extends from the oblique line of the thyroid cartilage to the 6th tracheal ring. Occasionally a pyramidal lobe is found arising from the upper border of the isthmus which lies on the 2nd and 3rd tracheal rings.

Relations

SUPERFICIAL—three strap muscles (not thyrohyoid), the cervical fascia and sternocleidomastoid.

MEDIAL—larynx and trachea and, more posteriorly, the pharynx and oesophagus. The recurrent laryngeal nerve lies between the trachea and the oesophagus. The external laryngeal nerve descends to reach cricothyroid.

POSTERIOR—contents of carotid sheath, the parathyroid glands, and the prevertebral fascia.

Blood supply

This is from superior and inferior arteries lying close to the external and the recurrent laryngeal nerves respectively as the vessels approach the gland. Venous blood passes via the superior and middle thyroid veins to the internal jugular vein and via the inferior thyroid veins by a common trunk to the left brachiocephalic vein.

Nerve supply

This is from sympathetic nerves from plexuses around superior and inferior thyroid arteries.

Lymph drainage

This is to pretracheal, paratracheal and inferior deep cervical nodes.

Histology

Cubical epithelium lines follicles which are filled with amorphous colloid (containing thyroglobulin). Between the follicles vascular connective tissue contains groups of parafollicular (C) cells which secrete thyrocalcitonin.

Embryology

The gland develops from a midline ventral diverticulum between 1st and 2nd pharyngeal arches which grows caudally to meet an outgrowth of the 4th pharyngeal pouch. The lower end of the diverticulum proliferates and forms the glandular tissue. The rest of the diverticulum (the thyroglossal duct) atrophies, but remnants may persist. The contribution from the 4th pouch may be responsible for the C-cells and the production of thyrocalcitonin. The stored thyroglobulin is changed into thyroxine. This secretion passes back through the follicle cells, is absorbed into the capillaries, and is a vital regulator of cellular metabolism, under the humoral control of the pituitary gland (hypophysis cerebri).

Over activity of the thyroid gland gives rise to thyrotoxicosis, whereas under activity produces cretinism in a child or myxoedema in later life.

During operations on the thyroid gland the recurrent and external laryngeal nerves are at risk and may be damaged. This may result in transient or more permanent hoarseness because of paralysis of the laryngeal muscles. Reduction of the blood supply to the closely-applied parathyroid glands or inadvertent removal of the glands may produce post-operative hypoparathyroidism and tetany.

THE PARATHYROID GLANDS

These are four small endocrine glands situated behind the lateral lobes of the thyroid gland within its capsule. There is a superior and an inferior pair. Each gland is oval in shape, about 3–6 mm across and may be mistaken for a fat globule. They have a rich blood supply from the superior and inferior thyroid arteries. They consist of columns of chief cells separated by blood spaces. The cells have dark-staining nuclei and a chromatin network. Around puberty, eosinophil-staining cells appear.

The superior and inferior glands develop from the 4th and 3rd pharyngeal pouches respectively. The glands secrete parathormone and play an important role in calcium metabolism.

THE TRACHEA (Figs. 6.3, 29.1)

The trachea is part of the respiratory passage. It descends from the larynx (at the level of the 6th cervical vertebra), through the neck and thorax to its bifurcation into the two bronchi at the sternal angle (the level of the lower border of the 4th thoracic vertebra). It is about 10 cm long and 2 cm in diameter. It is covered by the pretracheal fascia. (See also p. 413.)

The walls of the trachea are formed of fibrous tissue reinforced by 15–20 incomplete cartilaginous rings united behind by fibro-elastic tissue and smooth muscle where the trachea rests on the oesophagus. It is lined internally with respiratory epithelium. The rings are of hyaline cartilage.

Relations (in the neck): the trachea lies anterior to the oesophagus with the recurrent laryngeal nerve laterally in the groove between. It lies behind the cervical fascia and the infrahyoid muscles, and is crossed anteriorly by the isthmus of the thyroid gland and the jugular venous arch. Laterally are the lateral lobe of the thyroid gland, the inferior thyroid artery and the carotid sheath.

Relations (in the thorax): see page 88.

Blood supply: the inferior thyroid vessels.

Lymph drainage: to the pre- and paratracheal and tracheobronchial nodes.

Nerve supply: branches of the sympathetic trunk; parasympathetic branches from the vagus nerve via the recurrent laryngeal nerve and the pulmonary plexuses. Sensory fibres enter the recurrent laryngeal nerves.

In upper respiratory tract obstruction a tracheostomy (opening into the trachea) may be required. Usually the preferred site is at the 3rd and 4th tracheal rings and the isthmus of the thyroid gland often has then to be divided.

THE OESOPHAGUS (Figs. 6.3, 29.1)

This part of the digestive tube begins just behind the cricoid cartilage (at the level of the 6th cervical vertebra) as a continuation of the pharynx. It descends through the neck, thorax and diaphragm, inclining slightly to the left, to enter the stomach at the level of the 10th thoracic vertebra. Its upper end is about 15 cm from the incisor teeth and it is about 25 cm long. (See also pp. 88 and 123.)

Relations (in the neck): the oesophagus lies anterior to the prevertebral fascia and muscles and posterior to the trachea, with the recurrent laryngeal nerves in the grooves between. Laterally are the lateral lobe of the thyroid gland and the common carotid artery. The thoracic duct ascends for a short distance along the left side.

Relations (in the thorax and abdomen): see pages 88 and 123.

Blood supply: branches from the inferior thyroid artery, the aorta and the left gastric artery. Venous blood passes to the inferior thyroid, azygos and left gastric veins. The communication of the azygos and left gastric veins in the walls of the oesophagus forms an important portal-systemic anastomosis (p. 142).

Lymph vessels: pass to the deep cervical lymph chain, the posterior mediastinal lymph nodes and along the left gastric artery to the coeliac nodes.

Nerve supply: sympathetic—thoracic sympathetic chains and greater splanchnic nerves; parasympathetic—from the vagus nerve via the oesophageal plexus; recurrent laryngeal nerves to the upper part.

Histology: The oesophagus has four coats. An inner mucosa of stratified squamous epithelium with numerous mucous glands which extend into the vascular submucous coat except at the upper and lower ends. The muscular coat contains inner circular and outer longitudinal layers; it is composed of striated muscle in its upper one-third and smooth muscle below. The outer fibrous coat is a layer of areolar tissue with many elastic fibres. Peritoneum covers part of the abdominal oesophagus.

— 30 —

The vessels, nerves and lymph drainage

VESSELS

THE SUBCLAVIAN ARTERY (Figs. 6.3 and 30.1)

On the right side the artery is formed from the brachiocephalic artery behind the sternoclavicular joint. It passes laterally behind scalenus anterior and the prevertebral fascia, arches over the apex of the lung and crosses the 1st rib, at the outer border of which it becomes the axillary artery. On the left side the artery arises from the arch of the aorta, ascends through the mediastinum, arches over the lung and enters the arm.

Relations (of the right artery)
Medial to scalenus anterior it lies behind the right common carotid artery, vagus nerve and internal jugular vein, and in front of the suprapleural membrane, pleura and lung. The right recurrent laryngeal nerve and ansa subclavia (p. 430) hook around it. Behind scalenus anterior, which separates it from the phrenic nerve, the artery lies on the cervical pleura. Lateral to the muscle, the artery lies on the 1st rib with its vein and the clavicle in front, and the lower trunk of the brachial plexus and scalenus medius behind.

Relations (of the left artery)
In the thorax, the artery lies behind the left common carotid artery and the left vagus and left phrenic nerves. The oesophagus, trachea and left recurrent laryngeal nerve lie medial to it, and the left pleura and lung lie lateral to it. In the root of the neck its relations differ from those of the right, in that (*a*) the thoracic duct and phrenic nerve cross anterior to it, (*b*) the left recurrent laryngeal nerve does not hook around it (Figs. 6.1, 6.3).

Branches
The vertebral, internal thoracic and thyrocervical branches arise in the root of the neck medial to scalenus anterior. On the right, the costocervical trunk is formed behind the muscle; on the left it usually arises medial to the muscle.

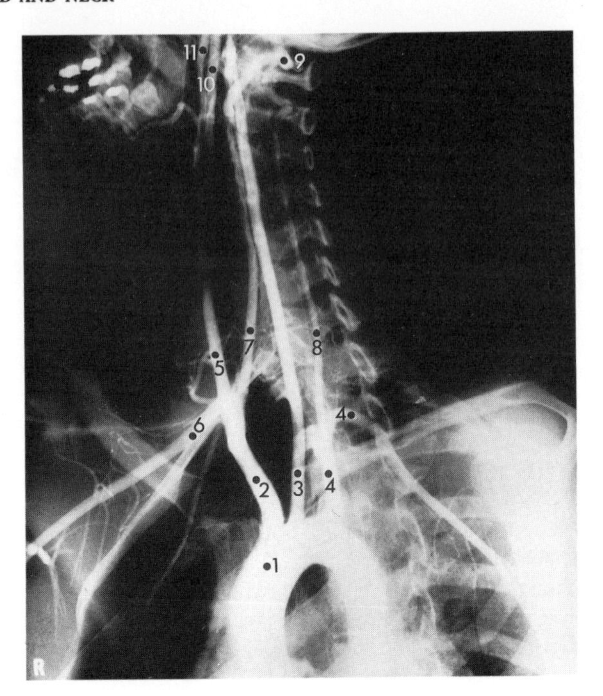

Fig. 30.1 Arteriogram of the aortic arch and its main branches.
1. Aortic arch
2. Brachiocephalic artery
3. Left common carotid artery
4. Left subclavian artery
5. Right common carotid artery
6. Right subclavian artery
7. Right vertebral artery
8. Left vertebral artery
9. Loop of left vertebral artery as it passes around the lateral mass of the atlas
10. Right internal carotid artery
11. Right external carotid artery

(i) **vertebral artery**—passes upwards and medially to reach the transverse process of the 6th cervical vertebra. It ascends through the foramen transversarium of the upper six vertebrae, turns medially behind the lateral mass of the atlas and, piercing the dura, enters the skull through the foramen magnum. It unites with its fellow of the opposite side in front of the brain stem and forms the basilar artery.

Relations: the artery is surrounded by a sympathetic plexus derived from the inferior cervical ganglion. In the lower part of the neck the artery lies on the prevertebral fascia and crosses over the inferior cervical ganglion. It lies behind the common carotid artery and is crossed over by the inferior thyroid artery and the thoracic duct or right lymph trunk. Within the vertebrae it is accompanied by the vertebral vein. As it winds around the lateral mass of the atlas, in the groove on the posterior arch, it crosses the floor of the suboccipital triangle and passes below the edge of the posterior atlanto-occipital membrane. The dorsal ramus of the 1st cervical nerve emerges between the artery and the atlas. The vessel has a very tortuous course. Inflammation of the vertebral synovial joints may produce excess bone and compress the vessel, thus reducing the blood flow to the hindbrain and upper spinal cord.

Branches: in the neck it supplies the spinal cord and vertebral muscles. Its branches within the posterior cranial fossa supply the upper part of the spinal cord and the hindbrain (see p. 490).

(ii) **internal thoracic artery**—see page 58.

(iii) **thyrocervical trunk**—is a short vessel which soon divides into inferior thyroid, transverse cervical and suprascapular arteries. The **inferior thyroid artery** passes upwards and medially and reaches the inferior pole of the thyroid gland. It lies behind the common carotid artery and sympathetic trunk, and in front of the vertebral artery and prevertebral fascia. It supplies the thyroid gland, the trachea, oesophagus, larynx and pharynx. The **transverse cervical** and **suprascapular arteries**, passing laterally over the scalene muscles, the phrenic nerve and prevertebral fascia, join the anastomosis around the scapula and supply the muscles of the shoulder girdle.

(iv) **costocervical trunk**—passes backwards over the dome of the pleura and reaches the neck of the 1st rib where it divides into **deep cervical** and **highest intercostal arteries**. The former passes above the neck of the rib and supplies the postvertebral muscles. The latter descends behind the pleura, in front of the neck of the rib and gives rise to the 1st and 2nd posterior intercostal arteries.

THE SUBCLAVIAN VEIN

This is the continuation of the axillary vein and extends from the outer border of the 1st rib to the medial border of scalenus anterior, where, joining the internal jugular vein, it forms the brachiocephalic vein. Lateral to scalenus anterior the vein lies in front of its artery; it then passes in front of the muscle, crossing over the phrenic nerve. The vein lies behind the clavicle. Its main tributary is the external jugular vein. A bicuspid valve is usually present just lateral to the entry of this vein.

THE COMMON CAROTID ARTERY (Figs. 6.3, 29.1, 30.1)

On the right side the artery arises from the brachiocephalic artery behind the sternoclavicular joint. It ascends through the neck and divides into internal and external carotid arteries at the level of the upper border of the thyroid cartilage. On the left, the artery arises from the aortic arch and ascends through the thorax and neck to bifurcate at the same level. The carotid sinus is a dilatation of the upper end of the artery and is sensitive to blood pressure changes. The carotid body is a small neurovascular structure on the posterior wall of the sinus; it functions as a chemoreceptor. Both structures are richly innervated by the glossopharyngeal nerve. The bifurcation of the common carotid artery is a common site for degenerative arterial disease to occur and local thrombosis or an embolus from this region can have serious effects on cerebral function.

Relations in the neck, of the right and left arteries
Each artery lies within the carotid sheath with the vagus nerve behind and internal jugular vein laterally. The sheath is covered anterolaterally by the infrahyoid and sternocleidomastoid muscles. Posteriorly the artery lies on the right subclavian artery at its origin, and then on the prevertebral fascia and muscles. Medially at first are the trachea and the oesophagus (with the recurrent laryngeal nerve between) and then the larynx and pharynx. The thyroid gland and its vessels may be anterior or medial to the artery.

Relations in the thorax, of the left artery
The artery lies anterior to the vagus and phrenic nerves and the left subclavian artery, to the left of the trachea and brachiocephalic artery, and is covered anterolaterally by the left pleura and lung (Figs. 6.1 and 6.3).

Branches
None, other than its terminal branches.

THE EXTERNAL CAROTID ARTERY

This artery supplies the structures in the upper part of the neck and the face. It arises from the bifurcation of the common carotid artery at the level of the upper border of the thyroid cartilage and ascends to reach and pass through the parotid gland. The artery ends behind the neck of the mandible by dividing into maxillary and superficial temporal arteries.

Relations
In the neck the artery lies on the lateral wall of the pharynx, at first anteromedial and then lateral to the internal carotid artery, separated from it by styloglossus and stylopharyngeus, the glossopharyngeal nerve and the pharyngeal branches of the vagus nerve. It lies deep to sternocleidomastoid and is crossed from above downwards by stylohyoid, the posterior belly of digastric, the hypoglossal nerve, and the lingual and facial veins. Within the parotid gland the artery lies medial to the retromandibular vein and the facial nerve.

Branches (in order of origin)
Superior thyroid artery—descends with the external laryngeal nerve to the upper pole of the thyroid gland. Its superior laryngeal branch pierces the thyrohyoid membrane and supplies the larynx. Further branches supply the pharynx, thyroid gland and infrahyoid muscles.

Ascending pharyngeal artery—ascends on the lateral wall of the pharynx and gives branches to the pharynx, tonsil, soft palate and auditory tube.

Lingual artery—arises at the level of the hyoid bone and passes forwards deep to hyoglossus into the substance of the tongue and then to its tip.

Relations: before passing deep to hyoglossus, it lies on the middle constrictor and here forms an upward loop which is crossed by the hypoglossal nerve. The artery supplies the suprahyoid muscles, the tonsil, the soft palate, the submandibular and sublingual glands and the tongue.

Facial artery—arises just above the lingual artery and arches over the submandibular gland. It then turns around the inferior border of the mandible (4 cm in front of the angle) and takes a tortuous course across the face to the medial angle of the eye.

Relations: medial to the submandibular gland, the artery lies on the middle and superior constrictor muscles, separated by the latter from the tonsil. Lateral to the gland it is related to the body of the mandible. On the face, the artery lies between the superficial and deeper muscles.

Branches: in the neck the artery supplies the tonsil, soft palate and submandibular gland, and on the face it supplies the upper and lower lips and the facial musculature.

Occipital artery—arises opposite the facial artery and passes backwards along the lower border of the posterior belly of digastric, and medial to the mastoid process. It turns upwards over the back of the skull and reaches the vertex.

Relations: with digastric it crosses the carotid sheath and lies in the medial of the two grooves on the medial side of the mastoid process. On reaching the back of the skull it passes between semispinalis and splenius capitis and pierces trapezius. The greater occipital nerve lies medial to it on the back of the skull. The artery supplies the upper part of the neck and the back of the scalp.

Posterior auricular artery—passes backwards along the upper border of the posterior belly of digastric, then lateral to digastric and on to the lateral aspect of the scalp. It supplies the auricle and adjacent scalp.

Superficial temporal artery—is formed behind the neck of the mandible in the parotid gland and ascends over the posterior end of the zygomatic arch on to the lateral aspect of the scalp, where it divides into anterior and posterior branches. It supplies the face (by its transverse facial branch), the auricle and the scalp.

Maxillary artery

This arises behind the neck of the mandible in the parotid gland. It passes forwards through the infratemporal fossa and pterygomaxillary fissure and enters the pterygopalatine fossa where it divides into terminal branches.

Relations

It lies at first between the neck of the mandible and the sphenomandibular ligament and is crossed by the inferior alveolar nerve. It then usually passes on to the lateral surface of lateral pterygoid, lying deep to temporalis, and runs forwards on lateral pterygoid, reaching the pterygomaxillary fissure and fossa. In the pterygopalatine fossa it is related to the maxillary nerve and the pterygopalatine ganglion.

Branches

(i) to the ear.

(ii) **middle meningeal artery**—ascends through the foramen spinosum into the middle cranial fossa where it divides into anterior and posterior branches. The anterior branch passes medial to the pterion (p. 342) and often grooves the bone. It may be damaged in fractures of the skull and bleed intracranially, so producing an extradural clot with a life-threatening rise in intracranial pressure. The surface marking of the middle meningeal artery is 3.5 cm behind and 1.5 cm above the zygomaticofrontal suture and surgical access to the vessel can be achieved by drilling through the bone over this spot, the pterion. The artery supplies the dura and bone of the middle cranial fossa and the vault of the skull.

(iii) **inferior alveolar artery**—passes through the mandibular canal and is distributed to the gums and teeth.

(iv) branches to the muscles of mastication.

(v) terminal vessels accompany the branches of the pterygopalatine ganglion and supply the nose, palate and pharynx. The infra-orbital branch gives off the superior alveolar arteries to the upper teeth.

THE INTERNAL CAROTID ARTERY

This supplies the greater part of the brain and the contents of the orbit. It arises from the bifurcation of the common carotid artery at the level of the upper border of the thyroid cartilage. It ascends to the base of the skull and, passing through the carotid canal in the petrous temporal bone, enters the middle cranial fossa. In the fossa it passes upwards and forwards through the cavernous sinus, pierces the roof of the sinus, and turns backwards on itself before dividing into anterior and middle cerebral arteries, medial to the anterior clinoid process.

Relations

The artery is surrounded by a sympathetic plexus formed from fibres of the superior cervical ganglion. In the neck the artery lies within the carotid sheath with the internal jugular vein laterally, and the vagus nerve behind. The pharynx lies medially, and the prevertebral fascia and muscles posteriorly. The external carotid artery is at first anteromedial to it and then lateral, separated from it by styloglossus and stylopharyngeus, the glossopharyngeal nerve and the pharyngeal branches of the vagus. Laterally, the artery is crossed by stylohyoid, the posterior belly of digastric and the hypoglossal nerve.

In the petrous temporal bone the artery lies below the floor of the middle ear and, as it enters the middle cranial fossa, passes over the foramen lacerum.

Within the cavernous sinus, it lies in the carotid groove of the body of the sphenoid, covered by the venous endothelium of the sinus. On its lateral side it is crossed by the abducent nerve and further laterally the oculomotor, trochlear, ophthalmic and maxillary nerves lie in the lateral wall of the sinus. It pierces the roof of the sinus medial to the anterior clinoid process and here lies lateral to the optic chiasma and nerve.

Branches

(i) to the **pituitary gland** and the **meninges**.
(ii) **ophthalmic** and **posterior communicating arteries**.
(iii) **anterior** and **middle cerebral arteries** (p. 488).

VENOUS DRAINAGE OF THE HEAD AND NECK

This may be subdivided into: (1) Veins of the brain and the intracranial venous sinuses (p. 493). (2) Veins of the face and scalp. (3) Veins of the neck. The arrangement of the veins is more variable than that of the arteries but most vessels are tributaries of the jugular system.

THE VEINS OF THE FACE AND SCALP

The **facial vein** is formed at the medial angle of the eye by the union of the supraorbital and supratrochlear veins draining the scalp. It passes obliquely backwards in the superficial fascia of the face, posterior to the facial artery, and is joined by the anterior branch of the retromandibular vein below the mandible. The facial vein drains the anterior part of the scalp and the superficial structures of the face. It communicates with the cavernous sinus by the ophthalmic veins, with the pterygoid venous plexus by the deep facial vein and with the diploic veins.

The **pterygoid venous plexus** is situated around the lateral pterygoid muscle and drains via the maxillary vein to the retromandibular vein. The plexus drains the infratemporal region and communicates with the cavernous sinus, and the ophthalmic and facial veins.

The **retromandibular vein** is formed behind the neck of the mandible by the union of the superficial temporal and maxillary veins. It descends through the parotid gland lateral to the external carotid artery and medial to the facial nerve. Below the gland it divides into anterior and posterior divisions. The anterior joins the facial vein below the angle of the mandible, the posterior joins the posterior auricular vein and forms the external jugular vein. The veins of the back of the scalp drain into the suboccipital venous plexus and thence into the vertebral veins which are tributaries of the brachiocephalic veins.

The **superficial temporal** and **posterior auricular veins** drain the lateral aspect of the scalp and the external acoustic meatus; most of the blood passes to the external jugular vein.

THE VEINS OF THE NECK

The **internal jugular vein** drains blood from the intracranial region, the head and neck. It arises at the jugular foramen on the base of the skull as a continuation of the sigmoid venous sinus, passes downwards through the neck and, behind the medial end of the clavicle, is joined by the subclavian vein so forming the brachiocephalic vein. The vein has a dilatation at each end, the superior and inferior jugular bulbs, and a fairly constant pair of valves above the lower bulb. In patients with heart failure the jugular venous pressure is raised, and by careful positioning of the patient the height and wave form of the pulse in the internal jugular vein can be observed and recorded in the neck.

Relations: the vein lies within the carotid sheath with the vagus nerve behind and the internal or common carotid arteries medially. The deep cervical lymph nodes lie around it. At the base of the skull it lies lateral to the last four cranial nerves. Posteriorly it lies on the sympathetic chain, prevertebral fascia and muscles, the phrenic nerve and, inferiorly, the subclavian artery. It is crossed over laterally by the accessory nerve passing downwards and laterally, then by the posterior belly of digastric passing upwards and backwards, and the omohyoid muscle passing downwards and backwards. Further laterally are the styloid process and its muscles, the sternocleidomastoid muscle and the medial end of the clavicle from above downwards.

Tributaries

(i) **inferior petrosal sinus**.

(ii) branches from the **pharyngeal plexus** of veins.

(iii) **facial vein** having drained blood from the face and scalp, passes backwards and downwards over the external and internal carotid arteries and the hypoglossal nerve. It enters the internal jugular vein opposite the hyoid bone.

(iv) **lingual vein**.

(v) **superior** and **middle thyroid veins**.

(iv) on the left the **thoracic duct** may enter the vein and on the right the **right lymph duct**.

The **external jugular vein**, a superficial vein, is formed behind the angle of the mandible by the union of the posterior auricular and the posterior branch of the retromandibular vein. It descends over sternocleidomastoid, pierces the cervical fascia above the midpoint of the clavicle and enters the subclavian vein. The vein may be used to assess venous pressure.

The **anterior jugular vein** begins below the hyoid bone near the midline and passes downwards and laterally deep to sternocleidomastoid. It enters the external jugular vein behind the clavicle. The **jugular arch** unites the two anterior veins just above the sternum.

NERVE SUPPLY

The **cutaneous nerve supply** is summarised in Figure 30.2.

The trigeminal nerve: see page 404.

THE CERVICAL PLEXUS OF NERVES

This is formed from the ventral rami of the upper four cervical nerves. The plexus lies on scalenus medius and is covered anteriorly by scalenus anterior, the prevertebral fascia and the internal jugular vein.

Branches

(i) **lesser occipital nerve** (C2 & 3) is distributed to the skin over the posterior aspect of the auricle and mastoid region.

(ii) **great auricular nerve** (C2 & 3) supplies the skin over the inferior part of the auricle and the parotid region of the face.

(iii) **transverse cervical nerve** (C2 & 3) supplies the skin over the anterior and lateral parts of the neck.

(iv) **supraclavicular nerves** (C3 & C4) supply the skin above and below the clavicle.

(v) **communications** with the hypoglossal nerve—fibres from C1 are carried by the hypoglossal nerve to geniohyoid and thyrohyoid, and also form the superior root of the ansa cervicalis. The superior root descends on the internal carotid artery and forms a communicating loop (the ansa cervicalis) with the inferior root (from C2 & 3) which descends on the internal jugular vein. The ansa supplies the infrahyoid muscles.

(iv) *other muscle branches*—supply the prevertebral muscles, sternocleidomastoid, trapezius and levator scapulae.

(vii) **phrenic nerve** (C3, 4 & 5, mainly 4)—descends through the neck and thorax to supply the diaphragm.

Relations (in the neck)

The phrenic nerve winds around the lateral border of scalenus anterior on to the anterior surface. It is here covered by the prevertebral fascia and crossed by the internal jugular vein, the transverse cervical and suprascapular arteries, and the subclavian vein. On the left, the nerve is more medially placed in the root of the neck and crosses the subclavian artery. The nerves pass in front of the internal thoracic arteries as they enter the thorax.

Relations (in the thorax)
See page 85.

Because the phrenic and supraclavicular nerves come from the same segments of the cord, pain which originates in the region of the diaphragm may be referred by the brain to the point of the shoulder.

Dorsal rami of cervical nerves supply the regions on each side of the midline posteriorly. The largest nerve, the *greater occipital*, arises from C2 and supplies much of the back of the scalp.

The brachial plexus: see page 235.

THE AUTONOMIC NERVOUS SYSTEM OF THE HEAD AND NECK

SYMPATHETIC

The cervical sympathetic trunks pass vertically upwards to the base of the skull, lying on the prevertebral fascia and muscles, over the transverse processes of the cervical vertebrae and deep to the carotid sheath. Each trunk has three ganglia and is continuous below with the thoracic ganglionated trunk. Preganglionic fibres arise mainly in the first thoracic segment. Postganglionic fibres are distributed as vascular, spinal and visceral branches.

The **superior cervical ganglion** lies opposite the 2nd and 3rd cervical transverse processes and behind the angle of the mandible. Its postganglionic fibres pass—

(i) with the external carotid artery and its branches.
(ii) with the internal carotid artery and its branches.
(iii) to the pharyngeal plexus.
(iv) to the upper four cervical nerves—rami communicantes.
(v) in a cardiac branch to the cardiac plexus.

The **middle cervical ganglion** lies at the level of the 6th cervical vertebra and cricoid cartilage. Postganglionic fibres pass—

(i) with the inferior thyroid artery.
(ii) to the 5th and 6th cervical nerves—rami communicantes.
(iii) in a cardiac branch to the cardiac plexus.

The **inferior cervical ganglion** is a small ganglion lying opposite the 7th cervical vertebra. It may combine with the 1st thoracic ganglion and form the stellate ganglion lying on the neck of the 1st rib. Postganglionic fibres pass—
(i) with the vertebral artery.
(ii) to the 7th and 8th cervical nerves—rami communicantes.
(iii) in a cardiac branch to the cardiac plexus.

There is a fine communication, the ansa subclavia, between the middle and inferior ganglia. It hooks around the subclavian artery.

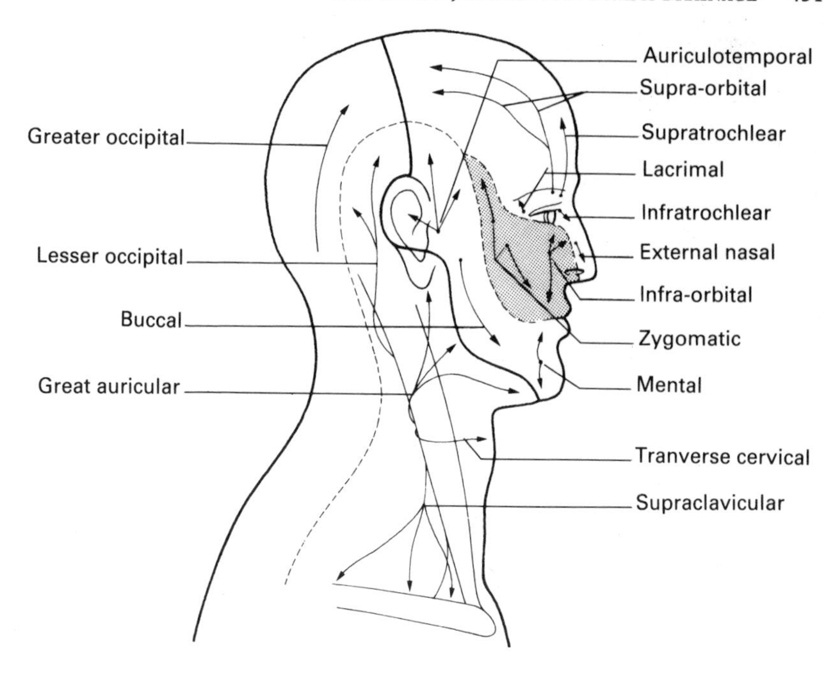

Fig. 30.2 Diagrammatic lay-out of the cutaneous innervation of the head and neck. The trigeminal nerve supplies the area in front of the thick line and the cervical nerves supply the remaining area.

Injury to the sympathetic trunk or ganglia in the neck or upper thorax causes on the same side: 1. pupillary constriction, because the dilator pupillae is paralysed; 2. drooping of the upper eyelid (ptosis), because some smooth muscle in the lid is paralysed; 3. flushing of the face, because of vasodilatation; 4. lack of sweating, because the secretomotor fibres are lost. This combination of signs is called Horner's syndrome.

PARASYMPATHETIC

Preganglionic fibres pass in the oculomotor, facial, glossopharyngeal and vagus nerves and synapse either in four small ganglia in the head (the ciliary, pterygopalatine, submandibular and otic) or in the walls of the organs they supply in the neck, thorax and abdomen. The named ganglia, as well as containing parasympathetic synapses, give uninterrupted passage to sympathetic and somatic fibres.

The **ciliary ganglion** lies in the orbit, lateral to the optic nerve. The parasympathetic fibres are derived from the oculomotor branch to inferior oblique. Postganglionic fibres pass to the sphincter pupillae and ciliary muscles. Sympathetic fibres come from the superior cervical ganglion via the plexus on the ophthalmic artery and are distributed to the eyeball. Sensory fibres from the eyeball pass to the nasociliary nerve.

The **pterygopalatine ganglion** lies in the pterygopalatine fossa, connected to the maxillary nerve by small branches. The parasympathetic fibres are from the facial nerve via the greater petrosal branch which becomes the nerve of the pterygoid canal. Postganglionic fibres supply the lacrimal gland (through the zygomaticotemporal and lacrimal nerves) and glands of the nose, palate and nasopharynx (through nasal, palatine, and pharyngeal branches of the ganglion).

Sympathetic fibres come from the superior cervical ganglion via the plexus on the internal carotid artery. The deep petrosal nerve (sympathetic) joins the greater petrosal nerve to form the nerve of the pterygoid canal which enters the ganglion. The fibres are distributed to the nose, palate and nasopharynx. Sensory fibres from the nose, palate and nasopharynx may pass in the branches of the ganglion to the maxillary nerve.

The **submandibular ganglion** lies lateral to hyoglossus, suspended from the lingual nerve by small branches. Its parasympathetic fibres come from the facial nerve via its chorda tympani branch and then along the lingual nerve to the ganglion. Postganglionic fibres supply the submandibular and sublingual glands and other glands in the floor of the mouth.

Sympathetic fibres come from the superior cervical ganglion along the facial artery and pass to the glands in the floor of the mouth. Taste fibres may pass through the ganglion.

The **otic ganglion** lies on tensor veli palatini medial to the mandibular nerve. The parasympathetic fibres come from the glossopharyngeal nerve via the tympanic plexus and its branch, the lesser petrosal nerve. Postganglionic fibres pass with the auriculotemporal nerve to the parotid gland. Sympathetic fibres come from the superior cervical ganglion along the middle meningeal artery and pass to the parotid gland. Motor branches from the mandibular nerve pass through the ganglion, without synapsing, and supply tensor veli palatini and tensor tympani.

Parasympathetic fibres from the glossopharyngeal nerve supply directly, or through the pharyngeal plexus, the glands of the oropharynx and the posterior one-third of the tongue. Parasympathetic fibres from the vagus supply the glands of the laryngopharynx, larynx, oesophagus and trachea. In all these situations, the ganglion cells are on the viscus.

THE LYMPH DRAINAGE OF THE HEAD AND NECK

The lymph nodes of the head and neck are divided into two groups,

(*a*) the circular chain around the base of the skull.

(*b*) the deep and superficial cervical chains—the deep around the internal jugular vein and the superficial around the external and anterior jugular veins.

The circular chain

The circular chain consists of seven named groups of lymph nodes:

(i) **occipital**—around the occipital artery; drain the posterior part of the scalp and adjacent part of the neck.

(ii) **retro-auricular**—on the mastoid process; drain external acoustic meatus, the back of the auricle and adjacent scalp.

(iii) **superficial parotid**—in front of the tragus; drain the external acoustic meatus, the front of the auricle and the adjacent scalp.

(iv) **deep parotid**—deep to the parotid fascia and within the gland; drain the anterior half of the scalp, infratemporal region, the orbit, the lateral part of the eyelids, the upper molar teeth, the external acoustic meatus and the parotid gland.

(v) **retropharyngeal**—between the pharynx and the atlas; drain the upper part of the pharynx and related structures.

(vi) **submandibular**—between the mandible and the submandibular gland; drain the anterior nasal cavities, the tongue, teeth and gums, submandibular and sublingual glands and all the face except the lateral part of the eyelids and the medial part of the lower lip and chin.

(vi) **submental**—behind the chin and on mylohyoid; drain the tip of the tongue, the floor of the mouth, the lower lip and chin.

Efferent vessels from all these groups, except the superficial parotid nodes, pass to the deep cervical lymph chain. The superficial parotid group pass to the superficial cervical lymph nodes.

The deep cervical chain

The deep cervical lymph chain lies around the internal jugular vein. The upper part of the chain is extended medially towards the retropharyngeal group of nodes. The lower part of the chain is extended laterally above the clavicle. Other named nodes of the chain are the jugulodigastric, juglo-omohyoid, pretracheal and paratracheal:

(i) **jugulodigastric node**—behind the angle of the mandible above the posterior belly of digastric; drains the tonsil and lateral part of the tongue.

(ii) **jugulo-omohyoid node**—lies in the angle between the internal jugular vein and the superior belly of omohyoid; drains the tongue via the submental and submandibular nodes.

(iii) **para- and pretracheal nodes**—lie along the inferior thyroid vessels and drain the trachea and thyroid gland: efferent vessels may descend to the tracheo-bronchial nodes in the mediastinum.

The deep cervical lymph chain drains most of the circular chain of nodes and receives direct efferents from the salivary and thyroid glands, tongue, tonsil, nose, pharynx and larynx. Their efferents form the jugular lymph trunk which joins the thoracic duct on the left, and on the right it either joins the right lymph duct or opens separately into the internal jugular or subclavian veins. The clinician must be constantly aware that the first sign of cancer arising in the oropharynx, middle ear and buccal cavity is frequently a hard enlargement of the lymph nodes in the upper deep cervical group.

The superficial chains

The superficial lymph chains lie along the external and anterior jugular veins. The former drains the superficial parotid nodes and the side of the neck and joins the subclavian lymph trunk. The latter drains the superficial structures on the front of the neck and joins the external jugular chain or the lower deep cervical chain directly.

PART SEVEN

Central nervous system

— 31 —

Introduction

The central nervous system consists of the brain and spinal cord. The emerging cranial and spinal nerves pass through the surrounding meninges and cerebrospinal fluid and form the peripheral nervous system. The autonomic nervous system has both central and peripheral components and is concerned with the innervation of smooth muscle, cardiac muscle and glands.

The brain is situated within the cranium and comprises a forebrain (prosencephalon), a midbrain (mesencephalon) and a hindbrain (rhombencephalon). The last is continuous through the foramen magnum with the spinal cord. The forebrain is subdivided into a telencephalon consisting of the two cerebral hemispheres laterally, and a central part (the diencephalon) composed mainly of the two thalami and the hypothalamus. The hindbrain is subdivided into the pons, medulla and cerebellum.

The midbrain, pons and medulla are collectively known as the **brain stem**; they form a central axis connecting the diencephalon to the spinal cord and give attachment posteriorly to the cerebellum.

The nervous system is made up of innumerable nerve cells (**neurons**), each with its branches, the **dendrites**, and an **axon**. Most of the neuronal cell bodies in the central nervous system are found in the cortex of the cerebrum and cerebellum, and in deeply placed aggregations called **nuclei**. Some neurons found outside the brain and spinal cord are grouped into **ganglia** (dorsal root, cranial nerve and autonomic). Many of the axons are myelinated and when massed together in the central nervous system form the tracts of the white matter. Neuroglia and ependyma are described on page 14.

The neurotransmitter substances in the central nervous system include noradrenalin, dopamine, GABA and substance P; numerous other substances may also be involved.

ATTACHMENT OF CRANIAL NERVES (Fig. 31.1)

The olfactory nerve (1) in bundles enters the olfactory bulb, and the optic nerves (2) lead back to the optic chiasma. Both olfactory bulbs and tracts, and the optic nerves are part of the primitive forebrain. The oculomotor nerves (3) emerge from the interpeduncular fossa behind the mamillary bodies. The trochlear

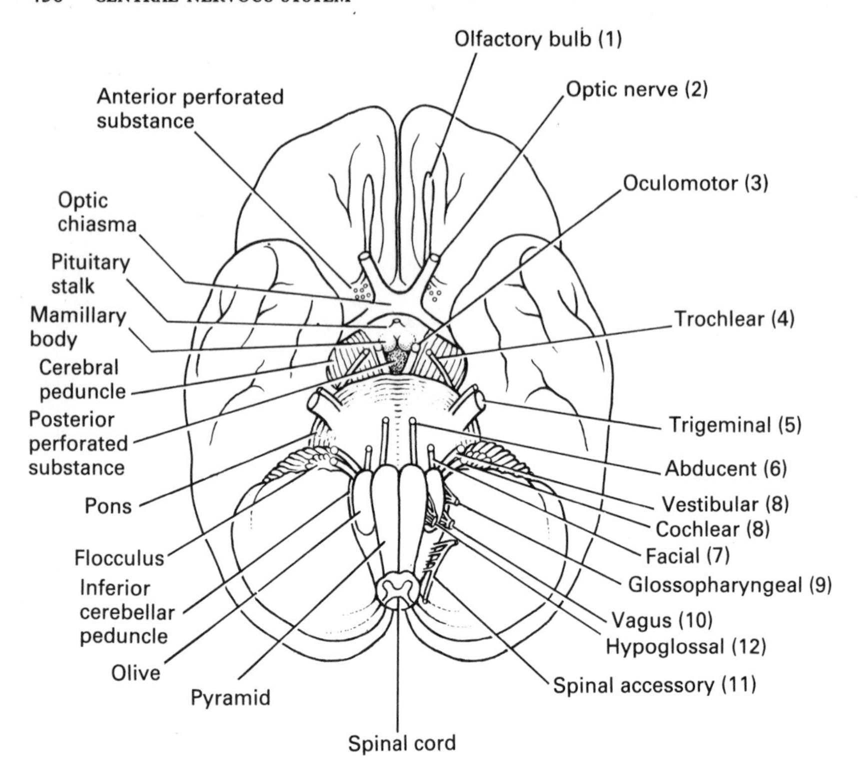

Fig. 31.1 Base of the brain showing the attachments of the cranial nerves. The trochlear nerve is seen winding round the cerebral peduncle from its attachment on the dorsal surface.

nerves (4) emerge dorsally below the caudal (inferior) colliculus and wind around the cerebral peduncles, lying between the temporal lobe and the peduncle. The trigeminal nerves (5) are attached to the lateral surface of the pons where motor and sensory roots may be distinguished. The abducent nerves (6) emerge between the pyramids and pons, the facial (7) and vestibulocochlear (8) nerves emerge in the cerebellopontomedullary angle. The glossopharyngeal (9), vagus (10) and cranial accessory (11) nerves emerge between the olives and the inferior cerebellar peduncles, the hypoglossal nerves (12) emerge between the olives and pyramids.

DEVELOPMENT OF THE CENTRAL NERVOUS SYSTEM

(Figs. 2.2, 31.2 and p. 40).

The nervous system develops from a dorsal longitudinal **neural plate** of ectoderm. The lateral margins of the plate become raised and its central part is

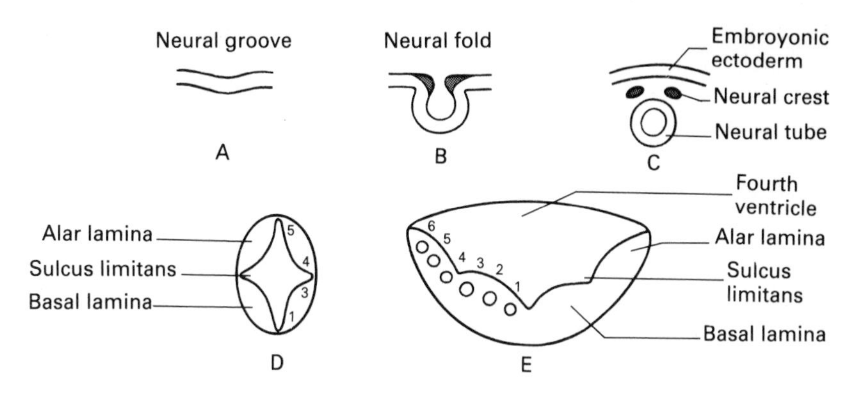

Fig. 31.2 The development of the central nervous system. **A. B.** and **C.** show the formation of the neural tube and the neural crest; **D.** shows the arrangement of the alar (sensory) and basal (motor) laminae in the spinal cord and the lower medulla; **E.** shows the re-arrangement of the alar and basal laminae in the floor of the 4th ventricle. (Nuclei: 1 = somatic efferent; 2 = special visceral efferent; 3 = general visceral efferent; 4 = general and special visceral afferent; 5 = general somatic afferent; 6 = special somatic afferent.)

depressed to form a **neural groove**. The groove deepens and the margins of the plate meet and fuse dorsally, forming the **neural tube**. The cephalic part of the tube is wider and forms the brain, the remainder becomes the spinal cord. Incomplete closure of the tube occurs occasionally in the lower part, giving rise to a condition known as spina bifida. This may be associated with the nonfusion of the neural arch of the related developing vertebrae and failure of development of the tissues over them. More rarely the tube fails to close in the cerebral region.

The junctional zone between neural ectoderm (forming the neural plate) and the embryonic ectoderm gives rise to the **neural crest tissue**. As the neural tube is separated from the embryonic ectoderm the neural crest tissue is also drawn in and forms a longitudinal plate of tissue on each side of the neural tube. From these longitudinal plates develop, amongst other things, the spinal, cranial and autonomic ganglia and the medulla of the suprarenal gland.

The ectodermal plate shows a marked increase in the number of its cells as the groove and tube are formed. Differentiation occurs into an **ependymal layer** lining the cavity of the tube, an intermediate **mantle layer** with numerous neuroblasts, and an outer **marginal layer** composed largely of fibres derived from the neuroblasts. The ependyma remains as the lining of the cavities of the central nervous system, the mantle layer becomes the future grey matter and the marginal layer the white matter.

Regional proliferation of neuroblasts causes changes in the size and shape of the tube. Two constrictions around the neural tube in the brain region divide it into three separate parts, the **primitive vesicles** of the forebrain, midbrain and hindbrain.

The tube becomes ventrally flexed at the junction of the brain and spinal cord (cervical flexure) and also in the midbrain region (mesencephalic flexure). Later a dorsal (pontine) flexure develops in the hindbrain region. The dorsal wall of the hindbrain vesicle is much stretched and forms the inferior part of the roof of the fourth ventricle. The **cerebellum** develops from the cranial (rostral) edge of the roof of the hindbrain vesicle, at first growing into the vesicle cavity but later becoming everted and undergoing external enlargement. Rapid multiplication of the neurons in the ventral wall of the hindbrain vesicle produces the pons.

In the spinal region the wall of the neural tube becomes thinned dorsally and ventrally, and an internal lateral longitudinal furrow, the **sulcus limitans**, divides each lateral wall into **basal** (ventral) and **alar** (dorsal) **laminae** (plates): From the basal laminae develop neurons subserving motor function and from the alar, sensory neurons.

The neurons of the spinal cord are arranged in four longitudinal columns which form the ventral (**anterior**) **horn of grey matter**. Behind this, also in the basal lamina and bordering the sulcus limitans, is the visceral motor column. The alar lamina also possesses visceral and somatic sensory columns, the latter forming the dorsal (**posterior**) **horn of grey matter**. The visceral sensory column together with the visceral motor column form the **lateral horn of grey matter**, found in the thoracic, lumbar and sacral regions. The nerve fibres form motor and sensory tracts but their dorsoventral arrangement is not so well-defined as that of the basal and alar lamina from whose neurons many are derived.

In the region of the pontine flexure of the hindbrain the dorsal part of the neural tube is thin, and the two alar laminae are more widely separated and lateral to the corresponding basal laminae. The columnar arrangement of the neurons remains although any one column may be absent or ill-defined in parts. In the region of the brain stem additional special visceral (branchial) efferent and afferent columns are formed, each lying between a somatic and a visceral column. The special visceral efferent column is motor to the muscles derived from the pharyngeal arch mesoderm and the afferent column is concerned with the sensation of taste (Fig. 31.2).

Development of the forebrain

The basal plate is absent rostral to the midbrain, and the forebrain is derived from a continuation of the alar plate. The ventral wall and floor of the forebrain vesicle become the diencephalon, the cavity becomes the third ventricle, its rostral wall becoming the thin lamina terminalis.

Two large outgrowths develop from the sides of the forebrain vesicle, each destined to become a cerebral hemisphere. Each has a cavity, the lateral ventricle which is continuous with the cavity of the primitive vesicle through an interventricular foramen. The lateral wall of the diverticulum undergoes considerable thickening and the lateral ventricle comes to lie near the midline. Masses of neurons in the wall of the diencephalon form the thalamus and hypothalamus, and in the hemisphere, the neurons on the floor of the lateral ventricle form the basal nuclei. The cleft between the thalamus medially and the hemisphere laterally is filled up. Many neurons migrate to the surface of the hemisphere and form the cortex (grey matter). The increasing surface is thrown into folds (convolutions, gyri) and grooves (sulci, fissures).

The medial surface of the developing hemisphere dorsal to the interventricular foramen becomes the hippocampus and that ventral to it the piriform cortex. The hippocampus and the piriform cortex are known as the allocortex (the neurons are usually covered by a thin layer of fibres). The rest of the cortex is called isocortex (the neurons being near the surface).

The hemisphere enlarges laterally and caudally and then ventrally, the lateral ventricle thus becoming elongated and C-shaped. The hippocampus curves around its convexity and the basal nuclei lie in its concavity. Fibres projecting from the isocortex to lower centres form part of a well-defined bundle, the internal capsule, which passes through the tissue in the concavity and splits the basal nuclei into a lateral part (lentiform nucleus) and a medial part (caudate nucleus), the latter lying adjacent to the thalamus. The internal capsule also contains fibres projecting from the thalamus to the cortex.

Commissures consist of fibres joining corresponding areas in the two sides of the brain. Commissural fibres between the temporal poles (piriform cortex and some neocortex) pass through the lamina terminalis forming the ventrally placed anterior commissure. Dorsally in the lamina the fornix commissure develops between the two hippocampal areas. The large neocortical commissure, the corpus callosum, develops in the fornix commissure and soon overgrows it, covering over the roof of the diencephalon. As the corpus callosum increases in size the dorsally placed hippocampus becomes much attenuated and remains only as a thin layer of cells on top of the corpus callosum. That part of the hippocampus carried backwards and then ventrally persists in the adult, and lies in the floor of the inferior horn of the lateral ventricle.

Association fibres join together different areas of the same hemisphere. Most are quite short, but not all.

The roof of the diencephalon remains thin ependyma and is invaginated by a network of blood vessels. With them it forms the choroid plexus of the third ventricle. The medial wall of the hemisphere, between the hippocampus and the interventricular foramen also remains thin ependyma and is invaginated by a similar network of blood vessels. With them it forms the choroid plexus of the lateral ventricle. As the hemisphere grows the line of invagination follows the C-shaped pattern of the lateral ventricle already described and forms the choroid fissure. The cells of the choroid plexus secrete the cerebrospinal fluid which fills the cavities of the brain and spinal cord, and bathes their outer surfaces.

Small forward diverticula from the diencephalon form the optic stalks, giving rise to the optic nerves and part of the eyeball. The pineal gland grows dorsally from the diencephalon, and the infundibular contribution to the pituitary gland grows downwards.

Sources of information on the structure and function of the central nervous system

Some of the functions of component parts of the central nervous system will be considered under their relevant sections. However, the evidence on which these conclusions are based is drawn from six main sources:

1. Pathological—studying the effects of disease or operative removal of any region.
2. Physiological—demonstrating resting electrical activity and any changes in this activity associated with changes elsewhere in the nervous system or in the body at large.
3. Anatomical—dissection, macroscopic observations, histological examination of stained material, histochemical and immunohistochemical techniques, and electron microscopy.
4. Morphological and embryological—comparison of homologous parts of the central nervous system developed to a variable extent in different animals may give pertinent information about their function.
5. Stimulation—the study of the effects of electrical stimulation of the central nervous system in the conscious patient has been carried out extensively.
6. Neuropharmacological and psychopharmacological—studying the effects of chemical substances on the patterns of behaviour.

— 32 —

The forebrain

THE CEREBRAL HEMISPHERES (Figs. 32.1, 32.2, 32.3)

The cerebral hemispheres lie on each side of the midsagittal plane separated by a fold of dura mater called the falx cerebri. They are covered superiorly by the bones of the skull vault and lie inferiorly on the bones of the anterior and middle cranial fossae and on the upper surface of the tentorium cerebelli. The hemispheres are joined together by a broad band of fibres, the **corpus callosum**. The corpus callosum overlies the diencephalon and the dorsum of the midbrain and is separated from them by a horizontal cleft, the transverse fissure. The basal nuclei are deeper structures of the hemispheres lying near its medial side, being closely applied to the C-shaped lateral ventricle, but lateral to the diencephalon.

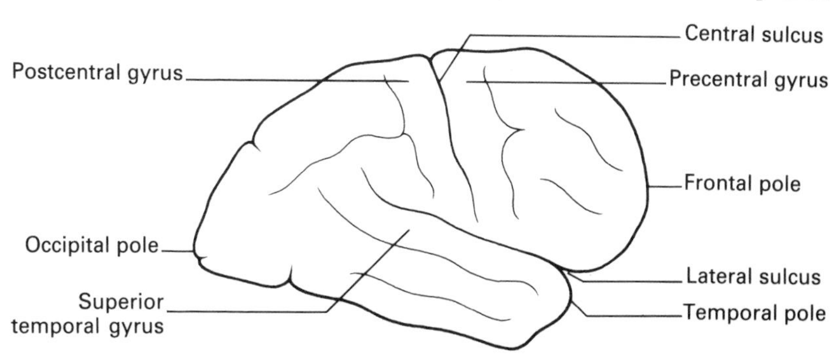

Fig. 32.1 The right cerebral hemisphere, lateral aspect.

The **hemispheres** have an outer layer of grey matter called the **cerebral cortex** which varies in thickness from 1–4 mm at different sites. The cortex is greatly convoluted: grooves, the **sulci**, separate ridges of brain tissue, the **gyri**. About two-thirds of the total surface area of the cerebral cortex is hidden from view in the sulci.

Each hemisphere has frontal, occipital and temporal poles; lateral, medial and inferior surfaces; and frontal, parietal, occipital and temporal lobes. The lobes are named after the closely related bones of the skull. The frontal and occipital poles are the rostral and caudal extremities of the hemisphere. The inferiorly-placed temporal pole lies in the middle cranial fossa under the frontal lobe.

On the convex **lateral surface** the deep **lateral sulcus** extends upwards and backwards and divides the frontal from the temporal lobes. The margins of the sulcus overlap the deeply placed **insula**, an area of cortex which has become submerged during the enlargement of the hemispheres. The **claustrum** is a thin sheet of grey matter lying deep to the insula, and closely linked to the insular cortex (see Fig. 32.4). The **central sulcus** passes downwards and forwards from the superior border towards the lateral sulcus and separates **precentral** and **postcentral gyri**, concerned with motor and somatosensory functions respectively. These gyri are limited by pre- and postcentral sulci. Sulci above and parallel to the lateral sulcus divide the surface into superior, middle and inferior frontal gyri. Sulci below and parallel to the lateral sulcus divide the surface into the superior, middle and inferior temporal gyri. A part of the superior temporal gyrus is concerned with auditory function.

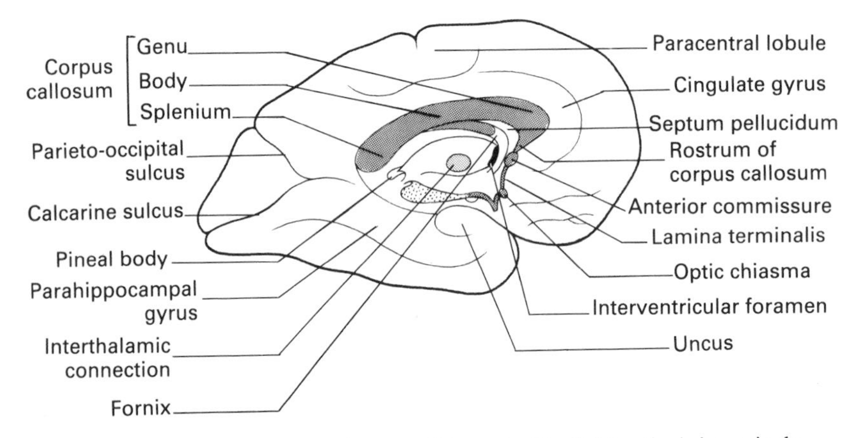

Fig. 32.2 The left cerebral hemisphere, medial aspect. The lightly stippled area is the cut brain stem.

A prominent feature of the flat **medial surface** is the cut surface of the corpus callosum. The **cingulate gyrus** and sulcus lie above and parallel to the corpus callosum. The central sulcus extends for a short distance on to this surface and the cortex around it is known as the **paracentral lobule**; it also is concerned with motor and somatosensory functions. The deep **calcarine sulcus** extends horizontally backwards from the splenium of the corpus callosum to the occipital pole. The **parieto-occipital sulcus** passes obliquely upwards from about its middle to the superior margin of the surface. The cortex overlying both margins of the posterior half of the calcarine sulcus, and that along the inferior margin of its anterior half, is concerned with visual functions.

The **inferior surface** is separated from the midbrain by a deep **hippocampal sulcus**. The **parahippocampal gyrus** borders the sulcus and is separated laterally by the longitudinal **collateral sulcus** from the rest of the cortex. Anteriorly the parahippocampal gyrus is continued into the **uncus**, which is a short medially projecting recurved part. The uncus is bounded laterally by the **rhinal sulcus**, which is usually continuous posteriorly with the collateral sulcus. The olfactory bulb and tract lie on the undersurface of the frontal lobe, and the sulcus in which they lie separates the gyrus rectus medially from the short irregular gyri laterally. The olfactory tract divides posteriorly into medial and lateral olfactory striae, and the area of cortex between the striae gives passage to a number of vessels entering the brain substance. It is known as the **anterior perforated substance**.

Grey matter of the hemispheres

Microscopic examination of the cerebral isocortex shows that the neurons are not distributed evenly nor are they all the same size and shape. Two populations of neurons can be identified.

1. Pyramidal neurons have branching apical and basal dendrites of varying sizes and an axon which arises at the axon hillock on the base of the pyramid. The axon usually enters the white matter and divides nearer to its termination. The apical dendrite often forms a fine plexus near the pial surface of the cortex.
2. Nonpyramidal (granular, stellate) neurons are usually smaller, irregularly multipolar. Their branching axons usually end within the cortex.

Most areas of the isocortex have their neurons arranged in six layers. Layer 1 (molecular) is subpial and is composed of a few nonpyramidal neurons and the apical dendrites of deeper pyramidal cells. Layers II and IV are the external and internal granular layers; they have numerous nonpyramidal cells. Layers III and V are the external and internal pyramidal layers. The very large pyramidal (Betz) cells of the motor areas are found in layer V. The innermost (fusiform) layer VI is composed of nonpyramidal neurons and is adjacent to the white matter. The pattern of the entering fibres also varies from area to area. Two tangentially placed plexuses are present, the outer in layer IV and the inner in layer V. The outer layer is very prominent in the visual cortex around the calcarine sulcus and the region is called the **striate area**.

In the sensory areas the nonpyramidal cells are most numerous. The motor areas have a preponderance of the pyramidal cells and are sometimes referred to as the agranular cortex. Using histological criteria it is possible to identify the **sensori-motor cortex** around the central sulcus, the **visual cortex** around the calcarine sulcus and the **auditory cortex** on the superior temporal gyrus. Other cortical areas are less well defined.

White matter of the hemispheres

The nerve fibres of the hemispheres are of three groups: commissural, association, and projection.

Commissural fibres unite corresponding areas of the cortex of the two hemispheres across the midline. They comprise the corpus callosum, and the anterior and fornix (hippocampal) commissures. The habenular and posterior commissures belong to the diencephalon.

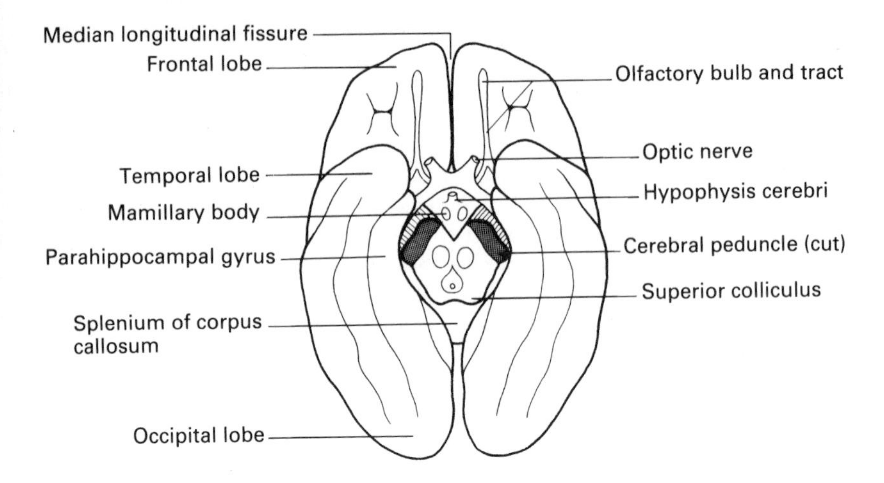

Fig. 32.3 The inferior aspect of the brain after removal of the hindbrain. The anterior perforated substance lies posterior to the division of the olfactory tract and the posterior perforated substance is just behind the mamillary bodies.

The **corpus callosum** is a broad band of fibres passing between corresponding cortical areas of the two hemispheres. It lies at the base of the median longitudinal fissure and above the diencephalon and midbrain. In midsagittal section it is the shape of a hook lying horizontally with its bend anteriorly and its point downwards. The pointed portion is known as the rostrum, the bend as the genu, the horizontal part as the trunk (body) and the expanded posterior end as the splenium. The callosum extends laterally into each hemisphere; the anterior fibres pass forwards into the frontal pole and are known as the forceps minor. A thick posterior bundle of fibres, known as the forceps major, passes backwards into the occipital poles. The rostrum of the corpus callosum fuses inferiorly with the lamina terminalis. A bundle of fibres within this lamina, the **anterior commissure**, unites the piriform areas (p. 449) and the olfactory tracts of the two sides. The **fornix (hippocampal) commissure** is found on the undersurface of the corpus callosum where the two crura meet and form the fornix. Fibres here pass across the midline between the two hippocampi.

Association fibres may be long or short, uniting adjacent or widely separated gyri of the same hemisphere.

Projection fibres ascend from or descend to lower parts of the central nervous system. Many form a well-defined layer, the **internal capsule**, between the lentiform nucleus laterally and the thalamus and caudate nucleus medially. Superiorly the fibres of the internal capsule fan out as the corona radiata, interdigitating with the fibres of the corpus callosum. In horizontal section the internal capsule is V-shaped. It possesses an anterior limb (between the caudate nucleus and the lentiform nucleus and crossed by fibres and grey matter uniting the two structures), an apex (the genu) pointing medially, and a posterior limb lying between the thalamus and the lentiform nucleus (Fig. 32.4).

The anterior limb carries (*a*) frontopontine fibres from the frontal lobe to the pons, and (*b*) fibres from the thalamus (medial and ventro-anterior nuclei) to the frontal lobe. The posterior limb carries from before backwards, (*a*) pyramidal fibres from the motor cortex which pass to the cranial nerve nuclei (cortico-nuclear fibres situated at the genu) and to the spinal nuclei (corticospinal fibres), (*b*) somatosensory fibres passing from the thalamus (ventroposterior nucleus) to the postcentral (somatosensory) cortex, (*c*) temporopontine fibres from the temporal lobe to the pons, (*d*) the auditory radiations passing from the medial geniculate body under the lentiform nucleus, to the superior temporal gyrus, (*e*) the visual radiations passing from the lateral geniculate body around the lateral aspect of the posterior horn of the lateral ventricle to the visual cortex.

The course of the fibres is such that many cross the midline (decussate) and end on the opposite (contralateral) side. Some fibres however end on the same (ipsilateral) side. The motor areas of each hemisphere control the voluntary muscles of the contralateral side of the body and the sensory areas receive information from the contralateral side.

THE BASAL NUCLEI (GANGLIA) (Figs. 32.4, 32.5)

These nuclear masses lie within the substance of the cerebral hemispheres. The caudate and lentiform nuclei are connected across the anterior limb of the internal capsule and, because of the striated appearance so produced, are collectively known as the **corpus striatum**. They form part of the extrapyramidal system and are involved in the regulation of movement. Diseases of the basal nuclei may produce rigidity and abnormal movements (spontaneous tremors) which affect muscle activity, particularly that involved in posture, manual tasks and speech. Defects in the production of certain transmitter substances have been described and symptoms can often be alleviated by supplying such substances as L-dopa to the patient.

The **caudate nucleus** is C-shaped and lies within the concavity of the lateral ventricle. It has an expanded head, a narrow body and a tail. The globular head lies above the anterior perforated substance in front of the thalamus, medial to the anterior limb of the internal capsule, and bulges medially into the anterior horn of the lateral ventricle. The body passes backwards round the thalamus in the floor of the lateral ventricle. The narrow tail curls downwards and then forwards within the concavity of the ventricle and comes to lie in the roof of its inferior horn. The thalamostriate vein runs in a groove between the thalamus and the caudate nucleus and, near the interventricular foramen, it joins the choroidal vein forming the internal cerebral vein. The caudate nucleus receives fibres from the cortex and sends efferent fibres mainly to the globus pallidus part of the lentiform nucleus.

The **lentiform nucleus** is roughly lens-shaped with its less convex surface facing laterally and its more convex surface applied to the genu of the internal capsule. Anteromedially it is united across the anterior limb of the internal capsule to the head of the caudate nucleus and posteromedially separated by the

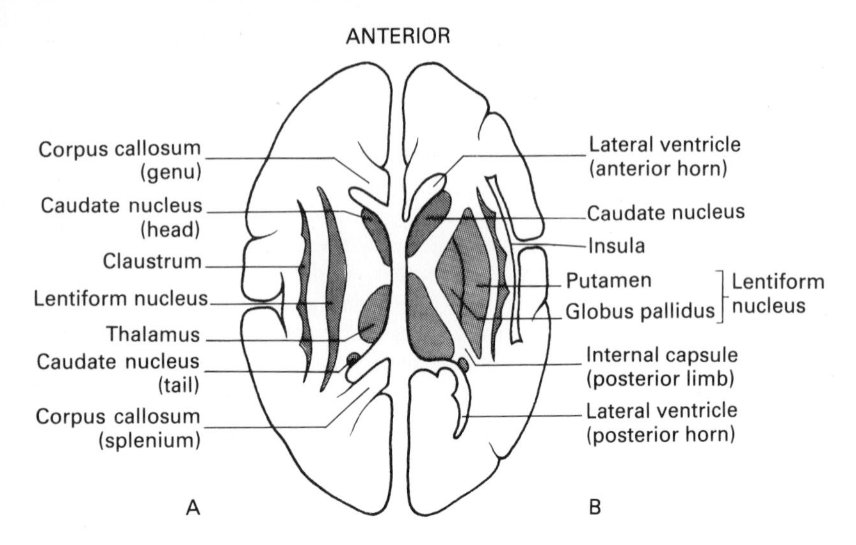

Fig. 32.4 Horizontal sections through the forebrain: **A.** at the level of the genu and splenium of the corpus callosum and **B.** through the interventricular foramen.

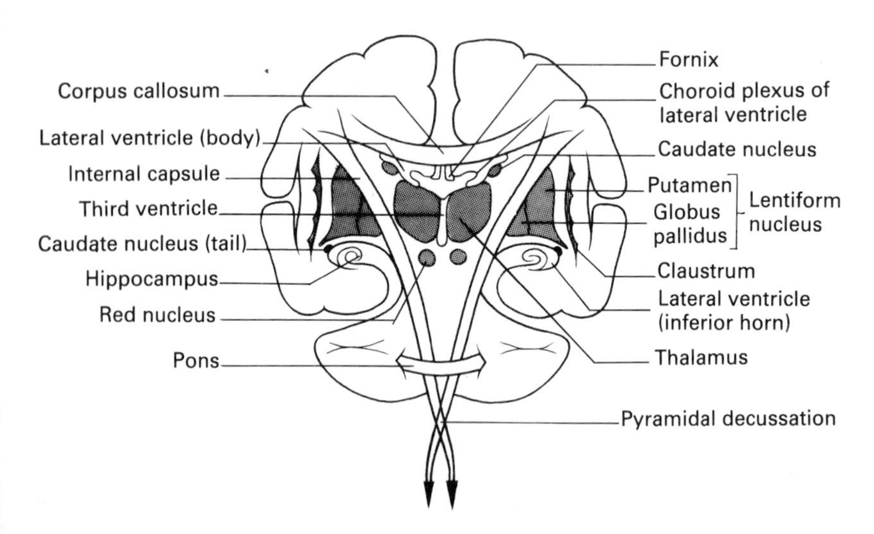

Fig. 32.5 Diagrammatic layout of the main relations of the pyramidal tracts (oblique coronal section).

posterior limb of the capsule from the thalamus. Superomedially it is covered by the internal capsule. Inferiorly, the anterior commissure and white matter of the temporal lobe separate it from the inferior horn of the lateral ventricle. The lentiform nucleus is divided into an outer putamen and an inner globus pallidus.

The **putamen** receives fibres from the caudate nucleus and the cortex; its efferents pass to the globus pallidus. The **globus pallidus** receives fibres from the caudate nucleus, the putamen and the substantia nigra and its efferent fibres pass to the thalamus (ventrolateral and ventroanterior nuclei), hypothalamus, subthalamus, substantia nigra and reticular formation.

THE LIMBIC SYSTEM

This system of interconnected centres surrounds the regions where the hemispheres join the diencephalon. Some of the centres effect emotional and visceral aspects of behaviour and have also been implicated in memory processes. The system includes structures with close olfactory connections (paraterminal gyrus and septal nuclei, the piriform cortex and the amygdaloid body), and others (the complex of hippocampus, fornix, hypothalamus, cingulate gyrus and parahippocampal gyrus). Lesions of the hippocampus affect short term memory, especially spatial memory, and lesions of the hypothalamus may produce visceral effects (see below). Efferent fibres from many of these limbic centres pass to the reticular formation of the brain stem and so influence lower centres.

The **olfactory bulb** lies on the undersurface of the frontal lobe and lies over the cribriform plate of the ethmoid bone. It receives the olfactory nerves after they have pierced the plate and a number of relays occur in the bulb. The olfactory tract passes backwards from the bulb along the surface of the brain and divides into medial and lateral olfactory striae in front of the anterior perforated substance. The medial stria has two components. One passes through the anterior commissure to the olfactory bulb of the opposite side and the other passes to the cortex of the medial part of the hemisphere just anterior to the lamina terminalis. This area is known as the **paraterminal gyrus** and it contains the **septal nuclei**. The lateral stria passes to the **prepiriform cortex**, i.e. the frontal lip and the floor of the lateral sulcus in front of the uncus. It is probable that direct olfactory impulses extend no further than this area. The prepiriform cortex is continuous posteriorly, over the anterior part of the insula, with the uncus and parahippocampal gyrus. The cortex of the anterior insula, the uncus and the anterior part of the parahippocampal gyrus is known as the **piriform cortex**. The amygdaloid body lies deep to the uncus and receives a small contribution from the lateral olfactory stria.

The **amygdaloid body** is an oval nuclear mass situated at the end of the tail of the caudate nucleus and overlying the anterior extremity of the inferior horn of the lateral ventricle. It is in contact with and receives fibres from the uncus and lateral olfactory stria and communicates with its fellow of the opposite side through the anterior commissure. Most of its efferent fibres pass in the stria terminalis, a narrow bundle lying alongside the thalamostriate vein, to the hypothalamus and septal nuclei.

The hippocampus is a nuclear mass, covered by a thin layer of white matter, the alveus, lying mainly in the floor of the temporal (inferior) horn of the lateral ventricle. (It should not be confused with the parahippocampal gyrus, p. 444.)

Its grooved anterior part is wide and narrows posteriorly. Axons from each hippocampus lie on its ventricular surface, the alveus, and then form the fimbria and the crus of the **fornix**. The two crura meet on the undersurface of the corpus callosum where some fibres decussate and form the commissure of the fornix. Most fibres in the fornix diverge anteriorly and pass forwards round the interventricular foramina and then downwards into the hypothalamus and the mamillary bodies. The fornix is the main efferent system of the hippocampus.

A midline triangular partition, the **septum pellucidum**, passes between the columns of the fornix and the undersurface of the corpus callosum. It is formed of two laminae enclosing a narrow cavity. The septum separates the anterior horns of the two lateral ventricles.

THE DIENCEPHALON

This medial part of the forebrain comprises the structures surrounding the third ventricle, namely, the two thalami, the hypothalamus, epithalamus, and subthalamus. The ventricle is limited anteriorly by the thin **lamina terminalis**, the rostral limit of the primitive neural cavity.

THE THALAMUS (Figs. 32.4, 32.5)

The two thalami are nuclear masses situated on the sides of the superior part of the third ventricle. Each thalamus is egg-shaped, having a larger posterior end. Its long axis is almost at right angles to that of the brain stem. The narrow anterior end reaches to the interventricular foramen. The posterior end projects backwards and laterally over the midbrain as the **pulvinar**. There are two small swellings on the inferior surface of the pulvinar, the **medial** and **lateral geniculate bodies**.

The arched upper surface is related to the stria medullaris (the line of attachment of the roof of the third ventricle) medially, part of the floor of the transverse fissure, the thalamostriate vein and the stria terminalis, the floor of the lateral ventricle, and the body of the caudate nucleus laterally. A prominent anterior tubercle is found near the interventricular foramen and a dorsal groove passes backwards from the tubercle.

The inferior surface lies on the hypothalamus, subthalamus and midbrain from before backwards. The medial surfaces of the two thalami are parallel and border the third ventricle, usually being united across it by a glial mass, the connexus interthalamicus. The lateral surface is applied to the posterior limb of the internal capsule which separates it from the lentiform nucleus.

THALAMIC NUCLEI

A band of white matter, the **internal medullary lamina**, passes downwards and medially from the groove on the superior surface through the anterior two-thirds of each thalamus. It splits anterosuperiorly to enclose the anterior thalamic nucleus which forms the anterior tubercle. Medial to the lamina are the medial (dorsomedial) nucleus, the central (centromedial) nucleus and, near the wall of the ventricle, the midline nuclei. Lateral and inferior to the lamina are the lateral and ventral nuclear masses. The lateral and medial geniculate bodies are thalamic nuclei which have been cut off from the back of the ventral nucleus but may be grouped together with it as part of a ventral nuclear complex. A thin nuclear layer over the lateral surface of the thalamus is known as the reticular nucleus. Amongst the fibres of the internal lamina are nuclei which together are called the intralaminar nuclei. The central nucleus is partly surrounded by these laminar fibres.

In summary, there are four large nuclear masses, the ventral nuclear complex, the lateral, anterior and medial nuclei, and three nuclear layers, the thalamic reticular nucleus laterally, the midline nuclei medially and the intralaminar group (including the central nucleus) centrally. Many of these nuclei are interconnected.

1. The **ventral nuclear complex**
   (*a*) The **ventral nucleus**—includes ventroposterior, ventrolateral and ventroanterior nuclei. The ventroposterior nucleus receives the ascending sensory systems in the medial, spinal and trigeminal lemnisci, and many of its efferents pass to the somatosensory cortex. The ventrolateral nucleus receives fibres from the superior cerebellar peduncle and the globus pallidus; its efferents pass to the motor cortex. The ventro-anterior nucleus also receives fibres from the globus pallidus and the ascending reticular formation but other connections are ill-defined. Efferent fibres pass to the motor and premotor cortex.
   (*b*) The **medial geniculate body**—is a relay station in the auditory pathway, receiving fibres from the lateral lemniscus. Its efferents pass as the auditory radiation to the superior temporal gyrus.
   (*c*) The **lateral geniculate body**—receives fibres from the optic tract, and its efferents pass as the visual radiation to the visual cortex. The central part of the nucleus is laminated and connected with the foveal (macular) area of the retina.
2. The **lateral nuclear complex**—includes the pulvinar, and its afferent fibres are from other thalamic nuclei especially the lateral geniculate body and other ventral nuclei. Its efferent fibres are distributed to the association areas of the cortex in the occipital, parietal and temporal lobes.
3. The **anterior nucleus**—receives afferent fibres from the mamillary body in the mamillothalamic tract, and its efferents pass to the cingulate gyrus.
4. The **medial (dorsomedial) nucleus**—probably receives afferent fibres from the hypothalamus, globus pallidus and some thalamic nuclei. Its efferents pass to the frontal cortex.
5. The **thalamic reticular nucleus**—receives afferents from other thalamic nuclei. The efferents pass in a diffuse manner to all parts of the cortex.

6. The **midline nuclei**—may receive fibres from the hypothalamus.
7. The **intralaminar nuclei** (including the central nucleus)—these receive afferent fibres from the ascending reticular formation and other thalamic nuclei. Many of the ascending fibres are concerned with pain sensations. The efferents pass to the corpus striatum and thence to the cortex and some thalamic nuclei.

From its numerous connections, the thalamus is seen to be a large relay station for all sensory tracts, and some forms of crude sensation may here reach conscious level. It also takes part in extrapyramidal pathways involving the cerebellum and basal nuclei as well as the cerebral cortex.

THE HYPOTHALAMUS

This forms the floor and inferior part of the lateral wall of the third ventricle. It is separated on the ventricular wall from the thalamus by the **hypothalamic sulcus** and is continuous posteriorly with the subthalamus. Anteriorly is the optic chiasma and behind this an elevation, the tuber cinereum, gives rise to the stalk, the infundibulum, of the hypophysis cerebri. Two small elevations, the mamillary bodies, lie behind the tuber cinereum and separate it from the posterior perforated substance and the cerebral peduncles.

The anterior columns of the fornix fan out in the hypothalamus dividing each side into medial and lateral parts. The medial parts have a number of named nuclear groups: the more obvious being the pre-optic, supra-optic and paraventricular nuclei above and behind the optic chiasma. Efferent fibres from these nuclei pass to the posterior lobe of the hypophysis cerebri and are involved in hormonal secretions.

Afferent fibres pass to the hypothalamus from the orbital surface of the frontal cortex, the septal nuclei, the hippocampus (through the fornix) and the amygdaloid body (through the stria terminalis). Efferent fibres as well as passing in the supra-opticohypophyseal tract to the posterior lobe of the pituitary gland, also pass in the median forebrain bundle, and dorsal longitudinal fasciculus to the midbrain reticular nuclei. Fibres from the mamillary bodies pass to the anterior nucleus of each thalamus (mamillothalamic tracts) and the midbrain (mamillotegmental tracts). There is also a portal system of veins between some hypothalamic nuclei and the pituitary gland.

The hypothalamus plays an important rôle in the regulation of the autonomic nervous system, the anterior part controlling mainly the parasympathetic system, and the posterior region the sympathetic system. It also influences most endocrine glands and, through the supra-optic and paraventricular nuclei, the rate of water secretion by the kidneys.

THE HYPOPHYSIS CEREBRI

The hypophysis cerebri (pituitary gland) is an endocrine gland situated in the hypophyseal fossa of the body of the sphenoid bone between the two cavernous sinuses. It is somewhat spherical in shape and is connected to the tuber cinereum by the infundibulum. The infundibulum evaginates a sheet of dura, the diaphragma sellae, which covers the organ. The hypophysis receives its blood supply from two main sources: (1) the internal carotid artery, (2) vessels in the tuber cinereum. A portal circulation exists between the hypothalamus and the hypophysis. Venous blood eventually drains to the cavernous sinus.

The gland possesses anterior, middle and posterior lobes. The anterior has acidophil, basophil, and chromophobe cells. Their secretions influence the thyroid and adrenal glands, the gonads, and also include growth and lactogenic factors. Secretions from the intermediate (middle) portion of the gland influence 'pigment reactions'. The posterior lobe consists of modified neurons and neuroglial cells. It secretes hormones which have passed down the axons of cells situated in the supra-optic and paraventricular nuclei of the hypothalamus. The posterior lobe hormones raise blood pressure, decrease urine production and contract the uterus.

The hypophysis cerebri develops by the fusion of a diverticulum from the diencephalon forming the posterior lobe, and a diverticulum from the stomadeum (Rathke's pouch) forming the anterior and middle lobes. Tumours of the hypophysis cerebri may produce pressure effects on the adjacent optic chiasma; they usually also secrete excessive amounts of one or more circulating hormones.

The **epithalamus** is situated in the posterior wall of the third ventricle between the posterior ends of the thalami. It comprises the habenular nuclei and commissure, the posterior commissure and the pineal gland. The **habenular nuclei** receive the stria medullaris, a narrow bundle of fibres passing backwards over the thalamus from the septal nuclei. Its efferents pass through the habenular commissure to the interpeduncular nucleus of the subthalamus. The **posterior commissure** unites the two superior colliculi, the pretectal regions and the medial longitudinal bundles of the two sides. Neurons in the pretectal region are involved in the light reflex (see p. 481). The **pineal gland** develops as an outgrowth from the posterior wall of the third ventricle. It is an oval mass lying between the superior colliculi. Its functions are probably concerned with the secretion of serotonin and other substances which influence cyclical activities. The pineal body is frequently calcified and is a useful radiological landmark. Its displacement may indicate an intracranial space-occupying lesion.

The **subthalamus** adjoins the midbrain, lying behind the hypothalamus and above the posterior perforated substance between the cerebral peduncles. It contains the subthalamic and interpeduncular nuclei which receive fibres from the globus pallidus and the habenular nuclei. Its efferent fibres pass to the reticular formation of the brain stem.

THE LATERAL VENTRICLE (Figs. 32.4, 32.5, 32.6, 32.7)

This is the cavity of the cerebral hemisphere and lies towards its medial side and below the corpus callosum. Each ventricle is united to the cavity of the dien-

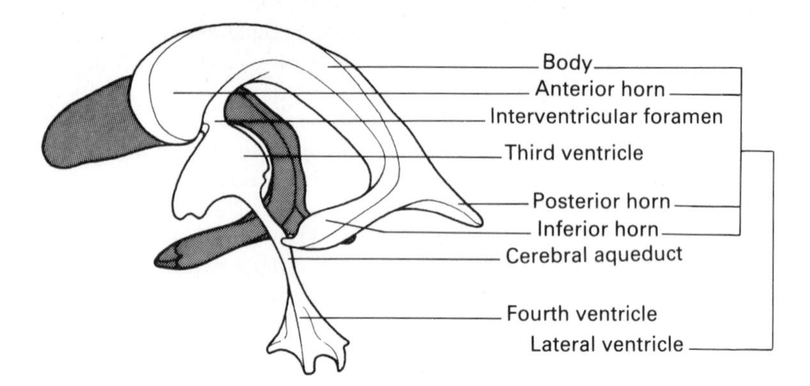

Fig. 32.6 Diagram of a model of the ventricular system of the brain. The right lateral ventricle is stippled.

cephalon (third ventricle) through an **interventricular foramen**. Each lateral ventricle is basically C-shaped with an anterior concavity. For descriptive purposes it is divided into an anterior horn, projecting into the frontal lobe, a body arching backwards, and an inferior horn projecting into the temporal lobe. From the posterior convexity a posterior horn, of variable form, extends backwards into the occipital lobe.

The **anterior horn** is that part of the ventricle in front of the interventricular foramen. It is separated from its fellow of the opposite side by the septum pellucidum. Its roof and anterior wall are formed by the corpus callosum, and inferolaterally the head of the caudate nucleus projects into the cavity.

The **body** is roofed by the corpus callosum, and its floor, which meets the roof laterally, is formed by the body of the caudate nucleus laterally and the thalamus medially, with the stria terminalis and the thalamostriate vein in the

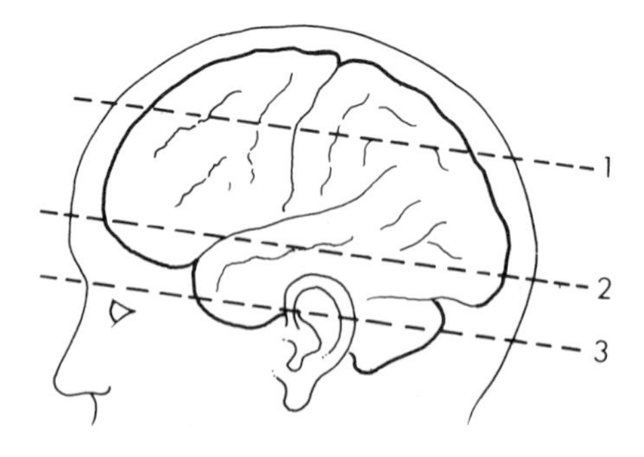

Fig. 32.7 Diagram of surface marking of the brain on the lateral aspect of the head. The 3 lines indicate the plane of the subsequent 3 CT scans. The line of reference for the 3 head scans is one joining the outer angle of the eye and the external acoustic meatus.

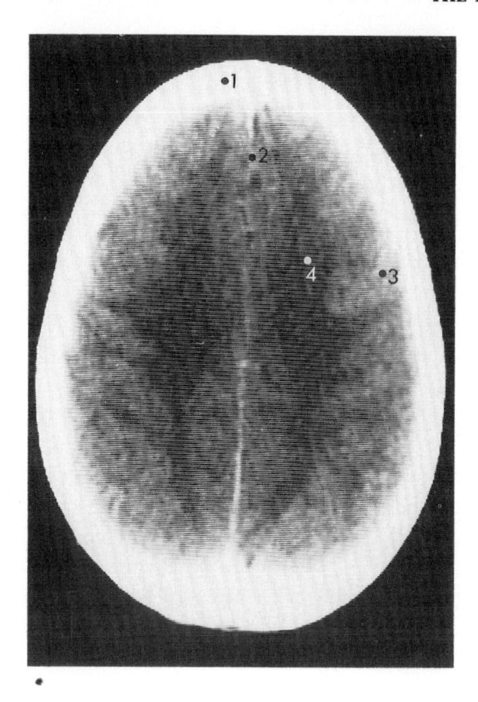

Fig. 32.8 CT scan of the head through level 1 of Figure 32.7. The orientation of most head scans is as if the observer was looking down from above; the right hemisphere being on the right side of the image. (See also the caption to Figure 6.2, and compare with the orientation of CT scans through the trunk.)
1. Frontal bone
2. Midsagittal plane
3. Cortical grey matter
4. White matter

groove between. Medially are the septum pellucidum and the fornix, below which the choroid plexus is invaginated into the cavity.

The **posterior horn** is covered superiorly and laterally by the radiation of the corpus callosum which separates the ventricle from the optic radiation. Inferomedially there are two longitudinal ridges in the cavity, an upper formed by the fibres of the corpus callosum, and a lower one by the infolding of the cortex of the calcarine sulcus.

The floor of the **inferior horn** is formed by the hippocampus, and its lateral wall by the fibres of the corpus callosum. Its roof is formed posteriorly by the white matter of the temporal lobe and the internal capsule, the tail of the caudate nucleus, the stria terminalis, and anteriorly by the amygdaloid body. The medial wall is invaginated by the choroid plexus.

The **choroid plexus** of the lateral ventricle is invaginated into the cavity between the interventricular foramen above and the tip of the inferior horn below. The arched line of invagination is known as the **choroidal fissure**. The medial ependymal wall of the ventricle is attached to the edge of the fornix and

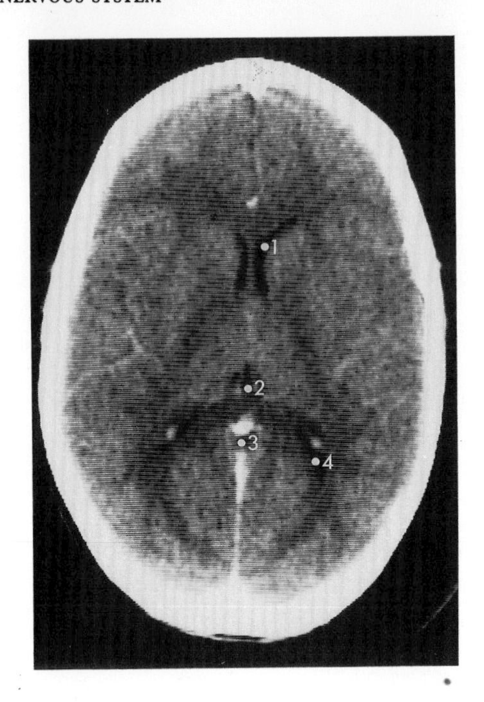

Fig. 32.9 CT scan of the head through level 2 of Figure 32.7. (See also caption to Figure 32.8).

1. Anterior horn of right lateral ventricle
2. Third ventricle
3. Cerebral aqueduct
4. Posterior horn of right lateral ventricle.

the fimbria on the one hand, and to the region of the stria terminalis and the thalamostriate vein on the other.

THE THIRD VENTRICLE (Figs. 32.2, 32.5, 32.6, 32.7)

This is the cavity of the diencephalon. It communicates anterosuperiorly with the lateral ventricles through the interventricular foramina and with the fourth ventricle through the cerebral aqueduct. It forms a midline vertical slit between the two thalami superiorly and has the hypothalamus inferiorly. The cavity is crossed by the connexus interthalamicus (when this is present).

It is bounded anteriorly by the lamina terminalis with the anterior commissure. Its floor is formed, from before backwards, by the optic chiasma, tuber cinereum, mamillary bodies and posterior perforated substance. The cavity is recessed into the optic chiasma and the infundibulum. The short posterior wall is formed by the pineal body with the habenular and posterior commissures. The cavity is recessed above and within the pineal stalk. Postero-inferiorly the

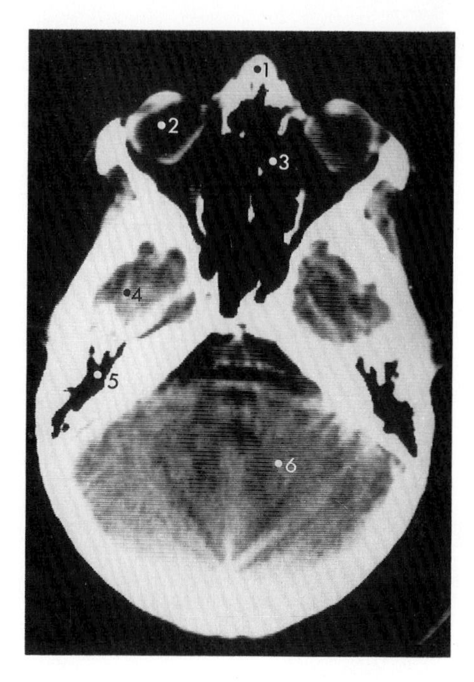

Fig. 32.10 CT scan of the head through level 3 of Figure 32.7. (See also caption to Figure 32.8).
1. Nose
2. Left orbit
3. Right nasal cavity
4. Left temporal lobe of the brain
5. Inner ear
6. Cerebellum

ventricle communicates with the aqueduct, and superiorly the ependymal roof is invaginated and forms the choroid plexus of the third ventricle.

It is possible to replace the cerebrospinal fluid in the ventricles with air so that their shape can be investigated radiologically. Air introduced through a lumbar puncture needle will quickly find its way into the ventricular system. By suitable positioning of the patient's head, the ventricles can be demonstrated. The position of a space-occupying lesion may be identified by the displacement of the ventricles as demonstrated by these radiological techniques.

— 33 —

The brain stem and cerebellum

BRAIN STEM

The brain stem (Fig. 33.1), formed of the midbrain, pons and medulla from rostral to caudal, is continuous rostrally with the diencephalon and caudally with the spinal cord. Dorsally the cerebellum lies in the posterior cranial fossa.

The brain stem contains the nuclei and central connections of the third to the twelfth cranial nerves, and nuclei associated with the visual and auditory systems and with the extrapyramidal system. It gives passage to ascending and descending fibres between the forebrain and spinal cord, and afferent and efferent cerebellar fibres. The region also contains a large number of scattered

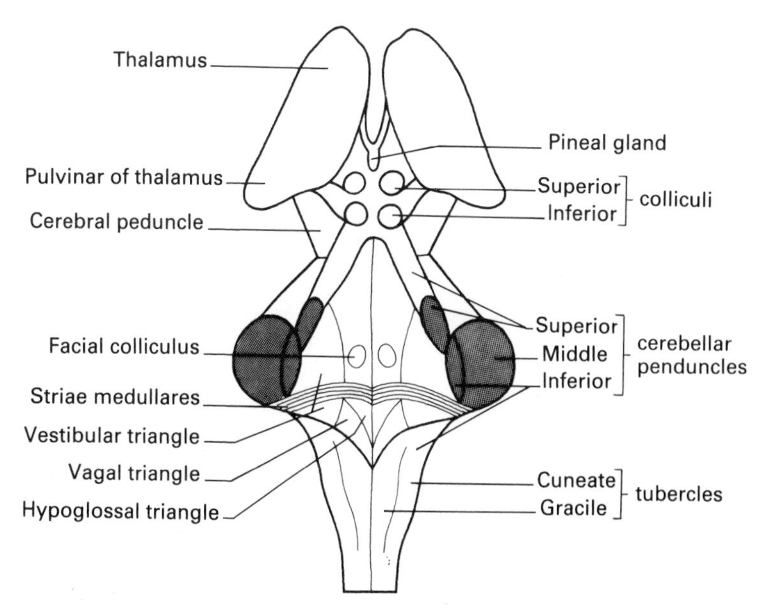

Fig. 33.1 Diagram showing some of the structures on the dorsal surface of the brain stem. The cerebellum has been removed, thus exposing the floor of the 4th ventricle.

nuclei situated between the previous structures and collectively known as the reticular formation (p. 476).

THE MIDBRAIN (Figs. 33.2, 33.3)

This is a short narrow region which joins the forebrain to the hindbrain through the opening in the tentorium cerebelli. It is largely overhung by the cerebral hemispheres. Its dorsal surface possesses four small elevations, the paired superior and inferior colliculi. The superior colliculi lie between the geniculate bodies of the two thalami. On each side a ridge, the brachium of the inferior colliculus, passes from the inferior colliculus to the medial geniculate body. Similar but less well defined ridges join the lateral geniculate bodies to the superior colliculi. The ventral surface of the midbrain has a pair of converging fibre bundles, the cerebral peduncles. Between them the surface is pierced by a number of blood vessels and is known as the **posterior perforated substance**. The cavity of the midbrain is the narrow **aqueduct** which unites the third and fourth ventricles. It is surrounded by the peri-aqueductal grey matter. The part of the midbrain dorsal to the aqueduct contains the colliculi, and is known as the **tectum**. The part ventral to the aqueduct, but excluding the cerebral peduncles, is known as the **tegmentum**. The peduncles form the **base**.

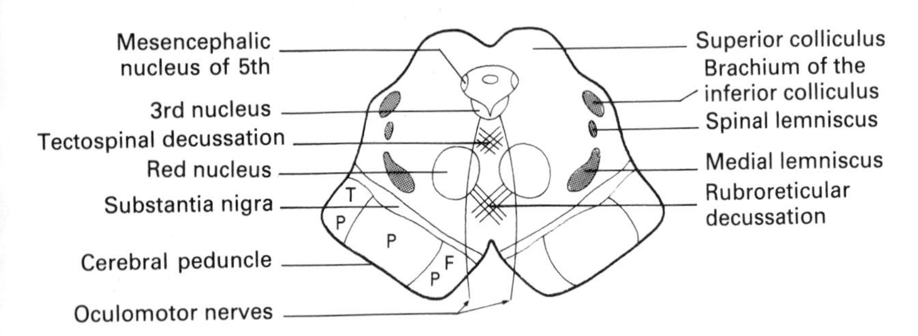

Fig. 33.2 Diagram of a cross-section through the superior colliculi of the midbrain. P = pyramidal fibres; FP and TP = frontopontine and temporoparietopontine fibres.

Nuclei

(*a*) The **oculomotor nucleus** has somatic and visceral parts and lies in the upper part of the midbrain ventral to the aqueduct. It consists of (i) central and lateral (somatic) nuclei supplying all extra-ocular muscles except superior oblique and lateral rectus, (ii) parasympathetic nuclei (visceral nuclei) supplying the sphincter pupillae and ciliary muscles. The oculomotor fibres pass anteriorly through the tegmentum and leave the brain stem between the cerebral peduncles.

(*b*) The **trochlear nucleus** lies on the side of the midline in the lower midbrain ventral to the aqueduct. The fibres first pass dorsally and decussate

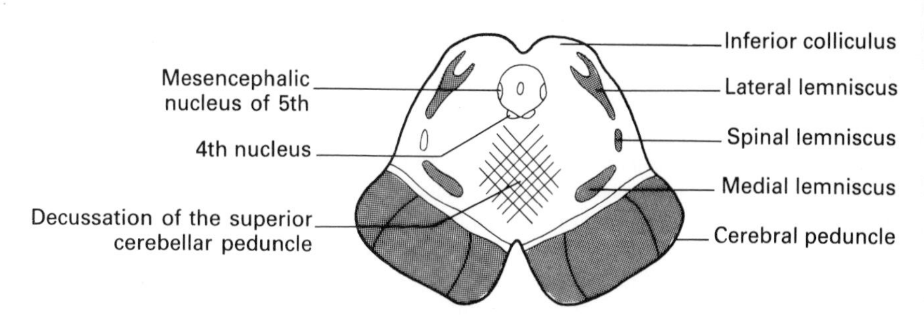

Fig. 33.3 Diagram of a cross-section through the inferior colliculi of the midbrain.

below the inferior colliculi before emerging and passing ventrally again around the cerebral peduncles. The fibres supply superior oblique muscle.

(*c*) The **mesencephalic nucleus** of the trigeminal nerve lies lateral to and along the whole length of the peri-aqueductal grey matter. The tract lies lateral to the nucleus. It carries proprioceptive impulses mainly from the muscles of mastication.

(*d*) The **red nuclei** are a pair of oval nuclear masses most obvious in the upper tegmentum. They receive afferents from the cerebellum through the superior cerebellar peduncles, and from the globus pallidus. Their efferents pass in the rubroreticular and rubrospinal tracts to the lower brain stem reticular formation and the spinal cord. The nuclei form part of the extrapyramidal system.

(*e*) The **substantia nigra** is an area of pigment containing neurons separating the tegmentum from the cerebral peduncles of each side. It is also part of the extrapyramidal system and receives fibres from the basal nuclei (strionigral tract) and its efferents pass to the reticular formation.

(*f*) The **superior colliculi** receive fibres from the optic tracts and also from the visual cortex. They communicate, through the medial longitudinal bundle, with motor nuclei of the head and neck, and through the tectospinal tract with cells of the spinal cord. They are concerned with reflex movements of the head and neck in response to visual stimuli.

(*g*) The **inferior colliculi** receive some auditory fibres from the lateral lemniscus. Some efferent fibres enter the reticular formation while others pass through the inferior brachium to the medial geniculate body. They may be concerned with reflex movements of the head and neck in response to auditory stimuli.

Fibre tracts

The **cerebral peduncles** are motor tracts which are the continuation downwards of the internal capsules. They contain the pyramidal (corticospinal) fibres passing from the motor cortex to cells of the spinal cord, and on either side of these the corticopontine fibres from the frontal, temporal and parietal cortex to the pontine nuclei. Scattered throughout the peduncles are corticonuclear (corticobulbar) fibres passing from the motor cortex to the cranial nerve nuclei of the opposite side.

Ascending in the lateral tegmentum are the medial, trigeminal and spinal lemnisci passing to the thalamus (ventroposterior nucleus). The **medial lemniscus** carries somatosensory fibres from the opposite side of the trunk and limbs; the **trigeminal lemniscus** carries somatosensory fibres from the opposite side of the face, and the **spinal lemniscus** carries some of the pain and temperature fibres from the opposite side of the body. The **lateral lemniscus** is present in the lower midbrain and passes dorsally to reach the **inferior colliculus** and the medial geniculate body. It carries auditory fibres from the cochlear, trapezoid, and lateral lemniscal nuclei.

The **medial longitudinal fasciculus** lies in the tegmentum adjacent to the 3rd and 4th cranial nerve nuclei. Many of its fibres carry impulses from the vestibular nuclei to the cranial nuclei (especially those concerned with the positioning of the eyeball) and to the upper spinal cord (concerned with the position of the head in space). Alongside the medial longitudinal fasciculus are fibres which descend from the hypothalamus to the lower autonomic nuclei.

Three tracts decussate (cross the midline) in the midbrain:

(i) the superior cerebellar peduncles, passing from the cerebellum to the red nucleus, thalamus and globus pallidus, decussate in the lower midbrain.
(ii) the rubroreticular and rubrospinal tracts, passing from the red nucleus to the cells of the spinal cord, decussate in the upper midbrain.
(iii) the tectospinal tracts, passing from the superior (and inferior) colliculi to the motor nuclei of the brain stem and spinal cord, decussate dorsal to the rubroreticular decussation.

The midbrain contains many large and important ascending and descending fibre tracts; its cranial nerve nuclei control most of the eye muscles. The red nucleus and the substantia nigra are involved in extrapyramidal activities, and the tectal nuclei in visual and auditory reflexes. Many of the nuclei of the ascending reticular formation activate (arouse) the cerebral cortex and have facilitatory or inhibitory effects on other centres.

THE PONS (Figs. 33.4, 33.5)

Anteriorly the pons forms a prominent convexity between the midbrain and the medulla, and is composed mainly of transversely running fibres which pass backwards and laterally to become the middle cerebellar peduncles. The trigeminal nerve emerges from the middle of the pons at the junction of the middle cerebellar peduncle and the pons. The abducent, facial and vestibulocochlear nerves emerge, from medial to lateral, at the lower border of the pons. The posterior surface of the pons forms the upper part of the floor of the fourth ventricle. The ventricle separates the pons anteriorly from the cerebellum posteriorly.

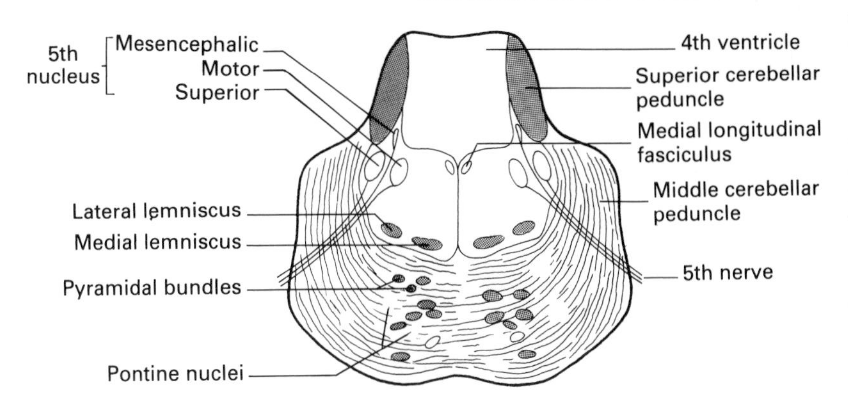

Fig. 33.4 Diagram of a cross-section through the middle of the pons, showing the trigeminal nuclei.

In transverse section the pons has a ventral (basilar) part and a dorsal (tegmental) part. The **basilar part** of the pons is continuous with the cerebral peduncles of the midbrain. In the pons the corticospinal (pyramidal) fibres are split into small scattered bundles and the corticopontine fibres ending in numerous **pontine nuclei**. Pontocerebellar fibres arise from these nuclei, cross the midline and sweep dorsally to form the middle cerebellar peduncle. The **tegmental part** is continuous with the midbrain tegmentum and contains the nuclei of the trigeminal, abducent, facial and vestibulocochlear nerves.

Nuclei

(*a*) The **trigeminal group of nuclei** is situated in the midpons region in the dorsolateral angle of the tegmental part. The large cells of the motor nucleus (supplying the muscles of mastication) lie medial to the smaller cells of the superior (principal) nucleus. The mesencephalic nucleus, with its afferent fibres on its lateral side, extends rostrally into the midbrain. The spinal nucleus, with its afferent fibres on its lateral side, extends into the spinal cord. The lowest fibres of this spinal tract end in the dorsal horns of the upper cervical segments. Fibres from the sensory nuclei form the trigeminal lemniscus which joins the medial and spinal lemnisci of the opposite side as they pass to the thalamus. The superior nucleus is concerned with most somatosensory information from the face and scalp, other than proprioception. This passes to the mesencephalic nucleus. Pain, hotness and coldness information passes mainly to the spinal nucleus, however the arrangement is probably not as clear-cut as it has been described.

(*b*) The **abducent nucleus** supplies the lateral rectus and is situated dorsally near the midline deep to the floor of the fourth ventricle. The fibres pass ventrally to emerge through the groove between the lower border of pons and the medullary pyramid.

(*c*) The **facial motor nucleus** supplies mainly the muscles of facial expression and is situated in the lateral tegmentum. The superior part of the salivary nucleus

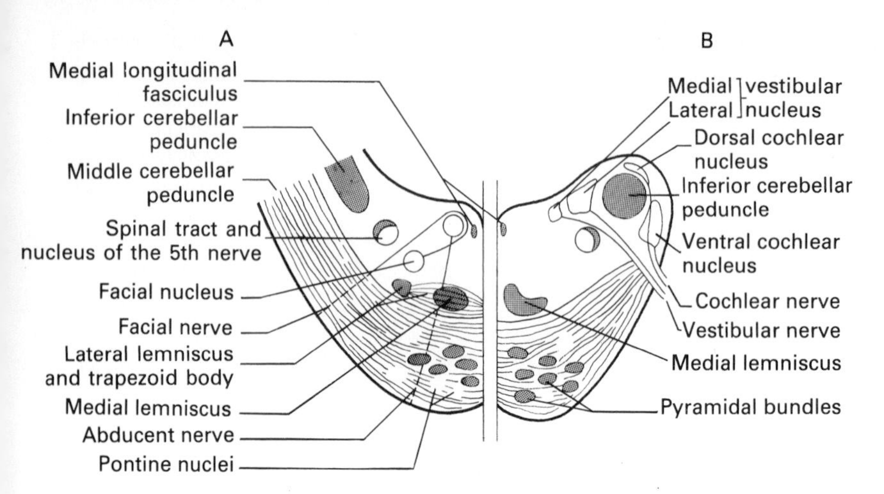

Fig. 33.5 Diagram of cross-section through (**A**) the lower pons (**B**) the lateral recess of the 4th ventricle.

supplies parasympathetic fibres to the submandibular and sublingual glands and is associated with the facial nucleus. The nucleus for the taste fibres of the anterior two-thirds of the tongue, the **nucleus of the tractus solitarius**, lies lateral to the motor nucleus. The motor fibres make a dorsal loop over the abducent nucleus before leaving the lower pons anterolaterally in company with the parasympathetic and taste fibres.

(*d*) The **dorsal** and **ventral cochlear nuclei** receive auditory fibres from the inner ear, and lie in the lower part of the pons on the inferior cerebellar peduncle. The nuclei lie in the floor of the lateral recess of the fourth ventricle. Most of their axons cross the midline in a compact bundle, the **trapezoid body**, which forms the dorsal part of the basilar pons. Some secondary cochlear fibres from the cochlear nuclei end in nuclei near the trapezoid body, known as trapezoid nuclei. Fibres in the trapezoid body turn rostrally and form a tract, the **lateral lemniscus**, which ascends through the pons and lower midbrain to the inferior colliculus and medial geniculate body.

(*e*) The **vestibular group of nuclei** receive vestibular fibres from the inner ear and lie under the lateral part of the floor of the fourth ventricle in both the pons and medulla. The large lateral part gives rise to the vestibulospinal tract, the fibres of which pass uncrossed to the motor cells of the spinal cord. Fibres from all parts of the nuclei pass to the medial longitudinal fasciculus (and medial lemniscus) of both sides. Many fibres pass uncrossed in the inferior cerebellar peduncle to the anterior lobe, vermis and flocculonodular lobe of the cerebellum. The nuclei also receive fibres from the fastigial nuclei of the cerebellum.

Fibre tracts

The trapezoid body and lateral lemniscus have been described with the cochlear nuclei. The **medial longitudinal fasciculus** lies dorsally near the midline in the floor of the brain stem cavity. It receives fibres from the motor nuclei supplying the extra-ocular muscles, the vestibular nuclei and the motor cells of the cervical spinal cord. The two bundles communicate with each other through the posterior commissure. They are associated with reflex control of eye and head movements. The medial lemniscus becomes more laterally and ventrally placed as it ascends the pons; the spinal lemniscus lies lateral to it. The anterior spinocerebellar tract ascends through the lateral tegmentum, enters the superior cerebellar peduncle and passes to the cerebellum.

THE MEDULLA OBLONGATA (Figs. 33.6, 33.7, 33.8)

The medulla oblongata (the bulb) narrows rapidly between the pons and the foramen magnum where it becomes continuous with the spinal cord. The upper part (the open medulla) mostly forms the lower half of the floor of the fourth ventricle. Posteriorly there is the thin ependymal roof of the ventricle (the inferior medullary velum) below, and the cerebellum above. The lower part (closed medulla) has a narrow central canal surrounded by grey matter and outside this by white matter. The basic structure of the lower medulla is similar to that of the spinal cord.

The **anterior median fissure** begins at the lower edge of the pons and continues into the cord. In the upper medulla, it is flanked by the **pyramids**—containing bundles of corticospinal (pyramidal) fibres. Lower down most of these fibres cross the fissure (the pyramidal decussation) as they pass downwards, posteriorly and laterally into the cord. Lateral to the pyramid is an oval mass, the **olive**, and posterolateral to the olive is the **inferior cerebellar peduncle** passing upwards into the cerebellum. The **posterior median sulcus** runs downwards from the lower recess of the fourth ventricle into the cord. On each side of it is a small ridge—the **gracile tubercle**—and lateral to it, the **cuneate tubercle**.

The rootlets of the hypoglossal nerve leave the medulla in the groove between

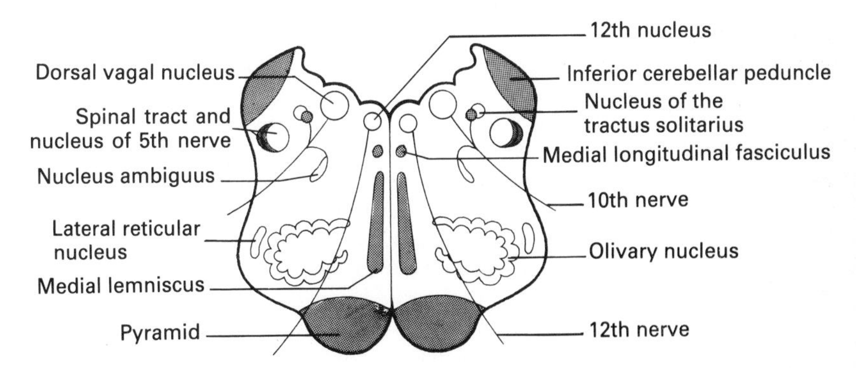

Fig. 33.6 Diagram of a cross-section through the open medulla at the level of the olivary nucleus.

the pyramid and the olive, and the rootlets of the glossopharyngeal and vagus nerves leave in the groove between the olive and the inferior cerebellar peduncle.

Internal structure

The central canal is wide above, where it forms part of the fourth ventricle, but narrows inferiorly. The embryological pattern of alar and basal plates is found in the lower medulla but is modified above where the alar plates lie lateral to the basal plates, the roof plate being much thinner. The roof plate, which remains as the ependymal inferior medullary velum, is invaginated by blood vessels and forms the choroid plexus of the fourth ventricle. Dorsally the cerebellum develops in the ventricular roof.

Nuclei (see Fig. 31.6)

(*a*) The **spinal nucleus** of the trigeminal nerve lies posterolaterally and extends throughout the medulla; inferiorly it is continuous with the posterior horn of the spinal cord.

(*b*) The **hypoglossal nucleus** (somatic efferent) lies in the floor of the fourth ventricle near the midline and extends caudally into the closed medulla. It supplies the muscles of the tongue.

(*c*) The fibres travelling in the glossopharyngeal and vagus nerves arise in a complex group of nuclei. The most rostral fibres go into the glossopharyngeal nerve and the most caudal run for a short distance with the accessory nerve. Lateral to the hypoglossal nucleus is the **dorsal vagal nucleus** (visceral efferent). Most of its fibres travel in the vagus. Some cells (the inferior part of the salivary nucleus) associated with the upper end of this nucleus send their axons into the glossopharyngeal nerve. They are the preganglionic secretomotor fibres to the parotid gland and synapse in the otic ganglion. Lateral to the motor nucleus is the **dorsal sensory vagal nucleus** (visceral afferent). Some of its fibres come from the glossopharyngeal nerve, but most come from the vagus. Lateral to it is the **tractus solitarius** and its **nucleus** (special visceral afferent—taste) receiving fibres from the glossopharyngeal nerve (posterior one-third of the tongue) and the vagus nerve (region of the epiglottis). Fibres from the chorda tympani branch of the facial nerve, conveying taste sensations from the anterior two-thirds of the tongue, end in the rostral extension of this nucleus. Situated ventrally between the hypoglossal and dorsal vagal nuclei is a group of cells forming the **nucleus ambiguus** (special visceral efferent). Their axons pass to the muscles of the palate, pharynx and larynx in the glossopharyngeal, vagus and cranial accessory nerves. These muscles are derived from pharyngeal arch mesenchyme and may become paralysed in diseases of the medulla (bulbar paralysis).

(*d*) The **gracile** and **cuneate (dorsal column) nuclei** are situated posteriorly in the lower half of the medulla, the gracile being the more medially placed. They receive sensory fibres from the posterior column of the spinal cord; their efferents pass ventrally, cross the midline as internal arcuate fibres, form the **medial lemniscus** and pass rostrally to the thalamus (ventroposterior nucleus). The **accessory cuneate nucleus**, situated more laterally, receives proprioceptive fibres from the neck region; its efferents pass uncrossed in the inferior cerebellar peduncle to the cerebellum.

(*e*) The **olivary nucleus** is large, irregular and cup-shaped, forming a prominence on the lateral surface. Its open part faces medially and from it efferent fibres cross the midline as internal arcuate fibres, and enter the inferior cerebellar peduncle. Lateral to the olivary nucleus is the lateral reticular nucleus which sends fibres into both inferior cerebellar peduncles.

The vestibular and cochlear nuclei are considered with the pons (p. 464).

White matter

The **pyramid** is a large bundle of fibres forming the prominence on each side of the midline on the anterior surface of the medulla. It is a convergence of the numerous bundles traversing the basal part of the pons. Inferiorly most of the fibres of the two pyramids decussate, passing posterolaterally across the medulla to form the lateral corticospinal tract which passes to the motor cells of the spinal cord. The uncrossed pyramidal fibres continue as the anterior corticospinal tract. As the decussating fibres pass laterally they cut across the anterior horn of grey matter. Corticonuclear fibres, running with the pyramidal fibres, end in the cranial nerve motor nuclei of the opposite side, e.g. the hypoglossal nuclei.

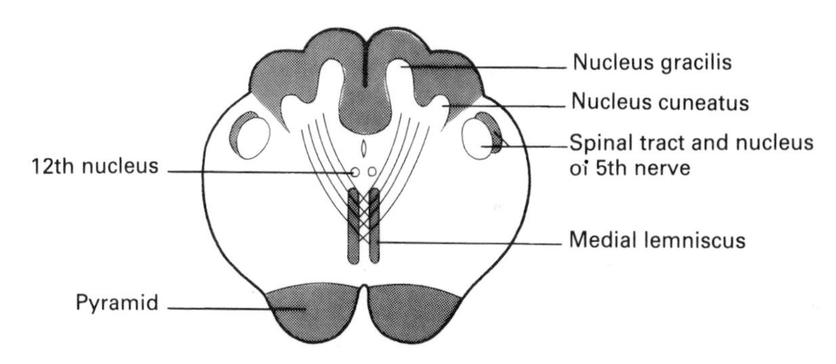

Fig. 33.7 Diagram of a cross-section through the closed medulla at the level of the sensory decussation.

Many of the fibres crossing the midline in the region of the open medulla are called **internal arcuate fibres**. Most go to the thalamus (passing from the gracile and cuneate nuclei through the medial lemniscus) or to the cerebellum (passing from the olivary nuclei and the reticular formation through the inferior cerebellar peduncle).

The **medial longitudinal fasciculus** descends near the midline dorsally, separated from the medial lemniscus by the tectospinal tract. All three tracts lie alongside the midline. The **spinal tract of the trigeminal nerve** overlies the lateral aspect of its nucleus; it descends from the pons and is continuous with the dorsolateral tract of the spinal cord. The **posterior spinocerebellar tract** ascends from the spinal cord and passes uncrossed in the inferior cerebellar

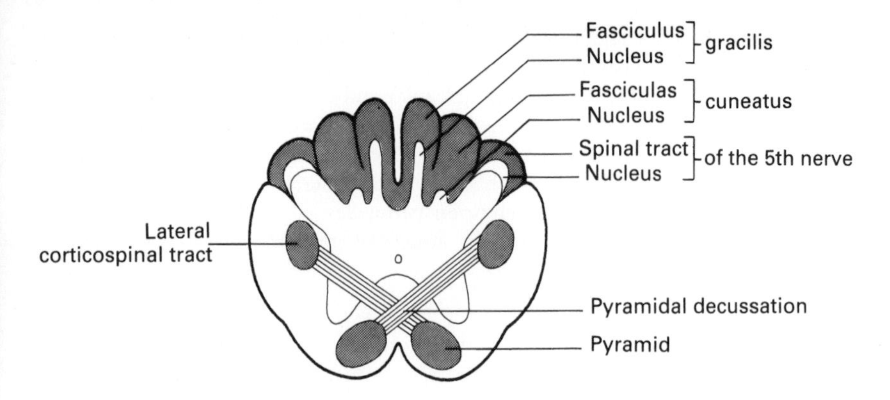

Fig. 33.8 Diagram of a cross-section through the closed medulla at the level of the motor decussation.

peduncle to the cerebellum. The **anterior spinocerebellar tract** ascends through the medulla, pons and midbrain and reaches the cerebellum through the ipsilateral superior cerebellar peduncle.

Other tracts passing through the medulla include the spinothalamic tracts which ascend from the spinal cord to the thalamus (ventroposterior nucleus), the vestibulospinal tract which descends from the lateral vestibular nucleus, and the rubroreticular tract which descends from the red nucleus. Both these descending tracts end in the region of the anterior horn neurons.

THE CEREBELLUM

The cerebellum fills the larger part of the posterior cranial fossa and is attached to the back of the brain stem by three paired fibre bundles. These are the superior, middle and inferior cerebellar peduncles passing to the midbrain, pons and medulla respectively. The cavity of the fourth ventricle lies between the cerebellum and the inferior medullary velum dorsally, and the pons and medulla ventrally; laterally the cavity passes over the inferior cerebellar peduncle and forms the lateral recess of the ventricle.

The cerebellum is divided into a midline irregular **vermis** and two laterally placed **hemispheres**. The ends of the vermis overlie the superior and inferior medullary vela roofing the fourth ventricle. When the hemisphere is cut vertically, the deepest sulcus, the **fissura prima**, can be followed across the upper surface of the cerebellum. It divides the vermis and hemispheres into **anterior** and **middle lobes**, the anterior lying above and in front of the fissure. A further subdivision of the organ, the **flocculonodular lobe**, lies caudal to the **posterolateral fissure** across the lower surface of the cerebellum. The nodule is the most caudal part of the vermis and the flocculi are narrow lateral extensions from it into the cerebellopontomedullary angles.

The cerebellum has an outer covering of grey matter, the **cortex**, which is greatly convoluted, the visible surface representing only about a sixth of the cortical area. The sulci are parallel and the ridges between them are known as **folia**. The cellular arrangement of the cerebellar cortex is very uniform, there being three layers—inner granular, middle Purkinje and outer molecular. Most afferent fibres to the cerebellar cortex end in the granular layer and in the outer molecular layer where the profuse dendritic network of the Purkinje cells is situated. The axons of the large, flask-shaped Purkinje cells pass to the central nuclei of the cerebellum. These nuclei of each side comprise the large laterally placed **dentate nucleus** bordered medially by the **globose** and **emboliform** nuclei (together known as the nucleus interpositum) and the medially placed **fastigial** (roof) **nucleus**. The Purkinje cells of the vermis project on to the fastigial nuclei, and an adjacent cortical band of each hemisphere (the paravermal zone) projects to a nucleus interpositum. The remainder of each hemisphere projects to a dentate nucleus.

AFFERENT CONNECTIONS

The cerebellum has four main groups of afferent connections:

1. Cerebral cortical fibres

These synapse in the pontine nuclei whose axons pass in the middle cerebellar peduncle of the opposite side to all parts of the cerebellum save the flocculo-nodular lobe.

2. Olivary fibres

These cross the medulla as the upper group of internal arcuate fibres and pass through the inferior cerebellar peduncle to all parts of the cerebellum. The olivary nuclei receive afferent fibres from the motor cortex, the red nucleus, peri-aqueductal grey matter of the midbrain and the spinal cord.

3. Spinal fibres

(*a*) The posterior spinocerebellar tract, originating in the cells of the thoracic nucleus, pass uncrossed through the inferior cerebellar peduncle to the vermis and paravermal zone of the anterior and middle lobes.

(*b*) Fibres ascending in the dorsal column of the upper part of the spinal cord synapse in the accessory cuneate nucleus whose axons pass uncrossed in the inferior cerebellar peduncle to the vermis and paravermal zone of the anterior and middle lobes.

(*c*) The anterior spinocerebellar tract originates in the posterior horn of the opposite side and passes in the superior cerebellar peduncle mainly to the vermis of the anterior lobe.

(*d*) Fibres from the anterolateral part of the spinal cord synapse in the lateral reticular nucleus whose axons pass in the inferior cerebellar peduncle of the opposite side to all parts of the cerebellum.

4. Vestibular fibres

These arise from the vestibular nuclei and pass mainly to the flocculonodular lobe.

EFFERENT CONNECTIONS

Efferent cerebellar fibres from the **dentate nucleus** and **nucleus interpositum** enter the superior cerebellar peduncle and decussate in the lower midbrain. They end in the red nucleus, thalamus (ventrolateral nucleus) and the globus pallidus. From the latter two regions impulses pass to the motor cortex. Some fibres end in reticular nuclei of the pons and medulla. Fibres from the **fastigial nuclei** pass in both superior and inferior cerebellar peduncles to the vestibular nuclei and reticular nuclei of the pons and medulla.

From the above information, the fibre content of the cerebellar peduncles may be summarised:

1. The **superior cerebellar peduncle** contains efferent fibres from all cerebellar nuclei and also the anterior spinocerebellar tract.
2. The **middle cerebellar peduncle** carries pontocerebellar fibres.
3. The **inferior cerebellar peduncle** carries the posterior spinocerebellar tract, fibres of the accessory cuneate and lateral reticular nuclei, afferent and efferent vestibular connections and olivocerebellar fibres.

The cerebellum, through its corticopontine connections, receives information on corticospinal activity. Simultaneously, through its proprioceptive spinal and vestibular connections it is able to compare this information and, through its efferent tracts, influence the cortical and subcortical motor centres. Thus the cerebellum can modify the pattern of a movement as it is taking place, ensuring maximum efficiency with minimum effort. Disease of the cerebellum interferes with the co-ordination of voluntary movement; patients show an intention tremor.

The different parts of the body are represented on the cortex of the cerebellum mainly in an ipsilateral manner. Vestibular activity is confined to the most caudal part (the flocculonodular lobes) and also to the most rostral part. The head, neck and trunk muscles are represented on the vermis and paravermal zones. The large part of the hemispheres is concerned with the more peripheral limb muscles. The signs of disease are most obvious on the ipsilateral (same) side as the cerebellar lesion. This is in sharp contrast to cerebral disease where the signs appear on the contralateral (opposite) side of the body.

THE FOURTH VENTRICLE (Figs. 32.6, 33.1)

This is the cavity of the hind brain. It communicates superiorly with the third ventricle through the cerebral aqueduct and is continuous inferiorly with the central canal of the spinal cord. When looked at from behind, the cavity is somewhat diamond-shaped, bounded superolaterally by the superior cerebellar peduncles, inferolaterally by the inferior cerebellar peduncles and inferiorly by the gracile and cuneate tubercles.

The **roof** of the cavity extends backwards into the cerebellum; superiorly it is formed by a glial sheet (the **superior medullary velum**) stretching between the superior cerebellar peduncles and inferiorly by a thin ependymal sheet (the **inferior medullary velum**) between the inferior cerebellar peduncles. The lateral angles of the roof are evaginated to form the **lateral recesses** which pass forwards round the inferior cerebellar peduncles. The inferior velum is invaginated by the blood vessels and pia mater and forms the choroid plexus of the ventricle. A lateral opening in the tip of each lateral recess and a median opening in the inferior medullary velum allow cerebrospinal fluid to pass from the ventricular system to the subarachnoid space.

The diamond-shaped **floor** of the ventricle is formed above by the pons and below by the medulla. A **median sulcus** grooves its length and a number of ridges, the **striae medullares**, pass transversely across it into the lateral recess. On either side of the median sulcus, below the striae medullares, a small depression, the **inferior fovea**, is present. Two diverging sulci descend from the fovea dividing the area below the striae into a medial **hypoglossal triangle**, an intermediate **vagal triangle** and a lateral **vestibular area**. They overlie their respective cranial nerve nuclei.

Just above the striae medullares an elevation, the **facial colliculus**, is present on each side of the median sulcus. It is formed by facial nerve fibres arching over the abducent nucleus. Lateral to this is a rostral extension of the vestibular area.

— 34 —

The spinal cord and reticular formation

THE SPINAL CORD (Fig. 34.1)

The spinal cord lies within the vertebral canal and extends from the foramen magnum to the level of the first or second lumbar vertebra after which a fibrous remnant, the filum terminale, descends to be attached to the back of the coccyx. The cord is about 45 cm long. It is cylindrical in shape, flattened slightly antero-posteriorly, and has cervical and lumbar enlargements where the nerves supplying the upper and lower limbs originate. The enlargements lie opposite the lower cervical and lower thoracic vertebrae. Since the spinal cord is shorter than the vertebral canal, the nerve roots descend with increasing obliquity to reach the appropriate intervertebral foramen where the dorsal and ventral roots join and form a spinal nerve which passes through the foramen. The collection of lower lumbar, sacral and coccygeal nerve roots below the spinal cord is known as the **cauda equina**. The tapered lower end of the spinal cord is continued as a pia-glial filament, the filum terminale, and this is attached to the coccyx. The

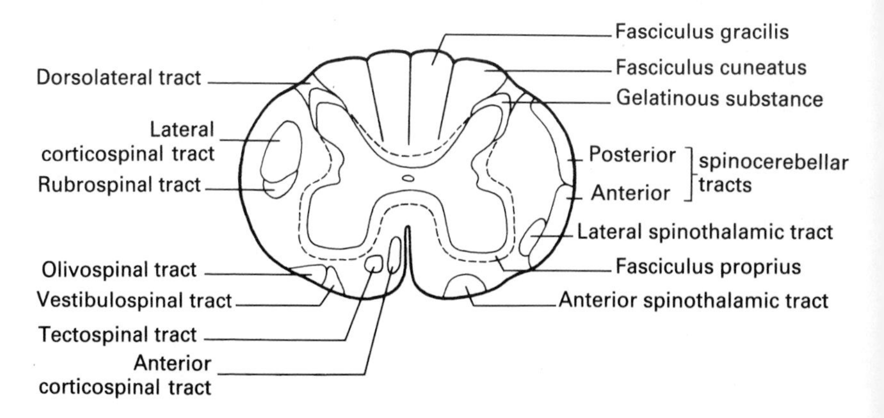

Fig. 34.1 Diagram of a cross section of the spinal cord, showing the main ascending pathways on the right and the main descending pathways on the left side of the diagram.

cord has an **anterior median fissure** and a **posterior median sulcus**. On its sides the rootlets of the spinal nerves emerge from **anterolateral** and **posterolateral sulci**.

Grey matter

The grey matter of the spinal cord is arranged around the central canal and projects towards the anterolateral and posterolateral sulci to form paired **anterior** and **posterior horns**. The cells of the former are mainly motor in function and of the latter mainly sensory. A small celled region, the **gelatinous substance**, overlies the posterior horn. There is proportionately more grey matter in the cervical and lumbar enlargements corresponding to the large nerve roots supplying the limbs. The neurons supplying the limb muscles are found in large groups (nuclei) laterally in the anterior horns. Also in the upper cervical region are the nuclei of the phrenic and spinal accessory nerves. In the thoracic and sacral regions a small lateral projection, the **lateral horn**, is present; its cells give rise to preganglionic autonomic nerve fibres. In the thoracic region a large group of large, closely packed cells, the **thoracic nucleus**, is present at the base of the posterior horn. Proprioceptive fibres from muscle spindles end on the cells of this nucleus and the efferent fibres form the posterior spinocerebellar tract mainly of the same side. However, the majority of the nerve cells in the grey matter are intercalated (intermediate) neurons.

White matter (Fig. 34.1)

The anterior and posterior horns and the emerging spinal rootlets divide the white matter into three columns (funiculi) on each side. An anterior funiculus lies between the anterior horn and the median fissure, a lateral between the anterior and posterior horns, and a posterior between the posterior horn and the posterior septum. The anterior median fissure incompletely separates the anterior columns, the remaining communication being called the anterior white commissure. The posterior median septum completely separates the posterior columns. Each posterior column is further divided in the cervical and upper thoracic regions into a **fasciculus cuneatus** laterally and a **fasciculus gracilis** medially. Between the gelatinous substance and the surface of the cord is the **dorsolateral tract**. The region closely surrounding the grey matter is known as the **fasciculus proprius**.

The fibres of the white matter are grouped into tracts (Fig. 34.1) which ascend and descend in the cord. Most of these tracts cannot be identified in normal material and pathological and experimental lesions have helped to define some of them. Many of the fibres have their cell bodies within the spinal cord or brain. The entering dorsal root fibres of spinal nerves divide into a large ascending branch and a small descending branch. Both of these give off innumerable collateral branches which take part in intrasegmental and intersegmental reflexes.

The **posterior column** contains:

(*a*) Mainly ascending fibres associated with light touch, pressure, vibration, and conscious proprioception. Their cell bodies lie in the dorsal root ganglia and they pass to the gracile and cuneate nuclei in the medulla.

(*b*) Intersegmental and intrasegmental fibres.

The **lateral column** contains:

(*a*) Peripherally placed, mainly ascending fibres;

(i) anterior and posterior spinocerebellar tracts—associated with unconscious proprioception. The anterior tract, whose cell bodies lie in the posterior horn of the opposite side, enter the cerebellum in the superior cerebellar peduncle. The posterior tract has its cell bodies in the thoracic nucleus of the same side and passes uncrossed to the cerebellum in the inferior cerebellar peduncle.

(ii) dorsolateral fasciculus—a mixed bundle of fibres. Some axons arise in the dorsal root ganglia (possibly carrying pain and thermal impulses) but most axons probably arise in or near the gelatinous substance (propriospinal fibres). All the axons run for short distances before ending in the posterior horn.

(iii) lateral spinothalamic tract—associated with pain and temperature sensations. Its cell bodies lie in the posterior horn of the opposite side and its fibres having crossed in the anterior white commissure, ascend to the thalamus in the spinal lemniscus.

(*b*) Intermediately placed, mainly descending fibres;

(i) lateral corticospinal (crossed pyramidal) tract—motor fibres whose cell bodies lie in the cerebral motor cortex. The fibres have crossed in the medulla and descend to the grey matter of the cord. Many of the fibres in the lateral corticospinal tract end at the base of the posterior horn where they synapse with cells whose short axons end amongst the anterior horn cells.

(ii) reticulospinal and rubrospinal tracts—contain extrapyramidal motor fibres descending to the cells of the spinal cord. They come from the red nucleus in the midbrain and from scattered reticular nuclei.

The **anterior column** contains:

(*a*) Anterior corticospinal (uncrossed pyramidal) tract. This is found in the cervical and upper thoracic regions and is formed of descending fibres from the motor cortex. They cross almost horizontally in the anterior white commissure to the contralateral anterior horn region.

(*b*) Vestibulospinal, olivospinal and tectospinal tracts are extrapyramidal fibres passing to the anterior horn region from nuclei in the brain stem.

(*c*) Anterior spinothalamic tract carries touch fibres. Its cells lie in the posterior horn of the opposite side and its fibres pass in the spinal lemniscus to the thalamus.

The fibres of the fasciculus proprius surround the grey matter and form intrasegmental and intersegmental connections in the spinal cord.

The anterior and lateral spinothalamic tracts together are sometimes called the **spinal lemniscus**.

THE RETICULAR FORMATION

The reticular formation is the name given to a large number of ill-defined groups of neurons situated between the larger nuclear masses of the brain stem and around the grey matter of the spinal cord.

With expanding knowledge of the nervous system the widespread activity of the reticular formation has become recognised and the accurate mapping of its nuclear groups and fibre tracts has been undertaken. Over a hundred nuclear groups have been described, the larger groups (such as the lateral reticular nucleus of the medulla) being more easily recognised. The neurons are of a variety of sizes having prolific dendritic trees and axons with numerous collateral branches. Ascending axons pass through the dendritic fields of descending neurons, probably influencing their activity. Similarly, descending axons tend to pass through the dendritic fields of ascending neurons.

Ascending neurons of the reticular formation receive collaterals from all afferent pathways, especially those of the trigeminal nerve, and also from the cerebellum. They project to the intralaminar thalamic nuclei and from there project diffusely to all parts of the cerebral cortex. There is a marked lack of organisation in this system, stimulation producing no specific local response but giving rise to widespread cortical activity (cortical arousal) and wakefulness. Enlarging lesions of the midbrain or increasing pressure on the midbrain reduce this cortical activity and lead to progressive stupor.

Descending neurons receive fibres from the motor cortex, either directly from the corticonuclear and corticospinal pathways or via the basal nuclei. They also receive cerebellar fibres, especially from the anterior lobe and fastigial nuclei. The descending pathways have both ipsilateral and contralateral effects on the spinal anterior horn cells which may inhibit or facilitate spinal motor activity governing, in particular, postural adjustments.

— 35 —

The afferent and efferent pathways

Axons run for variable distances within the spinal cord and brain stem. Most are short intrasegmental and intersegmental fibres but many run for long distances uniting the brain and lower spinal segments. In the central nervous system there appears to be some functional localisation of the fibres. Short intrasegmental and intersegmental fibres lie close to the grey matter of the cord in the fasciculus proprius; many ascending fibres lie in the posterior column and the periphery of the lateral column; and many descending fibres lie in the anterior column and the intermediate part of the lateral column.

Most of the information entering the central nervous system is employed in reflex, automatic and unconscious ways, only a small amount reaching consciousness. Similarly on the motor side of the central nervous system, only a small proportion of the descending pathways are directly involved in voluntary, conscious activity.

The pathways are so arranged that each cerebral hemisphere receives information from the opposite side and also controls muscles of the opposite half of the body. The primary sensory and motor areas of the cortex show a marked degree of anatomical localisation. In the sensorimotor cortex related to the central sulcus, the lower limb is represented near the top, the trunk and upper limb about the middle, and the head and neck are at the lower parts of the gyri. Though the limbs and trunk are represented in an inverted position, the head and neck are not inverted.

The areas of cortex represented are proportional to the acuteness of the sensory detection and to the complexity of the muscle activity. The more 'important' parts, e.g. lips, tongue, thumb and index finger, have larger representations than the buttock, foot and back. The macula of the eye has a larger representation on the calcarine gyri than all the peripheral retina.

In most humans, one hemisphere appears to be dominant. This usually determines the 'handedness' of the individual and is the side of the brain most concerned with speech processes and memory. Lesions of the nondominant hemisphere give rise to difficulties in spatial conception and sometimes neglect of the paralysed parts of the body.

AFFERENT PATHWAYS (Fig. 35.1)

The afferent pathways of the central nervous system are involved in the conduction of information about environmental conditions received by the peripheral nerve endings. Some of these receptors are simple, free nerve endings while others are very elaborate end organs (such as the encapsulated corpuscles and the muscle spindles). The extent to which a particular 'sensation' is dependent on a single form of nerve ending or on the summation of information from a variety of endings is in many cases undecided. The stretch receptors of muscle are specific in function while the free nerve endings of the cornea, mostly identical in structure, can distinguish touch, pain, hotness and coldness.

Most of the somatosensory pathways ascending the spinal cord and brain stem may be grouped under three headings: (*a*) pathways for touch and pressure,

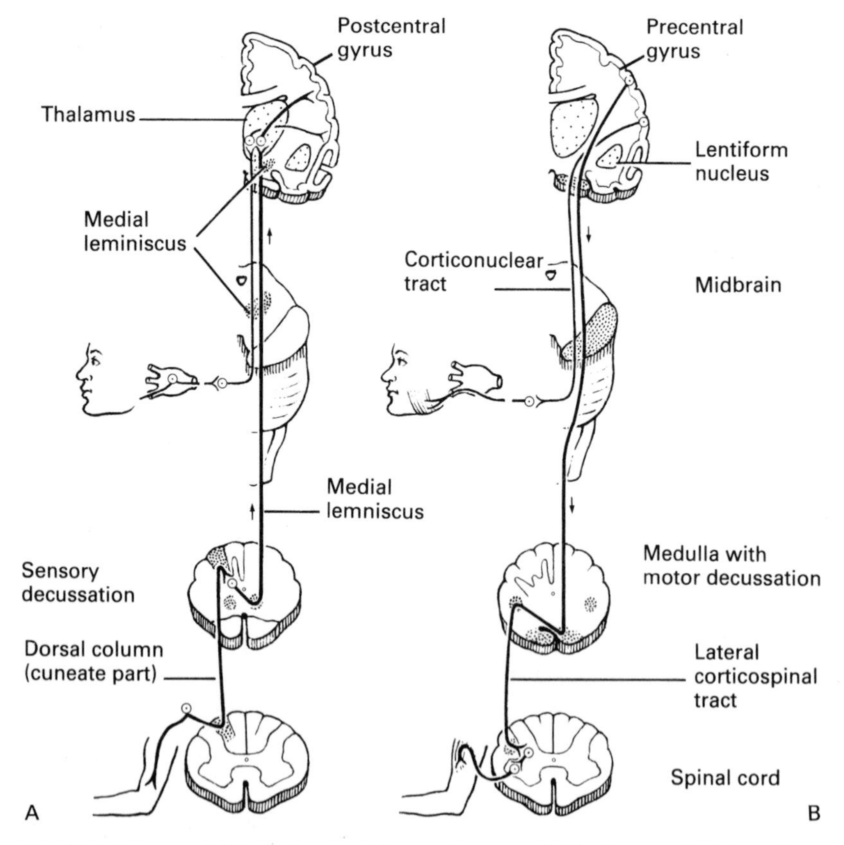

Fig. 35.1 Diagram of (**A**) the lemniscal (sensory) tracts and (**B**) the pyramidal (motor) tracts.

(*b*) pathways for pain, hotness and coldness, (*c*) proprioceptive pathways. Those pathways reaching the cerebral cortex are usually formed of a chain of three sets of neurons.

In addition, some of the cranial nerves contain pathways for olfaction, vision, taste, balance and hearing, the so-called special senses.

Touch and pressure fibres in a spinal nerve have their cell bodies in the dorsal root ganglion. The central branches ascend in the posterior column of the spinal cord and synapse in the gracile or cuneate nuclei in the medulla. In the posterior column, fibres from the lower limb lie medial to those from the trunk, upper limb and neck.

The second set of neurons has its cell bodies in the gracile and cuneate nuclei. The axons swing round the central grey matter, cross the midline and then ascend as the medial lemniscus to the thalamus (ventroposterior nucleus). The third set of neurons has its cell bodies in the thalamic nucleus and the axons ascend through the internal capsule to the cortex of the postcentral gyrus.

Other touch fibres enter the cord and synapse in the posterior horn. The second neurons cross the midline and ascend in the anterior spinothalamic tract to the ventroposterior thalamic nucleus. The third set of neurons projects to the postcentral gyrus.

Touch fibres in the trigeminal nerve have their cell bodies in the trigeminal ganglion and their central branches end in the superior sensory trigeminal nucleus. The second set of axons cross the midline, join the medial side of the medial lemniscus and end in the ventroposterior thalamic nucleus. The third set of axons project to the lower part of the postcentral gyrus.

Pain and temperature fibres in a spinal nerve have their cell bodies in the dorsal root ganglion. The central fibres enter the cord and ascend for a short distance in the dorsolateral tract before synapsing in the posterior horn. The second set of axons cross the cord in the anterior white commissure and form the lateral spinothalamic tract. The fibres ascend the cord, pass through the brain stem as the spinal lemniscus and synapse in the thalamus (ventroposterior nucleus). A third set of axons pass through the internal capsule to the postcentral cortical gyrus.

Pain and temperature fibres in the trigeminal nerve have their cell bodies in the trigeminal ganglion. Their central branches enter the pons and descend into the upper cord as the spinal tract of the trigeminal nerve (cf. the spinal dorsolateral tract). Their fibres synapse in the spinal nucleus of the trigeminal nerve (cf. the posterior horn), whose axons cross the midline and join the spinal lemniscus. The fibres end in the ventroposterior thalamic nucleus.

The impulses of some pain-carrying fibres are relayed in the reticular formation to the intralaminar thalamic nuclei and thence to most areas of the cortex. This nonlemniscal pathway is a much more diffuse projection system. Response to a painful stimulus is both immediate (along fast conducting fibres) and delayed (along slow conducting fibres). The delayed response is long lasting.

Some **proprioceptive information** may reach a conscious level (cerebral cortex) but most is utilised by the cord and cerebellum and never reaches consciousness. Conscious proprioceptive fibres reach the cerebral cortex by the same pathways as do touch and pressure fibres in the posterior columns.

Unconscious proprioceptive fibres in a spinal nerve have their cell bodies in the dorsal root ganglion and their impulses reach the cerebellum by one of three pathways.

(*a*) Branches of the first neuron may ascend in the posterior column of the spinal cord to the accessory cuneate nucleus in the medulla. Fibres arising here pass in the ipsilateral inferior cerebellar peduncle to the cerebellum (vermis and paravermal zone of the anterior and middle lobes). This is an uncrossed pathway for impulses from the upper limb and upper trunk.

(*b*) The first neuron may synapse in the posterior horn of grey matter. A second neuron arising here crosses immediately and ascends in the anterior spinocerebellar tract. The tract reaches the cerebellum (vermis of the anterior lobe) via the superior cerebellar peduncle. This is mainly a crossed pathway.

(*c*) The first neuron may synapse in the thoracic nucleus of the posterior horn of grey matter. A second axon arising here passes to the ipsilateral posterior spinocerebellar tract and thence through the inferior cerebellar peduncle to the cerebellum (vermis and paravermal zones of anterior and middle lobes). This is mainly an uncrossed pathway.

Proprioceptive fibres are also present in the trigeminal nerve, especially the mandibular division. The cell bodies of the first set of neurons are in the trigeminal ganglia. The mesencephalic trigeminal nucleus is concerned with passing on proprioceptive impulses to the cerebellum and thalamus but the paths in the brain stem are ill-defined.

Pathways of the special senses

Olfactory impulses from the sensory mucosa in the upper part of each nasal cavity pass along the olfactory nerves to an olfactory bulb. Impulses are transmitted along the olfactory tracts and their subsequent passage is ill-defined, but it seems unlikely that primary olfactory projection areas extend beyond the prepiriform cortex (p. 449).

Visual impulses from the retina pass to the optic nerve. The two optic nerves undergo partial decussation at the optic chiasma; the nasal fibres of each hemiretina cross to the opposite side. The optic tracts, which pass backwards from the chiasma, thus contain ipsilateral temporal fibres and contralateral nasal fibres. Each tract skirts the midbrain and passes mainly to the lateral geniculate body of the thalamus. Some fibres, however, pass beyond the thalamus to the superior colliculus and pretectal region of the midbrain.

Fibres arising in the lateral geniculate body form the optic radiation which passes through the retrolentiform part of the posterior limb of the internal capsule, then lateral to the posterior horn of the lateral ventricle, and reaches the visual cortex. This borders the calcarine sulcus on the medial surface of the occipital lobe and is called the striate cortex.

Impulses arising from parts of the retinae receiving light from the left visual fields are relayed in the right geniculate body to the visual (striate) area of the right occipital pole. The lower halves of the retinae receive impulses from upper visual fields and the impulses are projected through the geniculate bodies to the lower lips of the calcarine sulci of both sides.

From the colliculi, impulses pass down the cord in the tectospinal tracts. In the **light reflex**, impulses from the pretectal nuclei pass to the visceral part of the oculomotor nucleus and so, via the ciliary ganglion, to the circular muscle of the iris—thus reducing the size (aperture) of the pupil. In the **accommodation reflex**, impulses return from the visual cortex to the visceral and medial rectus parts of the oculomotor nucleus. As a result, the pupil is reduced in size, the lens becomes more convex and the eyes converge slightly.

Taste impulses carried in the seventh, ninth and tenth cranial nerves pass to the nucleus of the tractus solitarius. Further connections are ill-defined but probably pass via the thalamus to the insular cortex. Many connections are made with centres controlling visceral activities.

Auditory impulses from the spiral organ in the cochlear duct pass through the bipolar neurons of the cochlear nerve to the dorsal and ventral cochlear nuclei in the floor of the lateral recess of the fourth ventricle. Secondary neurons arising here cross the midline mostly as a ventrally placed decussating bundle, the trapezoid body. These fibres synapse in the trapezoid nuclei (embedded in the trapezoid body) and in the nuclei of the lateral lemniscus. Some fibres cross the brain stem dorsal to the trapezoid body and synapse in the contralateral trapezoid nucleus. The axons from these nuclear groups turn rostrally, and form the lateral lemniscus which ascends to the medial geniculate body and the inferior colliculus. Neurons from the medial geniculate body pass in the posterior limb of the internal capsule to the cortex of the superior temporal gyrus. Most auditory impulses pass up to the opposite hemisphere but some pass to the same side.

Vestibular impulses arising in the specialised organs of balance in the saccule, utricle and semicircular ducts pass through the bipolar neurons of the vestibular nerve to the vestibular nuclei situated in the floor of the fourth ventricle. These nuclei comprise lateral, medial, superior and inferior groups. The lateral has secondary connections with the spinal cord through the vestibulospinal tract and all nuclei have connections which pass in the inferior cerebellar peduncle to the flocculonodular lobe of the cerebellum. Connections between the nuclei and the medial longitudinal bundle are concerned with reflex movements of the eyes, head and neck. Other fibres lead to the thalamus and thence to the cortex (superior temporal gyrus).

EFFERENT PATHWAYS (Fig. 35.1)

Efferent pathways passing from the cerebral cortex to the brain stem and spinal cord form two complementary systems, the pyramidal and the extrapyramidal.

Pyramidal fibres (corticospinal) arise mainly from the pyramidal cells of the motor cortex. The fibres pass through the internal capsule, the middle part of the cerebral peduncle and are split into smaller bundles by the transversely running fibres of the basilar part of the pons. The bundles reunite in the upper medulla to form the prominent pyramid. Most of the fibres decussate just above the junction of the medulla and spinal cord and form the lateral corticospinal tract. A smaller number of fibres do not cross in the lower medulla but descend as the anterior corticospinal tract and decussate caudally, at the appropriate segmental level. Some efferent fibres (corticonuclear) leave the main tract in the brain stem and end in the cranial nerve motor nuclei. A number of areas other than the precentral gyrus have motor functions. Part of the inferior frontal gyrus of the dominant side probably controls the motor aspects of speech. Other areas co-ordinate movements of the eye and of the hand.

Many efferent fibres (corticonuclear, and lateral and anterior corticospinal) probably end in intermediate neurons in the region of the cranial nerve motor nuclei or of the anterior horn cells. The intermediate neurons serve as a second neuron in the pyramidal pathways. Some pyramidal fibres synapse directly with the final neurons. The final neuron (lower motor neuron) is the cranial motor neuron or anterior horn cell whose axon passes to the motor end plates of striated muscle.

The **extrapyramidal system** is the name given to groups of fibres passing from the motor cortex via subcortical nuclear masses to the lower motor neurons. The subcortical nuclei are found in the basal nuclei and the subthalamus of the forebrain; the substantia nigra, red nucleus and reticular nuclei of the midbrain; the reticular and lateral vestibular nuclei of the hindbrain. This system has obvious cerebellar connections. The main descending pathways are the rubroreticulospinal and vestibulospinal tracts.

Other descending pathways leave (*a*) the tectum of the midbrain (especially the superior colliculi) and descend in the tectospinal tracts, (*b*) the olivary nuclei of the medulla and descend as the olivospinal tracts.

Cerebellar efferent pathways are considered on page 470.

The efferent pathways of the **autonomic nervous system**, innervating smooth muscle, cardiac muscle and glands, are ill-defined. Certain cortical areas, particularly the inferior surface of the frontal lobe and the hippocampus, are closely linked with the hypothalamus from which autonomic fibres descend. Sympathetic fibres originate in the posterior parts of the hypothalamus and descend to the thoracolumbar outflow (T1-L2), synapsing on the lateral horn cells of these segments. Parasympathetic fibres originate in the anterior parts of the hypothalamus and descend to the craniosacral outflow. The cranial fibres synapse in the 3rd, 7th, 9th and 10th cranial nerve nuclei and the sacral fibres synapse in the lateral horn cells of the 2nd, 3rd and 4th sacral segments.

— 36 —

The meninges and blood supply

Three membranes form a protective covering around the brain and spinal cord, an outer dense fibrous dura mater, an intermediate delicate avascular arachnoid and an inner delicate vascular pia mater. The dura and arachnoid are separated by a potential subdural space and the arachnoid and pia by a wider subarachnoid space containing cerebrospinal fluid. The arachnoid and pia can be looked on as different parts of one enclosing membrane, the pia-arachnoid.

THE DURA MATER

Cranial

The cranial dura mater is in two layers. The outer (endocranial) layer is the periosteum on the inner surface of the cranial bones and is thus continuous with the periosteum on the outer surface of the skull through the foramina and sutures. The inner (cerebral) layer of dura mater is continuous with that of the spinal cord. In most areas the two layers are firmly adherent but in places they are separated to enclose the venous sinuses. The inner layer is folded to form the falx cerebri and the tentorium cerebelli, and is stretched between the clinoid processes as the diaphragma sellae.

The **falx cerebri** lies in the midsagittal plane and separates the two cerebral hemispheres. It is attached to the vault of the skull from the crista galli to the internal occipital protuberance. It is sickle-shaped, narrower at its anterior end, and posteriorly blends with the upper surface of the tentorium cerebelli. The inferior sagittal sinus lies within the lower free border and the superior sagittal sinus within the attachment of the fold to the vault of the skull.

The **tentorium cerebelli** roofs in the posterior cranial fossa and separates the cerebral hemispheres from the cerebellum. Anteriorly is a hole for the midbrain. It thus has attached and free borders. The attached border extends from each posterior clinoid process along the superior border of the petrous temporal bone and then round to the internal occipital protuberance. The free border is at a higher level than the attached; it runs around the posterior and lateral aspects of the midbrain and can be followed forwards to the anterior clinoid processes. The superior petrosal and transverse venous sinuses lie along the attached border

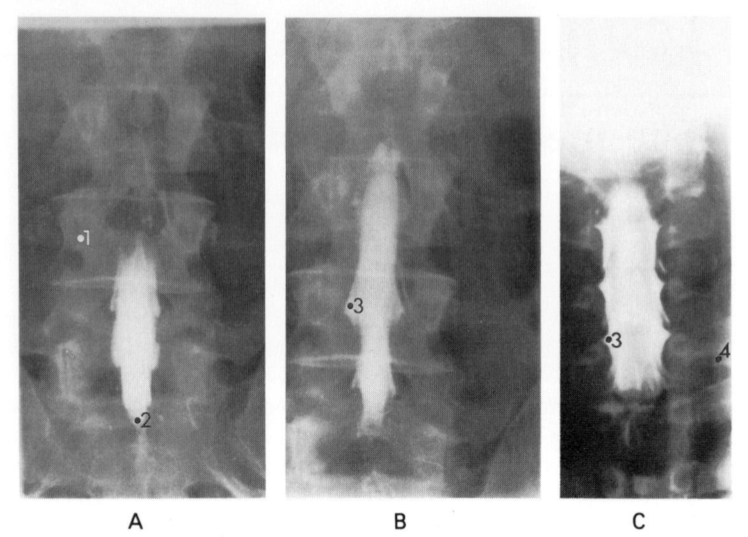

Fig. 36.1 Myelogram. Contrast material has been injected into the subarachnoid space through a lumbar puncture needle (p. 486) which has then been withdrawn. By tilting the patient this material, which is more dense than CSF, can be used to outline the various parts of the subarachnoid space. **A.** Lower end of subarachnoid space. **B.** Lumbar region. **C.** Cervical region.

1. 4th lumbar vertebra
2. Lower part of subarachnoid space
3. Meningeal sheath of nerve root
4. 1st rib

of the fold. In the midline the straight sinus passes backwards in the line of union of the falx cerebri to the tentorium cerebelli.

The **diaphragma sellae** is attached to the four clinoid processes and forms a roof for the hypophyseal fossa. In the midline it is evaginated and closely surrounds the pituitary gland. The medial wall of the cavernous sinus forms the lateral wall of the pituitary fossa.

The cerebral layer of dura forms a sheath around the cranial nerves as they leave the skull. The sheath so formed around the optic nerve extends to the sclera.

The cranial dura mater receives its blood supply from anterior ethmoidal, middle meningeal, internal carotid and vertebral arteries. It is innervated by meningeal branches of the trigeminal, glossopharyngeal, vagal and upper three cervical nerves from before backwards.

The dura lining the vault is much less firmly attached to the bones than is the dura over the base of the skull. Fractures of the base usually rupture the dura, and cerebrospinal fluid can escape. Fractures of the vault may injure the middle meningeal vessels but the blood clot is extradural and the dura usually remains intact.

Spinal

The spinal dura mater is continuous at the foramen magnum with the cerebral layer of cranial dura and extends downwards to the level of the second sacral vertebra. It lies within the vertebral canal separated from its walls by extradural fat and the internal vertebral venous plexus (p. 49). It is pierced by the ventral and dorsal roots of the spinal nerves and the filum terminale, and ensheaths these structures and the mixed spinal nerves as far as the intervertebral foramina. The spinal nerves and filum terminale help to stabilise the dural sac within the vertebral canal (Fig. 36.1).

THE ARACHNOID

This is closely applied to the cranial and spinal dura mater and fine trabeculae pass from it to the pia mater. In places the arachnoid herniates through the dura mater, forming arachnoid villi which come into contact with the endothelium of the cranial venous sinuses, especially the superior sagittal sinus and its lateral recesses. The clusters of villi seen in later life are known as arachnoid granulations. Through these villi cerebrospinal fluid is returned to the blood in the venous sinuses.

THE PIA MATER

This is closely applied to the surface of the brain and spinal cord. It extends into the sulci of the brain and the anterior median fissure of the cord and invests the cranial and spinal nerves, their roots and the filum terminale. On each side of the spinal cord the pia forms a serrated fold, the **ligamentum denticulatum**. The ligament runs the whole length of the cord and is attached to the dura mater, between the spinal nerves, by its serrations which pierce the arachnoid. The ligament helps to stabilise the spinal cord within the dural sheath.

Grey and white matter are absent over the roof of the third and fourth ventricles and the medial wall of the lateral ventricles. In these regions the pia mater lies in contact with the ependymal lining of the ventricles. The two layers fuse and are invaginated into the cavities by blood vessels. The folds so formed are known as **choroid plexuses**. The ependymal cells take on a secretory appearance and are concerned with the production of cerebrospinal fluid. A forward projecting fold of pia extends into the transverse fissure below the corpus callosum and is known as the **tela choroidea**. Choroidal arteries and the great cerebral vein lie in the tela. Lateral and inferior extensions of the tela invaginate the lateral and third ventricles forming their choroidal plexuses. Median and lateral apertures in the pia and ependyma of the roof of the fourth ventricle allow cerebrospinal fluid to pass from the ventricular system to the subarachnoid space.

The **subarachnoid space** is filled with cerebrospinal fluid and is narrowest over the cerebral hemispheres. In some regions, where the arachnoid stretches over subdivisions of the brain, larger spaces (**cisterns**) are formed. The larger of these are the **cerebellomedullary cistern** (**cisterna magna**) in the angle between the cerebellum and the medulla, the **interpeduncular** between the

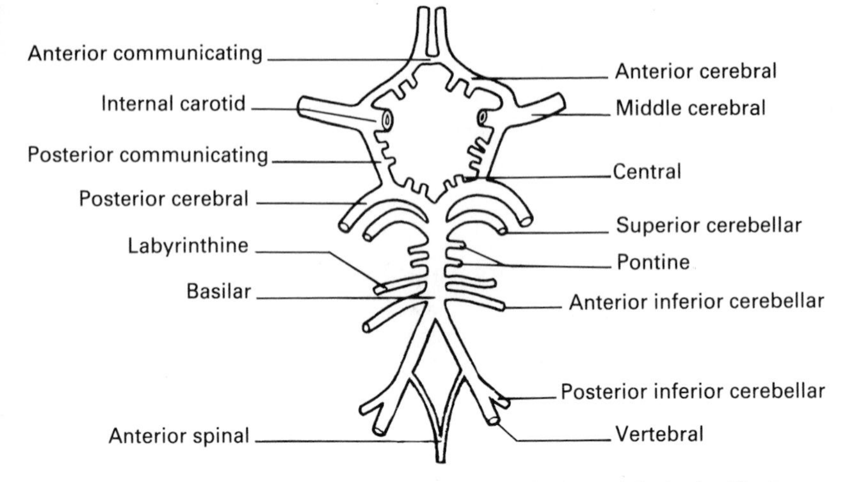

Fig. 36.2 Diagrammatic layout of the arterial circle on the base of the brain. The larger branches are shown.

cerebral peduncles, and the **pontine** in front of the pons. The cerebellomedullary cistern can be drained by a hollow needle inserted through the posterior atlanto-occipital membrane and the dura. The **lumbar cistern**, caudal to the end of the spinal cord, contains the cauda equina. A hollow needle can be pushed forwards in the midline between the spines of the lower lumbar vertebrae and after passing through the interspinous ligaments, it enters the extradural space. The needle then pierces the dura and arachnoid and enters the subarachnoid space. In this way a sample of cerebrospinal fluid can be aspirated for examination, the procedure being known as a lumbar puncture. A local anaesthetic can be introduced through such a needle into the subarachnoid space to produce regional anaesthesia; the procedure is known as spinal anaesthesia. Air, or other contrast media, may be introduced into the subarachnoid space to help radiological diagnosis (Fig. 36.1). The subarachnoid space extends along the bundles of olfactory nerves which pierce the cribriform plate, and along the optic nerve as far as the eyeball. Some effects of increased intracranial pressure can be observed by inspection through an ophthalmoscope of the optic disc.

Circulation of the cerebrospinal fluid

Most of the cerebrospinal fluid is produced by the choroid plexuses of the ventricles by active secretion of the modified ependymal cells. It passes through the cavities of the brain and then through the apertures in the roof of the fourth ventricle into the subarachnoid space. The fluid circulates around the outside of the brain and spinal cord, and is returned to the blood stream either through the arachnoid villi and granulations or directly to the small veins. Some fluid may be absorbed into perineural lymph vessels. Interference with the circulation of cerebrospinal fluid produces a condition known as hydrocephalus. If the blockage

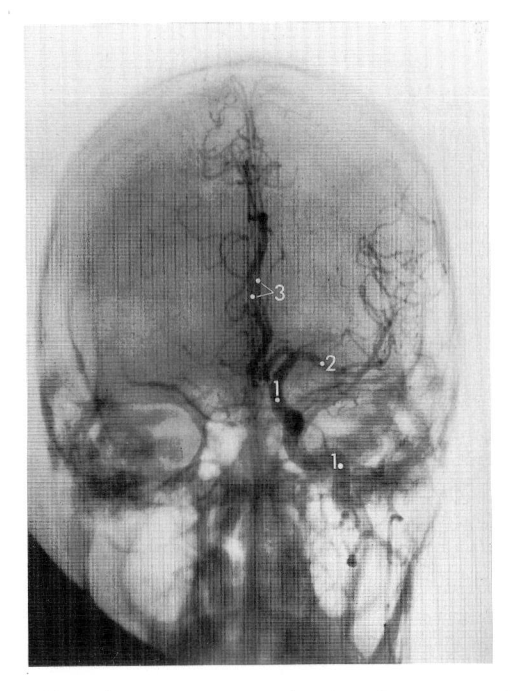

Fig. 36.3 Left carotid arteriogram, anteroposterior view. Contrast material has been introduced into the common carotid artery in the neck.
1. Internal carotid artery
2. Middle cerebral artery
3. Anterior cerebral arteries

is within the ventricular system (such as with pressure from a tumour) the ventricle will become dilated proximal to the level of obstruction—a condition known as internal hydrocephalus. Blockage of the arachnoid granulations (as with infection) produces communicating (external) hydrocephalus.

THE BLOOD SUPPLY OF THE BRAIN

Owing to the unyielding nature of the skull and spinal dura, and the relative incompressibility of the brain and cerebrospinal fluid, the vascular system is very susceptible to changes in intracranial pressure. The pattern of the vessels and the rate of flow through them can be investigated radiologically after the injection of contrast medium into the large arteries in the neck. Local cerebral ischaemia (see p. 18) occurring as a result of cerebral or general degenerative arterial disease may produce variable degrees of muscle weakness, speech disturbance, somatosensory and visual loss, the condition being known as a 'stroke'. Similar symptoms may be produced by pressure on a cortical area

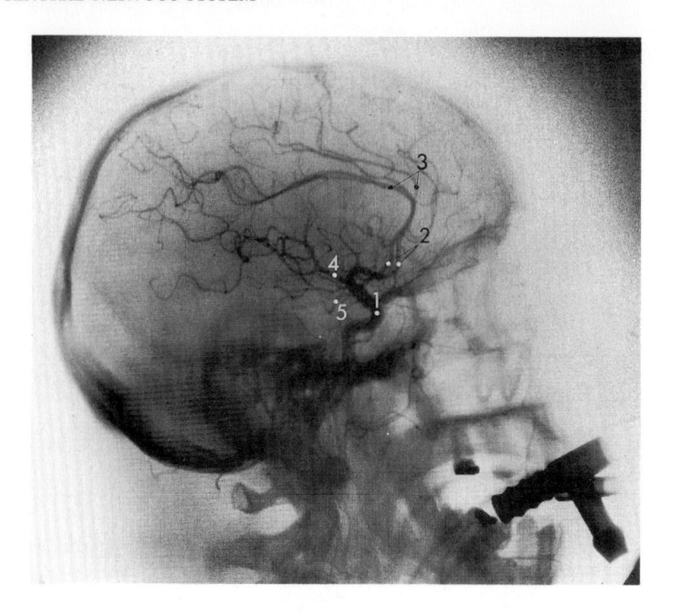

Fig. 36.4 Left carotid arteriogram, lateral view. Arterial phase. Contrast material has been introduced into the common carotid artery in the neck. Part of the anaesthetic tube is seen in the lower right corner.

1. Internal carotid artery
2. Anterior cerebral arteries
3. Branches of anterior cerebral artery above the corpus callosum
4. Middle cerebral artery
5. Posterior communicating artery

resulting from injury, local inflammation, or a tumour. Occasionally, an aneurysm (p. 169) of a cerebral artery in the subarachnoid space ruptures (a subarachnoid haemorrhage) or a meningeal vessel may be torn by a fracture of the skull. In both cases the patient shows the signs of increasing intracranial pressure.

ARTERIAL SUPPLY (Figs. 36.2, 36.3, 36.4)

The brain receives its blood from the two internal carotid and the two vertebral arteries.

Internal carotid artery (see also p. 426)
This enters the cranial cavity through the carotid canal in the petrous temporal bone, traverses the cavernous sinus, piercing its roof near the anterior clinoid process, and then divides into anterior and middle cerebral arteries (Fig. 36.4).

Branches

(i) **anterior cerebral artery**—passes forwards towards the midline and then upwards around the genu of the corpus callosum. It divides into two terminal

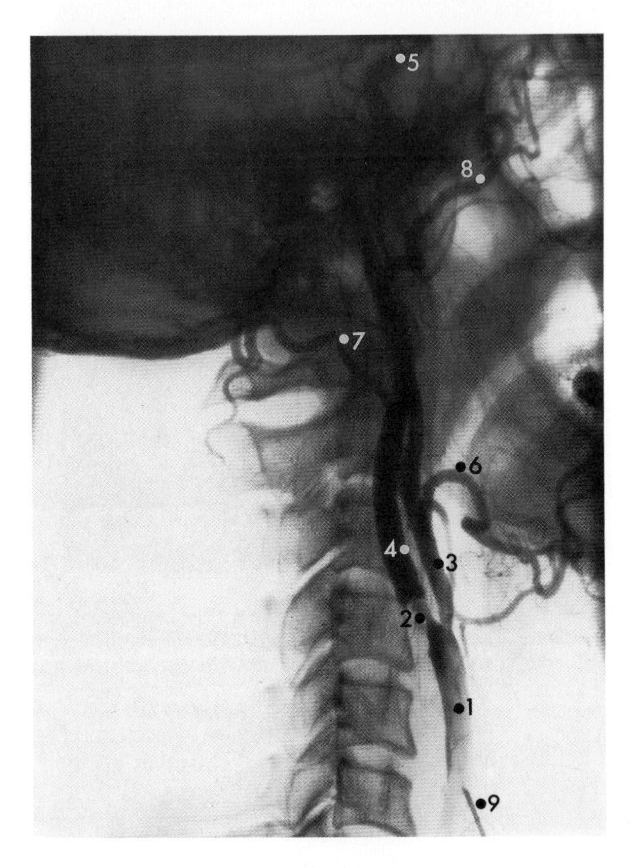

Fig. 36.5 Lateral view of carotid arteriogram showing gross stenosis and deformity at the origin of the internal carotid artery due to arterial disease.

1. Common carotid artery
2. Stenosis of internal carotid artery
3. External carotid artery
4. Internal carotid artery
5. Internal carotid artery in region of the cavernous sinus
6. Facial artery
7. Occipital artery
8. Maxillary artery
9. Needle in common carotid artery for injecting contrast material.

branches which run backwards on the corpus callosum and in the cingulate sulcus. The artery supplies the medial aspect of the frontal and parietal lobes as far back as the parieto-occipital sulcus, and a small part of the adjacent lateral surface. The two anterior cerebral arteries are united near their origin by the short **anterior communicating artery**.

(ii) **middle cerebral artery**—passes laterally into the lateral sulcus in which

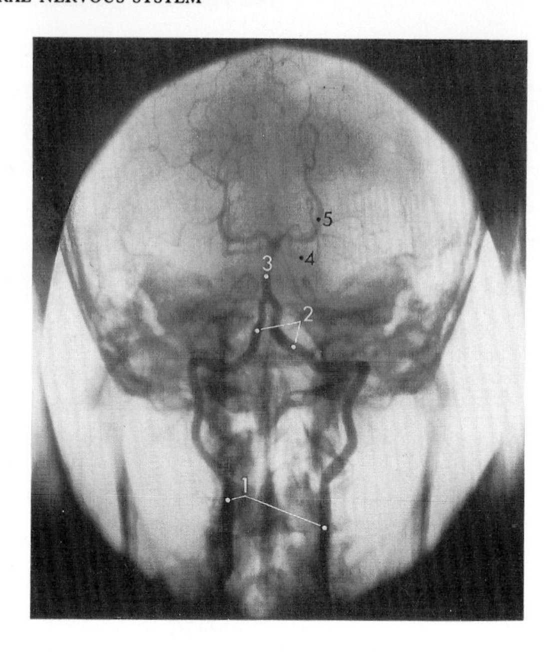

Fig. 36.6 Vertebral arteriogram, anteroposterior view. The contrast material has been introduced into a vertebral artery. The contralateral artery is also demonstrated since the material is introduced under pressure and some passes retrogradely down to the other side.

1. Vertebral arteries (in foramina transversarii)
2. Vertebral arteries inside skull
3. Basilar artery
4. Superior cerebellar artery
5. Posterior cerebral artery

it crosses the temporal pole and insula and turns upwards and backwards across the lateral aspect of the hemisphere. The artery supplies the inferior surface of the frontal lobe, the insula and all save the periphery of the lateral surface of the hemisphere.

(iii) **ophthalmic artery**—see page 362.

(iv) **anterior choroidal artery**—arises in the subarachnoid space, passes backwards with the optic tract and contributes to the blood supply of the choroid plexus in the inferior horn of the lateral ventricle.

(v) **posterior communicating artery**—passes backwards in the subarachnoid space to the posterior cerebral artery.

(vi) branches to the pituitary gland, meninges and trigeminal ganglion.

Vertebral artery (see also p. 422)

This enters the cranial cavity through the foramen magnum, passes forwards, upwards and medially on to the basal part of the occipital bone and, uniting with its fellow of the opposite side at the lower border of the pons, forms the basilar artery (Figs. 36.2, 36.6, 36.7).

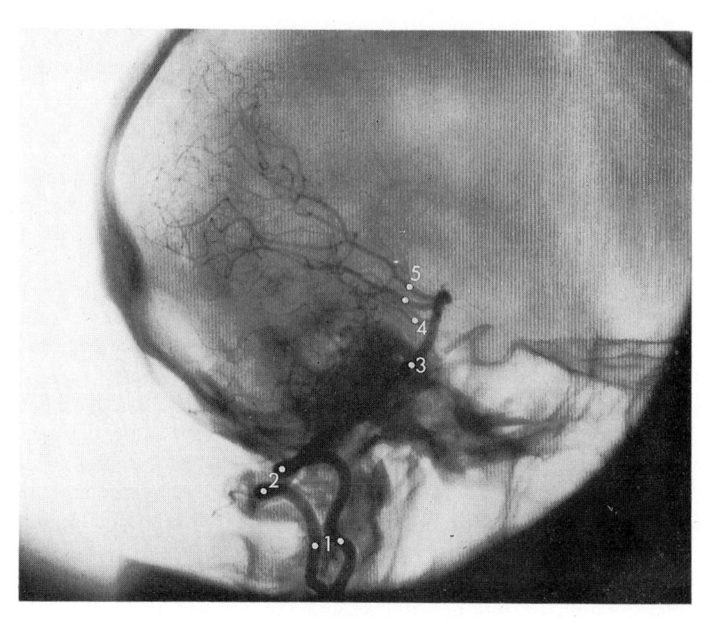

Fig. 36.7 Vertebral arteriogram, lateral view (see Fig. 36.6).
1. R & L vertebral arteries
2. Vertebral loops round lateral mass of the atlas
3. Basilar artery
4. Superior cerebellar artery
5. Posterior cerebral arteries

Branches

(i) **anterior spinal artery**—a single vessel is formed by the union of branches from both vertebral arteries. It descends in front of the medulla and reaches and supplies the spinal cord.

(ii) **posterior inferior cerebellar artery**—passes posteriorly and supplies the posterolateral part of the medulla, the lower cerebellum, a choroidal branch to the fourth ventricle, and a posterior spinal branch which descends to the spinal cord.

Basilar artery

This ascends in front of the pons and divides near its upper border into the two posterior cerebral arteries.

Branches

(i) **anterior inferior cerebellar artery**—passes backwards to its named area.

(ii) **pontine branches.**

(iii) **labyrinthine artery**—passes through the internal acoustic meatus to the inner ear.

(iv) **superior cerebellar artery**—passes backwards around the cerebral peduncles and supplies the superior surface of the cerebellum.

(v) **posterior cerebral artery**—passes backwards around the cerebral peduncles and supplies the medial surface of the occipital lobe. Branches supply the inferior surfaces of the occipital and temporal lobes and the adjacent lateral surface. Its **posterior choroidal** branch passes into the tela choroidea and supplies the choroid plexuses of the third and lateral ventricles.

CIRCLE OF WILLIS

The principal arteries supplying the forebrain form an anastomotic ring, the circle of Willis, around the optic chiasma and the pituitary gland (Fig. 36.2). It is formed by the stem of the anterior and middle cerebral arteries of each side anterolaterally, the posterior cerebral arteries posteriorly, the anterior communicating artery uniting the two anterior cerebral arteries, and the posterior communicating arteries uniting the corresponding posterior cerebral arteries to the middle cerebral or internal carotid arteries. The pattern of this anastomosis, however, is variable. A number of branches from the circle and its larger tributaries pass upwards into the nuclear masses of the forebrain, entering the brain mainly through the anterior and posterior perforated substances. These perforating vessels supply the internal capsule and the adjacent thalamus and basal nuclei. The anterior cerebral artery supplies the upper part of the sensorimotor cortex (leg and foot areas). The middle cerebral artery supplies the remainder of the sensorimotor cortex and also the auditory, speech and gustatory areas. The posterior cerebral artery supplies the visual area (the striate calcarine cortex).

VENOUS DRAINAGE

The veins draining the hemispheres may be divided into superficial and deep groups which anastomose freely and drain to the cranial venous sinuses.

Superficial group

(i) the superior cerebral veins, passing to the superior sagittal sinus.

(ii) the superficial middle cerebral vein passing in the lateral sulcus to the cavernous sinus.

(iii) the inferior cerebral veins passing to the transverse sinus. Superior and inferior anastomotic veins join the superficial middle cerebral vein to the superior sagittal and transverse sinuses respectively.

Deep group

(i) the anterior cerebral vein, draining the region supplied by its artery, joins

(ii) the deep middle cerebral vein, lying deeply in the lateral sulcus and forms the basal vein. This passes backwards round the midbrain and joins the great cerebral vein near the pineal gland.

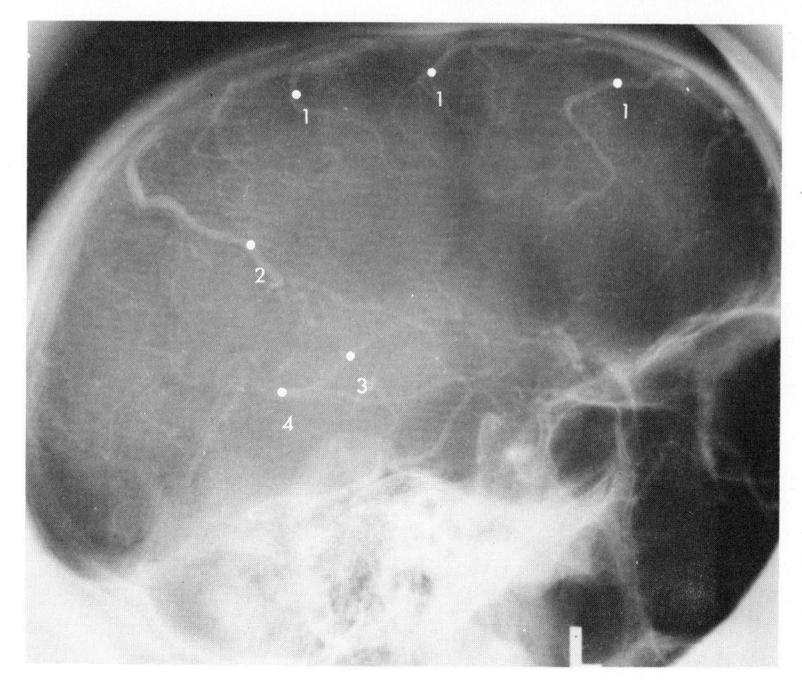

Fig. 36.8 Left carotid arteriogram—later exposure. Venous phase (see Fig. 36.4).
1. Superior cerebral veins
2. Superior anastomotic vein
3. Internal cerebral vein
4. Origin of great cerebral vein

(iii) each internal cerebral vein receives the thalamostriate vein, draining the caudate nucleus and thalamus, and the choroidal veins of the lateral and third ventricles.

(iv) the **great cerebral vein**, formed when the two internal cerebral veins join, lies in the tela choroidea in the transverse fissure and opens into the straight sinus.

THE CRANIAL VENOUS SINUSES

These lie either between the two layers of cranial dura mater or within a fold of the cerebral layer. The sinuses receive venous blood from adjacent parts of the brain and communicate via other sinuses and emissary veins with the exterior of the skull. The sinuses may be unpaired or paired.

UNPAIRED SINUSES

(i) ***superior sagittal sinus***—passes backwards in the upper border of the falx cerebri and reaches the internal occipital protuberance where it usually joins the right transverse sinus. Its walls are invaginated by many arachnoid granulations (p. 485), as are the lateral recesses of the sinus.

(ii) **inferior sagittal sinus**—passes backwards in the lower border of the falx cerebri and joins the straight sinus at the attachment of the falx to the tentorium.

(iii) **straight sinus**—runs downwards within the tentorium cerebelli to the internal occipital protuberance, usually joining the left transverse sinus.

(iv) **basilar sinus**—is a network of sinuses lying on the basal part of the occipital bone and uniting the two inferior petrosal sinuses.

(v) **intercavernous sinuses**—unite the cavernous sinuses anterior and posterior to the pituitary gland.

PAIRED SINUSES

(i) **transverse sinus**—commences at the internal occipital protuberance, the right usually continuous with the superior sagittal sinus and the left with the straight sinus. Each runs along the attached border of the tentorium, grooving the occipital and parietal bones, to the lateral part of the petrous temporal bone where it becomes the sigmoid sinus. The two transverse sinuses often communicate at the internal occipital protuberance and form the confluence of the sinuses.

(ii) **sigmoid sinus**—is a continuation of the transverse sinus and is formed at the lateral end of the superior border of the petrous temporal bone. It curves downwards and then forwards, grooving the inner surface of the mastoid process, and reaches the jugular foramen where it becomes the internal jugular vein.

(iii) **cavernous sinus** (Fig. 36.9)—lies on the body of the sphenoid bone lateral to the pituitary gland. It is traversed by the internal carotid artery, the oculomotor, trochlear and abducent nerves and the ophthalmic and maxillary branches of the trigeminal nerve. It extends from the apex of the petrous temporal bone and the opening of the carotid canal forwards to the superior orbital fissure. Its roof is continuous with the diaphragma sellae which lies under the hypothalamus and it is pierced in front by the internal carotid artery. Its lateral wall is related to the temporal lobe of the brain and within this wall lie the oculomotor, trochlear, ophthalmic and maxillary nerves. Posteriorly the wall is related to the trigeminal ganglion. The internal carotid artery passes upwards and forwards through the sinus and pierces its roof near the anterior clinoid process. The abducent nerve passes forwards and downwards through the sinus lateral to the artery. The maxillary nerve is only in the sinus for a short distance. All the nerves and the artery, though inside the dural cleft, are outside the endothelial lining of the sinus.

The sinus receives the ophthalmic veins, the sphenoparietal sinus, the superficial middle cerebral vein and smaller veins from the adjacent parts of the brain. It drains via emissary veins to the pterygoid venous plexus, and via the superior and inferior petrosal sinuses to the transverse sinus and internal jugular vein respectively. The two sinuses communicate via the intercavernous sinuses.

(iv) **inferior petrosal sinus**—arises from the cavernous sinus, descends over

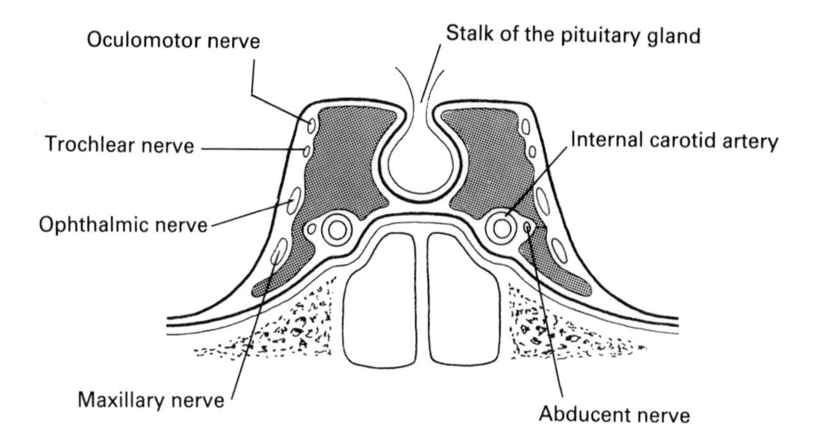

Fig. 36.9 Coronal section through the pituitary fossa showing the structures related to the cavernous sinus (shown in stipple).

the petro-occipital suture, passes through the jugular foramen and joins the internal jugular vein.

(v) **superior petrosal sinus**—passes along the superior border of the petrous temporal bone uniting the cavernous and transverse sinuses.

(vi) **sphenoparietal sinus**—runs along the edge of the lesser wing of the sphenoid bone to the cavernous sinus.

(vii) **occipital sinus**—passes from the transverse sinus near the internal occipital protuberance, around the foramen magnum to the sigmoid sinus.

Emissary veins

These pass through most foramina of the skull and unite the intracranial venous sinuses with the veins outside the skull. They include the condylar, parietal and mastoid veins, and veins passing through the foramen ovale, foramen lacerum and hypoglossal foramen.

Diploic veins

These are present in the frontal, parietal and occipital bones. They drain the diploë and communicate with the veins of the scalp and the dura.

BLOOD SUPPLY OF THE SPINAL CORD

Arterial supply—is from a single **anterior spinal artery** (derived from both vertebral arteries) and two **posterior spinal arteries** (branches of the posterior inferior cerebellar arteries). The anterior artery descends in the anterior median fissure and each posterior artery behind the posterior spinal nerve rootlets. These three arteries are reinforced in different regions by branches passing through the intervertebral foramina from the vertebral, deep cervical, posterior intercostal, lumbar and lateral sacral arteries. The lumbar arteries provide a rich supply to the lumbar enlargement.

Venous drainage—is via midline anterior and posterior spinal veins, and smaller channels near the emergence of the rootlets. They drain to the internal vertebral venous plexus (p. 49).

PART EIGHT

Clinical examination

— 37 —

The anatomy of clinical examination

METHODS

Clinical (bedside) examination involves the assessment of each region of the body by inspection, palpation, percussion and auscultation. To this is added the assessment of active and passive movement at the joints. In this chapter on the anatomy of clinical examination, structures that can be inspected and palpated have been outlined on the model together with some deeper structures, a number of which can be assessed by percussion and auscultation.

Inspection requires a warm, appropriately secluded area so that the region being studied is fully exposed and in optimal lighting. The patient is usually lying on his back on a couch and a doctor stands on the right side firstly having washed and warmed his/her hands. The skin and contours are examined from different angles to identify superficial structures, vascular pulsation, and movements such as those related to respiration, swallowing and coughing.

Palpation of superficial structures delineates their shape, surface and consistency. Bony contours can be examined when they are not covered by excess muscle and soft tissue, and some superficial nerves and glands are palpable. Lymph nodes such as those inferior to the angle of the mandible, in the groin and in the axilla are often palpable and regular examination of these sites is required to appreciate the range of normality.

Movement against resistance will make a muscle belly more prominent. It will thus be more easily palpable, and structures deep to it will become less prominent. A superficial artery is most easily palpated when it can be compressed on an adjacent bony surface. The pulps of two or more fingers are placed together along the line of the artery and the distal finger pressed on to the bone to compress the vessel. The proximal finger(s) resting gently on the skin are then used for palpation. The heart beat can usually be felt by placing the flat of the right hand on the left anterior chest wall. The flat of the hand will also demonstrate chest movements during respiration. Palpation of the abdomen allows a number of abdominal viscera to be felt.

If the middle finger of the right hand is tapped on a table top first over a central unsupported area and then over one of its legs, the difference between the hollow and the firmer areas will be both heard and felt. If the palm of the left hand is now placed over the same two areas, and the right middle finger used to tap (staccato) the dorsum of the middle phalanx of the left middle finger, the differences of sound and feel will be magnified. This principle is extended into clinical examination by the process known as percussion. The palm of the left hand is placed over a body cavity or organ and tapping is carried out as already described. Air-filled organs such as the lungs and the gut sound more hollow than more solid organs such as the heart and liver. Tapping bony surfaces such as the clavicle and skull vault directly with the middle finger can be used to compare the two sides of the body.

Auscultation is the process of listening over the body. This was originally carried out by applying an ear directly over an area. Development of the stethoscope has simplified the manoeuvre and serves to conduct and localise underlying sounds. These sounds include the closing of heart valves, blood flow in some arteries and the movement of air in the trachea, lungs and gut. The bell of the stethoscope is small enough to give easy access to most areas of the body but the diaphragm is better for detecting low pitched sounds.

When assessing the extent of movement of a joint, active movement in all planes is first carried out by the subject. Observation of these movements indicates any limitation which might suggest discomfort and, after discussion with the subject, the examiner undertakes passive movement to assess whether this is greater than active or in any way abnormal. The bones and soft tissues round the joint are then palpated. Resting a hand over a joint during movement may elicit a soft grating sensation known as crepitus, due to slight irregularity of the bony surfaces.

Tactile sensation is assessed by the light touch of a finger or a piece of cotton wool, and pain with a gently applied pin. Muscle tone, power, co-ordination and reflexes are measures of motor function.

GENERAL EXAMINATION

Examination of a subject in clinical practice commences with observations of general features obtaining an impression of the health of the individual. This is made from examination of the head, neck, hands, feet and the shape of the body without exposing the trunk. It is followed by regional examination of the subject from the head to the feet. This sequence partly follows the systematic pattern by which the examination is reported but detailed examination of the nervous system is undertaken as a separate entity so that sensory and motor function is compared throughout the body. It is important to develop a set routine for a complete examination, incorporating thoroughness and later developing speed of execution.

The general health of the subject is assessed by the facies, colour, nutrition, hair distribution and shape. Note obesity, and lax skin indicating recent weight loss. General pallor is related to reduced circulating haemoglobin (anaemia) and is best estimated by gently pulling down the lower eyelid and by so doing everting it; the colour of the inside of the lid is then assessed. Yellow pigmentation due to raised circulating bilirubin (jaundice) is assessed by the degree of staining of the white sclera of the eyeball. In this general clinical examination particular attention is given to the hands, the palms, the nails and the skin over the dorsum of hands and feet, providing evidence of nutritional and other pathological disturbances.

THE HEAD AND NECK

The reader can examine his/her own face in a mirror. Observe symmetry, and changes of facial expression, together with the shape of the eye (canthi, corneas, pupils), nose, mouth and ears. On opening the mouth observe the oral cavity with the teeth, tongue, palate and anterior pillars of the fauces. The oropharynx contains the tonsils and posterior pillars of the fauces. Tongue and palate move-

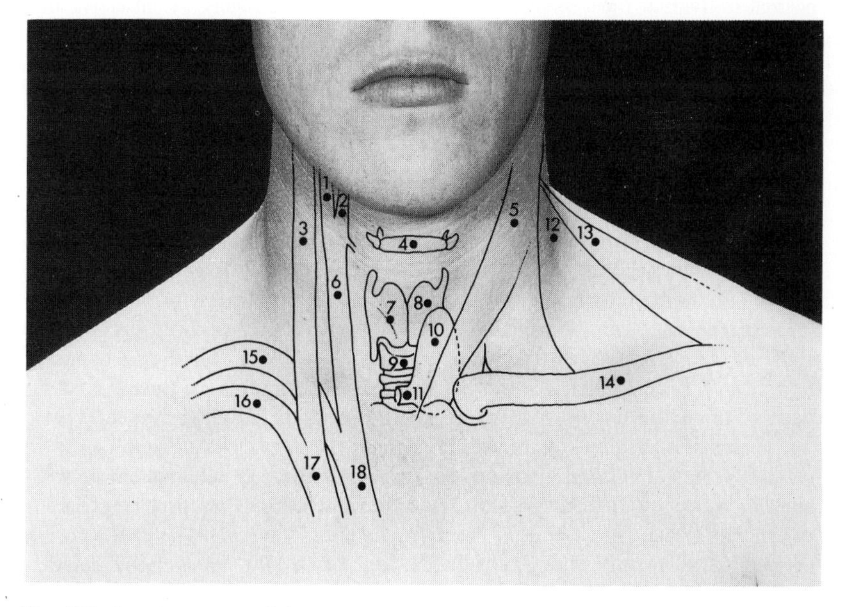

Fig. 37.1 Anterior aspect of the neck
1 Internal carotid artery
2 External carotid artery
3 Internal jugular vein
4 Body of the hyoid bone
5 Sternocleidomastoid muscle
6 Common carotid artery
7 Oblique line on right lamina of thyroid cartilage
8 Left lamina of thyroid cartilage
9 Cricoid cartilage
10 Left lateral lobe of the thyroid gland
11 Isthmus of the thyroid gland
12 Left posterior triangle of the neck
13 Trapezius muscle
14 Clavicle
15 Right subclavian artery
16 Right subclavian vein
17 Right brachiocephalic vein
18 Brachiocephalic artery

ments are demonstrated by asking the subject to stick out the tongue and to say "ah". Palpate the margins of the orbit noting the superior orbital notch, supraorbital ridge and the zygomaticofrontal suture. Pass on to the zygomatic arch, superior temporal ridge, the external occipital protruberance, superior nuchal line and the mastoid process. Note the feel of the hair of the eyebrows and head and its distribution.

The anterior of the neck (Fig. 37.1) is bounded superiorly by the lower border of the mandible and inferiorly by the upper border of the manubrium sterni and the clavicles which are subcutaneous throughout their length. The prominence of the thyroid cartilage can be seen in the midline, the thyroid angle being more acute and the prominence more accentuated in the adult male. The larynx can be seen to be elevated during swallowing. The anterior borders of the sternocleidomastoid muscles form a V on each side of the larynx, extending from the upper border of the manubrium sterni to the mastoid process. The carotid sheaths are situated on each side of the larynx mainly deep to the sternocleidomastoid muscle. The internal jugular veins lie lateral to the arteries. In the normal individual the venous blood pressure is approximately 11 cm above that in the right atrium. By positioning a subject with the head rotated to one side and resting on a pillow and the upper half of the body raised to an angle of approximately 30°, the character of the pulse wave in the internal jugular vein can be observed.

Features of the posterior aspect of the neck are the lateral borders, bounded by the trapezius muscles and the prominent spines of the 7th cervical and the lst thoracic vertebrae, the former being known as the vertebra prominens.

The body, angle and posterior border of the ramus of the mandible are superficial and palpable (Fig. 37.2). On opening the mouth the head of the mandible can be felt sliding forward below the zygomatic process. The tip of the transverse mass of the atlas can be palpated deeply, approximately midway between the angle of the mandible and the tip of the mastoid process. The hyoid bone and thyroid and cricoid cartilages can be palpated in the midline of the neck. If the body of the hyoid bone is held between the finger and thumb and pressed backwards and moved from side to side the greater horns can be felt rubbing on the cervical vertebral column. Much of the thyroid cartilage is palpable but only the anterior aspect of the cricoid and below this one or two tracheal rings before passing on to the soft isthmus of the thyroid gland. The remainder of the normal thyroid gland is not easily palpable. It is most easily palpated by standing behind a sitting subject.

The submandibular gland is situated inferior to the body of the mandible just anterior to the angle. It is most easily palpated bimanually, that is, its shape and consistency being assessed between the finger of one hand inside the mouth and the fingers of the other hand on the neck.

The order of palpation of cervical lymph nodes is not critical provided all groups are included. They may be examined from the front or behind as for the thyroid gland. If the former, simultaneous examination of the two sides can be undertaken with the pads of the middle two or three fingers of both hands. Examine in turn the occipital, postauricular, pre-auricular, submandibular and

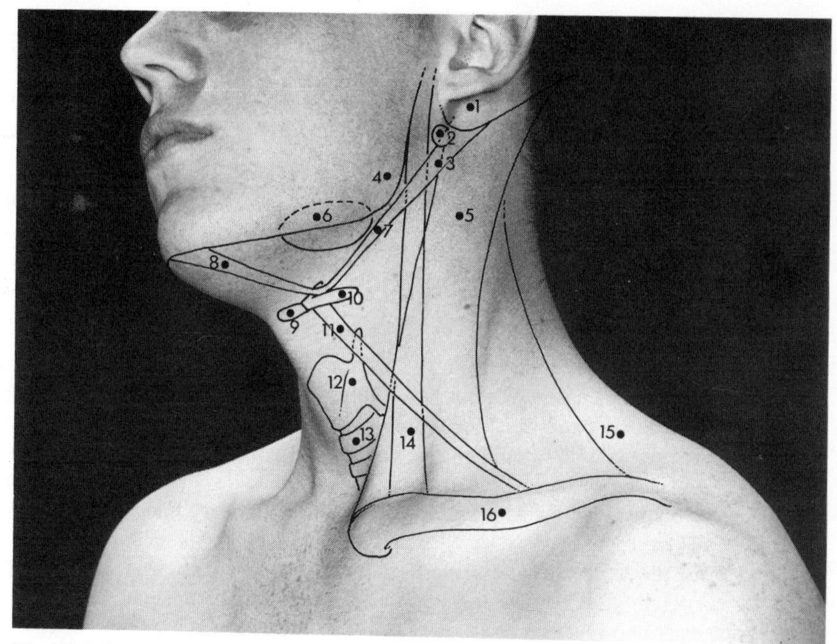

Fig. 37.2 Lateral aspect of the left side of the neck
1 Mastoid process
2 Transverse mass of the atlas
3 Posterior belly of the digastric muscle
4 Angle of the mandible
5 Sternocleidomastoid muscle
6 Submandibular gland
7 Intermediate tendon of the digastric muscle
8 Anterior belly of the digastric muscle
9 Body of the hyoid bone
10 Left greater horn of the hyoid bone
11 Omohyoid muscle
12 Thyroid cartilage
13 Cricoid cartilage
14 Left internal jugular vein
15 Trapezius muscle
16 Clavicle

submental nodes, followed by palpation downwards along the anterior border of the sternocleidomastoid muscle to the clavicle, thus palpating the internal jugular lymph chain. Passing backwards the supraclavicular nodes are examined and then, by ascending along the posterior border of the sternocleidomastoid muscle, the external jugular lymph chain. Additional nodes may be present anterior to the larynx and along the anterior jugular vein.

A large node can be missed deep to the lower end of the sternocleidomastoid muscle if not carefully looked for by gently squeezing the relaxed lower end of the muscle between finger and thumb. A sternocleidomastoid muscle can be tensed by turning the chin to the opposite side against the resistance of the palm of a hand. In this way it can be determined whether any palpable lymph node or other mass is situated deep or superficial to the muscle.

Palpable pulses in the head and neck include the subclavian, compressed on to the first rib behind the middle of the clavicle, the carotid bifurcation pressed posteriorly on to the vertebral column at the level of the hyoid bone and the facial artery, as it crosses the body of the mandible just anterior to the masseter muscle, this muscle border being demonstrated by clenching the teeth. The superficial temporal artery is palpated as it crosses the zygomatic arch just in front of the ear, and the occipital artery on the skull about half way between the external occipital protuberance and the mastoid process. Auscultation over the subclavian and carotid vessels may demonstrate normal or abnormal turbulent flow.

THE THORAX

Inspection

The anterior chest wall extends from the clavicles superiorly to the inferior costal margins made up by the lower six ribs and costal cartilages. In the thin male these bony landmarks and much of the rib cage are visible. The two sides of the chest are usually symmetrical. The rate, depth and character of respiration can be noted, as can be the apex beat of the heart over the left chest. The pectoralis major muscles are usually visible and form the anterior axillary folds. The deltoid muscles form the rounded contour of the shoulder. In the female the breasts extend from the second to the seventh intercostal spaces, and from the lateral sternal border to the anterior axillary line. The nipples in the male lie in the midclavicular line.

Palpation

The sternum can be felt from the jugular notch to the xiphisternum, the latter being of variable size and shape. The sternal angle is the junction of the manubrium and body of the sternum. The second costal cartilages articulate at this point and act as markers for counting the ribs. Chest movements can be magnified by placing the flat of the hands on the chest wall with the thumbs pointing towards each other. The hands are separated and the ribs are more horizontally placed in inspiration than in expiration. Feeling the reverberations with the flat of the hand over the chest wall when a subject talks is known as tactile vocal fremitus. This is accentuated in certain lung diseases.

The apex beat of the heart is situated in the left fifth intercostal space in the midclavicular line. In the male it is usually inside the nipple line. It can usually be felt by placing the flat of the hand over this site. The breast is palpated with the flat of the right hand, gently compressing it against the chest wall.

The axillary contents are palpated with the subject's arm slightly abducted, flexed at the elbow and the forearm resting across the chest. The examiner's left hand is used to examine the right axilla and the right hand the left axilla. During this manoeuvre the nonpalpating hand holds the elbow of the subject to support the weight of the arm. The apical and medial nodes are palpated with a cupped hand passing up the lateral aspect of the axilla and feeling the axillary contents against the chest wall as the hand is drawn downwards and medially. Other groups of lymph nodes are examined by compressing the axillary contents against the anterior and posterior axillary folds and laterally against the humerus. The axillary artery can also be felt in the last position.

Percussion

The lungs are resonant to percussion as can be demonstrated over the clavicles, the upper chest and axillary regions. Passing from the last region towards the midline on the left side a duller sound is heard and felt as the left lateral border of the anteriorly placed heart is encountered. The right border of the heart usually lies behind the sternum and is not easily detected in a normal person. It is generally easier to detect borders by percussion from resonant to dull. Percussion downwards from the clavicle on the right side will demonstrate the dull sound produced by the upper border of the liver at about the fourth intercostal space.

Auscultation

If the bell of a stethoscope is placed over the upper anterior chest and the axillary regions, air will be heard to enter the lungs in inspiration, the noise extending slightly into the beginning of expiration. The sound stops during the remainder of expiration. This normal pattern is known as vesicular breathing. If the stethoscope is placed over the larynx a noise will be heard during both inspiration and expiration with a break between. This is known as bronchial breathing. The latter pattern may also occur over the peripheral area of the chest in certain lung diseases. If the patient speaks during auscultation the sound is transmitted to the chest wall—this is known as vocal resonance and its character may be altered by lung disease. In disease there may also be added noises to the respiratory pattern.

On listening over the heart two distinct noises will be heard close together at the beginning of each beat. These are usually referred to as "lubb-dupp". The first is produced mainly by the closure of the mitral and tricuspid valves and the second by the aortic and pulmonary valves. Each valve is heard more clearly over specific areas of the anterior chest but the two valves in the systemic circulation provide most of the noise. The mitral valve is heard best over the apex beat. If this is not palpable, place the stethoscope in the left fifth intercostal space in line with the nipple. Auscultation of this area in the female may require raising a pendent breast. The aortic area is just to the right of the sternum in the first intercostal space and the pulmonary area at an equivalent site to the left of the sternum. The tricuspid area is over the right side of the sternum at the level of the fourth costal cartilage.

Abnormalities of the heart and its valves may alter the rate, rhythm and character of the sounds and produce added sounds to the cycle. Fig. 37.3 shows the surface markings of the lungs, heart and great vessels. Note that the apex of each lung extends above the midpoint of the clavicle and the left lung has a notch out of its lower anterior border to accommodate the presence of the heart. The arch of the aorta is situated in the superior mediastinum, that is, above a line joining the sternal angle to the lower border of the body of the fourth thoracic vertebra.

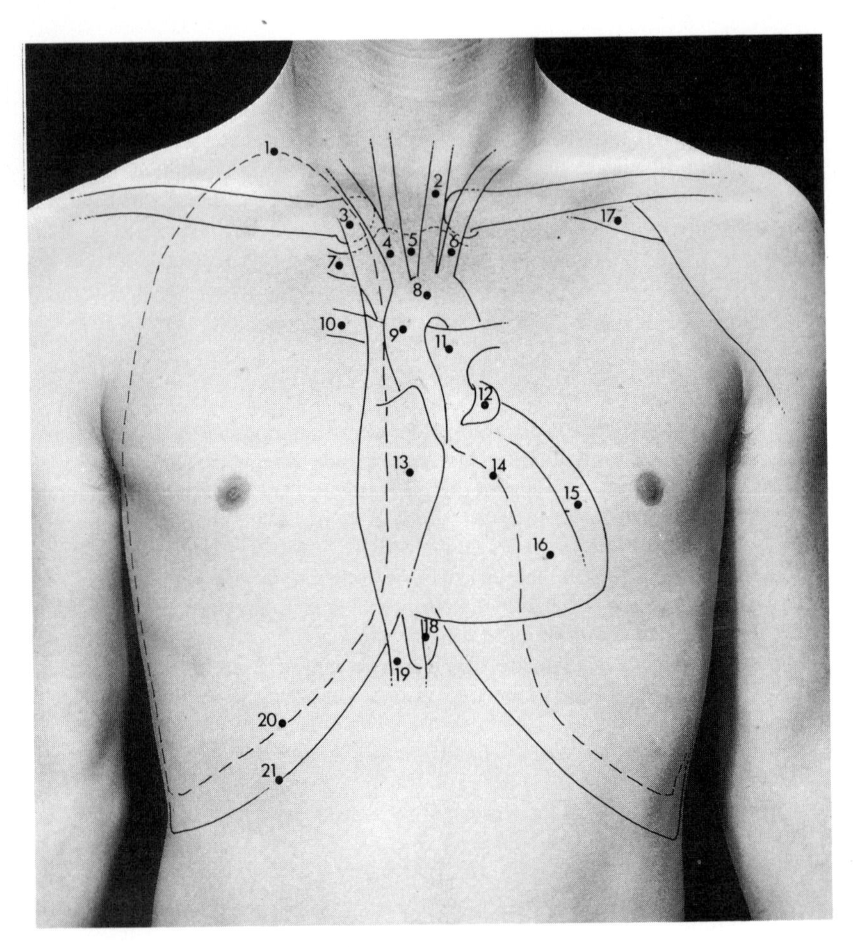

Fig. 37.3 Anterior aspect of the thorax

1 Apex of the right lung
2 Left common carotid artery
3 Right brachiocephalic vein passing behind the medial end of the clavicle
4 Brachiocephalic artery
5 Manubrium sterni
6 Left subclavian artery
7 First right costochondral junction
8 Aortic arch
9 Ascending aorta
10 Right pulmonary artery
11 Pulmonary trunk
12 Left auricular appendage
13 Right atrium
14 Cardiac notch in left lung and pleura
15 Left ventricle
16 Right ventricle
17 Deltopectoral groove
18 Xiphisternum
19 Inferior vena cava
20 Lower border of the right lung
21 Lower border of the right pleural cavity

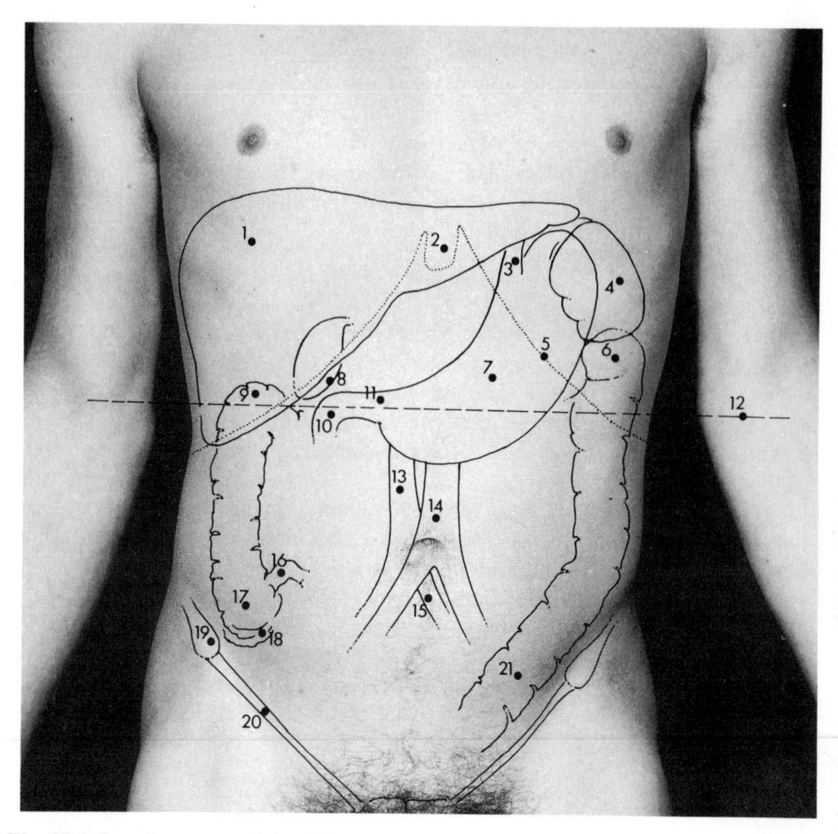

Fig. 37.4 Anterior aspect of the abdomen
1 Liver
2 Xiphisternum
3 Oesophagus
4 Spleen
5 Left costal margin
6 Splenic flexure of the colon
7 Body of the stomach
8 Fundus of the gall bladder overlain by the 9th costal cartilage
9 Hepatic flexure of the colon
10 Duodenum
11 Pylorus
12 Transpyloric plane
13 Inferior vena cava
14 Abdominal aorta
15 Left common iliac vein
16 Terminal ileum
17 Caecum
18 Appendix
19 Right anterior superior iliac spine
20 Inguinal ligament
21 Sigmoid colon

THE ABDOMEN

The anterior abdominal wall extends from the lower costal margin to the inguinal ligaments. The abdominal cavity extends upwards behind the lower ribs and downwards into the pelvis. Abdominal examination includes the perineum, its contents, and digital examination of the pelvis, per rectum and vaginam. The abdomen (Fig. 37.4) is divided by two horizontal planes (subcostal and trans-tubercular) and two vertical lines through the midinguinal point (this crosses the costal margin at the tip of the ninth costal cartilage). From above downwards there are three central regions (epigastrium, umbilical and suprapubic) and three lateral regions (hypochondrial, lumbar and iliac). The term transpyloric plane (halfway between the jugular notch and the symphysis pubis) is also used for descriptive purposes in clinical practice.

Inspection

The abdomen is examined with the subject lying flat on his back on a couch with a single pillow supporting the head and with the legs uncrossed. This is the recumbent (supine) position. The abdomen is fully exposed, the breasts being covered in the female and the pubic region in both sexes until specific examination of these areas. The size, shape and symmetry are noted and the position and form of the umbilicus.

At rest respiratory movement is predominantly diaphragmatic and the abdominal wall reflects this activity. The abdominal muscles tense during coughing and on raising the head off the couch. In a thin person central aortic pulsation and occasionally gut peristalsis may be visible. These movements are best observed by kneeling so that the examiner's eyes are at the level of the anterior abdominal wall. Not uncommonly abnormalities such as scars of previous operations, and inguinal hernias, which are protrusions of abdominal viscera through the superficial inguinal ring, may be observed. The latter are accentuated by coughing, this being referred to as an expansile cough impulse.

Palpation

Palpation should be with clean, warm hands, the observer standing on the right of the subject and the right hand carrying out most of the manoeuvres even in a left-handed person. The hand first gently rests on the four corner regions of the abdomen in a circular order. This gentle palpation assesses muscle tone in the abdominal wall and any differences between areas. A tense abdominal wall can be felt by asking the subject to cough or raise his head off the pillow.

In normal subjects examination produces no pain but abdominal disease may be accompanied by tenderness and this will induce protective voluntary tensing (guarding), or involuntary muscle tension (rigidity), of the abdominal wall. Following this initial assessment of abdominal tone, each region is examined in turn, palpating for abdominal viscera and abnormal masses.

It is customary to start with the right hypochondrium and then to proceed in an orderly fashion so that no regions are omitted. A possible scheme is to pass on to the epigastrium, followed by the left hypochondrium, umbilicus, left iliac, suprapubic, and right iliac regions, leaving the lumbar regions to the last since these require bimanual examination. This order is sometimes related to local findings, the suspected tender areas being best left to the last, but all regions must always be palpated. Palpation of posterior abdominal viscera and masses is by gentle depression of the relaxed anterior abdominal wall compressing viscera on to the vertebral bodies and muscles of the posterior abdominal wall. The shape of any palpable organ is defined by gentle finger palpation around the area.

In the right hypochondrium a liver edge may just be palpable in a thin individual. The liver descends on inspiration and this fact is made use of in identification. The flat of the right hand is placed across the right side of the abdomen with the index finger parallel to the costal margin. The right hand is depressed into the abdomen on expiration and pressure maintained during inspiration, the distending abdomen lifting the hand but any descending firm liver edge will be felt by the lateral border of the index finger. As enormous enlargement of the liver can extend even to the right iliac fossa, the hand is depressed firstly in this area and then ascends 1 to 3 fingers breadth at a time with each respiratory cycle, repeating the same manoeuvre until the costal margin is reached or the edge is located. An enlarged gall bladder moves with the liver and in line with the tip of the 9th costal cartilage (i.e. where the right vertical abdominal line and the lateral border of the rectus abdominis muscle cross the costal margin).

The epigastrium contains much of the stomach, which is not usually palpable, and the left lobe of the liver which if enlarged is examined as described above. In a thin person the pulsation of the aorta may be felt through the stomach and pancreas. The left hypochondrium contains the rest of the stomach, and the spleen may enlarge into this area. Initial splenic enlargement is below the costal margin in the anterior axillary line, but a very large spleen crosses the umbilical region and reaches the right iliac fossa. The mode of examination for the notch on the anterior edge is the same as that for the liver, working cranially along this line of enlargement.

The semisolid contents of the colon mean that it is often palpable in the normal subject. The transverse colon may be felt across the umbilical region. The sigmoid colon is felt in the left iliac fossa, and the caecum and a tender appendix are often palpable in the right iliac fossa. The ascending and descending colon pass through the right and left lumbar regions respectively. Other normal palpable organs are the aorta in the umbilical region, and a full urinary bladder and a pregnant uterus rising out of the pelvis into the suprapubic region. Palpation of the umbilicus with the tip of a finger may reveal a circular defect in the linea alba. This is usually insignificant but larger defects may transmit viscera producing an umbilical hernia, which will have a cough impulse.

The lower pole of a normal right kidney may be palpable and this is felt by backward pressure of the flat of the right hand in the right lumbar region, combined with forward pressure from the left hand, placed behind the abdomen opposite the right hand. The kidney descends on inspiration and can be felt between the two hands in this bimanual palpation. For bimanual examination of the left lumbar region, the right hand is applied across the anterior abdomen in the usual fashion. The left hand is either crossed behind the back of the subject, with the examiner kneeling, or placed behind the left lumbar region as the examiner leans over the subject.

Percussion

Percussion is used to define the lower borders of the liver and spleen, being undertaken along the defined lines of their enlargement. An enlarged urinary bladder or uterus is dull to percussion as are abdominal masses and fluid within the peritoneal cavity. The latter is known as ascites, and produces dullness in the flanks in the recumbent position. If the subject rolls on to one side, however, the upper dull area will become resonant due to shifting of the fluid by gravity; a feature known as 'shifting dullness'.

Auscultation

Bowel sounds can be heard with a stethoscope over all the abdomen. After a meal they are loudest and most frequent, but at other times it may be necessary to wait for a minute to hear them. Turbulence in the abdominal aorta or iliac vessels may produce sounds detectable with a stethoscope.

THE INGUINAL REGION, PENIS AND SCROTUM

The inguinal region (Fig. 37.5) must be examined in all clinical examinations, and the penis and scrotum in all males. Each inguinal ligament extends from the anterior superior iliac spine to the pubic tubercle. The spine is easily palpable but the tubercle, forming the lateral end of the superior margin of the body of the pubis, may be difficult to find in a fat subject. The superficial inguinal ring is above and medial to the tubercle and admits the tip of the little finger. In a fat male it is more easily located by gently invaginating the upper part of the scrotal skin with the little finger until the ring is located. The subject is asked to cough with the finger tip in place to assess whether any gut is being expressed, i.e. whether an inguinal hernia is present. Examination for inguinal lymph nodes is considered with the lower limb.

Examination of the penis is primarily defining whether the foreskin is retractable over the glans penis. Scarring of the foreskin can produce a circular contraction—a condition known as phimosis. Examination of the scrotum begins with observation of the skin and the size of the sac. Gentle palpation of the contents of the scrotum on each side through its wall will identify the testis, its upper and lower pole, and the posteriorly situated body of the epididymis. The spermatic cord can be palpated most easily in the upper part of the scrotum by rolling the cord against the pubic tubercle or between the finger and thumb.

PELVIC EXAMINATION

The pelvic contents can be palpated with a gloved finger passed per rectum or vaginam (PR or PV). On digital examination of the rectum the tone of the anal

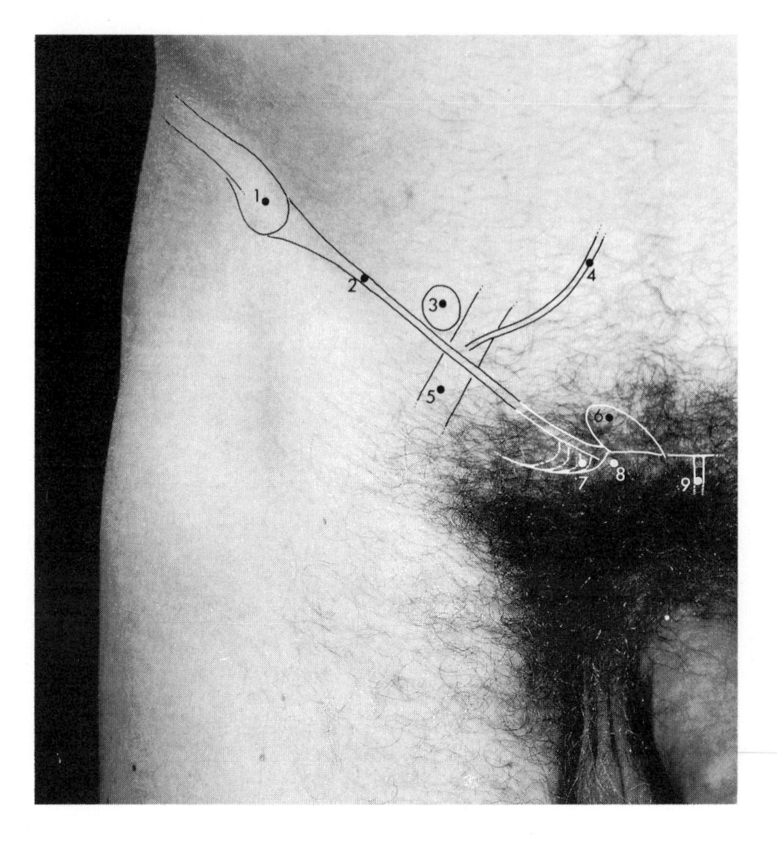

Fig. 37.5 Right inguinal region
1 Anterior superior iliac spine
2 Inguinal ligament
3 Deep inguinal ring
4 Inferior epigastric artery
5 Femoral artery
6 Superfical inguinal ring
7 Lacunar ligament
8 Pubic tubercle
9 Symphysis pubis

sphincter is firstly assessed. In the male the posterior aspect of the prostate can be palpated, and at a higher level through the anterior rectal wall the contents of the rectovesical pouch. The contents usually are loops of sigmoid colon or small gut. Vaginal examination allows palpation of the cervix and by depressing the left hand over the suprapubic region, bimanual examination usually enables the size and position of the uterus to be assessed. Ovaries lying on the broad ligament may also be palpable. The contents of the recto-uterine pouch (of Douglas) may be noted. Although such examinations are not necessarily very acceptable to clinician or patient they can elicit important information on pelvic viscera, and at least a PR is an essential part of every full clinical examination.

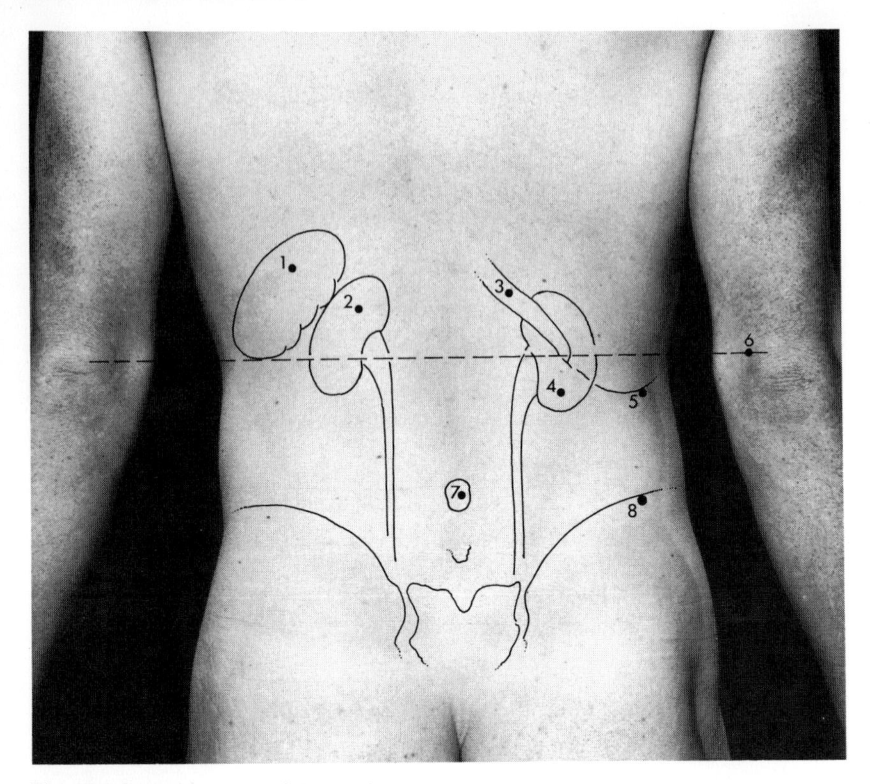

Fig. 37.6 Posterior aspect of the trunk
1 Spleen
2 Upper pole of the left kidney
3 Right 12th rib
4 Lower pole of the right kidney
5 Costal margin
6 Transpyloric plane
7 Spine of 4th lumbar vertabra
8 Iliac crest

THE DORSAL ASPECT OF THE TRUNK

Inspection and palpation of the posterior aspect of the body (Fig. 37.6) will demonstrate the back of the skull, the vertebral column, upper sacrum, coccyx and the posterior aspect of the hip bones. The prominent spine of the 7th cervical vertebra has already been referred to and the thoracic and lumbar spines are usually all palpable. The upper part of the sacrum may also be visible together with the iliac crests, and the dimples over the posterior superior iliac spines. The intercristal line, joining the highest point of the two iliac crests, passes through the 4th lumbar spinous process. The coccyx can be palpated in the midline as it approaches the anal margin. The gluteus maximus muscle forms the bulk of each buttock.

The spinal curvatures are usually in the sagittal plane. Lateral deviation from this plane is known as scoliosis. It is also accompanied by some rotation of the vertebrae in the affected areas. In the coronal plane the dorsal curvature is seen to be concave anteriorly, and the cervical and lumbar curvatures are convex. Exaggeration of the thoracic and lumbar curvatures are referred to respectively as kyphosis and lordosis. Note any deformity of these normal curvatures or of the spinal processes. Also note whether the spines remain in the sagittal plane during flexion and extension and observe the extent of all trunk movements. The scapulae lie over the second to seventh ribs. Their superior, medial and inferior angles and the spine may be palpated. The acromion extends anteriorly to form the bony tip of the shoulder, the deltoid muscle filling out its contour.

Examination of chest movements, percussion and auscultation follow the same pattern described for the anterior chest wall. The pleura extends down to the twelfth rib and the lungs to the tenth. Heart sounds are not usually audible since the heart is anteriorly situated.

THE UPPER LIMBS

Inspection and palpation of the upper limbs determines their symmetry, their shape, bony contours, and muscle bulk. The position of the nerves, arteries and veins is of particular clinical importance since they can be injured and the vessels are used for injections and insertion of catheters. Of particular note are the anatomy of the cubital fossa and the arrangement of structures around the wrist joint. The examination of the axilla has been described on page 504.

THE CUBITAL FOSSA

The fossa (Fig. 37.7) lies in front of the elbow joint. The lateral and medial epicondyles are both palpable, as is the ulnar nerve lying posterior to and closely applied to the medial epicondyle. The veins in the roof of the cubital fossa are variable in position. The cephalic and basilic veins usually lie along the lateral and medial borders respectively, united by a median cubital vein. These veins can be made to stand out by gripping around the arm or by placing a venous tourniquet on the arm. Repeated opening and closing of a subject's hand empties the venous blood from this lower area and further accentuates the cubital veins. The bicipital tendon, on its way to the bicipital tuberosity of the radius, passes distally through the fossa and can be easily palpated.

The brachial artery passes through the arm along the medial aspect of the biceps muscle and lies along the medial side of the tendon in the cubital fossa. The artery can be felt in the upper arm by compressing it on to the humerus in the groove between the biceps and brachialis muscles. It is also palpable in the cubital fossa. The arm needs to be extended to allow palpation against the lower end of the humerus. The artery is used at this site for auscultation during measurement of the blood pressure. Identification of the artery at this point is also important because it must not be inadvertently damaged during venepuncture. The site is frequently used for access in cardiac catheterisation. The

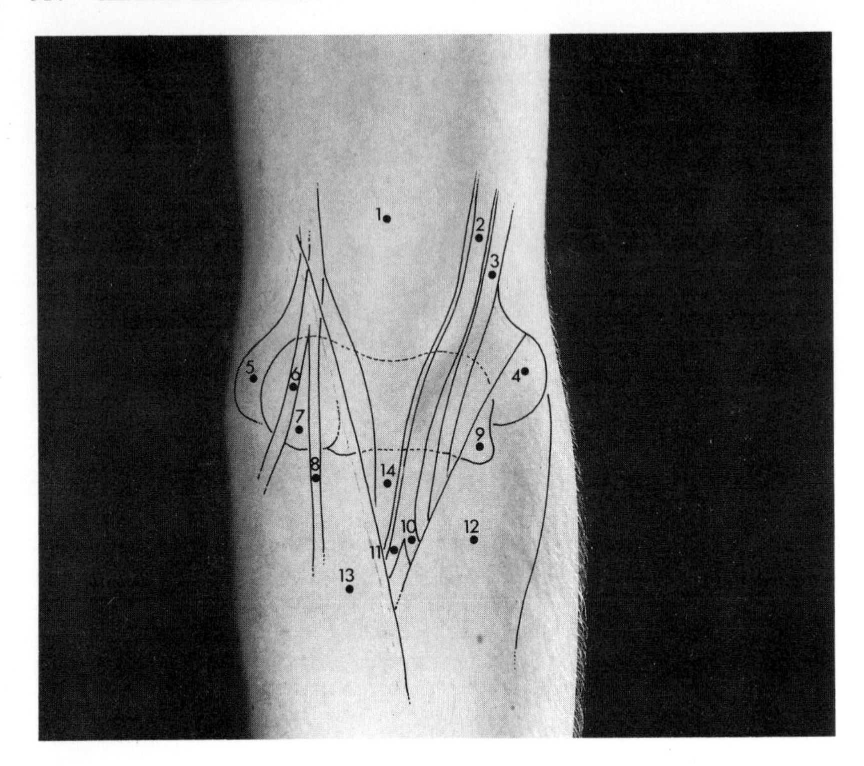

Fig. 37.7 Anterior aspect of right cubital fossa

1 Biceps muslce
2 Brachial artery
3 Median nerve
4 Medial epicondyle
5 Lateral epicondyle
6 Posterior interosseous nerve
7 Capitulum
8 Radial nerve
9 Trochlear
10 Ulnar artery
11 Radial artery
12 Pronator teres muscle
13 Brachioradialis muscle
14 Tendon of biceps muscle

median, radial and posterior interosseous nerves are not easily palpable around the elbow joint but their position is marked in Figure 37.7. The ulnar nerve has already been identified.

THE WRIST

In the anatomical position the radial and ulnar styloid processes are palpable on the sides of the wrist joint, the tip of the radial styloid being 1 cm distal to that of the ulnar. This relationship may change with fractures of the lower end of the radius. On pronation the lower end of the radius rotates around the ulnar and it is the anterior surface of the head of the ulnar which becomes palpable.

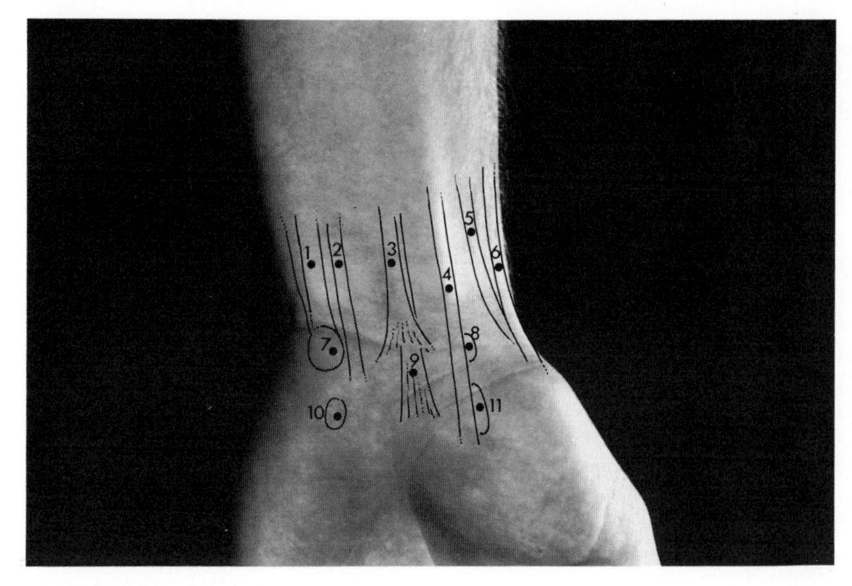

Fig. 37.8 Anterior aspect of the left wrist
1 Flexor carpi ulnaris tendon
2 Ulnar artery
3 Palmaris longus tendon
4 Flexor carpi radialis tendon
5 Radial artery
6 Abductor pollicis longus tendon
7 Pisiform bone
8 Tubercle of the scaphoid
9 Median nerve
10 Hook of the hammate
11 Ridge of the trapezium

On flexing the wrist against resistance the flexor carpi radialis and ulnaris tendons stand out on the sides of the anterior surface of the wrist (Fig. 37.8).

These tendons are attached respectively to the scaphoid and pisiform bones which are palpable. On deeper palpation just distal to the two bones the ridge of the trapezium (laterally) and the hook of the hamate (medially) are palpable. These four bony areas give attachment to and form the boundaries of the flexor retinaculum. In the flexed wrist the palmaris longus tendon may stand out between the carpal flexors as it passes into the flexor retinaculum. It may be absent in which case the underlying median nerve becomes palpable.

The radial artery lies subcutaneously along the lateral aspect of the anterior surface of the distal quarter of the radius. This site is routinely used to examine the pulse. The pads of three fingers are placed along the artery, the most distal being used to compress the artery on the radius, the other fingers assess the rate, rhythm, volume and character of the pulse.

Extension of the subject's thumb will demonstrate the tendons bordering the anatomical snuff box, namely the abductor pollicis longus with the extensor pollicis brevis anteriorly and the extensor pollicis longus posteriorly. The radial styloid and the scaphoid bone are palpable in the floor of the snuff box as is the radial artery as it crosses the wrist joint, passing deep to these three tendons. The cephalic vein passes subcutaneously across the snuff box and, together with a variable pattern of veins over the dorsum of the hand, can be distended as described above and used for venous access.

THE LOWER LIMBS

Inspection and palpation are used to assess symmetry, shape, bony contours, movement and muscle bulk. Areas of clinical importance are the hip joint, the femoral triangle, the knee joint and popliteal fossa, and the location of the structures around the ankle joint. The upper lateral aspect of the femur and the greater trochanter are palpable subcutaneously. The hip joint is about 4 cm inferior to the anterior superior iliac spine. The ischial tuberosities, which take the weight of the upper part of the body during sitting, are palpable at the lower border of the gluteus maximus muscles.

THE FEMORAL TRIANGLE

The triangle is bounded superiorly by the inguinal ligament, laterally by the sartorius muscle and medially by the medial border of the adductor longus muscle. None of these structures is easily palpated. A more useful landmark is the femoral (midinguinal) point situated half way between the anterior superior iliac spine and the middle of the symphysis pubis. This is the surface marking of the femoral artery which can be palpated by pressing backwards on to the head of the femur. The marking is particularly important to help find the artery in a fat subject. The femoral vein lies medial to the artery and both vessels are used for vascular access and for obtaining blood for analysis. The femoral nerve is situated lateral to the artery. The femoral canal is situated medial to the vein. The saphenous opening in the deep fascia of the thigh (the site of entry of the saphenous vein and of extrusion of a femoral hernia) is about 4 cm inferior and lateral to the pubic tubercle. This approximates to the groin crease. Inguinal lymph nodes may be palpable over the medial half of the inguinal ligament and vertically along the line of the femoral vein.

THE KNEE JOINT AND THE POPLITEAL FOSSA

The distal femur and proximal tibia are superficial and palpable on the lateral and medial aspects of the knee joint. The line of the joint can also be palpated and is identified more easily by gentle passive flexion and extension. Anteriorly the patella is palpable and the patellar tendon passing to the tibial tubercle. The latter and the anterior border and surface of the tibia are subcutaneous. The lateral ligament of the knee joint can be felt as it passes to the palpable upper end of the fibula. The common peroneal nerve can be found on deep palpation over the lateral aspect of the neck of the fibula as the nerve passes antero-inferiorly.

The hamstring tendons, forming the medial and lateral boundaries of the upper part of the popliteal fossa (Fig. 37.9), are easily palpable, particularly in

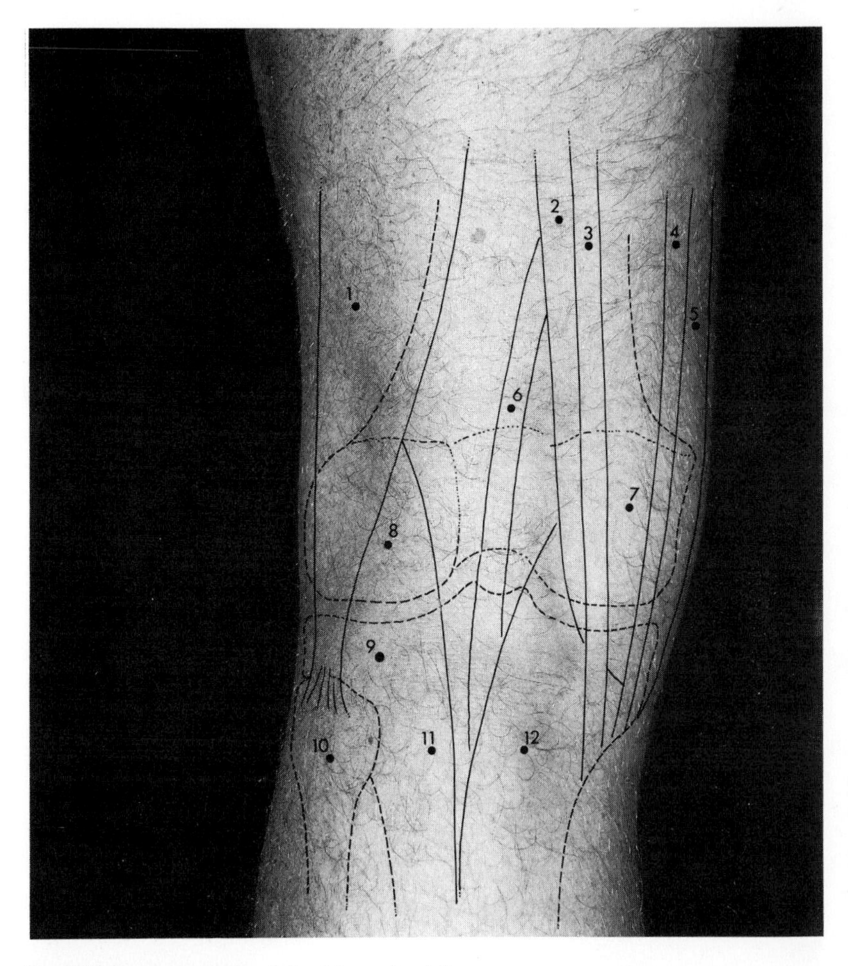

Fig. 37.9 Posterior aspect of the left popliteal fossa

1 Biceps femoris muscle
2 Semimembranosus muscle
3 Tendon of the semitendinosus muscle
4 Gracilis muscle
5 Vastus medialis muscle
6 Popliteal artery
7 Medial femoral condyle
8 Lateral femoral condyle
9 Lateral tibial condyle
10 Head of the fibula
11 Lateral head of the gastrocnemius muscle
12 Medial head of the gastrocnemius muscle

the flexed knee but the heads of the gastrocnemius muscle, forming the inferior margins of the fossa are less easily identified. The popliteal artery is sited deeply over the posterior aspect of the knee joint and adjacent bony surfaces. Palpation of the artery can be difficult. It is most easily compressed on to the upper end of the tibia between the heads of the gastrocnemius with the subject lying recumbent (supine), the knee slightly flexed and the muscles relaxed.

THE ANKLE JOINT (Figs. 37.10 and 37.11)

The lateral and medial malleoli are palpable on the sides of the ankle joint, the lateral passing more distally. The tendo calcaneus descends posteriorly to reach the calcaneus. The anterior tibial artery crosses the joint midway between the malleoli. It can be palpated at this point and more distally as the dorsalis pedis artery running parallel and just lateral to the extensor hallucis longus tendon. At the first site the artery is compressed against the joint and at the second site on to the adjacent medial tarsal and metatarsal bones. The tendons

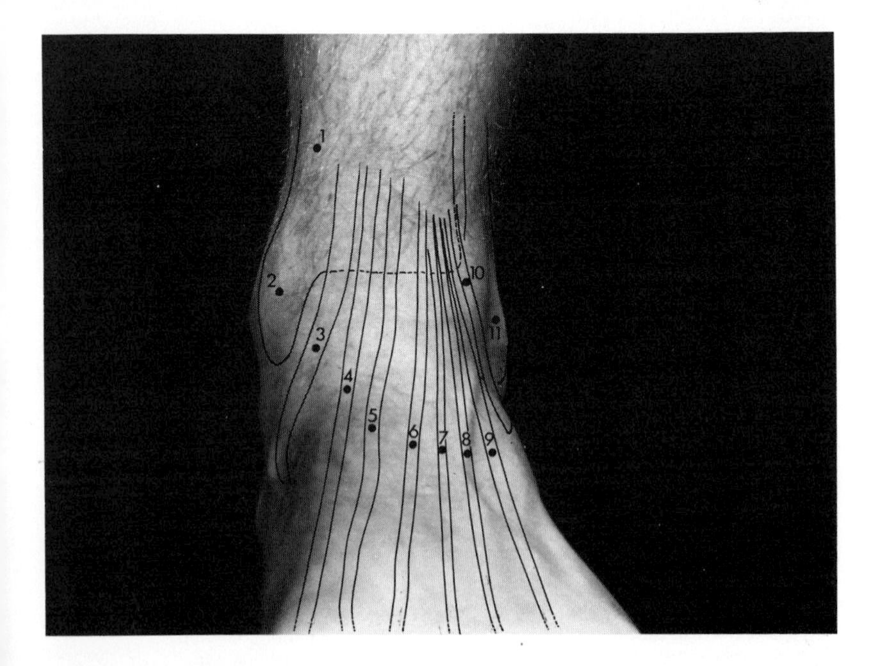

Fig. 37.10 Anterior aspect of the left ankle joint

1 Tibia
2 Medial malleolus
3 Tibialis anterior tendon
4 Extensor hallucis longus tendon
5 Dorsalis pedis artery
6–9 Tendons of the extensor digitorum longus muscle
10 Tendon of the peroneus tertius muscle
11 Lateral malleolus

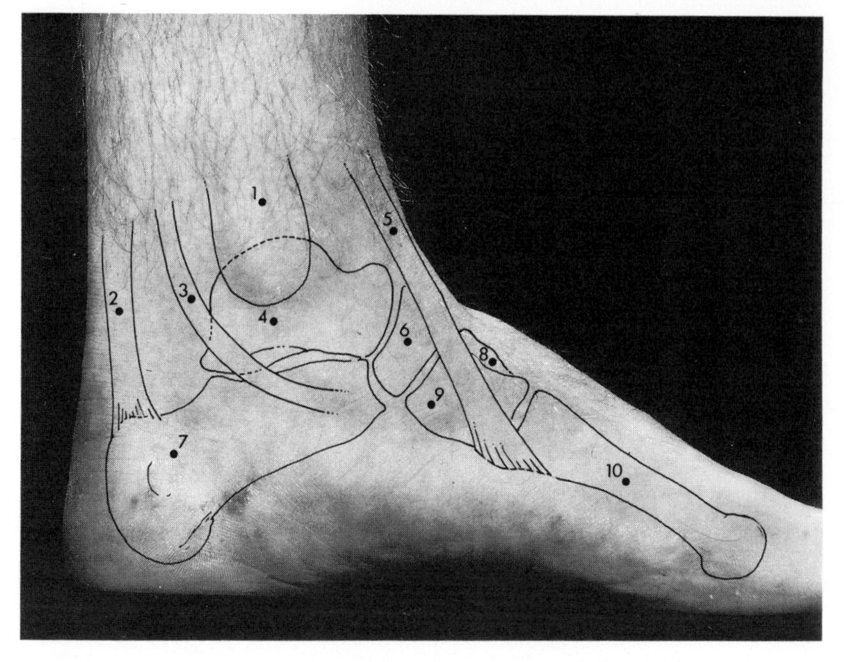

Fig. 37.11 Medial aspect of the left ankle joint
1 Medial malleolus
2 Tendo calcaneus
3 Posterior tibial artery
4 Talus
5 Tibialis anterior tendon
6 Navicular
7 Calcaneus
8 Intermediate cuneiform
9 Medial cuneiform
10 First metatarsal

of the peroneus longus and brevis are palpable proximal and distal to the lateral malleolus.

The tendons across the posterior aspect of the joint are less easily defined but an important marking is that of the posterior tibial artery. This passes midway between the medial malleolus and the medial prominence of the calcaneus, curving antero-inferiorly into the foot. It is compressed against the lower end of the tibia and the talus at this site. A perforating branch of the peroneal artery passes distally along the anterior surface of the lateral malleolus and may be palpated in the normal as well as the abnormal circulation.

The upright standing position places the lower limb under a high venous and lymph back pressure. Excess tissue fluid can collect—this being accentuated in a number of diseases. Excess tissue fluid is known as oedema and this can be demonstrated in the subcutaneous tissues as a persisting dent produced by digital pressure over the lower subcutaneous surface of the tibia.

Increased venous pressure can give rise to prominent tortuous thin-walled subcutaneous veins known as varicose veins. These are usually situated along the course of the long saphenous venous system. The long saphenous vein itself does not usually become varicose, particularly in the lower leg region. The surface marking of this vein at the ankle is 1 cm anterior to the medial malleolus. It ascends along the medial aspect of the lower leg and is in line with the posterior margin of the tibia at the level of the tubercle. It then passes posteromedial to the knee joint. From here the vein passes along the medial aspect of the thigh to the saphenous opening below and lateral to the pubic tubercle.

Further information on both the nervous and musculo-skeletal systems of the lower limb is obtained by examining the integrity of the gait of the subject. When the regional examination of the whole body is completed it may be necessary to have a more specialised examination of the nervous system.

Index